INDUSTRIAL RAILWAYS AND LOCOMOTIVES of ESSEX

Compiled by Robin Waywell and Frank Jux

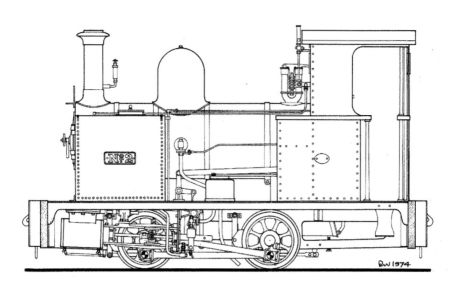

W. G. Bagnall 0-4-0 front tank locomotive 1424 of 1893, which worked the 3ft 6in gauge system at the London County Council's Northern Outfall Sewage Works for 60 years. The drawing is taken from an outline diagram on the makers' drawing sheet 4083, and completed by reference to photographs. Drawing by Roger West.

INDUSTRIAL RAILWAY SOCIETY

Published by the INDUSTRIAL RAILWAY SOCIETY

at 24, Dulverton Rd, Melton Mowbray, Leicestershire, LE13 0SF

www.irsociety.co.uk

© INDUSTRIAL RAILWAY SOCIETY 2011

ISBN (Hardbound) 978 1 901556 65 0

ISBN (Softbound) 978 1 901556 66 7

British Library Cataloguing-in-Publication Data
A catalogue record for this book is available from The British Library.

Produced for the IRS by Print Rite, Freeland, Witney, Oxon. 01993 881662

This book is copyright under the Berne Convention. Apart from any fair dealing for the purposes of private study, research, criticism, or review, as permitted under the Copyright Act, 1911, no portion may be reproduced by any process without the written permission of the publisher.

CONTENTS

Introduction		4
	Maps	14
	Locomotive Worked Sites	49
	Contractors' Locomotives	190
	Builders, Repairers & Dealers	261
	Preservation Sites	284
	Non-Locomotive Sites	301
Indexes:		
	Locomotives	315
	Locomotive Names	360
	Locations & Owners	365
Illustrations		373

Rear Cover Photograph
No.1(N1561 of 1870) at North Thames Gas Board, Beckton Gas Works, 26th October 1957.
Photograph by John Edgington.

INTRODUCTION

Essex is primarily an agricultural county, but due to its western boundary being determined by the River Lea, it includes part of what has long been, to all intents and purposes, a part of London. It has a long history of occupation, with Colchester being an early base for the Roman invaders, and claiming to be the oldest recorded town in the country. To the south, it is bordered by the River Thames, a fact having a major influence on its industrial history, as does its proximity to London. To the east, numerous inlets run inland from the North Sea, giving convenient access to small ships. However much of the riverside lands remained bleak with sparsely populated marshes and river walls necessary to keep out the tides. The damage caused by a failure of one of these walls resulted in the "Dagenham Breach" in 1707, which scoured out a large lake, and in 1845 there were plans to convert this into a new dock. Although work later started, the scheme was not a success, and the lake was eventually largely filled in by Samuel Williams & Co, who developed coaling piers and an industrial estate on the site. With its agricultural nature, Essex was not the scene of the great railway net-works of the Northern counties with their large industrial traffic base, and the Eastern Counties Railway was the main, and not very affluent, early railway promotion. The remainder of the railways came along with the later railway booms.

So far as industry is concerned, Essex has had a few special features. Firstly, it was the only one of the Home Counties to have had a modern blast furnace – installed at the Ford motor works at Dagenham when it was built in 1929-1930.

Secondly, it was, and continues to be, the main destination for most of London's rubbish. The marshes along the Thames from Barking to Tilbury were lined with dumps, on which at various dates were deposited every type of unwanted material from household and industrial waste to building rubble and river dredgings. Originally, household waste had been collected by "dust contractors", who made a profitable living by sorting reuseable items from it and selling them on. In the days when households depended on coal fires for heating, the refuse was screened to recover the unburnt cinders, which were used by brickmakers as a component of the famous "London Stock" bricks; while the organic contents were utilised by farmers as a fertiliser. The refuse dumps were naturally not welcome to anyone living nearby, and were consequently sited well away from habitation. In the 1920s there were many complaints regarding the very large dumps that had grown up near the new London County Council estate at Becontree, and controls were introduced to put them on a more sanitary basis, the operators having to cover the refuse with earth as it was deposited. The last firm to use rail haulage was W.R.Cunis at Rainham, where diesel locomotives worked over the undulating tracks on top of the rubbish, among flocks of attendant seagulls. Rubbish is now conveyed by river barge in containers and hauled from the wharves for tipping by rubber-tyred vehicles.

Other industries are more mundane. Among the earliest products were bricks made from the river clays, for which Grays was noted from at least the 1820s, being shipped from wharves on the river. London Stocks continued to be manufactured around Southend until recent times. Although being a smaller producer than the county of Kent, there were many small brickworks in Essex. Sand and gravel workings were also extensive, especially so since the advent of concrete buildings. Essex was also a substantial producer of cement. A small works was established at Dovercourt, near Harwich, in the 1860s, but the main manufacturing centre was on the chalk outcrop in the Grays area, which had been previously utilised for lime burning and for ships ballast. The banks of the Thames in Kent had long been a major centre for the industry, and it is somewhat surprising that it was not until the early 1870s that it gained a firm hold on the north bank, no doubt due to a booming market for cement, large quantities of which were exported throughout the world. The large extent of quarrying activity can be appreciated by visitors to the Lakeside Shopping Centre, which was established in the worked out quarries of the Tunnel Portland Cement Co. A remarkable feature was the inter-connection that existed between many of the latter-day works by a railway line to the south of the main line railway.

Another feature to be noted was the existence of a number of petroleum storage installations along the Thames. In the initial phase of the industry, petroleum was shipped in barrels or cans, and due to its flammable nature was considered a dangerous import. London's river authorities were not prepared to have such a product stored near to any built up area, and the sparsely populated Essex marshes were the nearest that it was allowed to come. Of the tank storage farms set up, that at Thames Haven was later a centre for oil refining, which expanded greatly in Great Britain after the Second World War.

The same sparse population of the marshlands was also an attraction for the military. Gunpowder magazines at Purfleet had long existed, and in 1849 work started on the construction on the first portion of land acquired at Shoeburyness of a gun testing range, in 1859 also becoming the base for the School of Gunnery. Much expanded over the years, the range was initially allied to the Royal Arsenal at Woolwich, and guns manufactured there were, from the 1870s, sent to Shoeburyness for trials in specially constructed barges, designed to carry their burden on railway wagons. The principle was similar to latter-day tank landing craft, wagons could be run on shore more or less where required, although generally at a special dock built for the purpose. A very extensive railway system was eventually built connecting the various parts of the ranges with the London Tilbury & Southend Railway.

The location of Docks had been moving further and further down river from London as ships became larger, the Victoria Dock being opened in 1855 near Silvertown. These were supplemented in 1880 by the Royal Albert Dock, and further expanded by the opening of the King George V Dock in 1921. These three were known collectively as the Royal Docks, and formed the largest impounded area of dock water in the world. However, these have now been superseded by the Tilbury Docks, where deeper water is available, which were conceived by the East & West India Dock Co in order to compete with rivals and to offer a dock that could be entered at whatever state of the tide. The first sod for the work was cut in July 1882, and the contractors, Kirk & Randall, commenced work. They soon found that site conditions were not as expected, and the Dock Company refusing to accept their request for an increase in the contract price to take account of this, the Dock Company turned them out and took over their plant in July 1884. New contractors Lucas & Aird were appointed a few months later and the dock was opened in April 1886. As a result of these problems the cost of the work soared from £1,100,000 to over £2,800,000 and worse still they were unable to attract the custom hoped for. The company was brought to the verge of ruin, and was no doubt relieved to be included in the amalgamation of the London Docks on the formation of the Port of London Authority on 31st March 1909. It is ironic that Tilbury now remains as the only one of the PLA's operating docks.

It will be noted that among the biggest industrial rail operators were the large gas works serving London. The largest and most prestigious were those at Beckton, the first attempt to consolidate gas production in very large modern plants. The concept was that of the Gas Light & Coke Company – founded in 1812 – and work on its construction started in 1868, with first production in 1870. With its allied by-product chemical works it expanded greatly over the years, with coal received by river and carried to the retort houses over an elevated railway system. It was large enough to require its own signalling system, and had large engineering workshops, which even built two locomotives in 1892. It was a fascinating railway that has been covered in a number of books and articles, and these should be consulted for further details. After the Second World War coal prices soared, and it became increasingly difficult for the gas companies – who employed a considerable amount of manual labour – to compete with electricity. Although a cheaper gas was produced in carburetted water gas plants, decreasing relative prices of feedstocks that became available from new oil refineries led to the use of these for production of town gas. The North Thames Gas Board agreed contracts to take refinery gas from Shell Haven oil refinery in 1956 and from Coryton in 1958. This was reformed at a new plant built at their Romford works, to which a pipeline was laid by William Press & Sons Ltd. The production of gas from coal was no longer an economic proposition, and the new plants gradually took over, and in 1959-1960 half of Beckton's retort houses were closed. With the importation of liquid methane from Texas in 1959 to a new terminal at Canvey Island, and the discovery of gas in the North Sea, the era of traditional production from coal came to an end, and Beckton shrunk to a shadow of its former self. The old works has been obliterated and most of the land redeveloped, a small part housing the Beckton Depot of the Docklands Light Railway.

From an early date a great deal of coal was brought to the Thames in sailing "collier" ships. These began to use steam propulsion in the 1840s, the screw-propelled "JOHN BOWES" of 1852 being a milestone in design of this type of ship. These modern ships needed better facilities for more rapid unloading than the traditional manual methods of the coal "heavers" then in use, and resulted in improved facilities. William Cory fitted out a large pontoon with a series of cranes for this purpose, which was moored in mid-stream and loaded the coal into lighters or storage facilities on board; while coal wharves and jetties were established to load the coal into railway wagons for onward despatch to retail or industrial users. Important examples of such jetties were those of Samuel Wiliams & Sons at Dagenham Dock, and Wm Cory & Son at Purfleet and at Gallions Wharf near the entrance to the Royal Docks. A large quantity of coal was also unloaded at power stations and at other industrial wharves and jetties and the eclipse of coal as the most popular fuel brought the closure of many of these enterprises based on coal, leaving the electricity industry as the main user. Even there gas and the policy to generate more electricity near coalfields has meant the closure of most of the smaller power stations, including the former "monster" station of the County of London Supply Co at Barking. Although the above may give a little of the background to the main industries of Essex, there were many other factories with long histories, especially in the Silvertown area.

Construction firms were large users of railways to convey spoil and materials. As will be seen, most of the known contracts using locomotives were in the late nineteenth century and later, and substantial numbers of locomotives were used on dock construction, where large quantities of spoil had to be moved. In between the wars much narrow gauge track and many locomotives were used on railway and housing contracts, many of which were funded by Government grants to alleviate unemployment. The work on the new Southend road assembled a large number of narrow gauge locomotives, and was perhaps the largest such fleet used in this country on roadworks, although greatly surpassed by the number brought to bear on the construction of the "Autobahn" system in Germany. The railway system used by C.J.Wills Ltd on the building of the Becontree Housing Estate for London County Council – their largest estate – was also noteworthy.

McEwan Pratt & Co of Wickford were relative pioneers in the construction of internal combustion engined locomotives and railcars, and reference should be made to the Society publication "The Railway Products of Baguley-Drewry and its Predecessors" by Allen Civil and Roy Etherington for details of their known output. Otherwise there were no major locomotive builders in the county apart from the Great Eastern Railway from its works at Stratford. The only major locomotive dealer was Thos W.Ward Ltd at Grays, where locomotives were invariably in stock ready for sale or hire, among a range of contractors plant and machinery. Their Titan Works had an interesting setting in a worked out section of Grays Chalk Quarries, and the site had formerly been the plant Depot of the contractor Sir John Jackson Ltd; rail access was through a tunnel under a road and via other parts of the old quarries, and Ward had an interesting network of lines around Grays, with a scrap yard on the Thames where numerous small and medium size ships were broken up.

Thus, although an agricultural county, Essex has had a wide range of industries, and a visit in the 1950s was a fascinating experience. As elsewhere in the United Kingdom, industry has declined in importance, and its use as an overflow area for London housing seems set to continue, with Government plans to expand housing on the riverside lands. Already new housing estates such as that at Chafford Hundred, near Grays, have been built on former industrial areas, and the waterfront that formerly saw the jetties of the Grays Chalk Quarries, early cement works, and Ward's shipbreaking yard, has seen the public house that once served the cement workers and watermen totally surrounded by new housing. However, it is difficult to disguise the evidence of quarrying, and we hope that this book will put on record some evidence of past activities in an interesting area.

The information relating to this county is set out in five sections as follows:-

1. Industrial locations where privately owned locomotives were used.

2. Known details of civil engineering contracts using locomotives, including contracts involving the construction of many sections of public railway.

3. Known details of builders, repairers and dealers in locomotives.

4. Preservation and Pleasure Railway locations where the gauge exceeds 1ft 3in.

5. Known details of non-locomotive worked systems which were of sufficient length to be of interest.

Indexes are included for the ease of referring to locomotives and locations listed within the volume.

I have over the years received help from enthusiasts and correspondents too numerous to mention who have answered questions or brought items to my attention and gladly given freely their time and information, for which I will always be grateful. Thanks are also due to the members of the Industrial Railway Society, whose observations over the years form the basis of the Society's records. The journals and newsletters of like-minded societies such as the Industrial Locomotive Society and the Narrow Gauge Railway Society have also formed a reliable source of research. In particular the advertisement extracts published by the ILS have given many leads to follow. Likewise the county record offices and reference libraries have proved their worth and I thank their attendants for their patience.

I am indebted to my co author Frank Jux who has helped research this volume and without his efforts it would be much the poorer, My special thanks also go to: Bob Darvill, Graham Feldwick, John Fletcher, Brian Gent, Gordon Green, Roger Green, Andrew Neale, Rob Pearman, Ken Plant, Ken Scanes, Clive Walters, Russell Wear, John Williams, Geoff Roughton, Maj A.S. Hill for records concerning MoD Shoeburyness, Adrian Corder-Birch for access to his extensive records of Essex brickworks, the late Harry Paar, the late Bill Williams and the late Eric Tonks who assisted over many years; also special thanks to Josephine Braines for proof reading, and lastly, I am indebted to Roger Hateley for his considerable help with formatting the text and producing maps together with much of the material in the contractors section and locomotive index and his encouragement in getting this volume into print.

Comments on and corrections to any information given in this book will be most welcome, as will any additional information that may have been omitted, which should be sent to:

Robin Waywell
December 2010

29 Caldbeck Close
Gunthorpe
Peterborough PE4 7NE

EXPLANATORY NOTES

GAUGE
The gauge of the railway is given at the head of the list. If the gauge is uncertain, then this is stated. Metric measurements are used where the equipment was designed to these units.

GRID REFERENCE
An indexed six-figure grid reference is given in the text, where known, to indicate the location of salient features of the site.

NUMBER, NAME
A number or name formerly carried is shown in brackets (   ). If it is an unofficial name or number, then inverted commas are used "   ".
In some cases, earlier or later names and/or running numbers are appended at the end of footnotes.

TYPE
The Whyte system of wheel classification is used wherever possible but when the driving wheels are not connected by outside rods but by chains or motors they are shown as 4w, 6w, 8w, etc. For ex-BR diesel locomotives the usual development of the Continental system is used. The following abbreviations are employed:-

	T	Side tank or similar. Tanks are invariably fastened to the frame.
	PT	Pannier Tank; a special type of side tank where the tanks are not fastened to the frame.
	ST	Saddle Tank; a round tank which covers the boiler top. This type includes the 'Box' and 'Ogee' versions popular amongst certain manufacturers during the nineteenth century.
	IST	Inverted Saddle Tank
	WT	Well Tank; a tank located between the frames below the level of the boiler.
	VB	Vertical boilered locomotive.
	F	Fireless steam locomotive.
	DM	Diesel locomotive; mechanical transmission.
	DE	Diesel locomotive; electric transmission.
	DH	Diesel locomotive; hydraulic transmission.
	DHF	Diesel locomotive; hydraulic transmission, flameproof.
	PM	Petrol or Paraffin locomotive; mechanical transmission.
	R	Railcar; a vehicle primarily designed to carry passengers.
	BE	Battery powered electric locomotive.
	BEF	Battery powered electric flameproof locomotive.
	R/R	Road-rail motive power.

CYLINDER POSITION

	IC	Inside cylinders
	OC	Outside cylinders
	VC	Vertical cylinders
	G	Geared transmission (used with IC, OC or VC)

FLAMEPROOF
Diesel or battery-electric locomotives that are flameproofed are shown with the suffix F to the type (e.g. 4wDMF).

STEAM OUTLINE
Diesel or petrol locomotives with a steam locomotive appearance added are shown as S/O.

MAKERS
The normal IRS abbreviations are used to denote makers. If any of these are unfamiliar to the reader, full details can be found in the index of locomotives (which are set out in the alphabetical order of these abbreviations).

MAKERS NUMBER AND DATE
The first column shows the works number, the second shows the date which appeared on the plate or the date the loco was built if none appears on the plate.

It should be noted that the ex-works date given in the locomotive index may be for a later year than that recorded as the building date.

Rebuilding details are denoted by the abbreviation 'reb', usually recording significant alterations to the locomotive.

SOURCE OF LOCOMOTIVE
'New' indicates that a locomotive was delivered from the makers to a location. A bracketed letter indicates that a locomotive was transferred to this location or subject to temporary transfer away. Full details, including the date of arrival, where known, appear in the footnotes below.

DISPOSAL OF LOCOMOTIVE
A locomotive transferred to another location is shown by a bracketed number and footnote, the date of departure being given in the footnote if it is known. In other cases the following abbreviations are used :-

	OOU	Loco noted to be permanently out of use on the date shown.
	Dere	Loco noted to be derelict and no longer capable of being used.
	Dsm	Loco both OOU and incomplete on the date shown.
	Scr	Loco broken up for scrap on the date shown.
	s/s	Loco sold or scrapped; disposal unknown.
	Wdn	Withdrawn from traffic.

Many sales of locomotives have been effected through dealers and contractors and details are given where known. If the dealers name is followed by a location, e.g. Abelson, Sheldon, it is understood that the loco went to Sheldon Depot before resale. If no location is given, the loco either went direct to its new owner or else definite information on this point is lacking. If a direct transfer is known to have been effected by a dealer, the word 'per' is used.

GENERAL ABBREVIATIONS

	c	circa; i.e. about the time of the date quoted
	reb	rebuilt
	w/e	week ending

FOOTNOTE ABBREVIATIONS
In addition to the abbreviations listed below, those used to denote the various locomotive builders are also used in footnotes where appropriate.

A&AEE	- Aeroplane & Armament Experimental Establishment
AMWD	- Air Ministry Works Department
APCM	- Associated Portland Cement Manufacturers Ltd
BAOR	- British Army of the Rhine
BPCM	- British Portland Cement Manufacturers Ltd
BSC	- British Sugar Corporation
CEA	- Central Electricity Authority
CEGB	- Central Electricity Generating Board
COD	- Central Ordnance Depot
CSD	- Central Stores Depot
CTRL	- Channel Tunnel Rail Link
GKN	- Guest Keen & Nettlefolds Ltd
GLC	- Greater London Council
HGTPD	- Home Grown Timber Product Department
HMSO	- Her Majesty's Stationery Office
ICI	- Imperial Chemical Industries Ltd
LATHOL	- London & Thames Haven Oil Wharves Ltd
LCC	- London County Council
LGOC	- London General Omnibus Company
LMOC	- London Motor Omnibus Company
LPTB	- London Passenger Transport Board
LT	- London Transport
MDHB	- Mersey Docks & Harbour Board
MoD	- Ministry of Defence
MoDAD	- Ministry of Defence Army Department
MoM	- Ministry of Munitions
MoS	- Ministry of Supply
MSC	- Manchester Ship Canal
MWB	- Metropolitan Water Board
NCB	- National Coal Board
PLA	- Port of London Authority
RAF	- Royal Air Force
RE	- Royal Engineers
REME	- Royal Electrical Mechanical Engineers
RNAD	- Royal Naval Armament Depot
RNPF	- Royal Naval Propellant Factory
ROD	- Railway Operating Department
ROF	- Royal Ordnance Factory
RTB	- Richard Thomas & Baldwins Ltd
UDC	- Urban District Council
USA/TC	- United States Army Transportation Corps
WD	- War Department
WDLR	- War Department Light Railways

There are references in footnotes to **Petrol Loco Hirers (Ltd)** and **Diesel Loco Hirers Ltd**. These companies specialised in the hire of Motor Rail locomotives and further details can be found in the entry for Simplex Mechanical Handling Ltd in the Bedfordshire Handbook.

MAIN LINE RAILWAY COMPANIES

BPGVR	- Bury Port & Gwendraeth Valley Railway
BR	- British Rail
CV&HR	- Colne Valley & Halstead Railway
ER	- Eastern Region
GER	- Great Eastern Railway
GNR	- Great Northern Railway
GWR	- Great Western Railway
LBSCR	- London Brighton & South Coast Railway
LD&ECR	- Lancashire Derbyshire & East Coast Railway
LMSR	- London Midland & Scottish Railway
LNER	- London & North Eastern Railway
LNWR	- London & North Western Railway
LSWR	- London & South Western Railway
LT&SR	- London Tilbury & Southend Railway
LYR	- Lancashire & Yorkshire Railway
MS&LR	- Manchester, Sheffield & Lincolnshire Railway
PD&SWJR	- Plymouth Devonport & South Western Junction Railway
RR	- Rhymney Railway
SER	- South Eastern Railway
SECR	- South Eastern & Chatham Railway
SR	- Southern Railway
TVR	- Taff Vale Railway

DOUBTFUL INFORMATION

Information which is known to be of a doubtful nature or subject to confirmation is denoted as such by the wording chosen or else printed enclosed in square brackets, sometimes with a question mark, e.g. [1910?].

LOCATION CODES

In the Industrial section within each county, each location entry is numbered in numerical sequence from 1. This number is prefixed by a letter indicating the key map on which the site appears. Similarly, the locations in the other sections are each numbered from 1 and are prefixed by first the map letter and then also by C, D, P or H (for Contractor, Dealer, Preserved or Non-loco systems (where H = 'Hand, Horse or Haulage') respectively. For example, EC25 denotes map E, and location 25 in the Contractors section. The letter X as a prefix indicates an entry which, for whatever reason, is not indicated on any key map. This includes contracts whose location is unspecified, and also lengthy pipeline laying contracts spanning several maps.

LEGEND FOR MAPS

─────────── Public Railway (Standard gauge)

═══════════ Industrial Railway (Standard gauge)

・・・・・・・・・・・ Narrow gauge railway or tramway

▬▬▬▬▬▬▬ Canal or waterway

MAPS

KEY MAPS
Arrangement of Key Maps
- A North Woolwich
- B Silvertown
- C Stratford
- D Lea Valley
- E Harlow to Shenfield
- F North West Essex
- G Chelmsford
- H Harwich
- J Southend-on-Sea
- K Corringham
- L Ockendon
- M Tilbury
- N Purfleet & Grays
- P Dagenham

SYSTEM PLANS
1. Wouldham Works
2. Collier's Works
3. Creekmouth
4. Ford Works
5. Globe Works
6. Titan Works
7. Romford Brewery
8. Kynoch Works
9. Shoeburyness Military Railway
10. Beckton Gas Works
11. Beckton By-Products Works
12. Royal Docks
13. Tilbury Docks
14. Rochford Nurseries
15. Rowhedge Ballast
16. Mucking Gravel Pits
17. Tunnel Portland Cement
18. Bretts & Aveley Pits
19. Dagenham Dock
20. Becontree Estate

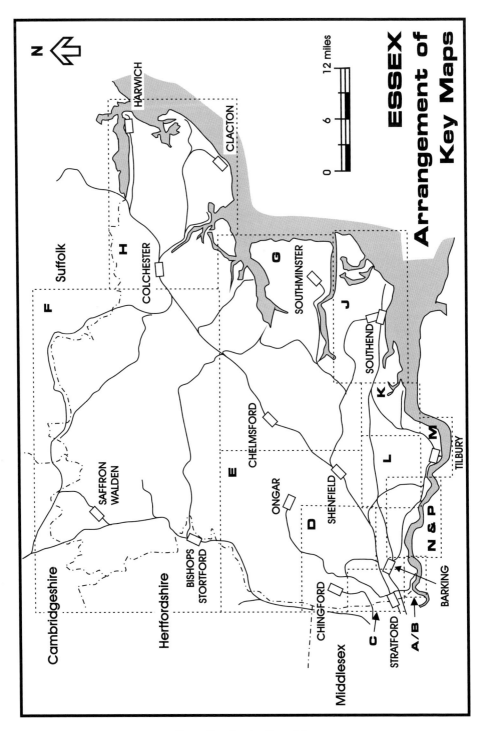

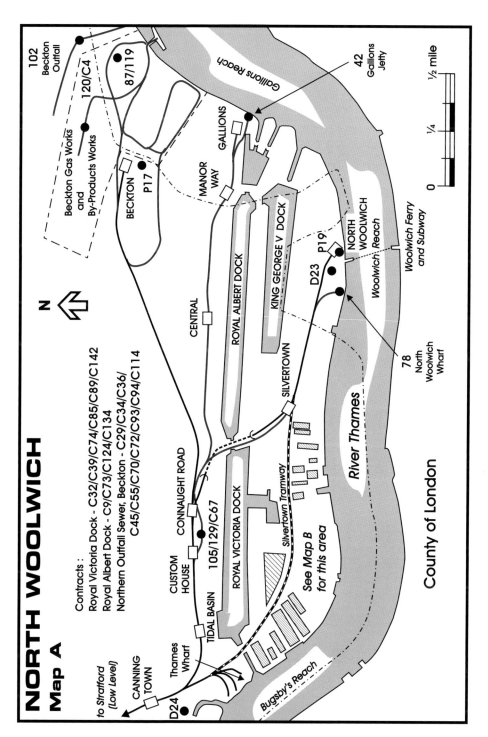

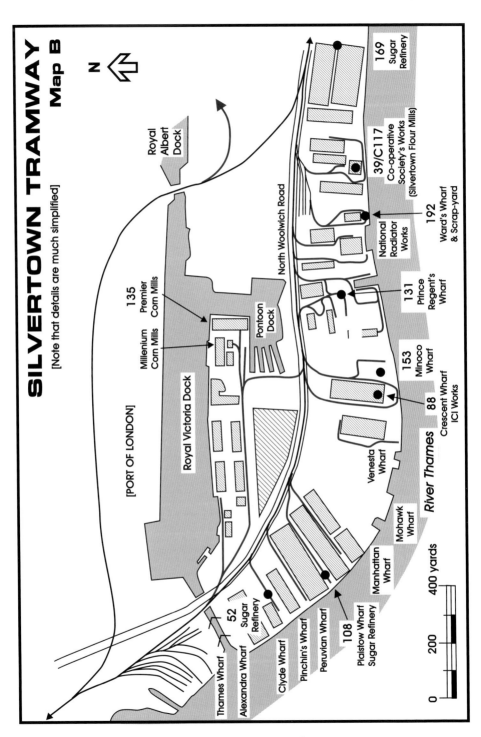

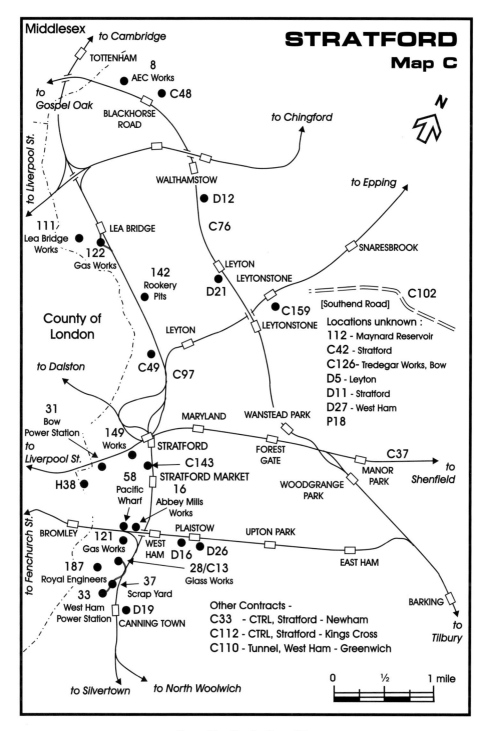

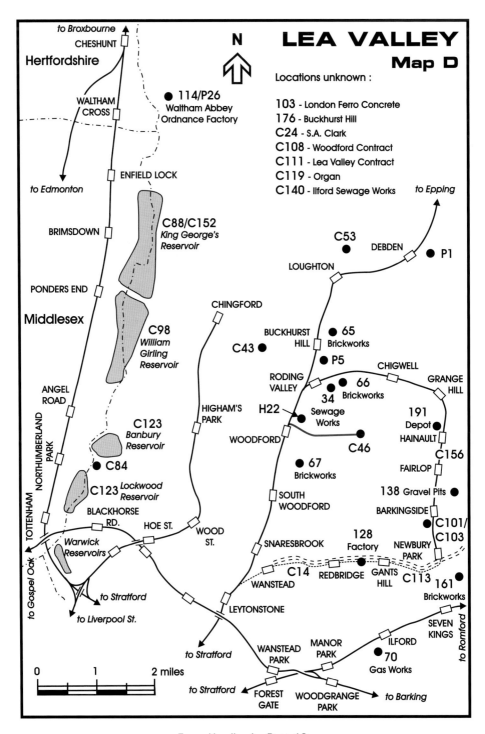

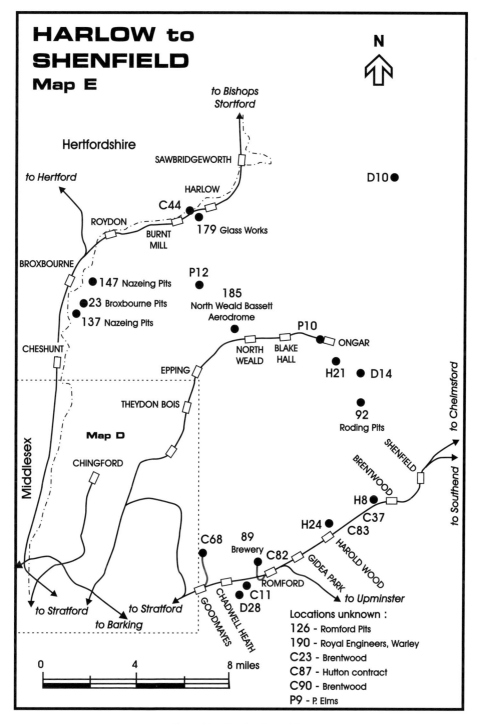

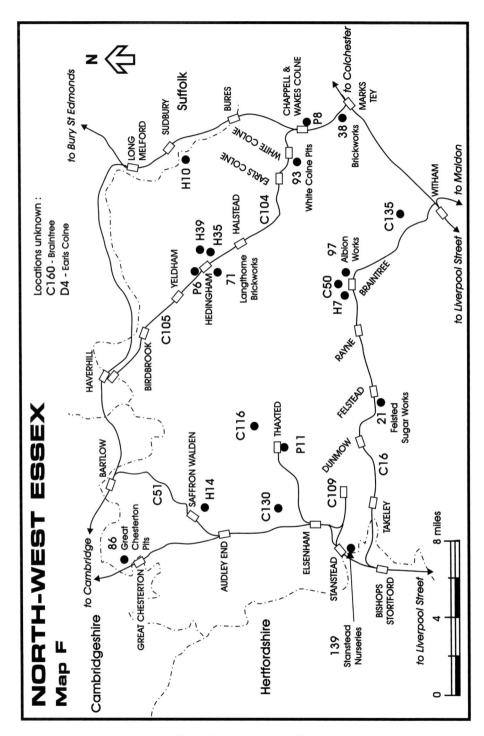

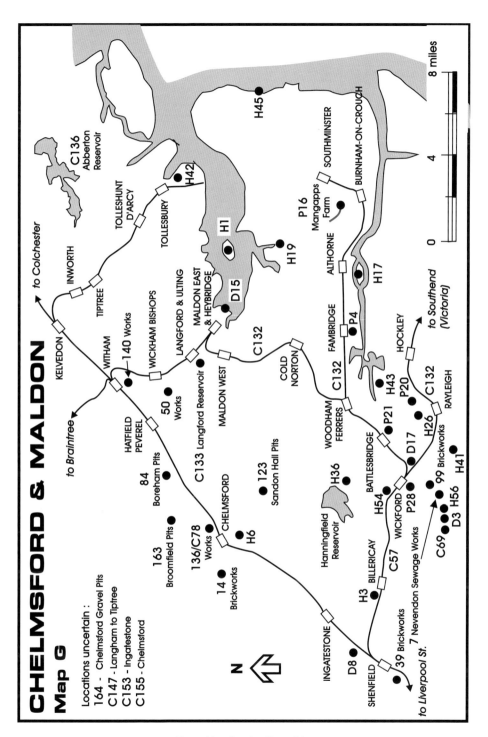

Map H HARWICH & CLACTON

Locations unknown:
198 - Colchester
C91 - Clacton
C137 - Stour scheme
C144 - Clacton
C162 - Colchester
D18 - Colchester
P7 and P29

Essex Handbook. Page 22

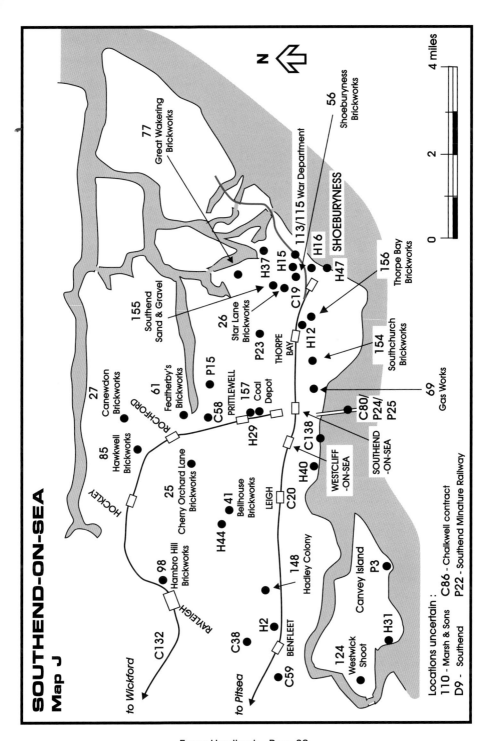

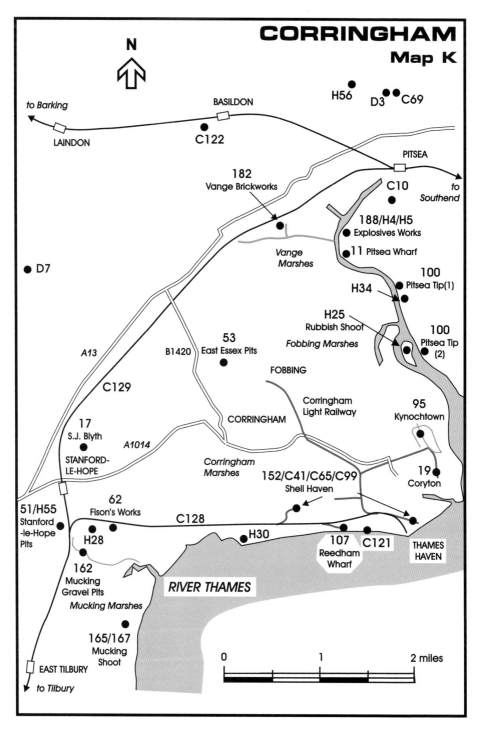

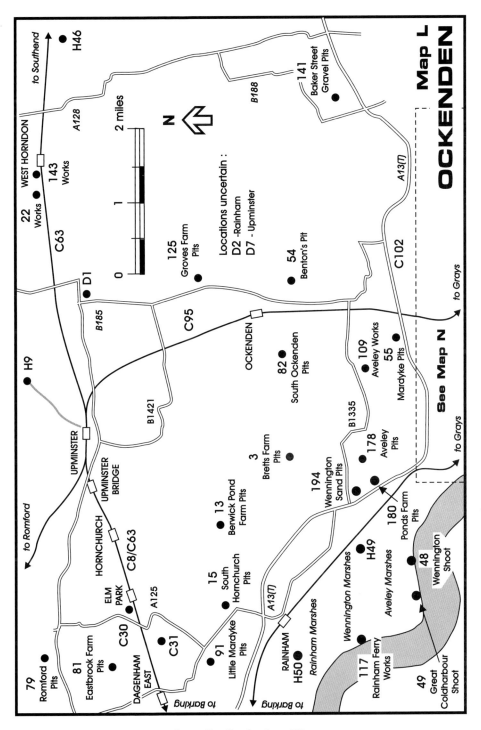

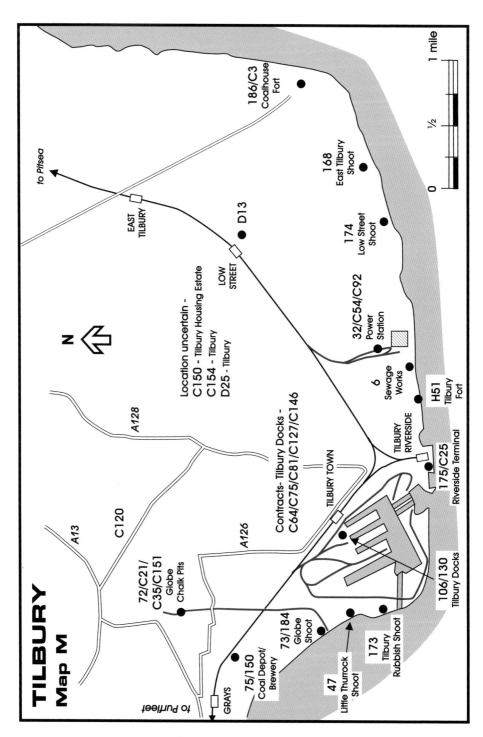

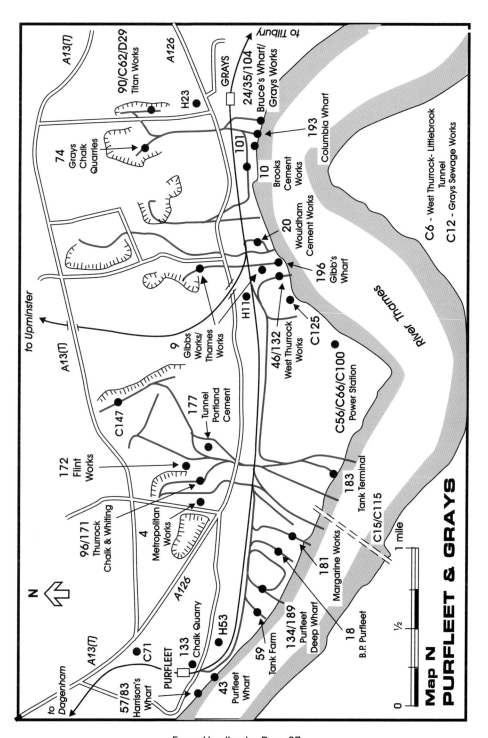

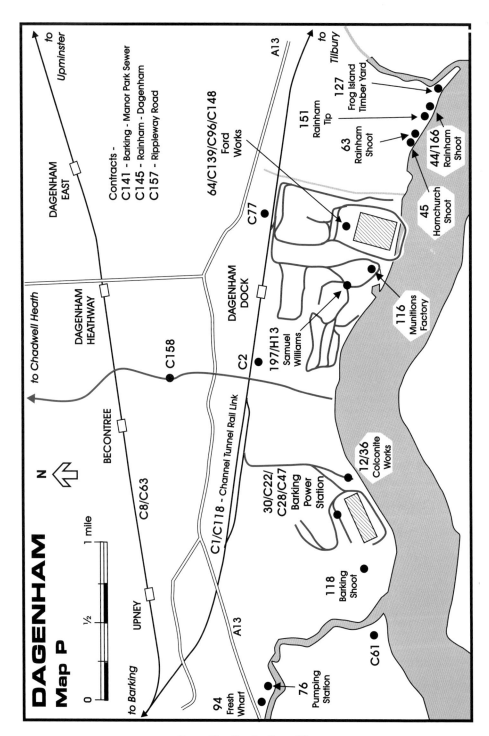

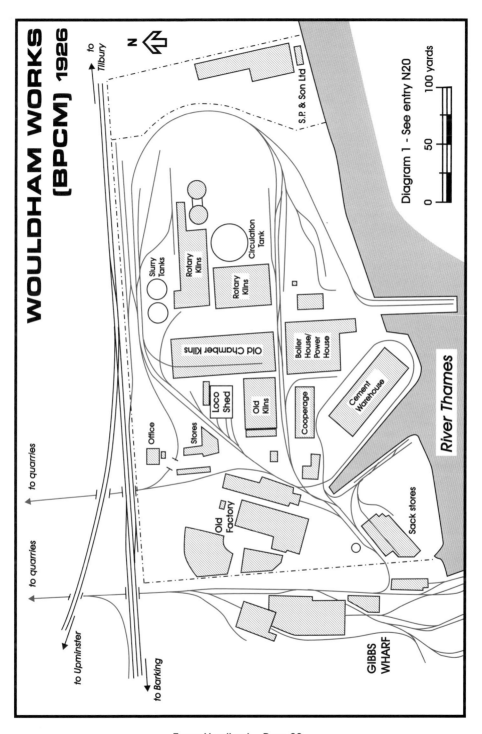

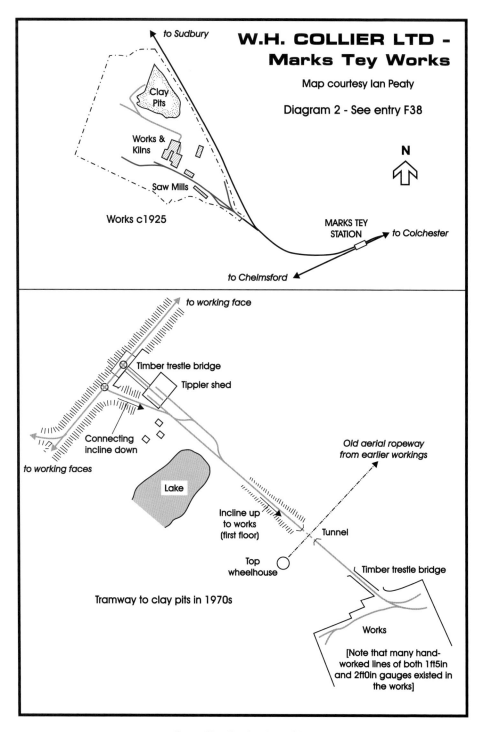

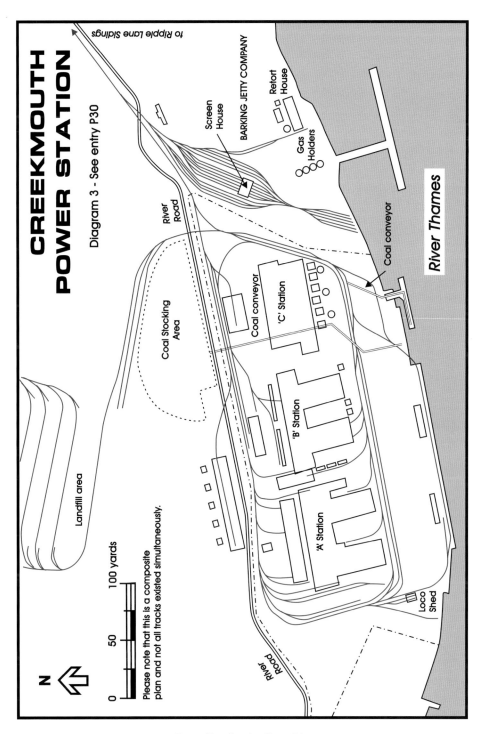

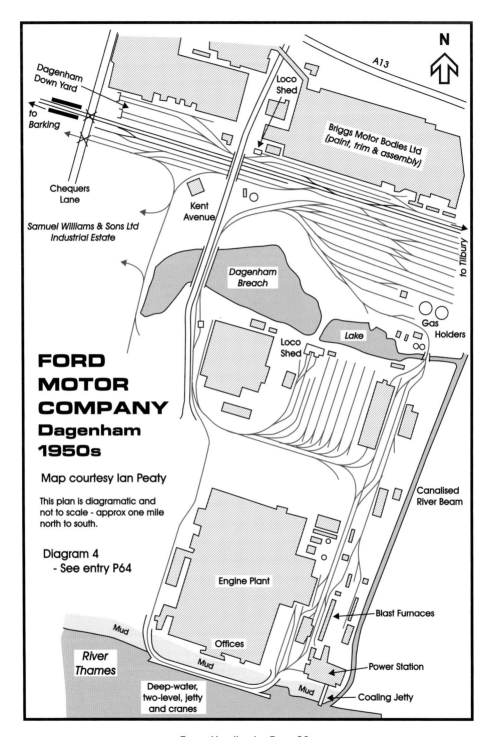

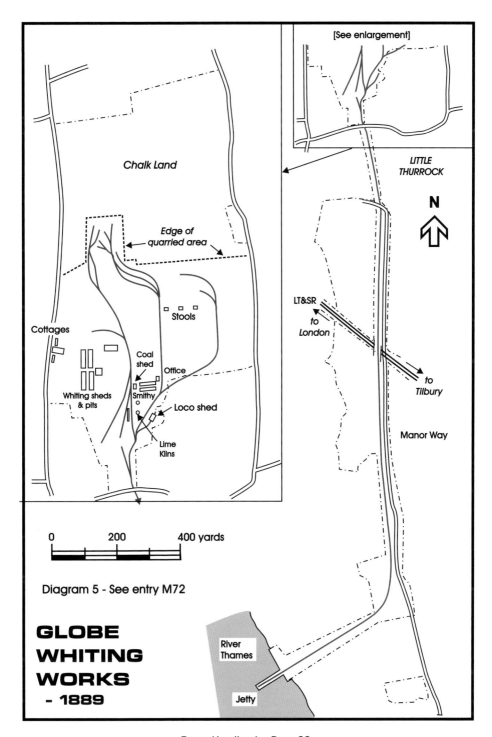

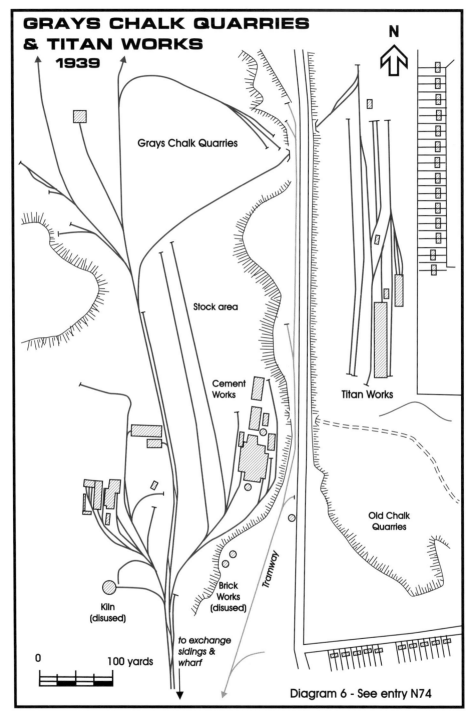

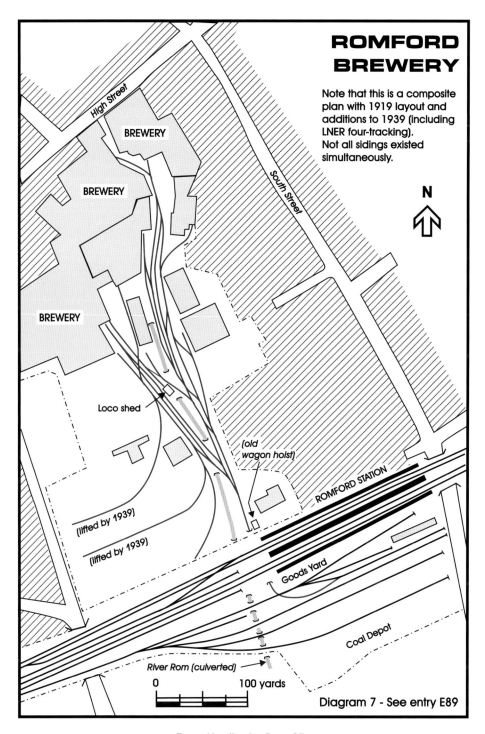

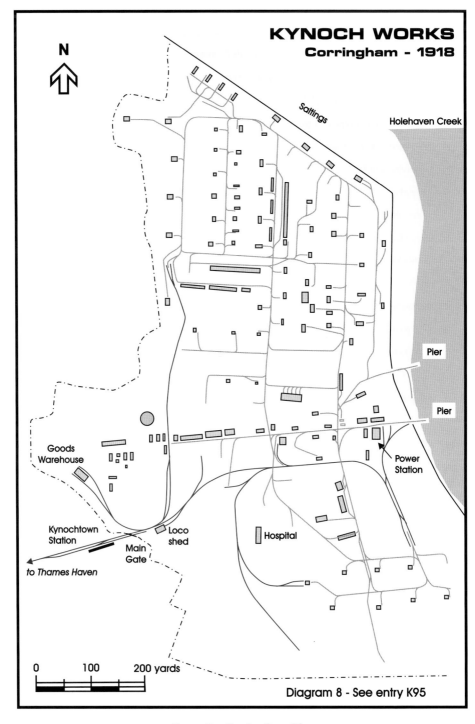

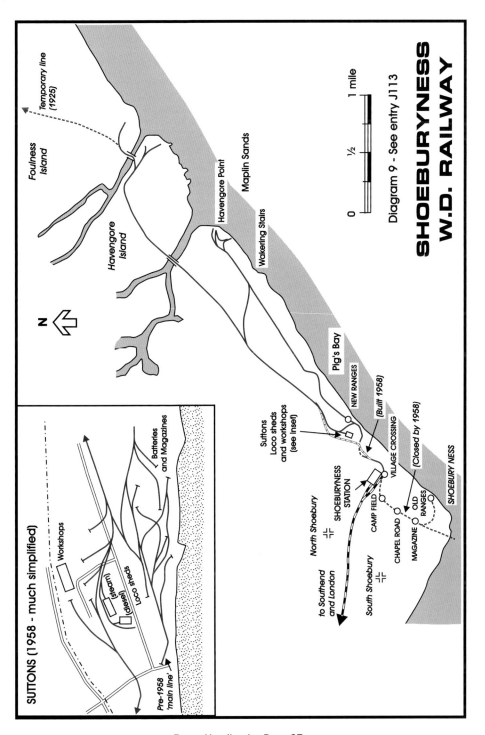

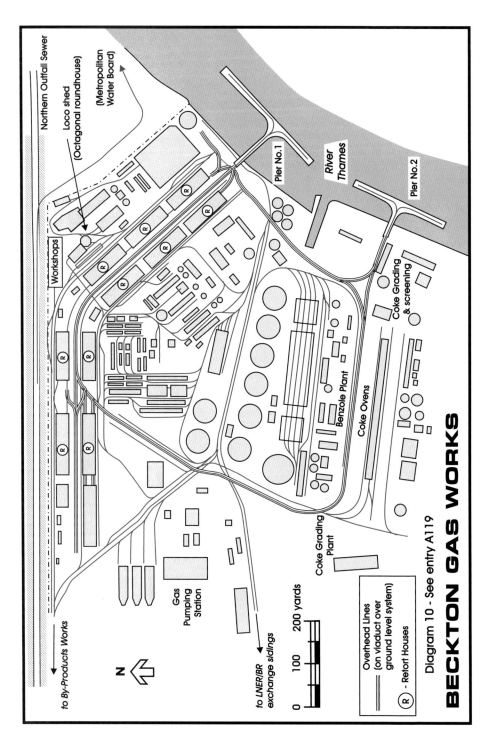

Diagram 10 - See entry A119

BECKTON GAS WORKS

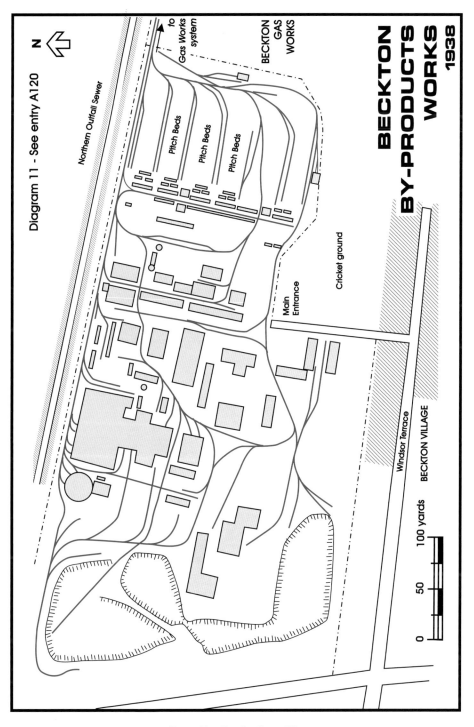

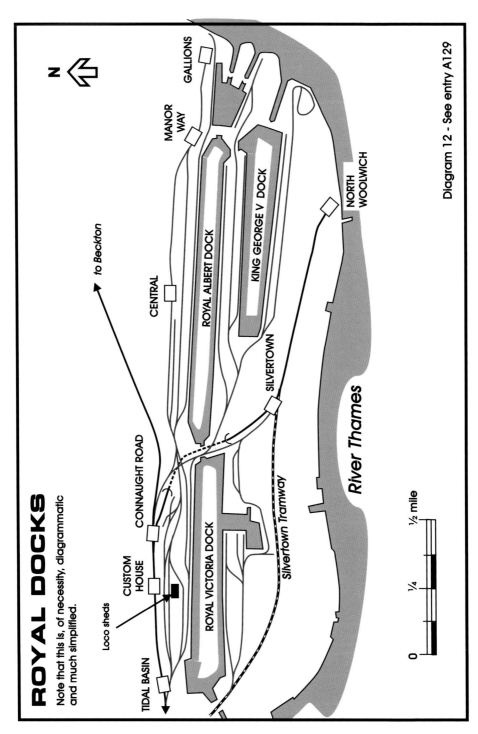

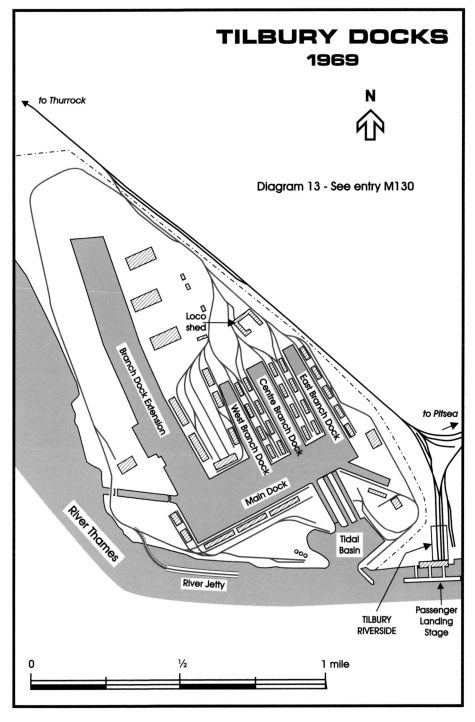

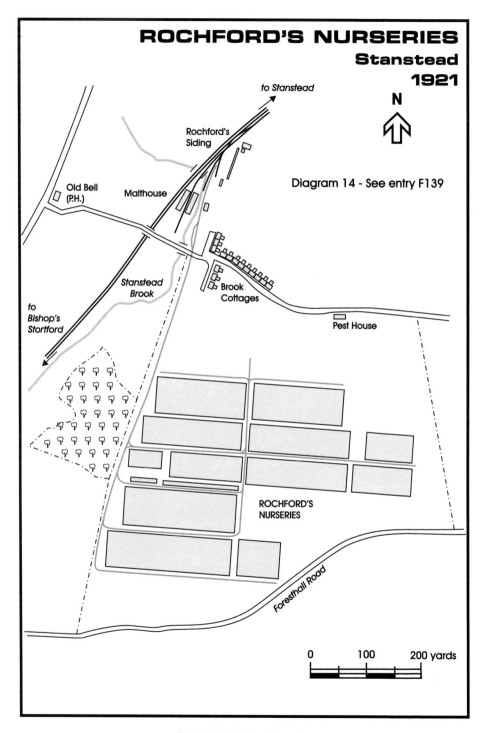

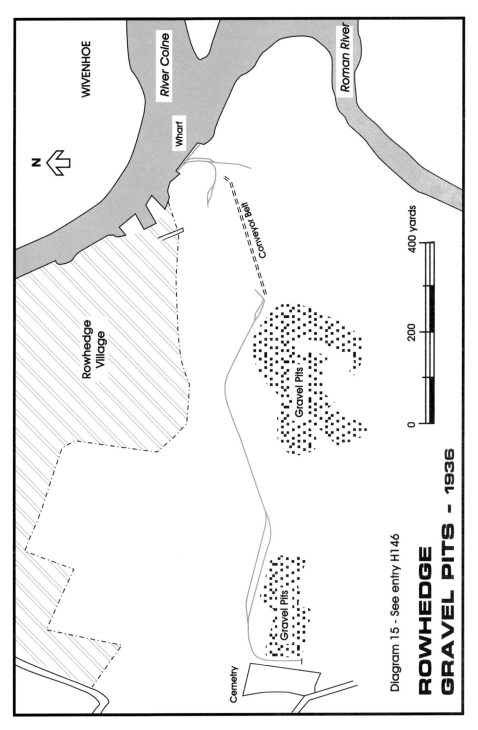

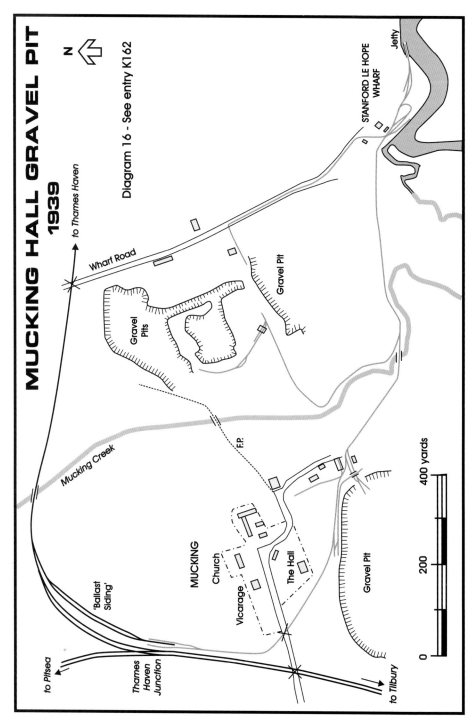

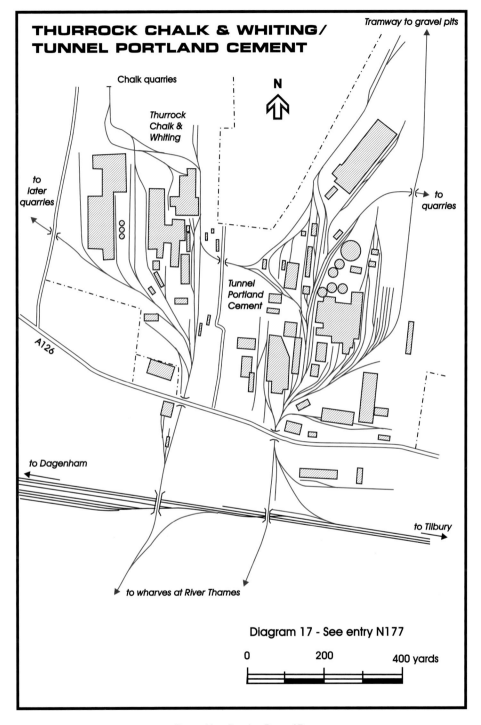

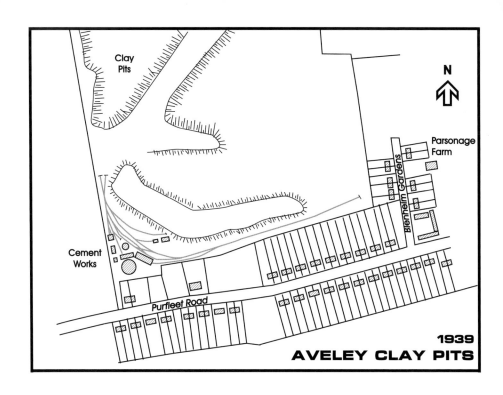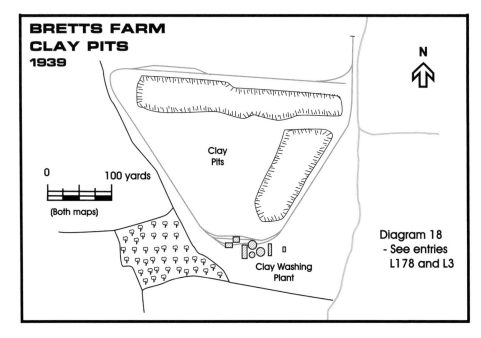

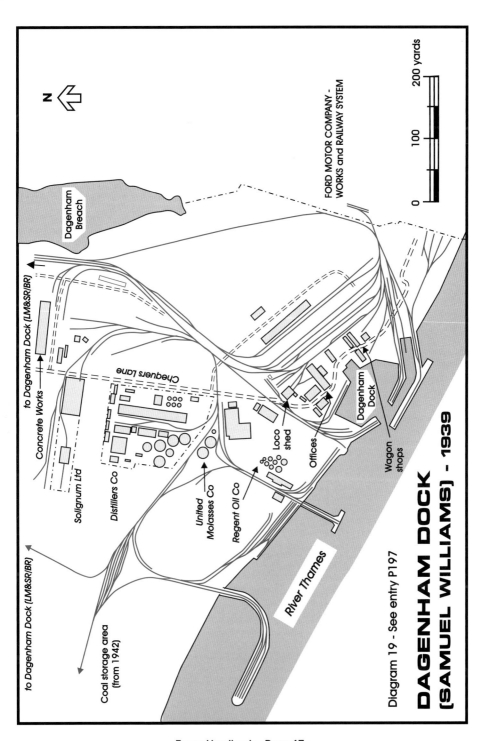

Diagram 19 - See entry P197

DAGENHAM DOCK (SAMUEL WILLIAMS) - 1939

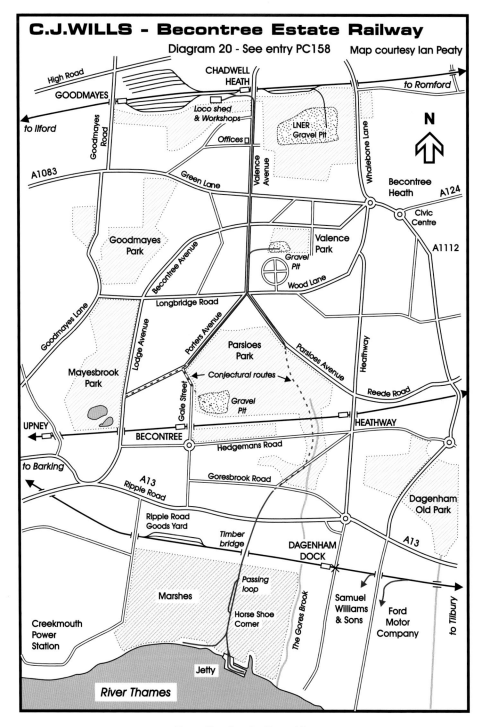

SECTION 1
INDUSTRIAL LOCATIONS

ADMIRALTY

ROYAL NAVAL ARMAMENT SUB-DEPOT, COPPERAS WOOD   H1
TM 205318

A sub-depot to RNAD Wrabness, located about two miles east of the main depot. Rail traffic ceased and depot closed at the end of 1961.

Gauge : 4ft 8½in

-	4wDM	FH	2496	1941	(a)	(1)

(a)   ex RNAD Ditton Priors, Shropshire by 2/1960.

(1)   to Lacmots Ltd, Queenborough, Kent, /1964.

ROYAL NAVAL ARMAMENT DEPOT, WRABNESS   H2
TM 161315

Armaments Depot, located on the River Stour about a mile west of Wrabness Station, with workshops used for the maintenance of mines; the depot covered 68 acres and had 11 magazines with a storage capacity of 6000 tons and was completed in 1920; standard gauge sidings served all the main buildings. Rail traffic ceased and depot closed 31/1/1965.

Gauge : 4ft 8½in

-		0-6-0F	OC	AB	1574 1918	New	(1)
-		4wDM		RH	210483 1942	(c)	(5)
YARD No.1339		4wDM		RH	207103 1941	(a)	(2)
-		0-4-0DM		RH	414301 1957	New	(3)
YARD No.43		4wDM		RH	221639 1943	(b)	(4)

(a)   ex Holton Heath Factory, Dorset, c/1953.
(b)   ex Lodge Hill Depot, Kent, 12/1961.
(c)   ex Philip & Son, Noss Works, Dartmouth, Devon, by 23/6/1948.

(1)   to RNAD, Dean Hill, Wiltshire.
(2)   to George Cohen, Sons & Co Ltd, dealers, c/1963;
      thence to Britannia Iron & Steel Works Ltd, Bedford, 5/1964.
(3)   to RNPF, Caerwent, Gwent, c/1964.
(4)   to W. Simmons & Son, dealers, Great Bentley, near Colchester, c/1964, by 3/1965.
(5)   s/s after 20/6/1956

ALPHA CEMENT LTD
Oxford & Shipton Cement Ltd until 6/1934

Subsidiary of **Associated Portland Cement Manufacturers Ltd** and **Tunnel Portland Cement Co Ltd** 1938 – 1948, and by APCM Ltd solely from 1949. In 1938 the ordinary share capital of Alpha Cement Ltd was acquired jointly by APCM Ltd and Tunnel Portland Cement Co Ltd, the latter company acquiring 26%. At the same time the three companies entered into a working agreement for closer co-operation in marketing cement products. APCM Ltd acquired Tunnel's 26% shareholding in 1/1949.

BRETTS FARM CLAY PITS, Aveley   L3

TQ 556820

Pits located about three miles south of Upminster. The company was listed in quarry directories after 1934 by 1939. The internal railway was replaced by conveyor belts by 1955; pits later worked out and flooded.

Gauge : 2ft 0in

-	4wDM	RH	177531	1935	New (b)	(1)
-	4wDM	RH	179866	1936	New	(2)
-	4wDM	RH	178003	1936	New	(6)
-	4wDM	RH	178004	1936	New	(6)
-	4wDM	RH	183744	1937	New	(5)
-	4wDM	RH	192861	1939	New	(4)
-	4wDM	RH	181824	1936	(a)	(3)

(a)   ex APCM, Shorne Wood Clay Pit, Cobham, Kent, after 10/5/1950, by 23/7/1951, converted from 2ft 6in gauge.
(b)   ex RH, Lincoln, 20/5/1936; earlier at the Public Works, Roads & Transport Exhibition, Royal Agricultural Hall, Islington, London, 13/11/1935 to 27/11/1935.
(1)   to Cliffe Works, Kent, c/1936, by 19/5/1938.
(2)   to APCM, Mason's Works, Ipswich, Suffolk, c/1953 (after 27/6/1953, by 15/4/1955).
(3)   to APCM Bevan's Works, Kent, after 7/8/1951, by c/1954.
(4)   to G. & T. Earle Ltd, Barrow Haven Clayfields, Lincolnshire, c3/1955.
(5)   to Rodmell Works, Lewes, Sussex, after 16/8/1939 by 5/12/1939.
(6)   to Cliffe Works, Kent, after 29/12/1936, by 24/3/1937.

METROPOLITAN WORKS, West Thurrock   N4

TQ 572782

Construction of the works was begun by the **Metropolitan Cement Company Ltd** (incorporated 19/4/1907); however the works was bought before completion by Alpha Cement Ltd in 5/1936. After 1949, rail traffic was worked by Thurrock Chalk & Whiting Co Ltd.

Gauge : 4ft 8½in

GOLIATH	6wVBT	VCG	AtW	110	1928	(a)	Scr 7/1952
IOTA	0-4-0T	OC	FE	211	1892	(b)	(1)
MARGAM	0-4-0ST	OC	MW	1306	1895	(c)	(2)
-	0-4-0DM		JF	21455	1936	(d)	(3)

(a)   ex Shipton-on-Cherwell Works, Oxfordshire, c/1940.
(b)   ex Stone Court Ballast Co Ltd, Kent, /1945
        via W.R. Cunis Ltd, Woolwich, London, for repair.
(c)   ex BPCM, Wouldham Works, /1948.
(d)   ex APCM, Highsted Chalk Pit, Kent, /1949.

(1)   returned to Stone Court Ballast Co Ltd, Kent, /1946, thence to BPCM Ltd, Penarth Works, Glamorgan.
(2)   returned to BPCM, Wouldham Works, /1949.
(3)   to APCM, Dunstable Works, Bedfordshire, c/1949.

ALRESFORD SAND & BALLAST CO LTD
ALRESFORD CREEK, near Colchester　　　　　　　　　　　　　　　　**H5**
　　　　　　　　　　　　　　　　　　　　　　　　　　　　　　　　TM 061200

Company incorporated 6/10/1932 with sand and gravel pits at Alresford Creek; by 1948 with pits at **Bradwell**, near Braintree, (TL 803229) also at **Ferriers Farm**, Bures (TL 894341), **Moverons**, Brightlingsea, (TM 068189) **Chase**, Asheldham, near Southminster, (TL 973018) and **Villa Farm**, Wivenhoe (TM 048223 approx); and by 1953 also with pits at **Wick Farm**, Stoke by Nayland (TM 009350) and Great Holland, near Clacton. Company still trading in 2010 with its registered office at Ford Lane, Alresford, near Colchester.

Gauge : 2ft 0in

-		4wPM	L	4449	1932	New	s/s
-		4wDM	RH	187099	1938	(a)	(1)

(a)　　ex Sidney-Wilmot Ltd, Thorington Street Gravel Pits, Suffolk, after 22/12/1942 by 25/6/1943.

(1)　　s/s after 9/3/1950

ANGLIAN WATER AUTHORITY, ESSEX SEWAGE DIVISION

MARSH FARM SEWAGE WORKS, Tilbury　　　　　　　　　　　　　　**M6**
Thurrock Urban District Council until 1/4/1974　　　　　　　　　　　TQ 656756

Sewage works served by an internal narrow gauge system. Railway closed in 1975 and dismantled.

Gauge : 2ft 0in

-		4wDM	FH	3787	1956	New	(1)

(1)　　to Alan Keef Ltd, Cote Farm, Bampton, Oxfordshire, 12/5/1980.

NEVENDON TREATMENT WORKS, Basildon　　　　　　　　　　　　**G7**
Basildon Development Corporation until 1/4/1974　　　　　　　　　TQ 738907

Sewage treatment works built by W.& C. French Ltd which was opened in 1958. It was served by an internal narrow gauge system with track set in concrete, apart from the elevated tipping dock and the use of 'jubilee' panels used to access the sludge beds. Sewage was treated using the latest technology, and the waste produced was a safe sludge dried and used for fertiliser. The railway was used to remove dried sludge from the sludge beds, which was a task for the warmer months of the year. Railway closed c/1975 but was still mainly complete in 1982; site later cleared for redevelopment.

Gauge : 2ft 0in

-		4wDM	RH	373359	1958	New	(1)

(1)　　to Pleasure-Rail Ltd, Knebworth House, Hertfordshire, c9/1980.

ASSOCIATED EQUIPMENT CO LTD
WALTHAMSTOW WORKS C8
TQ 357896

Company registered 13/6/1912, having a factory in Black Horse Lane, Walthamstow, served by a standard gauge siding from the Midland Railway Gospel Oak - Barking line, on the north side of the line west of Blackhorse Lane station. In 1905, as part of the early development of the London motor bus, the London Motor Omnibus Co Ltd (trading as "Vanguard") was registered 7/1/1905 and set up these workshops about 1906. The LMOC was acquired by Vanguard Motorbus Co Ltd (registered 4/1/1907) only to become part of London General Omnibus Co Ltd on 1/7/1908, and production of new buses commenced here with the first 'X' Type on 16/12/1909 and the more successful and numerous 'B' Type from 18/12/1910. Production reached a level of about 20 buses per week. In 1912 control of the LGOC passed to the Underground Electric Railways Co of London Ltd and the works was transferred to its subsidiary, Associated Equipment Co Ltd. The factory closed in 1927 when production was moved to the large new works at Southall, Middlesex. The Walthamstow works, including the rail system, was auctioned on 18/10/1927. (After the formation of the London Passenger Transport Board on 1/7/1933, AEC became an independent company).

Gauge : 4ft 8½in

   -                        0-4-0PM   McP
                                        BgC   702 1918 New   (1)

(1) to E.R. Cole dealer, Shepherds Bush, London, 10/1927;
thence to Calder & Co Ltd, Deptford Works, Rotherhithe, London, after 9/1928.

ASSOCIATED PORTLAND CEMENT MANUFACTURERS LTD
Formed 10/1919

THAMES WORKS, West Thurrock N9
APCM (1900) Ltd until 10/1919   TQ 595780
Gibbs & Co Ltd incorporated 8/5/1873 until 10/7/1900   TQ 595774
originally **Gibbs & Co**

A works, built 1872 by David and William Gibbs as **Thames Portland Cement Works**, to produce cement by the old process using brick built kilns. An experimental rotary kiln was tried in 1887. The works was served by a railway that ran south from it to a wharf on the Thames known as **Gibbs Wharf**. This line was used to transport clay and coal to the works and cement back for loading into barges. The works and quarries had ceased to be used for cement manufacture by 3/1910 but continued to produce sand & ballast. Later reopened 9/1911 and in /1916 were supplying chalk to Wouldham Works; continued whiting manufacturer and sand & ballast working until c/1920. Works taken over by **Thames Land Co Ltd** (see under Whitehall Securities Corporation Ltd for later history).

"The Engineer" 27/8/1875 wanted, contractors loco to hire for a year with option to purchase, a 4w loco to pull 50tons, Gibbs & Co Ltd, West Thurrock.

Gauge : 4ft 8½in

ALPHA	0-4-0WT	OC	[Lewin	c1876	New ?]	s/s /1922	
BETA	0-4-0WT	OC	[Lewin	c1876	New ?]	s/s /1922	
GAMMA	0-4-0T	OC	Lewin	c1877	New	s/s /1922	
DELTA	0-4-0T	OC	FE	150	1887	New(a)	(4)
IOTA	0-4-0T	OC	FE	211	1892	New	(3)

GUERNSEY		0-4-0ST	OC	MW	1241	1892	(b)	(1)
KAPPA		0-4-0ST	OC	CF	1164	1898	New(c)	(2)
BEATRICE		0-4-0ST	OC	[P?]			(d)	s/s
GLOBE No.1		[4wWT	G	AP]			(e)	s/s
GLOBE No.2		[0-4-2T	OC	KS	124	1897]	(e)	s/s

(a) here by 2/10/1896 (when boiler exploded).
(b) earlier John Mowlem & Co Ltd, Grays-Upminster-Romford (LT&SR) (1889-1893) contract, here by /1905.
(c) to Chas Wall Ltd, Grays, contractor, hire by 6/1910; returned.
(d) ex unidentified contractor at Salisbury, Wiltshire, on hire, /1910.
(e) ex Globe Portland Cement & Whiting Co, Greenhithe, Kent.

(1) to Bevans Works, Kent, /1914.
(2) to Brooks Works, by /1920.
(3) to Stone Court Ballast Co Ltd, Kent, /1925.
(4) to Swanscombe Works, Kent, 6/1927.

BROOKS WORKS, Grays   N10
APCM (1900) Ltd until 10/1919   Works TQ 606776
Hilton Anderson Brooks & Co Ltd until 10/7/1900   Quarry TQ 603785
Hilton Anderson Brooks & Co until 17/2/1893
Brooks Shoobridge & Co (formed 1870) until 1893

Edmund Wright Brooks together with his partner Frederick Hoffgaard Shoobridge, bought the disused Ulmate of Ammonia Works at Grays (established 1860, closed due to the collapse of the London finance house of Overend & Gurney and auctioned on 7/4/1869) and began manufacturing Portland Cement there in 1870. In that year the firm of Brooks, Shoobridge & Co was formed by the partners Edmund Wright Brooks, Frederick Hoffgaard Shoobridge, Bedford Marsh and Harris Hills, and in 1893 this firm amalgamated with Hilton Anderson & Co of Kent to form Hilton Anderson Brooks & Co; the works known as Anchor Works. Fredk H. Shoobridge left the partnership on 1/1/1879. The original tramway was worked by horses and may have been narrow gauge as narrow gauge lines existed in the works upon closure c1922. The works was later dismantled.

Gauge : 4ft 8½in

No.1		2-2-0WT	G	AP	1602	1880	New	Scr by/1914
No.2		2-2-0WT	G	AP	1780	1882	New	Scr by/1914
No.3	(AUNTY)	4wWT	G	AP	524	1870	(a)	Scr by/1914
No.4		0-4-0ST	OC	Lilleshall	187	1873	(b)	s/s
(8)	CLARENCE	0-4-0ST	OC	Ridley	13	1899	(c)	(1)
No.5	ALICE	0-4-0ST	OC	P	665	1897	New	(2)
No.6	ALLAN BRUCE	0-4-0ST	OC	AB	680	1890	New #	(3)
No.7	STUART	0-4-0ST	OC	AB	681	1890	New #	(4)
	KAPPA	0-4-0ST	OC	CF	1164	1898	(d)	(5)

\# supplied new through DK as agents.

(a) originally new 1/1870 to Wm Jay, Serpentine Dredging Contract, London, who advertised for hire an 8hp loco traction engine, 8/1/1874.
(b) ex Lincoln Wagon & Engine Co, /1890-1897, earlier Thomas Kinnersley & Co, Clough Hall Iron Works, Kidsgrove, Staffordshire, No.4.
(c) origins unknown.
(d) ex Gibbs Works, by /1920.

(1) to BPCM, Wouldham Works, after /1916.
(2) to BPCM, Wouldham Works, c/1920.
(3) to Grays Chalk Quarries Co Ltd.
(4) to Crown & Quarry Works, Kent, c/1921.
(5) to Burham Works, Kent.

RONALD L. BAKER (SHIPBREAKERS)
PITSEA WHARF, Marsh Road, Pitsea   K11

TQ 736859

A shipbreaking operation on the east bank of Vange Creek on a site which had earlier seen use of explosives storage (see British Explosives Syndicate Ltd in the Non-locomotive section). A locomotive was brought to this site and remained derelict for some time; it is not clear if any use was made of it, or track laid for its operation.

Gauge : 3ft 6in

    L C C No.3   4wDM   RH 235743 1945   (a) s/s after 5/1972

(a)   ex London County Council, Northern Outfall Works, Beckton, c8/1965.

BARKING JETTY CO LTD
COLCONITE WORKS, Creekmouth   P12

Barking Jetty Co Ltd incorporated 18/9/1934   TQ 469820

Coal Conversion Ltd registered 16/7/1928

Coal Conversion Ltd was set up to acquire from the Leeds Fireclay Company its interests in the Plassmann process for low temperature carbonization of coal and other fuels. The company erected a distillation works at Barking with three large retorts which went into production at the end of 1929. The works was served by a jetty 140ft long and extended about 1937 to 645ft for the unloading of coal both for domestic and industrial purposes, also for coke and slag. Coal Conversion Ltd set up a separate company called the Barking Jetty Company Ltd (in which Coal Conversion Ltd & Co- operative Wholesale Society Ltd had an equal share holding); this company acquired nine acres of land, the jetty, crane, conveying plant and other assets under agreement.

    Sir Alexander Gibb and Partners consulting engineers were responsible for supervising the installation of a new plant which would enable coal to be discharged from steamers into barges for conveyance up the river and into trucks for transport by rail; coal being brought by steamer from Northumberland & Durham. Coal Conversion Ltd sold their shareholding to Wm Cory & Son Ltd by 1/1946. The jetty company operated a branch from the south side of Ripple Lane Sidings on the LT&SR Barking -Dagenham line to their works and also to Barking Power Station. The railway had been laid in connection with the construction of the power station, which opened in 1925, but the Jetty Co did not operate rail traffic themselves until 1937. Works closed 1964, company went into voluntary liquidation 24/5/1965.

Gauge : 4ft 8½in

No.1	0-4-0ST OC	HL	3901	1937	New	(1)
No.2	0-4-0DM	DC	2251	1947		
		VF	D77	1947	New	(2)

(1)   scrapped on site by George Cohen, Sons & Co Ltd, Canning Town, 3/1962.
(2)   to Bristol Mechanised Coal Co Ltd, Filton, Bristol, 6/1965.

BARWYKE SAND & BALLAST CO
RAINHAM PITS   L13
                                                                                TQ 539837?

Rates records list these pits at Berwick Ponds Farm, also listed in quarry directory for 1933 but not thereafter.

Gauge : 2ft 0in

| | - | 4wPM | MR | 5086 | 1930 | (a) | (1) |

(a)   ex Petrol Loco Hirers, Bedford, 19/11/1932.

(1)   to Heathrow Sand & Gravel Co (Colnbrook) Ltd, West Drayton, Middlesex, 2/1934.

BEACH'S BRICKFIELDS LTD
RAINSFORD END BRICKWORKS, Chelmsford   G14
W. Beach & Son Ltd until 18/2/1946   TL 689071
W. Beach & Son until 23/12/1932

Brickworks established c1901. In the 1920's the works had two down-draft rectangular kilns each holding 38,000 bricks. In 1926 these kilns were changed from coal to oil firing and about two million sand-faced red facing bricks were produced per year. In 1942 the works was advertised for sale but remained unsold until 9/1945 when the brick works was sold to the London Brick Company. Beach's Brickfields Ltd was then incorporated and continued production until about 1966.

Gauge : 2ft 0in

| | - | 4wDM | OK | 6711 | 1936 | (a) | (1) |

(a)   origin unknown.

(1)   to Midland Brick Co Ltd, Long Buckby, Northamptonshire, by 1/1955.

BENT-MARSHALL (SOUTH HORNCHURCH) LTD
SOUTH HORNCHURCH GRAVEL PITS   L15
                                                                                TQ 522836

Company formed 23/1/1930 to take over gravel screening works on the Southend Road at South Hornchurch. Tramway closed c/1963.

Gauge : 2ft 0in

-	4wPM	FH	1709	1930	New	Scr /1956	
-	4wDM	FH	1763	1931	New	Scr /1956	
-	4wPM	FH	1894	1934	New	Scr /1956	
-	4wPM	FH	1889	1934	New	(1)	
-	4wPM	MR			(a)	Scr /1956	
-	4wPM	MR			(a)	Scr /1956	
-	4wPM	MR			(a)	Scr /1956	
-	4wDM	MR	7308	1938	(b)	s/s c/1963	
-	4wDM	MR	7377	1939	(c)	s/s c/1963	

(a)   identity and origin unknown, here by 27/6/1953.
(b)   originally George Wimpey & Co Ltd, Bishopton (ROF) contract, Renfrewshire.
(c)   ex George W. Bungay Ltd, dealers, Hayes, Middlesex; here by 1/1953;
      earlier J.N. Connell, dealer, Coatbridge, Lanarks;
      originally Glasgow Corporation Housing Dept, Robroyston contract, Glasgow.

(1)   to Lea Valley Sand & Ballast Pits Ltd, Cheshunt, Hertfordshire.

BERK SPENCER ACIDS LTD
ABBEY MILLS CHEMICAL WORKS, West Ham   **C16**
Berk Ltd   TQ 391829
F.W. Berk & Co Ltd until 1/1/1966
F.W. Berk & Co until 12/8/1891

This Vitreol works in Canning Road, West Ham was established by Thomas Bell & Co in the 1870s and taken over by Berk & Co c/1874. (T. Bell & Co Ltd went into voluntary liquidation on 5/12/1883). F.W. Berk & Co was founded in 1870 by Frederick William Berk and his brother Robert Berk in partnership and was incorporated on 12/8/1891. The company manufactured a wide range of chemical products and in 1960 F.W. Berk & Co Ltd & Borax (Holdings) Ltd acquired the whole share capitol of **Spencer Chapman and Messel Ltd** chemical manufacturers of Silvertown who in turn merged with the industrial acids business of **Berk Ltd** to become **Berk Spencer Acids Ltd in** 1966. This company built a new sulphuric acid plant at Stratford with a capacity of 120,000 tons which commenced production in 1965. Berk Ltd became a subsidiary of The Steetley Company Ltd in 1970. The works was served by standard gauge rail sidings near the intersection of the ex-LT&SR line and the ex-GER Stratford - North Woolwich line. Rail traffic ceased by 10/1972.

Gauge : 4ft 8½ in

         -                    4wDM   HU   c1928   New?   (1)

(a)   origin and identity uncertain, loco carried the number HT169, however this number has not been identified in surviving makers records.

(1)   to C.H. Higgins Ltd, Ilford, 10/1972.

S.J. BLYTH
SCRATTON ROAD SANDPITS, Stanford-le-Hope   **K17**
                                                                          TQ 687825
Mr S.J. Blyth of Hassenbook Hall, Stanford-le-Hope operated sand pits in Scratton Road, Stanford-le-Hope, he was also a director of Hawkwell Brickfields Ltd. The sand pits later became the site of school playing fields.

Gauge : 2ft 0in

                              4wDM   RH 166048   1934   (a)   (1)

(a)   ex Dobson, Ellis & Co, Stanford-le-Hope Sand and Ballast Pits, after 19/2/1935, by 19/6/1936.

(1)   to East Essex (Corringham) Sand Co Ltd, Corringham, after 20/6/1936, by 8/3/1938.

BP OIL LTD
PURFLEET WORKS N18
Shell-Mex & BP Ltd incorporated 12/11/1931 until ? TQ 567775

Oil storage and processing plant with wharves on the north bank of the River Thames served by standard gauge sidings connecting with the ex-LT&SR line east of Purfleet Station. Locos of Van de Berghs & Jurgens Ltd, who had premises adjoining this works, were occasionally borrowed to cover failures of the diesel locomotives. Works closed 4/1983.

Gauge : 4ft 8½in

		4wDM	FH	1853	1934	New	(1)
No.10		4wDM	FH	1977	1936	New	(4)
No.18		0-4-0DM	JF	22973	1942	(a)	(3)
No.13		4wDM	FH	2102	1937	(b)	(2)
		0-4-0DM	AB	395	1956	(c)	(5)
No.25		0-6-0DE	YE	2641	1957	(d)	(7)
No.20		0-4-0DM	JF	4210003	1949	New(e)	(6)

(a)   earlier ROF Bescot, Walsall, Staffordshire.
(b)   ex Hamble Refinery, Hampshire, 9/1950.
(c)   ex Thames Matex Ltd, West Thurrock, c/1969.
(d)   ex PLA, No.205, c/1970.
(e)   to Thames Matex Ltd, West Thurrock, c/1969; returned c3/1971.

(1)   to Potton Depot, Bedfordshire, by 4/1950.
(2)   to Trafford Park Depot, Lancashire, 2/1952.
(3)   to JF, 21/4/1950; thence to Hamble Refinery, Hampshire,
(4)   to Potton Depot, Bedfordshire, c/1958.
(5)   returned to Thames Matex Ltd, West Thurrock, c/1970.
(6)   to Thos. W. Ward Ltd, Silvertown, c5/1973.
(7)   to Foster Yeoman Ltd, Merehead Quarry, Somerset, c12/1983.

BP OIL CO LTD
CORYTON BULK TERMINAL K19
Mobil Oil Co Ltd until c2/1998 New Loco Shed TQ 744829
Vacuum Oil Co Ltd incorporated 13/5/1901, until 1/12/1955 Old Loco Shed TQ 742827
Cory Bros. & Co Ltd until 9/1950

In 1921 the Cory Bros, coal shippers and merchants, wished to diversify into the oil fuel trade. To do this they purchased the Kynochtown explosives factory and the associated Corringham Light Railway from Kynoch Ltd and converted the site into an oil fuel Depot. In 1951 the Vacuum Oil Company announced that they were to build a new oil refinery at Coryton this would process about 850,000 tons of crude oil per year and was completed in 1953. The company changed its name to the Mobil Oil Co Ltd on 1/12/1955, in 1964-1965 the refinery was expanded to increase its output by a third. Coryton was operated by BP from 1996, when Mobil's fuels operations in Europe were placed into a joint venture with BP. Following the merger in 1999 of Mobil & Exxon, the remaining interest in the refinery was sold to BP in 2000. On 1/6/2007 the plant was sold by BP to Petroplus Holdings AG for £714.6m and refining capacity had reached 10 million tonnes per annum. The plant was served by standard gauge rail sidings connecting with the BR (ex LT&SR) branch at Thames Haven which closed on 19/12/2008, rail traffic ceased.

The steam locos worked the Corringham Light Railway until its closure on 1/3/1952.

Reference : The Thames Haven Branch, Peter Kay, 1999.
Gauge : 4ft 8½in

		CORDITE	0-4-0WT	OC	K	T109	1884	(a)	Scr
		KYNITE	0-4-2T	OC	KS	692	1901	(b)	(1)
		-	0-6-0ST	OC	AE	1771	1917	(b)	(2)
		-	0-6-0ST	OC	AE	1672	1914	(c)	(2)
		-	4wDM		RH	386871	1955	New	(4)
		-	0-4-0DM		RH	418791	1958	New	(6)
		-	0-6-0DM		WB	3160	1959	New	(7)
		-	[0-4-0DH		NB]			(d)	(3)
		-	0-8-0DH		AB	506	1965	New	(5)
1			0-4-0DH		AB	506	1965		
				reb	AB	506/1	1969		
506/1				reb	AB		1982	(e)	(12)
2			0-4-0DH		AB	506	1965		
				reb	AB	506/2	1969		
506/2				reb	AB		1982	(e)	(12)
No.24			0-4-0DH		TH	239v	1971	(f)	(8)
H011	(08423)		0-6-0DE		Derby		1958	(g)	(9)
		-	0-6-0DH		TH	291v	1980	New(h)	(11)
		HAMBLE-LE-RICE	0-6-0DH		TH	294v	1981	(j)	
		(KENTISH MAID)	0-6-0DH		TH	295v	1981	(k)	(11)
(No 24	F-35)		0-6-0DH		HE	6950	1967	(k)	(10)

BR, D2224 on loan c/1964

(a) possibly ex Kynoch Ltd, with site.
(b) ex Kynoch Ltd, with site.
(c) ex William Jones Ltd, dealers, Greenwich, c/1933 (by 10/1933);
    originally WD, Shoeburyness.
(d) origin and identity unknown, here on trial/loan.
(e) ex AB, /1969; after 21/6/1969, rebuilds from 0-8-0DH AB 506/1965.
(f) ex Shell UK Ltd, Stanford-le-Hope, c7/1981.
(g) ex RMS Locotec Ltd, Dewsbury, West Yorkshire, on hire,
    9/1995 (after 2/6/1995 by 16/11/1995).
(h) to RMS Locotec Ltd, Dewsbury, West Yorkshire, for overhaul; returned, 11/2/1998.
(j) ex Hamble Oil Terminal, Hampshire, c4/1997 by 19/10/1998.
(k) ex Isle of Grain Bitumen Terminal, Kent, c10/1999 by 19/11/1999.

(1) withdrawn /1920s and remained derelict until scrapped, 3/1952.
(2) scrapped 8/1957 by Ray, of Southend.
(3) returned to makers after trial.
(4) to George Cohen Sons & Co Ltd, /1962, and at their Cargo Fleet Depot, Middlesbrough, Yorks (NR), by /1966.
(5) returned to AB; rebuilt /1969 to two 0-4-0DH locos, AB 506/1/1969 and AB 506/2/1969, both of which returned here..
(6) sold 24/3/1973, buyer unknown.
(7) to Resco (Railways) Ltd, Woolwich, London, c12/1979.
(8) to Gulf Oil Refining Ltd, Waterston Refinery, Dyfed, w/e 3/2/1984.
(9) to Faber Prest Ports Ltd, Flixborough, Lincolnshire, after 12/6/1998 by 11/7/1998.
(10) to Elsecar Steam Railway, Elsecar Heritage Workshops, Barnsley, Yorkshire, 19/5/2006.

(11) to C.F. Booth Ltd, Rotherham, South Yorkshire, 10/9/2009.
(12) to C.F. Booth Ltd, Rotherham, South Yorkshire, 16/9/2009

BRITISH PORTLAND CEMENT MANUFACTURERS LTD
Company registered 1/11/1911
WOULDHAM WORKS, West Thurrock   **N20**
Wouldham Cement Co Ltd until 12/1911   Quarries TQ 598786 & 597790
Wouldham Cement Co (1900) Ltd   Works TQ 598775
(incorporated 24/2/1900) until 23/6/1906
S. Pearson & Son Ltd until 26/2/1900
Wouldham Cement Co Ltd (incorporated 20/1/1865) until 24/8/1899
Lion Cement and Chalk Co Ltd incorporated 9/9/1874, until 21/9/1876
This cement works was established c1874 by the **Lion Cement Co** as Lion Works and was later purchased by Major Robertson (who also owned cement works at Wouldham near Rochester, Kent). The works was enlarged and traded as Wouldham Cement Co, Lion Cement Works. In 1899 the works was bought by S, Pearson & Sons Ltd to provide cement for their contract to build Dover Harbour. Pearson's greatly expanded the works which when purchased was producing 450 tons of cement per week, with the addition of six rotary kilns and a new block of chamber kilns production was increased to 2,500 tons per week. Wouldham Cement Company Ltd was taken over by the BPCM group in 1911 but continued to trade as an associate company of BPCM Ltd until 14/5/1926. Located on the south side of the LT&SR line west of Grays station (siding agreement dated 5/9/1902); the works had a jetty and wharf worked by steam cranes unloading coal and clay from colliers and barges and also for loading cement. Conveyor belt operation between the quarry and works commenced in 1961, rail traffic ceased and the works later closed in 1976.
Sale by auction by order of the Mortgagee 29/6/1877 – Newly erected Lion Cement and Whiting Works, including a line of railway connecting the chalk pit with the factory and wharf, with double winding gear for wagons worked by a seven - horsepower steam engine, by Clayton, Shuttleworth & Co.
The locomotive numbers were carried inside the cabs.
Gauge : 4ft 8½in

2		0-4-0ST	OC	HH			New	Scr by/1924
	BARKING	0-6-0ST	IC	MW	858	1882	(a)	(2)
2		0-4-0T	OC	FJ	168	1879	(b)	Scr by/1924
No.3	WOULDHAM	0-4-0ST	OC	P	771	1899	New	Scr 5/1958
No.6	ARAB	0-4-0ST	OC	P	800	1899	(c)	(9)
No.7	ALWILDA	0-4-0ST	OC	P	798	1899	(d)	(9)
	WARWICK	0-4-0ST	OC	HE	98	1873	(e)	Scr by/1924
	ANNIE	0-4-0ST	OC	MW	1135	1891	(f)	(3)
4	MARGAM	0-4-0ST	OC	MW	1306	1895	(g)	(8)
No.8	STANLEY	0-4-0ST	OC	P	1314	1913	New	Scr /1965
5	THOR (f. PETROS)	0-4-0ST	OC	HE	629	1896	(h)	(4)
	CLARENCE	0-4-0ST	OC	Ridley	13	1899	(i)	(1)
	CORSTON	0-6-0T	IC	MW	1196	1890	(j)	(5)
	FELSPAR	0-4-0ST	OC	MW	1846	1914	(k)	(7)
No.5	THOR	0-4-0ST	OC	AB	1391	1915	(l)	Scr c12/1965
"No.9"	LION	0-4-0ST	OC	P	665	1897	(m)	Scr 10/1958
"No.10"	WITHAM	0-4-0ST	OC	AB	1411	1915	(n)	(9)

"No.11"	BALDER		0-4-0ST	OC	AB	714	1893	(o)	(10)
	GEORGE		0-4-0ST	OC	AB	1281	1912	(p)	(6)
"No.12"	CAESAR		0-4-0ST	OC	AB	1816	1923	(q)	(9)
"No.13"	WEST THURROCK		0-4-0ST	OC	P	1707	1926	(r)	(9)
1"No.14"	WOULDHAM		0-4-0ST	OC	P	1701	1926	(s)	(10)
	GOLIATH		0-4-0ST	OC	AB	1741	1921		
					reb AB		1927	(t)	(9)
	5	HILTON	0-4-0ST	OC	P	633	1896	(u)	(11)
		ELIZABETH	4wDH		RR	10232	1965	New	(12)

(a)   ex S. Pearson & Son Ltd, West Thurrock Plant Depot, on loan.
(b)   ex Wouldham Cement Works Co, Kent, c/1899.
(c)   ex S. Pearson & Son Ltd, Banbury and Lockwood Reservoirs (1899-1904) contract, Walthamstow, No.97, by 3/1901.
(d)   ex S. Pearson & Son Ltd, Banbury and Lockwood Reservoirs (1899-1904) contract, Walthamstow, by 3/1901, No.95, to Thos. W. Ward Ltd, Grays for repairs, by 19/9/1954: returned, after 12/3/1955, by 14/7/1955.
(e)   ex S. Pearson & Son Ltd, Surrey Docks Extension (1895-1898) contract, London; here by /1902.
(f)   ex S. Pearson & Son Ltd, West Thurrock Plant Depot, /1909;
(g)   ex S. Pearson & Son Ltd, West Thurrock Plant Depot, No.39, /1909; to Alpha Cement Ltd, West Thurrock, /1947; returned, /1950.
(h)   ex E.W. Goodenough /1916, earlier J.S. Peters, Merstham, Surrey.
(i)   ex Brooks Works, Grays, after /1916.
(j)   ex S. Pearson & Son Ltd, Queen Mary Reservoir (MWB) (1919-1924) contract, Middlesex.
(k)   ex George Cohen, Sons & Co Ltd, Canning Town.
(l)   ex Vickers Ltd, Barrow, Lancashire, /1920.
(m)   ex APCM, Brooks Works, Grays, after/1924.
(n)   advertised for sale by Cohen 1/12/1933, earlier John Mowlem & Co Ltd, King George V Graving Dock (1931-1935) contract, Hampshire, /1935.
(o)   ex George Cohen Sons & Co, Port Talbot, /1937, earlier R.B. Byass & Co Ltd, Port Talbot, Glamorgan.
(p)   ex Thurrock Chalk & Whiting Co Ltd, /1945.
(q)   ex Alpha Cement Ltd, Kirton Lindsey, Lincolnshire, c9/1951, by 2/9/1951.
(r)   ex Whitehall Security Corporation Ltd, Grays, 4/3/1955.
(s)   ex Bevan's Works, Kent, 12/1957.
(t)   ex Johnson's Branch, Greenhithe, Kent, 4/1960, rebuilt from 3ft 9in gauge.
(u)   ex Kent Works, Stone, Kent, 4/11/1960.

(1)   to Kent Works, Stone, Kent, by 24/4/1943.
(2)   to S. Pearson & Son Ltd, West Thurrock Plant Depot, off loan.
(3)   to ICI Ltd, Silvertown Works.
(4)   to Seabrook & Sons, Grays, by 11/1918.
(5)   to G. & T. Earle Ltd, Melton Works, East Yorkshire, 7/1926.
(6)   returned to Thurrock Chalk & Whiting Co Ltd, /1945.
(7)   to George Cohen, Sons & Co Ltd, Canning Town.
(8)   loaned to Light Expanded Clay Aggregates Ltd, Grays, c/1952.
(9)   scrapped on site by W Rice, of Slades Green, Kent, c7/1961.
(10) scrapped on site by W Rice, of Slades Green, Kent, c/1961, after 7/8/1961.
(11) scrapped on site by W Rice, of Slades Green, Kent, 6/1963.
(12) to Swanscombe Works, Kent, 15/2/1978.

Gauge : 500mm
According to maker's records the following loco was delivered here but it may have worked at an associated quarry.

-	4wPM	H	985	1931	New	s/s

BRITISH SUGAR CORPORATION LTD
FELSTED WORKS   F21
Second Anglo-Scottish Beet Sugar Corporation Ltd (formed 10/1925) until 12/6/1936

TL 664209

This works (one of 15 such built following the passing of the Sugar Beet Subsidy Act in 1925) was built by Duncan Stewart & Co Ltd. It was served by standard gauge rail sidings connecting with the BR (ex GER) Bishop's Stortford-Braintree railway. Rail traffic ceased 1968 and works was later closed and demolished.

Gauge : 4ft 8½in

(No.3)		0-4-0ST	OC	AB	1886	1926	New	Scr 9/1968
No.5		0-4-0ST	OC	P	1725	1927	New	(1)
No.1		0-4-0ST	OC	P	1439	1916	(a)	(2)
	JAMES	0-6-0ST	IC	HC	1429	1920	(b)	(3)
-		0-4-0DM		RH	281269	1950	New	(4)

(a)   ex Colwick Factory, Nottinghamshire, c6/1953.
(b)   ex South Lynn Factory, Norfolk, c/1953.

(1)   to Colwick Factory, Nottinghamshire, c9/1951,by 13/10/1951.
(2)   to Colwick Factory, Nottinghamshire, by /1954.
(3)   derelict /1957; s/s c/1964, after 3/1964.
(4)   to Foley Park Factory, Worcestershire, 1/1969.

BROWN & TAWSE TUBES LTD
WEST HORNDON WORKS   L22
Company registered 1/6/1929   TQ 618880
Brown & Tawse Ltd
Brown & Tawse until 23/1/1917

Company established in 1881, by Peter Saunders Brown and James Tawse as Iron Steel and Metal Merchants with an address at Exchange Street in 1894, Dundee and Lime Street Square in the City of London. Standard gauge sidings served this steel stockholders on the north side of the ex-LT&SR Upminster - Southend line. Rail traffic ceased 2/1977.
Reference : Industrial Railway Record, No.86.
Gauge : 4ft 8½in

-		4wPM	Brown & Tawse		(a)	(1)
	BLUEBIRD	4wPM	MH		(b)	s/s /1971
-		4wDM	RH	200794 1940	(c)	s/s /1971
-		0-4-0DM	HC	D610 1939	(d)	Scr 12/1977
-		4wDM	RH	221639 1943	(e)	(2)

(a) built at West Horndon jointly by Brown & Tawse & Rotary Hoes with Brown & Tawse supplying the wagon chassis and Rotary Hoes supplying the Ford 10 engine, the chassis carried a Derbyshire Carriage & Wagon Co makers plate.
(b) ex scrap dealer; rebuilt at West Horndon with a petrol engine.
(c) ex Chas. Brand & Sons Ltd, Merton Plant Depot, Surrey, after 12/1955; originally Brand's Puriton (ROF) (1941-1942) contract, Somerset.
(d) ex T. Mitchell (Scrap Merchants & Plant Hire) Ltd, Bolton, Lancashire, /1965; originally Air Ministry, Hartlebury, Worcestershire.
(e) ex W. Simmons & Son, dealers, Great Bentley, c/1966; earlier Admiralty, Wrabness.
(1) to Rotary Hoes Ltd, West Horndon, after 5/1953.
(2) to Colne Valley Railway, Castle Hedingham, c9/1979.

BROXBOURNE SAND & BALLAST PITS LTD
BROXBOURNE PITS E23
Mr S.R. Sheppard trading as **Broxbourne Sand & Ballast Co** until 1/6/1933   TL 381058
Workings, served by a narrow gauge tramway, in Old Nazeing Road, Broxbourne, listed in quarry directories from 1933 until at least 1943/1944 but not listed in 1948. Motor Rail spares were ordered for here up to 2/5/1940. Company went into voluntary liquidation on 14/3/1966.
Gauge : 2ft 0in

		4wDM	RH	164337	1931	New	(a)	(1)
		4wPM	MR	4723	1938	New		(2)

(a) initially on hire (if not always);
to RH, Boultham Works, Lincoln, for overhaul, after 16/1/1934, by 5/9/1934; returned c10/9/1934; to RH, Boultham Works, Lincoln, for overhaul, 31/12/1937; returned after 20/1/1938, by 25/10/1940.
(1) to Eastwoods Cement Ltd, Barrington Works, Cambridgeshire after 2/4/1941 by 29/2/1945.
(2) to Wormley Sand Pits, Church Lane, Wormley, Hertfordshire.

ALEXANDER BRUCE (GRAYS) LTD
BRUCES WHARF, Grays N24
TQ 612776
Locomotive operated by this company of stevedores and wharf owners registered 21/7/1951 at 14 Buckingham Street, WC2. Originally the site of **Clark and Standfield** who applied in 1900 for permission to build a jetty and let it to Bruce. Standard gauge sidings served a Thames-side wharf with connections to the ex-LT&SR line just west of Grays Station. The company were also timber importers and manufacturers of railway sleepers.
Gauge : 4ft 8½in

L5 41		4wDM	FH	3885	1958	(a)	(1)

(a) ex W.R. Cunis Ltd, Rainham, c2/1976.
(1) to A.J. Birch & Son Ltd, Hope Farm, Sellinge, Kent, by 5/1986.

BUTTERLEY BUILDING MATERIALS LTD
Company registered 7/11/1985, subsidiary of **Hanson PLC**

CHERRY ORCHARD LANE BRICKWORKS, Rochford   J25
London Brick Co Ltd   TQ 859899
Milton Hall (Southend) Brick Co Ltd, until 1984
Thornback Brick Co Ltd, until 1926
James Thornback, until 1919
John Thornback, until c1910
W.J. Watts & Co Ltd, until 1904
William James Watts, c1890 until 1902

In 1/1926 the Milton Hall (Southend) Brick Co Ltd bought the works of the **Thornback Brick Co Ltd**, (incorporated 14/11/1919) in Cherry Orchard Lane, Rochford to replace the original works at Prittlewell, this works produced both stock and Red Bricks, manufacture of stocks ceased at the end of 1931. In 9/1932 the company bought land containing brick earth from Caleb Rayner at Townfield, Star Lane, Great Wakering, as the site for a new works. Clay pits and brickworks were served by a narrow gauge tramway some ½ mile in length. The railway remained in very limited use in 7/1993 but closed c17/2/1995. The works also had a hand-worked 2ft 6in gauge line running for 40 yards through the shrink-wrap plant, with flat wagons loaded/unloaded by forklift trucks.

Gauge : 2ft 0in

-	4wDM	RH	441951	1960	(a)	(2)
-	4wDM	RH	237916	1946	(b)	(1)
-	4wDM	RH	179881	1936	(b)	s/s /1966
-	4wDM	MR	21520	1955	(c)	(3)
-	4wDM	MR	8614	1941	(d)	(3)
-	4wDH	AK	10	1983	New	(7)
-	4wDM	MR	11111	1959	(e)	(4)
-	4wDM	MR	40S343	1969	(f)	(5)
-	4wDH (ex DM)	RH	283513	1949		
	reb	AK	20R	1986	(g)	(6)
-	4wDM	AK	40SD530	1987	New	(8)
-	4wDM	AK	28	1989	New	(8)
-	4wDM	AK	26	1988	(h)	(8)

Note that an uncertain number of internal combustion locos, s/s c1952, were present.

(a)   ex Thorpe Bay Brickworks, c/1963.
(b)   ex Star Lane Brickworks, /1965.
(c)   ex Star Lane Brickworks, /1966;
       to Star Lane Brickworks, /1980; returned.
(d)   ex M.E. Engineering Ltd, Cricklewood, Greater London, 24/9/1975;
       earlier G. Foster, Ramshaw, Co. Durham.
(e)   ex M E Engineering Ltd, Neasden, London, on hire, 5/1983.
(f)   ex Alan Keef Ltd, Cote, Oxfordshire, on hire 6/1983.
(g)   ex Alan Keef Ltd, Cote, Oxfordshire, 6/6/1986; earlier Great Bush Railway, East Sussex.
(h)   ex Star Lane Brickworks, for disposal;

(1)   to John S. Allen Ltd, dealers, Upminster, /1966; exported to Singapore, 5/1967.
(2)   to Star Lane Brickworks, c11/1969.
(3)   to Star Lane Brickworks, /1980.
(4)   returned to M. E. Engineering Ltd, Neasden, London, 6/1983.
(5)   to Star Lane Brickworks, on hire, 9/1983.
(6)   to Wickford Narrow Gauge Railway Group, Wickford, 4/12/1992.
(7)   to Wickford Narrow Gauge Railway Group, Wickford, 5/12/1992.
(8)   to Alan Keef Ltd, Lea Line, Ross-on-Wye, Herefordshire, c5/1998 (after 25/4/1998 by 3/6/1998).

STAR LANE BRICKWORKS, Great Wakering   J26
London Brick Co Ltd, 1984   TQ 934873
Milton Hall (Southend) Brick Co Ltd, 1932 until 1984

Brickworks and claypits originally served by an aerial ropeway in operation by 1933, later replaced by a narrow gauge tramway. Tramway closed c/1966 but re-opened c11/1969; extended c11/1984 to a new clay field to give a total length of ½ mile; finally closed on 10/10/1991. Works also served by internal hand-worked railways of 2ft 0in gauge (production area and kilns), 2ft 6in gauge (shrink-wrap plant) and 8ft 0in gauge (modern fast-firing kiln). Works closed 4/2005

Gauge : 2ft 0in

-		4wDM	RH	237916	1946	New	(1)
-		4wDM	RH	256194	1948	New	(3)
-		4wDM	MR	21520	1955	(a)	(4)
-		4wDM	RH	179880	1936	(b)	(2)
-		4wDM	RH	179881	1936	(b)	(1)
-		4wDM	RH	441951	1960	(c)	(5)
-		4wDM	AK	26	1988	New	(6)
-		4wDM	MR	40S343	1969	(d)	(4)

(a)   ex Great Wakering Brick Co Ltd, c/1963;
      to Cherry Orchard Brickworks, /1966; returned /1980.
(b)   ex Thorpe Bay Brickworks, c/1963.
(c)   ex Cherry Orchard Brickworks, c11/1969.
(d)   ex Cherry Orchard Brickworks, on hire, 9/1983.

(1)   to Cherry Orchard Brickworks, /1965.
(2)   to L.W. Vass Ltd, Ampthill Station, Bedfordshire, /1966.
(3)   to John S. Allen & Sons Ltd, dealers, Upminster, c/1966.
(4)   returned to Alan Keef Ltd, Cote, Oxfordshire, off hire c14/11/1983.
(5)   to Wickford Narrow Gauge Railway Group, Wickford, 19/11/1992.
(6)   to Cherry Orchard Brickworks for disposal.

The two locomotives listed below were delivered to the Milton Hall Brick Co, Southend-on-Sea; which works is unknown but may have worked at Star Lane Brickworks; remains of unconnected lengths of 3ft 0in gauge track were found inside the buildings at this works.
"Contract Journal", 12/8/1936 – for sale 7 wagons 3ft gauge, Southend- on- Sea Estates Co Ltd.
"Machinery Market", 15/8/1947 – for sale two Howard 3ton locos 3ft gauge with 25hp Dorman petrol engines, need overhaul – Milton Hall (Southend) Brick Co Ltd, Southend.

Gauge : 3ft 0in

-	4wPM	H	939	1928	New (a)	(1)
-	4wPM	H	964	1929	New (a)	(1)

(a) possibly new here or ex Prittlewell Brickworks.
(1) to Middlesex Contractors Plant Ltd, London, for sale 1/1948 & 9/1948, s/s.

CANEWDON BRICK, SAND & BALLAST CO
BRICKWORKS & GRAVEL PITS, Ballards Gore, near Rochford   **J27**
W.T. Lamb & Sons trading as Canewdon Brick, Sand & Ballast Co   TQ 903929
P.H. Cater, from 1920s until 1930s

Brickworks & gravel pits on the site owned by **Cater Bros (Scotts Hall) Ltd** (farmers) and listed in rates records from 1931 to 1939. The works ceased production at the outbreak of WW2 and site taken over by the War Dept. The Brickworks, plant and gravel pit including 116 acres of land were auctioned on 29/8/1947 on behalf of the land owners, Scott's Hall Estate.

Gauge : 2ft 0in

-	4wDM		RH	174528	1935	New	(1)

(1) to W.T.Lamb & Sons Ltd, Rayleigh Brickworks, after 10/2/1939 by 14/7/1944.

CANNING TOWN GLASS WORKS LTD
CANNING TOWN WORKS   **C28**
    TQ 390822

Company incorporated 17/3/1923 to take over works in Stephenson Street from British Glass Industries Ltd, works built by Henry Boot & Sons, closed, 3/1961 and demolished in 1962.

Gauge : 4ft 8½in

-	4wPM		MR	2034	1920	New	(3)
-	4wVBT	VCG	S	6994	1927	(a)	(1)
PEGGY	0-4-0ST	OC	AB	266	1894		
	Reb		Wake			(b)	(2)
No.4	0-4-0ST	OC	AB	1666	1920	(c)	(2)
-	4wDM		FH	3147	1947	New	(4)

(a) ex George Cohen Sons & Co Ltd, Canning Town, on hire.
(b) ex Gas Light & Coke Co, Bromley by Bow gas works, on hire, /1934.
(c) ex Gas Light & Coke Co, Bromley by Bow gas works, on hire.
(1) returned to George Cohen Sons & Co Ltd, Canning Town.
(2) returned to Gas Light & Coke Co.
(3) to Birds Commercial Motors Ltd, Stratford-on-Avon, Warwickshire, /1961.
(4) to United Glass Ltd, Ravenhead Works, St.Helens, Lancashire, c4/1961.

CARLESS SOLVENTS LTD
HARWICH REFINERY, Parkeston H29
Carless, Capel & Leonard Ltd incorporated 23/12/1948   TM 232323
A business started by Eugene Carless at Hackney Wick in London in 1859 as a distiller and refiner of mineral oils with partner W.G. Blagden; later joined by Mr Capel to became Carless & Capel between 1870 and 1872. In 1873, the business was purchased by John Hare Leonard to become Carless, Capel & Leonard; the firm invented the name "Petrol" used as a trade name to describe what was then, launch spirit. The firm acquired other works in London and refined by-products from gas works. Due to closure of coal gas works in the 1950s, the company built a refinery at Harwich in 1964 (site leased from 1/7/1963); by 1968 refining capacity had doubled and by 1970 storage capacity trebled. The works at Refinery Rd, Parkeston is served by standard gauge sidings on the south side of the ex-GER Manningtree - Harwich line.

Gauge : 4ft 8½in

No.35		0-4-0F	OC	RSHN 7803	1954	(a)	(1)
-		4wDM	R/R	S&H			
				WhC 4001	1966	(b)	(2)
BWC 687F	FP 41 CO	4wDH	R/R	NNM 73511	1979	New	(2)
Q240 JBV		4wDM	R/R	Unimog 092692	1982	(c)	

(a) ex North Thames Gas Board, Beckton, per C.& K. Metals Ltd, Barking, /1972.
(b) origin unknown; here by /1972.
(c) ex Frank Atkinson Ltd, High Bentham, North Yorkshire, by 9/2008.

(1) to S.A. Pye, Bramfield, Ipswich, Suffolk, 14/10/2000.
(2) s/s by 9/2008.

CENTRAL ELECTRICITY GENERATING BOARD
Central Electricity Authority until 1/1/1958
British Electricity Authority until 1/4/1955

BARKING POWER STATION, Creekmouth P30
County of London Electric Supply Co Ltd. from 23/4/1904 until 1/4/1948
Ripple Lane Sidings TQ 472832
Loco Shed TQ 469822

Coal-fired power station built on the north bank of the River Thames, east of Barking Creek. The site was purchased in 1914; design and construction was the responsibility of consulting engineers Sir Alexander Gibb & Partners on the site of an earlier rubbish tip. Work on the foundations began in 12/1922 with the first piles being driven on 2/1/1923. In addition to Essex the station was to supply power to part of Kent. This meant laying a high voltage cable under the Thames through a seven foot diameter tunnel half a mile long, bored specially for the purpose. Two vertical shafts were sunk to a depth of 100ft, one in the station area, the other in Woolwich Arsenal on the southern shore. The tunnel was formed by cast-iron rings bolted together; the gaps between this tube and the walls of the tunnel were grouted with cement injected under pressure. There were several extensions to the station some involving contractors Concrete Piling Ltd. The 'A' station opened on 19/5/1925; the 'B' station was built between 1931 and 1939 and the 'C' station between 1952 and 1954. The site was served by standard gauge rail sidings and a branch about a ½ mile long connected these to the ex-LT&SR Barking - Dagenham line on the south side of Ripple Lane Sidings. This line was laid in connection with the construction of the power station and was used for the delivery of general supplies, but almost all incoming coal arrived by sea rather than rail. The power station closed c1977 and was demolished during 1978.

Gauge : 4ft 8½in

1		0-4-0F	OC	HL	3595	1924	New	(2)
2		0-4-0F	OC	HL	3596	1924	New	Scr 8/1967
3		0-4-0PM		Bg	1386	1924	New	Scr
4		0-4-0PM		Bg	1387	1924	New	Scr
5		0-4-0ST	OC	HL	3653	1927	New	(5)
6		0-4-0ST	OC	HL	3791	1932	New	(6)
7		0-4-0ST	OC	HE	1872	1937	New	(7)
8		0-4-0ST	OC	RSHN	7150	1945	New	(8)
	SHIRLEY	0-4-0ST	OC	AB	1993	1932	(a)	(1)
	-	4wDM		FH	3294	1948	New	(10)
11257		0-4-0ST	OC	Hor	1111	1910	(b)	(3)
	-	4wVBT	VCG	S	9398	1950	(c)	(4)
(D2041)		0-6-0DM		Sdn		1959	(d)	(9)

BR, D3338 was on hire 11/9/1971

(a)   ex John Mowlem & Co Ltd, Chingford Reservoir (1935-1951) contract, /1943.
(b)   loaned from Holloway Bros.
(c)   ex S on demonstration 3/1952.
(d)   ex Rye House Power Station, Hertfordshire, c5/1971.
(1)   returned to John Mowlem & Co Ltd, Chingford Reservoir (1935-1951) contract, /1946.
(2)   to Gas Light & Coke Co, Beckton By-Products Works, after 3/1947.
(3)   to Holloway Bros, Littlebrook Power Station, Kent, contract, /1949-1950.
(4)   returned to S after demonstration,
       thence to NCB Desford Colliery, Leicestershire, on demonstration c17/3/1952.
(5)   to Bow Power Station, /1958.
(6)   to West Ham Power Station, /1958.
(7)   advertised for sale, 8/1958; sold for scrap, /1958.
(8)   to St John's Wood Power Station, London, c9/1958.
(9)   returned to Rye House Power Station, Hertfordshire. 8/1974.
(10)   to Stour Valley Railway Preservation Society, Chappel & Wakes Colne Station, 12/1978.

BOW POWER STATION   C31
London Power Co Ltd until 1/4/1948   TQ 378834
London Electricity Joint Committee (1920) Ltd until 1925
Charing Cross West End & City Electricity Supply Co Ltd until 1924
Charing Cross & Strand Electricity Supply Corporation Ltd until 1905

Construction started 1900, and the power station was commissioned in 1902. It was located on the south of the ex-GER main line west of Stratford station at Marshgate Lane, Bow. Siding agreement dated 7/2/1918, rail traffic ceased c2/1966, power station closed and demolished.

Gauge : 4ft 8½in

1		0-4-0ST	OC	P	936	1902	New	(2)
2		0-4-0ST	OC	KS	4168	1920	New	(1)
MET		0-4-0ST	OC	HL	2800	1909	(a)	(4)
5		0-4-0ST	OC	HL	3653	1927	(b)	(5)
ED 8		0-4-0ST	OC	AB	2353	1954	(c)	(3)

(a) ex Acton Lane Power Station, Middlesex by 1/1950.
(b) ex Barking Power Station, /1958.
(c) ex John Mowlem & Co Ltd, West Thurrock Power Station contract, off loan, c8/1965.

(1) to Battersea Power Station, London, 1/1937.
(2) to H. & J.R. Saunders & Co Ltd, Leyton Station, /1957, by 14/4/1957.
(3) to Goldington Power Station, Bedfordshire, c/1967.
(4) to Industrial Locomotive Preservation Group (Kent & East Sussex Railway), Tenterden, Kent, c3/1968.
(5) sold for scrap, 8/1968.

TILBURY POWER STATION   M32
TQ 659755

This power station was built by Holloway Bros. It was commissioned in 1956 with the 'B' station being due for completion in 1965. Coal was received at a jetty on the River Thames and was transported to a stocking site by conveyor.

Gauge : 4ft 8½in

No.1	0-4-0ST	OC	RSHN	7540	1949	(a)	(1)
ED 8	0-4-0ST	OC	AB	2353	1954	New	(2)
-	4wDM		RH	305314	1951	(b)	(3)

(a) ex BEA, Brunswick Wharf, London, c/1953.
(b) ex Drakelow Power Station, Derbyshire, c/1963.

(1) to CEA, Marchwood Power Station, Hampshire, /1955.
(2) to John Mowlem & Co Ltd, West Thurrock Power Station contract, on loan 9/3/1959.
(3) to Drax Power Station, North Yorkshire, c/1967.

WEST HAM POWER STATION, Bidder St, Canning Town   C33
TQ 389818

This power station was built by the County Borough of West Ham in 1904. It received coal by rail or river barges, and was later rebuilt and re-opened on 18/7/1951. The later 'B' station was constructed by Taylor Woodrow.

Gauge : 4ft 8½in

6	0-4-0ST	OC	HL	3791	1932	(a)	(1)
-	4wDM		RH	306092	1950	New(b)	(2)

(a) ex Barking Power Station, /1958.
(b) to Rye House Power Station, Hertfordshire, 7/1971; returned c10/1974.

(1) to G.J. Marshall & Co Ltd, Battersea, London, for scrap, 6/1967; scrapped 11/1967.
(2) to Rutland Railway Museum, Cottesmore, Rutland, 17/10/1981.

CHIGWELL URBAN DISTRICT COUNCIL
LUXBOROUGH LANE SEWAGE WORKS, Chigwell   D34
TQ 422926

Sewage works built in 1939 and later extended by W.& C. French Ltd, served by an internal narrow gauge system. Railway closed and dismantled.

Gauge : 2ft 0in

-	4wPM	L	34521	1949	(a)	(1)

(a)   ex W.& C. French Ltd, Buckhurst Hill.

(1)   to M.E. Engineering Ltd, Cricklewood, Middlesex, 8/3/1966; rebuilt to 4wDM and thence to Mordale Ltd, Little Woolden, Lancashire, 22/9/1966.

CLARK AND STANDFIELD
GRAYS ENGINEERING WORKS & IRON FOUNDRY   N35
TQ 612775

Firm listed in directories by 1884 specialised in building Floating Docks and in the salvage of sunken vessels using a device of their own invention called a gripping camel, attached to which were air-bags used when inflated to raise a sunken vessel. In 10/1888 the French steamer Ville de Calais sank in Calais harbour, during salvage the after part of the ship was being towed to London on 2/3/1890 when it sank in heavy seas off Margate. Mr Standfield senior and four crew were drowned whilst Mr Standfield junior and other crew members were saved.

The "Times" for 9/6/1900 carried an advert that Wheatley Kirk Price and Co were to auction on 13/6/1900 on behalf of Clark and Standfield the freehold premises known as Grays Engineering Works including wharves, dock, railway sidings together with the buildings including offices and erecting shop, fixed plant and machinery.

"The Engineer", 15/6/1900, Wheatley Kirk to auction 5/7/1900, re Clark & Standfield, Grays Engineering Works, Grays (given up manufacturing work) – plant including 6in std gauge loco.

Firm is said to have had a narrow gauge home made loco, details unknown.

Gauge : 4ft 8½in

No.20	0-4-0Tram	FE			(a) s/s after 7/1900

(a)   ex London & India Docks Co, c/1900.

GEORGE COHEN, SONS & CO LTD
JETTY WORKS, River Road, Barking   P36
TQ 469820

Works on site formerly occupied by **Metals & Ropes Ltd** until /1967 and earlier that of the Barking Jetty Co, rail traffic ceased. Site closed c/1972.

Gauge : 4ft 8½in

-	4wDM	RH	513142	1967	New	(1)
-	4wDM	RH	265611	1948	(a)	(2)
-	0-4-0DM	JF	19024	1930	(b)	(2)

(a) ex Eastern Gas Board, New Barnet Gas Works, Hertfordshire. c11/1965.
(b) ex Cransley Works, Northamptonshire, c7/1970.
(1) to Tinsley Works, Sheffield, 3/7/1968.
(2) to Bidder Street Scrapyard, /1972.

Gauge : 2ft 0in

	4wPM	JF	19231	1931	(a)	(1)

(a) ex ?, New to Constable Hart & Co Ltd, possibly Tovil Quarry, Maidstone, Kent.
(1) to M.E. Engineering Ltd, Cricklewood, Middlesex, 3/1948.

BIDDER STREET SCRAPYARD, Canning Town　　　　　　　　　　C37
TQ 391818

This scrapyard site is said to have operated from 1881. Adverts show the address as Quadrant Street, Canning Town. It was served by standard gauge sidings to the west of the GER Stratford - North Woolwich line north of Canning Town station. The premises were also used as a plant depot with many locos being bought and sold.

Note : the CP plant numbers allocated may not have been carried.

"Contract Journal" 11/4/1923 for sale or hire by dealer G. Cohen, 600 Commercial Road, E14, – 8x14 0-4-0ST MW, lying London.

"Contract Journal" 12/9/1923 for sale by dealer G. Cohen, 600 Commercial Road, E14, – 8in 0-4-0ST WB, 2ft gauge; also 8x12 MW 0-4-0ST std gauge, lying London.

"Contract Journal" 6/2/1929 for sale by G. Cohen, dealer, 600 Commercial Road, E14, – 0-6-0T HC 1069, 15½x20; 0-6-0ST AE 1603, 14x22; 0-6-0ST MW 1449, 14½x20; 0-6-0 ST AB 12x20; 0-4-0ST HC 14x20; 0-6-0ST AB/23, 14x22; 0-6-0ST YE/22, 14x20;　0-6-0ST AE 1518, 14x20; 0-6-0ST P, 14x20; 0-4-0ST BLW 14x20.

"Machinery Market" 29/4/1932 – for sale by G. Cohen, 600 Commercial Road, E14, - 16x20 0-6-0ST RS, 12x20 0-6-0ST AB, Sentinel – Cammell steam loco.

"Contract Journal", 17/5/1933 – for sale by dealer G. Cohen, 600 Commercial Road, E14, – including std gauge Sentinel – Cammell steam loco.

Gauge : 4ft 8½in

	SOMERFORD	0-6-0ST	OC	P	720	1898	(a)	(4)
CP 1578	CRYMLYN	0-6-0ST	OC	AE	1518	1907	(a)	(1)
CP 2065	PHOENIX	0-6-0ST	IC	MW	1449	1900	(b)	(2)
CP 119	FELSPAR	0-4-0ST	OC	MW	1846	1914	(c)	(3)
CP 231		4wVBT	VCG	S	6994	1927	(d)	(7)
CW 204	SWANSEA	0-6-0ST	IC	MW	595	1876	(e)	
		reb				1902	(e)	(17)
CW 207	SOUTHSEA	0-4-0ST	OC	HE	215	1879	(e)	(18)
CP 473	LORD MAYOR	0-4-0ST	OC	HC	402	1903	(e)	(6)
CP 467	SOMERTON	0-4-0ST	OC	HC	656	1903	(e)	(15)
CP 466	BARRY	0-6-0ST	IC	HC	440	1896	(e)	(11)
CW 201	CECIL LEVITA	0-6-0ST	IC	HE	1499	1926	(e)	(16)
CW 203	FORTH	0-6-0ST	OC	AB	1844	1924	(e)	(8)
CP 1939		0-6-0ST	OC	AB	1297	1912	(f)	(12)
	No.12	0-6-0ST	OC	RS	3094	1902	(g)	(19)
CP407	No.22	0-6-0ST	OC	RS	3120	1904	(h)	(5)

Essex Page 70

CP396 (CP391?)	NANCY	0-6-0ST	IC	P	1067	1905	(j)	(10)
	BASSETT	0-4-0ST	OC	Hor	1097	1910	(k)	(14)
	No.5	0-6-0ST	OC	RS	2844	1896	(l)	(9)
		0-4-0ST	OC	AB	1290	1912	(m)	(13)
	(CIRCE)	0-4-0DM		JF	19024	1930	(n)	s/s /1973
	-	4wDM		RH	237918	1945	(o)	Scr /1972
	(BASS No.4)	0-4-0DM		Bg	3589	1962	(p)	s/s c/1985
	-	4wDM		RH	265611	1948	(q)	s/s c/1974
	ELIZABETH II	4wDM		FH	3736	1955	(r)	s/s c/1985

(a) ex S. Pearson & Son Ltd, Queen Mary Reservoir (1919-1924) contract, Middlesex.
(b) ex Mersey Docks & Harbour Board, Lancashire, /1927.
(c) ex Shap Granite Co Ltd, Westmorland, c/1930.
(d) ex Chatham Dockyard, /1932?, advertised for sale 29/4/1932, here by 17/7/1933; to Canning Town Glass Works, hire and returned; originally Jersey Eastern Railway, BRITTANY, via A.O. Hill Ltd, shipbreakers, Admiralty Dockyard, Dover. (on 23/11/1929 A.O. Hill Ltd purchased the whole of the plant, permanent way and moveable effects of the Jersey Eastern Railway. On 11/4/1931 the RINGWOOD, a ship belonging to the Southern Railway Co, docked at Southampton from Jersey. According to the ship's manifest the cargo included what was described as a "chassis"; it is thought that this was the power unit from BRITTANY S 6994. No other shipment of any loco from Jersey has been traced. There were no regular sailings between Jersey & Dover other than the occasional shipment of granite).
(e) ex C.J. Wills & Sons Ltd, Becontree Housing (1920-1934) contract.
(f) originally Easton Gibb & Son Ltd, Rosyth, Fife.
(g) ex Port of London Authority, Royal Docks, /1931, 12.
(h) ex Port of London Authority, Tilbury Docks, /1933, 22.
(j) ex Madeley Collieries Ltd, Staffordshire, c/1933.
(k) ex John Mowlem & Co Ltd, Southampton, /1933.
(l) ex Port of London Authority, Royal Docks, /1937.
(m) ex Babcock & Wilcox Ltd, Renfrew, 4/1948.
(n) ex W. Cory & Son Ltd, Gallions Jetty, 10/8/1959;
to Cransley Depot, Northamptonshire, c1/1968; ex River Road Works, Barking, /1972.
(o) ex R.A. Main Ltd, Edmonton, Middlesex, c/1966.
(p) ex Bass, Mitchells & Butlers Ltd, Burton on Trent, Staffordshire, 9/1967.
(q) ex River Road Works, Barking, /1972.
(r) ex W.R. Cunis Ltd, Rainham, 1/1975.
(1) to Ely Beet Sugar Factory Ltd, Cambridgeshire, on hire by 11/1926, returned after 10/1931; to Ely again after 9/1934 following repairs at A. Dodman & Co Ltd, Kings Lynn.
(2) to Ely Beet Sugar Factory Ltd, Cambridgeshire, on hire /1927, returned /1928; to United Sugar Co Ltd, Bury St. Edmunds factory, Suffolk, on hire c9/1931, purchased c10/1934.
(3) to Thurrock Chalk & Whiting Co Ltd on hire, /1930 returned,
to Charles Wall Ltd, Grays, /1933-1934, returned,
to British Portland Cement Manufactures, Wouldham Works, returned,
thence to Holborough Works, Kent, by 21/3/1943.
(4) to William Jones Ltd, dealer, Greenwich, London.
(5) to Madeley Collieries Ltd, Leycett, Staffordshire, /1933.
(6) to Charles Wall Ltd, Grays, on hire /1934, returned, later Stanningley Depot, Leeds, Yorkshire.

(7) to William Jones Ltd for conversion work by 28/2/1935, returned?; to J.S. Peters, Merstham, Surrey, /1935.
(8) to Joseph Pugsley & Sons Ltd, dealers, Bristol; thence to John Mowlem & Co Ltd, Chingford Reservoir (1935-1951) contract.
(9) to Lanarkshire Steel Co Ltd, Motherwell, Lanarkshire, /1936.
(10) to Peckett & Sons Ltd, Bristol, 11/1936.
(11) to Edmund Nuttall, Sons & Co Ltd, Motspur Park-Chessington (SR) (1938-1940) contract, Surrey.
(12) to British Sugar Manufacturing Ltd, Wissington, Norfolk, on hire, /1928, returned /1929; to Berry Hill Collieries Ltd, New Haden Colliery, Staffordshire, hire, 4/1931 to 6/1931, advertised for sale lying at Stoke, 11/1933; to George Wimpey & Co Ltd, North Acton-Greenford contract, Middlesex, hire, by 6/2/1938; to WD Shoeburyness, 28/12/1939.
(13) to Burt Bolton & Haywood Ltd, Totton, Hampshire.
(14) to United Glass Bottle Manufacturers (Charlton) Ltd, Charlton, London.
(15) to Joseph Pugsley & Sons Ltd, dealers, Bristol, by 6/1935, thence to Davy & United Roll Foundry Ltd, Casebourne Works, Haverton Hill, North Yorkshire, by 4/1939.
(16) to Old Silkstone Colliery Co Ltd, Dodworth Colliery, South Yorkshire.
(17) to Charles Wall Ltd, Grays Works.
(18) to Cliffe Hill Granite Co Ltd, Markfield, Leicestershire.
(19) to Fountain & Burnley Ltd, North Gawber Colliery, West Yorkshire.

Gauge : 2ft 0in

DIANA	0-4-0WT	OC	HC	1132 1916	(a)	(1)
PN801	4wPM		Austro-Daimler		(b)	s/s
CP373	4wDM		OK		(b)	s/s

(a) ex William Muirhead Macdonald, Wilson & Co Ltd, Ilford contract, 7/1924.
(b) origin and identity unknown, for sale /1934.

(1) to A G Moore Ltd, Shieldmains Colliery, Drongan, Ayrshire, by 26/11/1930.

The following industrial locos are known to have been broken up for scrap.
Gauge : 4ft 8½in

No.5		2-4-0T	OC	Crewe		1849	(a)	Scr
No.6		2-4-0T	OC	Crewe		1849	(a)	Scr
No.7		2-4-0T	OC	Crewe		1847	(a)	Scr
	ELLESMERE	0-6-0ST	OC	AB	1158 1909		(b)	Scr
		0-4-2T	OC	HC	805 1907		(c)	s/s
	TERRIER	0-4-0ST	OC	HC	604 1902		(d)	Scr
	UNDAUNTED	0-4-0ST	OC	HL	3083 1914		(e)	Scr
10		0-6-0ST	IC	HC	1526 1924		(f)	Scr
		4wVBT	VCG	S	5735 1926		(g)	Scr
No.8	BOMBAY	0-6-0ST	IC	MW	1674 1906		(g)	Scr
No.9		0-6-0ST	IC	MW	1617 1903		(g)	
			reb	YE		1922		Scr
	LAUREL	0-4-0ST	OC	HL	3091 1914		(h)	Scr
15		0-6-0ST	IC	HC	1676 1937		(j)	Scr
No.10		0-4-0ST	OC	WB	1613 1900		(k)	Scr 23/5/1959
No.7		0-6-0ST	IC	MW	1488 1900		(l)	Scr
No.3	EDGWARE	0-6-0ST	IC	MW	2045 1926		(l)	Scr
14	MARYHILL	0-4-0ST	IC	P	1606 1923		(m)	Scr

(a)   ex London & India Docks Joint Committee, 8/1896.
(b)   ex S. Pearson & Son Ltd, Queen Mary Reservoir (1919-1924) contract, Middlesex, by 10/1925.
(c)   ex GWR 662, after 9/1925, form Rhymney Railway, Rail Motor No.1, (121).
(d)   ex J.S. Peters, Merstham, Surrey, /1928, possible hire to Thurrock Chalk & Whiting, /1930.
(e)   ex Royal Arsenal, Woolwich, London, 10/3/1954.
(f)   ex Samuel Williams (Dagenham Dock) Ltd, Dagenham Dock, /1957.
(g)   ex Samuel Williams (Dagenham Dock) Ltd, Dagenham Dock, 1/1957.
(h)   ex Royal Arsenal, Woolwich, London, 16/12/1957.
(j)   ex Samuel Williams (Dagenham Dock) Ltd, Dagenham Dock, c12/1958.
(k)   ex South Eastern Gas Board, East Greenwich Gas Works, c2/1959, by 22/3/1959.
(l)   ex Samuel Williams (Dagenham Dock) Ltd, Dagenham Dock, by 22/7/1959.
(m)   ex Samuel Williams (Dagenham Dock) Ltd, Dagenham Dock, /1960.

W.H. COLLIER & CO LTD
MARKS TEY BRICK, TILE & POTTERY WORKS   F38
Company registered 3/5/1935
W.H. Collier & Co until 1935   TL 910243
Colliers Ltd until 1914
William Homan Collier J.P. until 1910
John Wagstaff 1863 until 1879

Brickworks founded by John Wagstaff in 1863 and later sold to William Homan Collier in 1879. By 9/1988 the works was owned by Salvesen Brick Ltd, which became part of Chelwood Brick plc of Cheadle, Cheshire. Following a management buyout in 2005 the company now trades as W.H. Collier Ltd (registered 27/10/1961). An associated company was W. H. Collier & Co (Transport) Ltd registered 11/2/1937.

The brickworks operated a number of tramways of different gauges. A horse-worked standard gauge siding was laid from the adjacent GER Stour Valley line in 1898, being removed c/1960. A 1ft 8in gauge tramway was used in the pits, clay being transferred to an aerial ropeway until this was replaced by a railway incline which terminated inside the works. The tramway was replaced by lorries and closed 8/1978, although the section inside the works was modified to serve a new conveyor system. It remains in use at time of writing and is now hand-worked; the brick production areas and kilns served by 1ft 5in gauge and 2ft 0in gauge hand-worked lines. There is also a 3ft 6in gauge line for a traverser connecting a disused drying shed, a 3ft 11in/4ft 0in gauge line for a short traverser feeding kiln tunnel and serving the unloading bay area, a 4ft 2in gauge line for two traversers on the top floor of the 1950s drying area. Works still in production in 2010.
Gauge : 1ft 8in

-		4wDM (ex PM)	L	33937	1949	(a)	(2)
-		4wDM	LB	54183	1964	New	(1)
-		4wDM	OK	[6707	1936?]	(b)	Scr

(a)   ex John Hart & Co Ltd, New Barnet, Hertfordshire, by 9/1961;
      earlier National Smelting Co Ltd, Llansamlet, West Glamorgan.
(b)   origin and identity unknown, possibly ex Jaywick Miniature Railway, Jaywick Sands near Clacton.

(1)   to M.E. Engineering Ltd, Neasden, Greater London, 7/1979.
(2)   to Alan Keef Ltd, Cote Farm, Bampton, Oxfordshire, 18/9/1979.

CO-OPERATIVE WHOLESALE SOCIETY LTD
SILVERTOWN FLOUR MILLS　　　　　　　　　　　　　　　　　　　　B39
TQ 417798

Flour Mills in Thames Road in existence from c1901, the mill was leased to Spillers Ltd in 1945 to replace their mill which was destroyed through enemy action during WW2. Spillers rebuilt the mill at Royal Victoria Dock and commenced production in 1/1953. The society also operated a soap works in Thames Road.

Gauge : 4ft 8½in

		0-4-0ST	OC	P	1114	1907	New	(1)
	-	0-4-0ST	OC	P	1739	1928	New	(2)

(1) to E.R. Cole Ltd, London, dealer, by 3/1928, thence to Dunlop Co Ltd, Erdington, Birmingham, West Midlands, 5/1928, per R.H. Neal & Co Ltd, agents.
(2) to Thos. W. Ward Ltd, Titan Works, Grays, by 11/1949, thence to Inverkeithing Works, Fife, c/1949.

DANIEL CORNISH LTD
SHENFIELD BRICKWORKS, Brentwood　　　　　　　　　　　　　　　　　G40
Daniel Cornish & Co, until 6/5/1922　　　　　　　　　　　　　　　　　　TQ 610941
Colliers Ltd, until 1914
Shenfield and Cranham Brick & Tile Co Ltd, until 8/8/1910
Daniel Cornish trading as Daniel Cornish & Co, until 10/5/1906

Brickworks established in 1895. Daniel Cornish also owned brickworks at Hutton and Wickford. When the Shenfield and Cranham Brick Co Ltd acquired the business of the Cranham Brick & Tile Co Ltd (established 1899 by Joseph & Caleb Broodbank) and that of Daniel Cornish & Co in 1906; Daniel Cornish became a director of the new company which changed its name in 1910 to Colliers Ltd when William Homan Collier joined the company as a director. Colliers Ltd failed during WW1 and was dissolved 3/6/1921. Daniel Cornish Ltd went into voluntary liquidation 29/9/1944.

Gauge : 2ft 0in

	4wPM	FH	1636	1930	New	(1)

(1) to Edmonton Urban District Council, Deephams Sewage Works, Middlesex, by 9/1930.

EDWARD CORNISH LTD
BELLHOUSE BRICKWORKS, Eastwood, Southend　　　　　　　　　　　J41
Edward Cornish, c1902 until 1930　　　　　　　　　　　　　　　　　　　TQ 839888

Company formed 10/1/1930 to take over the business of Edward Cornish. Brickworks listed in trade directories from 1912, had a narrow gauge system from the clay pit to the works. Tramway superseded by dumpers 8/1968. Locos out of use by 1/1969. Company went into voluntary liquidation 10/10/1977.

Gauge : 2ft 0in

	4wPM	FH			(a)	(1)
	4wDM(exPM)	L	9256	1937	New(b)	(4)
	4wPM	L	37658	1952	New	(3)
	4wDM	L	+52031	1960	(c)	(2)

+carried works number 25031 in error.
- (a) origin and identity unknown.
- (b) rebuilt to DM c/1956.
- (c) ex M.E. Engineering Ltd, Cricklewood, Middlesex, after display at the Public Works Exhibition.
- (1) derelict 11/1960 and sold for scrap, /1963.
- (2) to M.E. Engineering Ltd, Cricklewood, Middlesex, c6/1970; thence to Miller & Baird Ltd, Porth contract, South Glamorgan.
- (3) to R.P. Morris, Longfield, Kent, /1970; thence to Fisons Ltd, Ashcott, Somerset, 18/2/1971.
- (4) to P.C. Vallins, Reigate, Surrey, for preservation, 11/1/1971, per R.P. Morris, Longfield, Kent.

WILLIAM CORY & SON LTD
GALLIONS JETTY　　　　　　　　　　　　　　　　　　　　　　　　　　　　　　A42
Wm Cory & Son until 15/10/1896　　　　　　　　　　　　　　　　　　　　TQ 443807

Company set up to adopt an agreement with Cory Francis Cory-Wright for the purchase of a number of coal merchants; these were Wm Cory & Son, Lambert Bros, D. Radford & Co, Beadle Bros Ltd, J.& C. Harrison, Green, Holland & Sons, Mann, George & Co and G.& J. Cockerell & Co Ltd. The company had coal handling wharves served by standard gauge sidings connecting with the Port of London Authority railways near Gallions Station, the locomotive shed was behind this station. Rail traffic ceased and system dismantled after 1967. The company also operated rubbish shoots along the Thames which had standard gauge tramways for transporting the rubbish from wharves to the tips, the rubbish being transported from London in barges and unloaded by grab cranes, (details given below).

Gauge : 4ft 8½in

		BELVEDERE	4wWT	G	AP	3888	1897	(a)	Scr /1929
		JETTY (KENT)	0-4-0ST	OC	AE	1578	1910	(b)	(1)
		DEPTFORD	0-4-0ST	OC	BH	1038	1893		
			reb	CF			1901		
			reb	YE		1923	1933	(c)	(2)
		WOOLWICH	4wVBT	VCG	S	7060	1927	New	(3)
		GREENWICH	4wVBT	VCG	S	7696	1929	New	(5)
		CHARLTON	4wVBT	VCG	S	8796	1933	New	(3)
		BELVEDERE	4wVBT	VCG	S	9365	1945	New	(4)
3		OBERON (OILER)	0-4-0DM		JF	19351	1931	(d)	Scr /1958
	No.2	CIRCE	0-4-0DM		JF	19024	1930	(e)	(6)
		PERSEUS	0-4-0DM		DC	2269	1949		
					VF	D98	1949	New	(9)
		PEGASUS	0-4-0DM		DC	2270	1949		
					VF	D99	1949	New	(7)
		PRIAM	0-4-0DM		DC	2566	1955		
					VF	D293	1955	New	(8)
		TEUCER	0-4-0DM		DC	2567	1955		
					VF	D294	1955	(f)	(10)

Essex Page 75

(a) ex Erith Wharf, Kent.
(b) ex Purfleet Wharf, /1932.
(c) ex Steetley Co Ltd, Coxhoe, Co. Durham, c/1943.
(d) ex Rainham Shoot, c/1947, after 16/2/1947.
(e) ex Erith Wharf, Kent, 2/1952.
(f) ex Purfleet Wharf, c/1962, after 9/1961.

(1) to Rochester Wharf, Kent.
(2) to Purfleet Wharf, c/1946 by 26/7/1947.
(3) to Rochester Wharf, Kent, by 4/1950.
(4) to Rochester Wharf, Kent, by 5/1950.
(5) to Rochester Wharf, Kent, 1/1952.
(6) to George Cohen, Sons & Co Ltd, Canning Town, 10/8/1959.
(7) to Rea Ltd, Bidston, Cheshire, /1962.
(8) to Erith Wharf, Kent, c/1967.
(9) to Renwick, Wilton & Dobson (Fuels) Ltd, Coal Concentration Depot, Exmouth Junction, Devon, 8/1967.
(10) to Rea Ltd, Birkenhead, Cheshire, /1968.

PURFLEET WHARF N43

TQ 554778

Construction of the wharf commenced c1904-1905 by the **Steamship Owners Coal Association Ltd**, this company was acquired by Wm Cory and Son Ltd in 1913.

Gauge : 4ft 8½in

ESSEX	0-4-0ST	OC	MW	1756	1910	New	(3)
JETTY	0-4-0ST	OC	AE	1578	1910	New	(1)
SCOT	0-4-0ST	OC	AB	1309	1913	New	(2)
ROMFORD	0-4-0ST	OC	HC	1561	1925	New	(4)
BEXLEY	0-4-0ST	OC	HC	911	1910	(a)	Scr 8/1954
DEPTFORD	0-4-0ST	OC	BH	1038	1893		
	reb		CF		1901		
	reb		YE	1923,	1933	(b)	(5)
PURFLEET	0-4-0ST	OC	MW	1973	1919	(c)	(6)
GALLIONS	0-4-0ST	OC	HC	1326	1919	(d)	(7)
TEUCER	0-4-0DM		DC	2567	1955		
			VF	D294	1955	New	(8)

The following cranes were noted here in 9/1945; Coles No.481, Grafton 473.

(a) ex Erith Wharf, Kent, /1941.
(b) ex Gallions Jetty, c/1946, (by 16/2/1947).
(c) ex Rochester Wharf, Kent, c/1950 (by 6/1950).
(d) ex Erith Wharf, Kent, c7/1954.

(1) to Gallions Wharf, /1932.
(2) to Erith Wharf, Kent, by 20/4/1935.
(3) to Erith Wharf, Kent, for repairs, 8/1943.
(4) to Hornchurch Shoot by 30/9/1945.
(5) to Cudworth & Johnson, Llay Hall Colliery, Denbighs, after 3/6/1952, by 8/1952.
(6) scrapped c/1956, after 9/1956.
(7) scrapped /1961, after 2/1961.
(8) to Gallions Jetty, c/1962, after 9/1961.

RAINHAM RUBBISH SHOOT   P44

Old Shoot TQ 505814
New Shoot TQ 509814

In 10/1924, a level crossing had been laid across the Manor Way with a line to a new shoot, 250 yards from the old tip and near to Rainham Creek. This divided into two sidings one 160 yards long to the south, the other 600 yards long running northwards parallel to the old shoot. An inspection of the tip in 1933 recorded that 14,000 tons of refuse had been received from Bethnal Green & Poplar Borough Councils. Barges transported the refuse from Bethnal Green via the Grand Union Canal and from Poplar via Northumberland Wharf at Blackwall, the material being unloaded by grab crane into wagons of 2½ ton capacity for tipping onto the shoot which was on 80 acres of marsh land.

The following cranes were noted here in 9/1945.

Grafton 471, a Coles crane and two unidentified cranes No 48 & 49.

Gauge : 4ft 8½ in

	HORNCHURCH	0-4-0ST	OC	TG	440	1907	New	(1)
	FERRO	0-4-0ST	OC	HC	857	1908	New(a)	Scr /1933
	BARKING	0-4-0ST	OC	HC	859	1908	New	(4)
	RAINHAM	0-4-0ST	OC	HC	696	1904		
		reb		Cory		1913	(b)	(6)
	LORD NELSON	0-4-0ST	OC	MW	706	1878	(c)	Scr /1933
No.2		0-4-0DM		JF	19024	1930	New	(5)
3	OILER	0-4-0DM		JF	19351	1931	New	(3)
	UTILITY	0-4-0ST	OC	HC	1206	1916	(d)	(2)
	(THURROCK)	0-4-0ST	OC	HC	1442	1921	(e)	(8)
	ROMFORD	0-4-0ST	OC	HC	1561	1925	(f)	(7)

(a)   consigned to Poplar Dock, London.
(b)   ex John Shelbourne & Co, with site, /1909, rebuilt from 3ft 0in gauge, /1913, to Thurrock Shoot, returned from Hornchurch Shoot, by 18/6/1950.
(c)   ex Royal Arsenal, Woolwich, London,/1919.
(d)   ex Erith Wharf, here by 10/4/1936.
(e)   ex Hornchurch shoot, after 30/9/1945, by 24/5/1947.
(f)   ex Hornchurch Shoot, here by 8/1950.

(1)   to Erith Wharf, Kent, /1932.
(2)   to Hornchurch Shoot, by 30/9/1945.
(3)   to Gallions Wharf, c/1947, after 16/2/1947.
(4)   scrapped /1949, after 5/1949.
(5)   to Erith Wharf, Kent, c/1950, after 5/8/1950, by 13/10/1951.
(6)   scrapped 6/1950, after 18/6/1950.
(7)   scrapped /1950, after 8/1950.
(8)   to Essex Welding Co, Romford, for scrap, by 14/4/1951.

HORNCHURCH RUBBISH SHOOT   P45

TQ 500815

Shoot owned by City of London Corporation but operated by Cory's was adjacent to the tip of Flower & Everett. Both shoots were very large the tips reaching a height of 90ft, a visit to these two shoots in 1929 found ten steam cranes in operation and ten locomotives. Cory's shoot was on 134 acres of marsh land and received refuse from the City of London which was transported in barges from Letts Wharf at Lambeth.

The following cranes were noted here in 9/1945 :
Grafton, 472, 1572, 2384, also a Smith Rodley diesel electric crane on the jetty.
Gauge : 4ft 8½in

RAINHAM	0-4-0ST	OC	HC	696	1904		
	reb		Cory		1913	(a)	(4)
THURROCK	0-4-0ST	OC	HC	1442	1921	(b)	(1)
PLUMSTEAD	0-4-0ST	OC	HC	1174	1915	(c)	(2)
ROMFORD	0-4-0ST	OC	HC	1561	1925	(d)	(5)
UTILITY	0-4-0ST	OC	HC	1206	1916	(e)	(3)
ESSEX	0-4-0ST	OC	MW	1756	1910	(f)	(3)

(a) ex Thurrock Shoot, by /1930.
(b) ex Thurrock Shoot.
(c) ex Erith Wharf, Kent, by 30/9/1945, derelict 5/1949.
(d) ex Purfleet Wharf by 30/9/1945.
(e) ex Rainham Shoot, by 30/9/1945.
(f) ex Erith Wharf, Kent, /1946.

(1) to Rainham Shoot after 30/9/1945 by 24/5/1947.
(2) scrapped, /1949, after 5/1949.
(3) scrapped, /1950 after 4/1950.
(4) to Rainham Shoot by 18/6/1950.
(5) to Rainham Shoot, by 8/1950.

WEST THURROCK RUBBISH SHOOT   N46
TQ 595773

Rubbish shoot on West Thurrock Marshes near to St Clement's Church in existence before 1923, site later built on by Thomas Hedley & Co Ltd.
Gauge : 4ft 8½in

RAINHAM	0-4-0ST	OC	HC	696	1904		
	reb		Cory		1913	(a)	(1)
THURROCK	0-4-0ST	OC	HC	1442	1921	New?	(2)

(a) ex Rainham Shoot.

(1) to Hornchurch Shoot, /1930.
(2) to Hornchurch Shoot.

H. COVINGTON & SONS LTD
LITTLE THURROCK MARSHES RUBBISH SHOOT   M47
H. Covington & Sons until 8/2/1910   TQ 623762

Land was leased on Little Thurrock marshes from 1894. Covington listed as lightermen & contractors had two jetties and a river frontage of 1300ft, rubbish was received by barge from London to be deposited on the marshes. Before the introduction of locomotives, operations commenced using a horse drawn railway system with side tip wagons loaded by steam crane. In 1926 refuse was being received in barges from Chelsea, Kensington & Greenwich Borough Councils. Shoot closed /1927, company went into voluntary liquidation on 1/12/1955.

"Contract Journal" 29/1/1947 for sale – one 9in WB steam & one HE (under reconstruction) steam, both std gauge – H. Covington & Sons Ltd, Chelsea, London.

Gauge : 4ft 8½in

MONMOUTH	0-4-0ST	OC	MW	941	1885	(a)	(1)
STRATFORD	0-4-0ST	OC	AB	992	1904	(b)	(1)
TAMAR	0-4-0ST	OC	HE	360	1884	(c)	Scr /1928
	0-4-0ST	OC	MW			(d)	Scr /1926

(a) ex Royal Engineers, Pwlholm, Monmouth, by /1927.
(b) ex Charles Wall Ltd, contractor, Grays, /1913.
(c) ex Sir J. Jackson, contractor, Grays, by 1/1923.
(d) origin and identity unknown.
(1) to Wennington Shoot, /1927.

WENNINGTON MARSHES RUBBISH SHOOT   L48
Wharf TQ 533790

Wharf under construction in 7/1926, refuse was transported to this shoot in barges from Chelsea, Kensington & Westminster Borough Councils. Shoot subsequently worked by W.R. Cunis Ltd, Covington's wharf being disused.

Gauge : 4ft 8½in

MONMOUTH	0-4-0ST	OC	MW	941	1885	(a)	Scr c/1938
(STRATFORD)	0-4-0ST	OC	AB	992	1904	(a)	(2)
THE CAPTAIN	0-4-0ST	OC	MW	1406	1898	(b)	Scr c/1939
(THOR)	0-4-0ST	OC	HE	629	1896		
	reb		Seabrooke		1921	(c)	(3)
FRANCES	0-4-0ST	OC	HC	435	1895	(d)	Scr c/1939
KENYA	0-4-0ST	OC	WB	2170	1921	(e)	(3)
FIREFLY	0-4-0ST	OC	MW	2044	1925	(f)	(1)

The following cranes were noted here in 9/1945.
Smith Rodley 1041, Grafton 975 & 2250

(a) ex Tilbury Shoot, /1927.
(b) ex J. Moffatt, Brentford (Great West Road) (1920-1925) contract, Middlesex, 2/1927.
(c) ex Seabrooke & Sons Ltd, Grays, /1929.
(d) ex C.J. Wills, Becontree Housing (1920-1934) contract, /1934.
(e) ex Pauling & Co Ltd, Crymlyn Burrows Plant Depot, Swansea, /1943.
(f) ex R. Leggott's Chalk Quarries Ltd, Ferriby, Lincolnshire, c5/1945, by 9/1945.
(1) to W.R. Cunis Ltd, Rainham, /1948.
(2) to W.R. Cunis Ltd, Rainham, by 1/7/1950.
(3) to W.R. Cunis Ltd, Rainham, after 1/7 /1950.

W.R. CUNIS LTD
GREAT COLDHARBOUR RUBBISH SHOOT, Aveley — L49

W.R. Cunis until 18/4/1906   Wharf TQ 525788
Shoot TQ 527791

Company with an office address of Seething House, Great Tower Street, EC, was registered to take over the business of a dredger owner, rubbish shoot proprietor, lighterman, contractor, steam tug & sailing barge owner, builder, repairer, sail maker & sand & ballast contractor carried on at Seething House, EC, and at Charlton, Rainham and elsewhere under the style of W.R.Cunis.

Rates records show that a dust shoot & wharf on ten acres of land was established at Coldharbour by 1907. The company transported rubbish by barge from Fulham, Deptford & West Ham to their wharf on the Thames at Aveley Marshes. A standard gauge railway was used to transport the rubbish from the wharf operated by three grab cranes to the tipping area on the marshes near Coldharbour Farm. The tip covered approximately 60 acres and was about 30-40ft high in 10/1924 when about 1,000 tons of refuse was received each week. In 1933 refuse was received in barges from Battersea, Fulham & Westminster Borough Councils with refuse from Deptford Borough Council transported via Copperas Wharf on Deptford Creek. By 1953 the combined tips of Cunis and that formerly operated by Covington covered 229 acres of marsh land. Rail traffic ceased 12/1974.

Gauge : 4ft 8½in

	-	0-4-0ST	OC	MW	825	1882	(a)	Scr c/1925
	BELVEDERE	0-4-0ST	OC	MW	1318	1896	(b)	Scr c/1939
	EDWARD	0-4-0ST	OC	MW	1242	1894	(c)	Scr c/1925
2		0-4-0ST	OC	AB	1652	1919	(d)	Scr c/1968
	-	0-4-0ST	OC	AE	1702	1915	(e)	(1)
	RALPH	0-4-0ST	OC	AB	1895	1926	(f)	Scr c3/1957
								(after 10/3/1957)
1		0-4-0ST	OC	AB	967	1903	(g)	Scr 4/1967
								(after 10/3/1957)
No.3		0-4-0ST	OC	AB	2167	1946	New	(2)
	FIREFLY	0-4-0ST	OC	MW	2044	1925	(h)	Scr /1958
								(after 10/3/1957)
	-	0-4-0ST	OC	HE	629	1896		
			reb	Seabrooke		1921	(j)	Scr 12/1954
	"COVINGTON"	0-4-0ST	OC	AB	992	1904	(j)	Scr /1955
	KENYA	0-4-0ST	OC	WB	2170	1921	(j)	Scr 12/1954
1		4wDM		FH	3791	1956	New	Scr c10/1974
2		4wDM		FH	3821	1957	New	Scr c10/1974
3		0-4-0DH		Cunis		1963	(k)	(3)
L4	46	4wDM		FH	3910	1959	(l)	(5)
L5	41	4wDM		FH	3885	1958	(l)	(7)
L3	ELIZABETH II	4wDM		FH	3736	1955	(m)	(4)
No.6		4wDM		FH	3994	1962	(n)	Scr /1975
L7	D2007 TRIBRUIT	0-6-0DM		HC	D917	1956	(o)	(6)
L8	D2008 GUINNION	0-6-0DM		HC	D918	1956	(o)	(6)

The following Grafton cranes were noted here in 9/1945.
Works Numbers 850, 1628, 1525, 1633, 2033, 2062

(a) originally Lucas & Aird, Hull & Barnsley Railway (1880-1885) contract, East Yorkshire..
(b) ex London County Council, Southern Outfall Construction.
(c) ex John Mowlem & Co Ltd, Plumstead Sewer (1909) contract, London, /1909.
(d) ex General Electric Co Ltd, Erith, /1932.
(e) ex Plumstead Shoot, London, /1934.
(f) ex Woolwich Wharf, London, by 5/1934.
(g) ex Plumstead Shoot, London, by 5/1934, to Woolwich Wharf, London, 7/1945; returned by 9/1948.
(h) ex H. Covington & Sons Ltd, Wennington, by 26/12/1948.
(j) ex H. Covington & Sons Ltd, Wennington, /1950.
(k) ex Woolwich Wharf, 11/1963 (rebuilt from 0-4-0ST AB 2167/1946 - see below)
(l) ex North Thames Gas Board, Beckton Gas Works, /1968.
(m) ex South Eastern Gas Board, Lower Sydenham Works, London, c12/1970.
(n) ex H.F.A. Dolman Ltd, Southend, c/1971;
     earlier North Thames Gas Board, Bromley-by-Bow Works.
(o) ex Port of Bristol Authority, Avonmouth Docks, Somerset, per T. Wakefield, scrap metal dealer, Romford, c12/1973.

(1) destroyed by German V2 rocket, /1945.
(2) to Woolwich Wharf, for conversion to diesel, c/1962. (see above)
(3) to Reeds Paper Mills Ltd, Greenhithe, Kent, 9/1968.
(4) to George Cohen, Sons & Co Ltd Sons & Co Ltd, Canning Town, 1/1975.
(5) to British Industrial Sand Ltd, Middleton Towers, Norfolk, c2/1975.
(6) to Howard-Doris Ltd, Strome Ferry, Ross & Cromarty, 5/1975,
     after repairs at British Rail Engineering, Stratford Works, 4/1975.
(7) to Alexander Bruce (Grays) Ltd, Bruces Wharf, Grays, c2/1976.

B. DANNATT LTD
HATFIELD PEVEREL WORKS G50
B. Dannatt until 28/9/1937 TL 813114

Pits in existence by 1939 until at least 1966. This company of sand and gravel merchants was listed in HMSO list of quarries with pits at Margaret Woods, Great Waltham near Chelmsford in 1931 until at least 1934. Bernard Dannatt died 5/8/1936.
Gauge : 2ft 0in

| - | 4wPM | L | 7622 | 1936 | New# | s/s |

\# recorded by Lister as second hand to Ernest Doe, Ulting Wick, Maldon.

DOBSON, ELLIS & CO
SAND & BALLAST PITS, Stanford-le-Hope K51
TQ 681816

Pits near Stanford-le-Hope and located 600yds from the station, listed in trade directories for 1929 and HMSO list of quarries for 1931, pits closed and site redeveloped as playing fields. Trade directories for 1929 list John Dobson as a builder of Butts Road, Stanford-le-Hope and William Ambrose Ellis as a barge owner of Stanford-le-Hope.

"Contract Journal", 5/6/1935, Fuller Horsey to sell 27/6/1935 re Dobson Ellis, Stanford-le-Hope sand & ballast pits (lease expired) – plant including two 2ft gauge, 10hp diesel locos.
Gauge : 2ft 0in

		4wDM	RH	164336	1932	New	(1)
		4wDM	RH	166048	1934	New	(2)

(1) to R.H.Neal & Co Ltd, Park Royal Depot, Middlesex, after 29/9/1934 by 13/8/1935, thence to John Nicholson & Sons Ltd, Hunslet Chemical Works, Leeds, Yorkshire, by 9/1940
(2) to S.J. Blyth, Scratton Road Sandpits, Stanford-le-Hope, after 19/2/1935, by 19/6/1936.

JAMES DUNCAN
CLYDE WHARF SUGAR REFINERY, Silvertown   B52
formerly **Duncan Bell & Elott**   TQ 400802

James Duncan of Greenock had established the manufacture of sugar from beetroot at Lavenham, Suffolk from 1869 to 1873. The enterprise in Suffolk failed due to the circumstance of transporting the juice to his refinery in London to be crystallized, and the insufficient supply of beetroots. In 1884 Bolton and Partners Ltd took over the works at Lavenham and introduced a new simpler and less costly process. Clyde refinery was established by 1862 and closed in 1886, the refinery was auctioned by Fuller, Horsey, Sons & Cassell on 16/3/1887 but was still for sale 6/1888; acquired by David Martineau & Sons in 1889 (incorporated 12/12/1889) and reopened 1890, but was destroyed by fire in 1893; the premises, plant and machinery were to be auctioned by order of the High Court of Justice (Chancery Division) on 6/10/1896. James Duncan died in 1905.

"The Engineer", 29/4/1887, Fuller Horsey to auction 18/5/1887 pending disposal of sugar refinery – surplus plant at refinery, Clyde Wharf, Victoria Docks, London – plant including loco traction engine by JF, std gauge, with extra set of road wheels. Later advert 16/9/1887 included twenty one 10ton railway trucks, still for sale on 14/8/1888.
Gauge : 4ft 8½in

		2-2-0WT G	JF	1347	1870	New(a)	s/s

(a) built new as a road engine.

DOCKLANDS LIGHT RAILWAY LTD
BECKTON DEPOT
                                                          TQ 442812
For locomotive listings see County of London Handbook.

EAST ESSEX (CORRINGHAM) SAND CO LTD
EAST ESSEX PITS, Corringham   K53
                                                          TQ 710834
Company formed 15/1/1930 and listed in HMSO list of quarries for 1934 and the quarry directory for 1939. In 1934 the company also operated pits at Linford near Stanford-le-Hope. The 1938 London telephone directory lists the pits as Alstons Corner, Corringham. Company went into voluntary liquidation on 9/5/1940.

Gauge : 2ft 0in

		4wDM	RH 166048 1934	(a)	(1)
		4wDM	(MR ?)	(b)	(2)

(a) ex S.J. Blyth, Scratton Road Sand Pits, Stanford-le-Hope, after 20/6/1936, by 8/3/1938.
(b) origin and identity unknown.
(1) to J.L. Eve Construction Co Ltd, contractor, Wimbledon Plant Yard, after 10/3/1938, by 4/3/1940.
(2) An unidentified 2ft 0in gauge "Simplex" diesel loco was auctioned with other plant on 12/7/1938.

EAST LONDON TRANSPORT LTD

BENTONS PIT, South Ockendon   L54
TQ 596820

Company had pits in Mollands Lane, South Ockendon. Loco haulage discontinued c4/1953. 1938 London telephone directory lists pits at Brook Farm Orsett. A later company registered 14/11/1953 at 3 Grosvenor Road, Ilford, to carry on business as quarry & mine owners and general carriers.

Gauge : 2ft 0in

		4wDM	OK 7268 1936	(a)	(1)
		4wPM	FH	(b)	s/s
		4wPM	FH	(b)	(1)
		4wPM	MR	(c)	(1)

(a) ex Thos. W. Ward Ltd, c/1944.
(b) ex Chas.T. Olley & Sons, South Ockendon, c/1948.
(c) origin and identity not known.
(1) to Mardyke Ballast Pits, East Purfleet.

MARDYKE BALLAST PITS, East Purfleet   L55
TQ 585795?

Pits located on the arterial road at West Thurrock and listed in HMSO list of quarries for 1948 and quarry directories until at least 1966, with pits in Grays Road, Purfleet later in use as a plant depot 100yds from Purfleet station, exact location unknown.

Gauge : 2ft 0in

		4wPM	Excelsior	(a)	(1)
		4wPM	FH	(b)	(2)
		4wPM	MR	(c)	(2)
		4wDM	OK 7268 1936	(b)	(2)

(a) origin unknown, loco described as a self propelled wagon, adverts of the day depict vehicles of this type by various makers including Jung. "Excelsior" was a registered trade name used by William Jones Ltd for a range of side tipping skip wagons and dump cars in adverts for the 1930s.
(b) ex South Ockendon Pits.
(c) ex South Ockendon Pits, loco carried a McAlpine plate.

(1) Scrapped after 9/1952, by 6/1956.
(2) derelict /1957; s/s by /1965.

EASTWOODS FLETTONS LTD
SHOEBURYNESS BRICKWORKS   J56
Eastwoods Ltd until 30/6/1927   TQ 937853
Eastwood & Co Ltd, until 1/3/1920
John Francis Eastwood trading as Eastwoods until 1902
Josiah Jackson trading as J. Jackson & Co until 1890s
Dale Knapping c1850s until 1878

Brickworks known as Model Field established in 1850s with standard gauge sidings adjacent to Shoeburyness LT&SR station. The brickworks had narrow gauge tramways connecting the works with clay pits. In 1962 Redland Holdings Ltd purchased Eastwoods brick and tile business and the works closed by 11/1964.

Gauge : 2ft 0in

| | | 4wDM | RH | 164337 | 1931 | (a) | s/s |
| | | 4wDM | MR | 8626 | 1941 | (b) | (1) |

(a) ex Barrington Works, Cambridgeshire, after 20/9/1946, by 15/7/1947.
(b) ex MR, 2/2/1950.

(1) to Orchard Farm Clay Pits, Iwade, Kent, by 7/1964.

ESSO PETROLEUM CO LTD
Company registered 6/4/1951
HARRISON'S WHARF BITUMEN TERMINAL, Purfleet   N57
Anglo American Oil Co Ltd until 6/4/1951
Ebano Oil Co until /1935   TQ 552782

The Ebano Oil Company was incorporated 7/12/1929 as manufacturers of asphalt, tar, pitch and other products produced from coal, oil and mineral substances; they had wharves on the north bank of the River Thames with standard gauge sidings which connected with the ex-LT&SR line just west of Purfleet Station; company went into voluntary liquidation on 19/6/1936. Rail traffic ceased.

Gauge : 4ft 8½in

		4wDM (ex PM)	MR	3896	1928	(a)	s/s c/1967
		0-4-0F OC	AB	1870	1925	(b)	(1)
		4wDM	MR	5755	1948	(c)	(2)
		0-4-0DM	HE	2067	1940	(d)	(3)
		0-4-0DE	YE	2686	1958	(e)	(4)
	2	0-6-0DH	HC	D1373	1965	(f)	(5)

(a) ex Fawley Refinery, Hampshire, by 3/1953.
(b) ex Pacific Wharf, West Ham, 22/10/1958.
(c) ex Fawley Refinery, Hampshire, after 30/12/1963, by 20/6/1965.
(d) ex Purfleet Tank Farm.
(e) ex Purfleet Tank Farm, /1974.
(f) ex Purfleet Tank Farm, by 6/1983.

(1) scrapped on site by Thos. W. Ward Ltd, Grays, 1/1965.
(2) to Thos. W. Ward Ltd, Columbia Wharf, Grays, for scrap, c/1967.
(3) returned to Purfleet Tank Farm, /1975.
(4) returned to Purfleet Tank Farm.
(5) returned to Purfleet Tank Farm, c/1986.

PACIFIC WHARF, West Ham C58
Redline Glico Ltd until c/1954 TQ 388828
Redline Motor Spirit Co Ltd from 13/3/1924 until 9/10/1931
Glico Petroleum Ltd from 9/5/1924
Gas Lighting Improvement Company Ltd incorporated 10/12/1888

Works established in Crows Road, West Ham. The Gas Lighting Improvement Company Ltd was established 12/1888 and changed its name in 5/1924 to Glico Petroleum Ltd. The company went into voluntary liquidation 5/10/1931 when the undertakings of the company were transferred by agreement to the Redline Motor Spirit Co Ltd, which in turn changed its name to Redline Glico Ltd, 9/10/1931. On 20/5/1927 fire broke out at the works in an area which contained the distilling plant for benzine motor spirit and turpentine, a storage tank exploded, the explosions were heard over two miles away when 8,000 gallons of benzine was destroyed; two men were injured and 150 evacuated, it took firemen five hours to extinguish the flames. Rail traffic ceased and rail connection removed, 10/1958.
Gauge : 4ft 8½in

LOCO No.1		0-4-0F	OC	AB	1870	1925	New	(1)

(1) to Bitumen Terminal, Purfleet, 22/10/1958.

PURFLEET TANK FARM N59
Anglo-American Oil Co Ltd (incorporated 27/4/1888) until 6/4/1951 TQ 563777

Oil storage facilities operated jointly by Anglo American Oil Co Ltd and **Tank Storage & Carriage Co Ltd** at Caspian Wharf to the south of the ex-LT&SR line about a mile east of Purfleet Station, served by standard gauge sidings which connected with this line.
Gauge : 4ft 8½in

ELSIE	2-4-0ST	IC	AE	804	1871	(a)	(1)
(PERFECTION)	0-4-0ST	OC	P	1571	1921	New	(4)
ANGLOCO	0-4-0ST	OC	P	1735	1927	(b)	(3)
-	0-4-0F	OC	AB	1603	1918	(c)	(2)
-	0-4-0DM		EE	1192	1941	(d)	
			DC	2161	1941		(8)
AMW No.246	0-4-0DM		JF	23003	1943	(e)	(6)
MOP No.7	0-4-0DM		JF	4210144	1958	(f)	(5)
-	0-4-0DE		YE	2686	1958	(g)	(9)
-	0-4-0DM		HE	2067	1940	(h)	(7)
2	0-6-0DH		HC	D1373	1965	(j)	(10)

(a) ex Bute Works Supply Co Ltd, dealers, Cardiff, after 3/1907; earlier GWR, 2, originally South Devon Rly, KING.
(b) ex Stanlow Refinery, Cheshire, /1938.
(c) ex Hickson & Partners Ltd, Middlesbrough.
(d) ex WD 70035, 2/1948.

(e) ex Shell-Mex & BP Ltd, Potton, Bedfordshire, by 23/6/1949.
(f) ex Ministry of Power, Portishead, Somerset, c/1962.
(g) ex Trafford Park Refinery, Manchester, /1964; to Harrisons Wharf, /1974; returned.
(h) ex Thos. W. Ward Ltd, 12/1967 (earlier WD, 849);
to Harrison's Wharf; and returned /1975.
(j) ex Salt End, Hull, Yorkshire (ER), 10/3/1977;
to Harrison's Wharf by 6/1983, returned c/1986.

(1) believed to be loco PERSEVERANCE advertised for sale by Stanlee Shipbreaking & Salvage Co Ltd, Dover, /1926 and thence to a Mr Hill, 25/3/1926. (possibly A.O. Hill Ltd, shipbreakers, Admiralty Dockyard, Dover).
(2) to Mode Wheel Refinery, Lancashire, /1936.
(3) to Purfleet Deep Wharf & Storage Co Ltd, by 23/6/1949.
(4) to Purfleet Deep Wharf & Storage Co Ltd, /1951.
(5) to Mode Wheel Refinery, Lancashire, 11/1963.
(6) to Thos. W. Ward Ltd, Grays, /1965.
(7) to Quainton Railway Society, Buckinghamshire, 29/4/1983.
(8) to Quainton Railway Society, Buckinghamshire, 10/10/1983.
(9) to North Downs Steam Railway, Kent, c/1987.
(10) to Esso Bitumen, Cattewater, Plymouth, Devon, after 11/7/1991, by 12/1991.

EXPLOSIVES & CHEMICAL PRODUCTS LTD
BRAMBLE ISLAND WORKS, Great Oakley, near Harwich   H60
TM 215265

Company registered 6/4/1905, established a factory on marsh land by Hamford Water, manufactured industrial explosives including gelignite. An extensive narrow gauge railway network was used to transport raw materials, intermediate and finished products between various buildings and magazines, with lines connected by small turntables. Mainl hand-worked, with 20 people employed to push bogies even in 1964. For the heavier operations, the locomotive was used when manpower was not sufficient. It was mainly used for transporting acids, either moving drums of acid (weighing up to 350kg each) or mobile bulk tanks which could contain 3 or 4 tonnes of acid. Use of the railway ceased c/1968-70, being replaced by tractors.

Gauge : 2ft 0in

| | | 4wDM | L | 50265 | 1958 | New(a) | s/s |

(a) supplied by Colchester Tractors Ltd.

FEATHERBY'S BRICKWORKS
Rochford   J61
Proprietor : Albert E. Mallandain by 1933   TQ 884900
Featherby's Brickworks Limited
G. Featherby Company (Rochford Essex) Limited until 10/11/1922
George Featherby & Co, until 14/11/1919
George Featherby, until 1902
Brickworks established c1894 at Roach End, Rochford. George Featherby in partnership with his son John traded as George Featherby & Co as makers of hand made red building and ornamental bricks until the partnership was registered as a limited company in 11/1919,

later liquidated 6/9/1922. A new company **Featherby's Brickworks Limited** (registered 10/11/1922) was formed to acquire the assets of the previous company from the liquidator by agreement of 20/11/1922, however this company also failed and a liquidator was appointed 13/3/1928, the company was wound up by court order 30/8/1932. An auction of the brickworks comprising of 22 acres, containing brick-earth, sand & ballast together with kilns, drying shed, pug mills, offices, plant and machinery and a large stock of bricks, took place on 11/7/1928 at the London Auction Mart, Queen Victoria Street, London EC, by order of the High Court of Justice. Albert Edward Mallandain was a director and major shareholder in G. Featherby Company (Rochford Essex) Ltd.

"Machinery Market", 16/12/1949 - for sale 2ft gauge 10hp petrol loco, Featherby's Brickworks, Rochford.

"Contract Journal" 15/2/1950 - for sale 10hp petrol loco previously used on chicken farm, Featherby's Brickworks, Rochford, Essex.

"Machinery Market" 11/5/1951 – for sale – six 2ft 6in gauge wagons – Featherby's Brickworks, Tinkers Lane, Rochford.

Gauge : 2ft 0in

		4wDM	RH 217999 1942	(a)	(1)
		4wPM		(b)	s/s

(a) ex RH, Lincoln, on hire, 6/10/1950; earlier MoS.
(b) origin and identity unknown.

(1) to J. Carter, North Fambridge, c/1969;
thence to Leighton Buzzard Narrow Gauge Railway Society, St Albans, Hertfordshire for overhaul, 10/1974; and to Leighton Buzzard, Bedfordshire, 6/1975.

FISONS LTD
Stanford-le-Hope   **K62**
Incorporated 23/7/1895   TQ 696815
Fertiliser factory served by standard gauge sidings adjacent to the ex-LT&SR Thames Haven branch south of Stanford-le-Hope village. Siding connection taken into use in 1959; rail traffic ceased and sidings disused (but still in situ) by 1995.
Gauge : 4ft 8½in

		0-4-0DM	JF 4220001 1959	New	(1)

(1) to Avonmouth Works, Bristol, c9/1982.

FLOWER & EVERETT LTD
Flower & Everett until 29/7/1912
RAINHAM RUBBISH SHOOT   **P63**
     TQ 502815
Company listed in directories as Dredging contractors was established c/1820 by Farnham and Henry Flower of Bow, East London; registered 29/7/1912 at 69 King William Street EC as lightermen, wharfingers, dredging contractors and removers of refuse and barge owners and listed in the 1938 London telephone directory with an address at 74 Bankside SE1. The shoot covered 90 acres of marsh land; refuse was received in barges from Westminster Borough Council, this contract terminated on 31/3/1932. In 1933 refuse was being transported from Fulham & Camberwell Borough Councils; material from the latter via

Glengall Wharf on the Grand Surrey Canal, this contract terminated 31/3/1934. In 10/1933 the journal "Public Cleansing" carried an article on Flower & Everett's shoot at Rainham which depicted an O&K Land Dredger in use, this machine excavated soil at the base of the tipping face to cover the refuse. The machine was fitted with a Deutz four stroke diesel engine and ran on track of 5ft 11in gauge and was supplied by William Jones Ltd of London; shoot closed after 1936 and track lifted. The company also owned and operated a shoot at Little Mussels Farm, Pitsea, later taken over by the Land Reclamation Company; the company later became **W.R.Cunis (Waste Disposal) Ltd**, dissolved 17/2/1998.

Gauge : 4ft 8½in

-	0-4-0ST	OC	MW	221	1866	(a)	s/s
-	0-4-0ST	OC	MW	809	1881	(b)	s/s
KIRKBY	0-4-0ST	OC	MW	1303	1895	(c)	s/s
WILLIAM	0-4-0ST	OC	MW	1450	1899	New	s/s
PRINCESS	0-4-0ST	OC	MW		1900	(d)	s/s
RAINHAM	0-4-0ST	OC	DK	6899	1901	(e)	(1)

(a) earlier John Aird & Sons Ltd, Staines Reservoirs (1898-1902) contract, Middlesex [and possibly John Aird & Sons Ltd, Avonmouth Dock (1902-1905) contract, Bristol?].
(b) originally Lucas & Aird, Hull & Barnsley Railway (1880-1885) contract, East Yorkshire.
(c) earlier W.H. Hutchinson, Leen Valley (GNR) (1898-1901) contract, Nottinghamshire.
(d) ex A.R. Adams & Son, dealers, Pill Bank Ironworks, Newport, Gwent, by 11/1927.
(e) ex Joseph Rank Ltd, Birkenhead, Cheshire, /1932.

(1) s/s after 4/1936.

FORD MOTOR CO LTD
DAGENHAM WORKS
P64
Ford Motor Co (England) Ltd until 7/12/1928   TQ 496825

A major motor car and commercial vehicle manufacturing works, construction began with the first sod cut in 5/1929. The works was built to the east of the Samuel Williams Dagenham Dock Industrial Estate, on the site of Flower & Everett's old rubbish tip which was levelled and the factory then built on thousands of reinforced concrete piles, some driven to a depth of 80ft; it also has its own deep water jetty. Production began in 9/1931 and the works was completed in 1935. The works had its own blast furnace, manufactured and installed by the Power-Gas Corporation of Stockton, this came into production in 1934 along with coke ovens and a foundry for the production of engine castings. The works was served by an extensive network of standard gauge lines (25 miles in the 1970s) and had a rail connection to the Samuel Williams estate. On the north side of the LT&SR line was the **Briggs Motor Bodies** works (TQ 496830) where one of Ford's locomotives was stationed to shunt the sidings there. The blast furnace was due to close in 1978 when a new plant for the manufacture of engines opened at Bridgend in South Wales; the foundry closed in 1984 and production of cars ceased with the Ford Fiesta in 2/2002. To replace the old assembly plant a new diesel engine complex has been built and the river jetty is used for the import & export of cars from 'drive-on drive-off' purpose- built ships. By 2004 the internal rail system much reduced in size was used principally to bring in raw materials and for the dispatch of completed engines to other Ford assembly plants. Records of Clayton Equipment Ltd show that spares were ordered in 1962 & 1963 for a transfer car which is believed to have been used in the foundry, further details unknown. (see also Trentham & Speight in contractors section).

Reference : "Railway Bylines" Ian P. Peaty, April 2004.
Gauge : 4ft 8½in

No.	Name	Type	Cyl.	Maker	Works No.	Year	New/Note	Disposal
1	ALICE	0-6-0ST	OC	AE	1460	1903	(a)	(2)
2	BURTON	0-4-0ST	OC	HL	2502	1901	(b)	(1)
-	*	0-4-0ST	OC	HE	515	1890		
		reb		HE		1915	(c)	(1)
-	-	0-4-0WE		WSO	1529	1930	New	Scr c/1979
1	-	Bo-Bo.DE		BTH		1932	New	(11)
2	-	Bo-Bo.DE		BTH		1932	New	s/s /1966
3	-	Bo-Bo.DE		BTH		1932	New	s/s /1966
No.6	(No.4)	0-6-0ST	OC	P	1861	1934	New	(13)
No.7	(No.5)	0-6-0ST	OC	P	1890	1936	New	(12)
No.6	*	0-4-0ST	OC	P	1908	1937	New	Scr c7/1965
No.7		0-6-0ST	OC	P	1938	1937	New	s/s /1967
No.8		0-6-0ST	OC	HC	1508	1924	(d)	Scr c10/1954
No.8		0-6-0ST	OC	P	2154	1954	New	(14)
-		4wPM		MH	L103	1930	New	s/s
-		4wDM		FH	1862	1934	New	s/s
-		4wDM		RH	210481	1941	New	Scr c/1952
	NOAH	0-4-0ST	OC	AB	859	1900	(e)	(3)
	SOUTHAMPTON	0-6-0ST	OC	HE	1647	1931	(f)	(4)
	ETTRICK	0-4-0ST	OC	HL	3721	1928	(g)	(5)
	-	0-4-0DM		P	5001	1956	(h)	(6)
No.1(No.9)		0-4-0DE		RH	412715	1958	New	(15)
	*	0-4-0ST	OC	AB	1979	1930	(j)	(7)
	A.E.CLOW	0-4-0DM		JF	4200018	1947	(k)	(8)
	-	0-4-0DH		RH	468048	1963	(l)	(9)
	-	4wDM		RH	476143	1963	(m)	(10)
7		0-6-0DM		DC	2606	1957		
				RSHD	7892	1957	(n)	Scr 7/1978
1	(No.6 until /1977)	0-6-0DM		DC	2611	1958		
				RSHD	7897	1958	(o)	(19)
3	(No.8 until 1/1978)	0-6-0DM		DC	2714	1961		
	P1062C			RSHD	8192	1961	(p)	Scr c/1988
2	(No.1 until /1977)	0-6-0DM		DC	2657	1960		
	P1381C			RSHD	8098	1960	(q)	(19)
7	P1212C							
	(No.2; No.10)	0-6-0DH		HC	D1376	1966	New	Scr 3/1981
3	(No.11)	0-6-0DH		HC	D1377	1966	New	(16)
5	(No.4, No.12)	0-6-0DH		HC	D1378	1966	New	Scr c/1979
6	(No.5 P1215C)	0-6-0DH		HC	D1396	1967	New	
		reb		HAB$	6385	1996		
4	(No.7 until /1978)	0-6-0DM		Don		1959	(r)	(18)
5		0-6-0DH		EEV	D1229	1967		
		reb		HE	8900	1977	(s)	(17)
	GT/PL/1 P260C	4wDM		Robel 21 11 RK1		1966		
No.1		0-4-0DH		EEV	D1124	1966	(t)	
	-	4wDH		TH	285v	1976	(u)	(20)
3	MALCOM (EDDIE)	0-4-0DH		S	10127	1963	(v)	
		reb		Wilmott #		2003		

Essex

*	Kept at Briggs Motor Bodies Ltd. (Ford subsidiary), Dagenham.
#	Wilmott Bros (Plant Services) Ltd, Ilkeston, Derbyshire.
$	HAB = Hunslet-Barclay Ltd, Kilmarnock, Ayrshire.

Note: A number of Class 03/04 0-6-0DM locos were hired from BR Stratford in the 1970s, locos seen included, 2164 in 2/1971, 2217 in 9/1971, 03168 in 12/1974 & 10/1977, 03081 in 2/1977, 03149 in 6/1977 & 10/1977, 03161 in 10/1977, 03389 in 6/1977.

(a) ex Thos. W. Ward Ltd, 3/5/1930, 39579; originally South Leicestershire Colliery Co Ltd, until 6/8/1929.
(b) ex Thos. W. Ward Ltd, /1931, 34507,
earlier Marston Thompson & Evershed Ltd, Burton, Staffordshire.
(c) ex L.J. Speight, contract on site, /1932;
earlier J. & B. Martin, Crayford Brick Sidings, Kent.
(d) ex John Mowlem & Co Ltd, Swynnerton (ROF) (1939-1945) contract, Staffordshire, MEECE, here by 12/1944.
(e) ex Abelson (Engineers) Ltd, Sheldon, Birmingham, hire, /1951.
(f) ex John Mowlem & Co Ltd, loan, c1/1953, by 19/3/1953.
(g) ex Abelson (Engineers) Ltd, Sheldon, Birmingham, hire, 1/1954.
(h) ex P, for trials, c6/1957.
(j) ex Thos. W. Ward Ltd, hire, 7/1960.
(k) ex Abelson (Engineers) Ltd, Sheldon, Birmingham, hire, 8/1960.
(l) ex RH, Lincoln, on hire 20/11/1963.
(m) ex RH, Lincoln, on hire 31/12/1963.
(n) ex BR, D2262, /1969.
(o) ex BR, D2267, /1970, to BR Swindon, Wiltshire, 19/5/1977 returned 8/11/1977.
(p) ex BR, D2333, 2/1970, to BR Swindon, Wiltshire, 5/1977 returned 23/1/1978.
(q) ex BR, Colchester, 2280, /1971, to BR Swindon, Wiltshire, 8/7/1977 returned 25/11/1977.
(r) ex BR, Colchester, 2051, 6/1973, to BR Swindon, Wiltshire, 10/1977, returned 23/2/1978.
(s) ex HE, 7/1978; earlier NCB, Nailstone Colliery, Leicestershire.
(t) ex RFS Engineering Ltd, Kilnhurst, South Yorkshire, 4/2/1993.
(u) ex RMS Locotec Ltd, Dewsbury, West Yorkshire, on hire 6/3/2002.
(direct from hire at Huntsman Tioxide, Lincolnshire).
(v) ex Wilmott Bros (Plant Services) Ltd, Ilkeston, Derbyshire, 6/6/2003.

(1) to Thos. W. Ward Ltd, for scrap 9/1934.
(2) to Naworth Colls. Co Ltd, Cumberland, via George Cohen, Sons & Co Ltd, /1935.
(3) returned to Abelson, (Engineers) Ltd, Sheldon, Birmingham, by /1953.
(4) returned to John Mowlem & Co Ltd, /1954.
(5) returned to Abelson, (Engineers) Ltd, Sheldon, Birmingham, by 12/1954.
(6) returned to P. /1958 (after 23/5/1958).
(7) returned to Thos. W. Ward Ltd, 13/9/1960.
(8) returned to Abelson, (Engineers) Ltd, Sheldon, Birmingham, c/1960, after 12/10/1960.
(9) returned to RH, Lincoln, 16/12/1963.
(10) to NCB Cwm Mawr disposal point, Carmarthenshire, 16/11/1965.
(11) to Kent & East Sussex Railway Preservation Society, Rolvenden, Kent, 7/1966.
(12) scrapped on site by Romford Scrap & Salvage Co, 10/1969.
(13) to Steel Breaking & Dismantling Co Ltd, Chesterfield, Derbyshire, for scrap, 4/1970.
(14) to Steel Breaking & Dismantling Co Ltd, Chesterfield, Derbyshire, for scrap, c1/1970.

(15) to Steel Breaking & Dismantling Co Ltd, Chesterfield, Derbyshire, for scrap, c11/1972.
(16) to HE, 21/7/1978; to Ford Motor Co Ltd, Bridgend, Glamorgan, 13/2/1980.
(17) to C F Booth, Rotherham, /1989, resold to Queenborough Rolling Mills, Sheerness, Kent.
(18) to Rother Valley Railway Ltd, Kent, 6/11/1997.
(19) to East Anglian Railway Museum, Chappel & Wakes Colne Station, 24/9/1998.
(20) returned to RMS Locotec Ltd, Yorkshire, 20/6/2002.

Gauge : 2ft 0in

		4wPM	BgE	2051	1931	New	(a)	(2)
		4wPM	MR	7001	1932	New	(b)	(1)

(a) built new as a 25hp Planet type with a Ford 4cyl engine and built for F. C. Hibberd & Co Ltd, Park Royal, London, loaned for exhibition in 1931.
(b) built new as a 25hp loco with a Ford engine, loaned to Ford Motor Co Ltd, 5/5/1932.

(1) returned to makers after loan.
(2) disposal unknown, s/s

W. & C. FRENCH LTD
BUCKHURST HILL BRICKWORKS
D65
TQ 422938
Company registered 23/3/1931
William and Charles French, early 1900s until c1930
Brick Works situated in Lower Queens Road and in operation from early 1900s until c1966; served by a narrow gauge tramway. Loco operation ceased c/1947.

Gauge : 2ft 0in

	4wDM	OK	5129	1933	(a)	(1)

(a) possibly new here.

(1) to Buckhurst Hill Plant Depot, c/1947.

LUXBOROUGH LANE BRICKWORKS, Chigwell
D66
TQ 424929
Harold E.B. Sheaves trading as Chigwell Brick Works
Brick Works in Luxborough Lane, Chigwell operated by William Cornish c1880's, and taken over by his son William Delhi Cornish; later operated by Harold E. B. Sheaves as Chigwell Brick Works from c1927 to at least 1937; the works made red bricks, tiles and drain pipes. Rail traction superseded by road vehicles. Works closed and demolished in 1969.

Gauge : 2ft 0in

LD 1198	4wDM	MR	7473	1940	(a)	(1)
LD 2979	4wDM	RH	191682	1938	(b)	s/s c/1969

(a) ex MR 12/2/1940, possibly new here.
(b) ex WD, 11/4/1957, here by 10/1958.

(1) to Buckhurst Hill Depot, by 10/1958.

WOODFORD BRICKWORKS

D67

TQ 412910

Brick Works off Chigwell Road established by William Cornish c1894, later operated by his son Herbert Robert Cornish from late 1920s to early 1930s. The works made Red facing bricks, roofing tiles and drain pipes. A narrow gauge rail system connected the clay pit and works. The line was in use 6/1964 but the works closed and was demolished in 1965.
Gauge : 2ft 0in

LD 1087   4wDM   MR   7486   1940   (a)   s/s

(a)   ex MR 22/4/1940, possibly new, here by 11/1958.

FRESHWATER SAND & BALLAST CO LTD

FRESHWATER BALLAST WORKS, Fingringhoe, near Colchester   **H68**
Freshwater Sand & Ballast Co until 3/10/1946
Freshwater Gravel Sand & Ballast Company Ltd until 1/1931   TM 049194

The Freshwater Gravel Sand & Ballast Company Ltd was registered 3/1/1929 to take over property at Wick Farm, Fingringhoe belonging to Ernest Edward Clinch, farmer and director of the company; the company went into liquidation in 1/1931. From 1931 **E.J.& W. Goldsmith Ltd** were sole proprietors of the Freshwater Sand & Ballast Co. The Freshwater Sand & Ballast Co Ltd was incorporated on 3/10/1946 to take over the existing business carried on by the Freshwater Gravel Sand & Ballast Co, one of the directors being Mr J. F. Barham (see also Peters & Barham and Barham & Tait Ltd). The company is shown HMSO list of quarries for 1948 with an address at Ocean House, 66-71 Lower Thames Street, E.C.3 and is listed in directories until 1955/56. Ballast works closed.
Gauge : 2ft 0in

-	4wDM	RH	164333	1931	New (a)	(1)
-	4wDM	RH	166010	1932	New	(2)
-	4wDM	RH	164347	1933	(b)	(3)
-	4wDM	RH	175408	1935	(c)	(4)
-	4wDM	RH	211608	1941	(d)	(4)

(a)   ex RH, Lincoln, 12/1/1932 new on one months trial, per Field & Ness, engineers, London (agents?); later purchased.
(b)   ex RH, Lincoln, 19/6/1934; originally a new loco hired to Tower Quarries Ltd, Whisby, Lincolnshire, 13/3/1933 to 20/11/1933.
(c)   ex Exors of John Heaver, Whyke Gravel Pits, Chichester, Sussex, after 11/9/1941, by 18/3/1947.
(d)   ex MoS.

(1)   s/s after 27/6/1946.
(2)   returned to RH, Lincoln, 9/7/1934; thence to George Marsh, 30 Sutton Street, Southend-on-Sea, 3/9/1934. Replaced by RH164347.
(3)   to Wennington Sand & Ballast Co Ltd, Wennington, after 7/11/1949.
(4)   to Wennington Sand & Ballast Co Ltd, Wennington.

GAS LIGHT & COKE CO
SOUTHEND-ON-SEA GAS WORKS
Southend-on-Sea & District Gas Co until /1931
Southend Gas Co until /1923

J69
TQ 895849

A gasworks was built on a four acre site at the east end of the town in 1854 by the Southend Gas Light & Coke Co and production commenced on 8/5/1855. Coal was supplied to this works via a pier immediately to the south. In 1877 the company became the Southend Gas Co and in 1902 this company built an elevated narrow gauge railway from the pier, over the Esplanade and into the east side of the works. Traffic handled was coal inwards and coke, tar and liquor outwards. This rail system was replaced by conveyor belts, closed and dismantled in 1935. The works remained in production to be taken over by North Thames Gas Board in 1949 and was finally closed in 1967.
Reference : Gas Light & Steam, Malcolm Millichip, British Gas, 1994.
Gauge : 2ft 1¾in

(PAULINE)	0-4-0ST	OC	KS	850	1903	New	s/s /1935
(BIM)	0-4-0ST	OC	KS	1099	1910	New	s/s /1935
(TIM)	0-4-0ST	OC	KS	4154	1920	New	s/s /1935

ILFORD GAS WORKS
Ilford Gas Company until /1922
Ilford Gas Light and Coke Co Ltd from 12/12/1881 until /1899

D70
TQ 433860

Ilford Gas Company was formed in 1839 and had a works off Ilford High Road on an Island in the River Roding. Coal was carted from Ilford Station and also arrived at a wharf on the River Roding, a narrow gauge internal tramway handled coal and other materials. The company was taken over by Gas Light & Coke Company in 1922, works closed 1923. Two 2ft gauge locos were for sale in an auction of plant by Gas Light & Coke Company on 3/7/1923.
Gauge : 2ft 0in

-		0-4-0WT	OC	HC	1034	1913	(a)	(1)
-		0-4-0ST	OC	KS	4020	1919	(b)	(2)

(a) ex Holloway Brothers [Rosyth Housing (1917-1918) contract, Fife ?], 10/1918.
(b) ex Kerr Stuart (built New for Ministry of Munitions).

(1) to William Jones Ltd, dealers, Greenwich, London, /1923;
thence to Tarmacadam Roads Ltd, Mostyn Works, Flintshire, by /1928.
(2) to Stewart & McDonnell, Kingston By- Pass contract, Surrey, by 5/6/1927.

MARK GENTRY
LANGTHORNE BRICKWORKS, Sible Hedingham
Orbell Cornish c1883 until 1884

F71
TL 769340

Mark Gentry traded as **Hedingham Brick & Tile Works** and from 1884 until 1911 as **Hedingham Brick Tile & Terra Cotta Works**. The works in Wethersfield Road was not in production in the period 1908 to 1910, had closed by 1912 (not listed in HMSO list of quarries for that year). Also by 1894 pits were operated at Highfields, Great Maplested until 1917, by 1918 this pit had been taken over by Rippers Ltd of Castle Hedingham but was not worked in that year. The loco listed below was constructed by Mark Gentry and his son in their own workshops and utilised a compound engine and vertical boiler manufactured by

A.G. Mumford Ltd, Culver Street, Colchester, marine engineers, and obtained second hand from a small launch. The loco weighed about one ton and was capable of hauling two trucks of brick earth at six miles per hour.
References : The British Clayworker 4/1898, 2/1905, 4/1908
                Model Engineer 4/1959
                East Anglian Magazine 1/1972
                Industrial Railway Record No.83 12/1979 and No.88 3/1981
                The Narrow Gauge No.198, /2007
Gauge : 2ft 0in

             -                 4wVBT   VC   1905   (a)   s/s

(a)   constructed on site, /1905.

THE GLOBE CEMENT, BRICK, WHITING, & CHALK CO
Whitehall, Grays   M72
                                                                                    TQ 624783

Globe Pit was first opened by Robert Ingram a brickmaker who installed a tramway to a Wharf at TQ 614773 on the River Thames in 1820, by 1849 the tramway had fallen out of use and was not in use when the LT&SR was constructed in 1854. On 22/7/1874 Robert Ingram applied to the LT&SR to make a tunnel under their railway to take a tramway along Manor Road from his brickfields to the Thames and in 1875 he was given permission to construct a bridge, however it has not been established if the line was ever built. In 7/1881 an agreement was made with the LT&SR for Messrs **Lockington & Mander** to erect a tramway bridge over their line, the tramway ran to a Wharf on the Thames at TQ 622763 and was known locally as the "Jumbo Line". The conveyance of land forming Globe Works passed to Mr Charles Wall in 1896 and Charles Wall Ltd were chalk quarry owners and manufacturers of Carbonate of Lime, a letter from the company in 1/1950 stated that the tramway closed in 7/1941.

"London Gazette" 3/1/1879, Notice is hereby given, that the Partnership heretofore subsisting between **James Barrows Lockington, Frederick George Norman Pochin** and **Charles John Mander**, carrying on business at Grays in the county of Essex under the style or firm of the Globe Cement, Brick, Whiting and Chalk Company, has this day been dissolved by mutual consent, so far as the said Frederick George Norman Pochin is concerned; and in future the business be carried on by the said James Barrows Lockington and Charles John Mander, under the same style as heretofore, by whom all debts due to and owing by the said late firm will be received and paid. – Dated this 2nd day of August, 1878.

"Gray's and Tilbury Gazette", 15/5/1886 carried a report that the works of the Tilbury Machinery and Ironware Company were undertaking repairs to a locomotive from the Globe brickfields which included the manufacture of new castings and turning of the tire's on the iron wheels.

The "Times" 6/12/1888 – Chalk and Ballast Quarry, Whiting Works, Capitol site for cement works and brickfield, close to Tilbury Docks, with good pier and wharf. To be Let or Sold, apply to J.B.L. 27 Gresham Street, E.C.

Fuller Horsey Sons & Cassell auctioned on 9/7/1889, the Globe Brick & Whiting Works, Whitehall, Grays which comprised an area of approximately eighty acres including workshops, stores, two lime kilns, five pug mills, tramway, pier and ballast wharf on the River Thames; the premises and jetty were leasehold for 99 years from 25/3/1876, loose plant included a **locomotive**.

Reference : Essex Record Office, D/DS 4/23.

The "Times" 29/6/1889 carried an advert for the auction on 9/7/1889 which stated, "May be viewed and particulars, plans and conditions of sale, Messrs **Mander & Watson**, solicitors, 9 New Square, Lincoln's Inn, WC.

Adverts in the "Times" show **Charles John Mander** solicitor of 9 New Square, Lincoln's Inn. **James Barrows Lockington** born Islington, traded as J. Lockington & Co, Commission Agents, 27 Gresham Street E.C.

Lockington & Mander formed the Tilbury Portland Cement Co Ltd in 1890, which by petition of James Barrows Lockington of 27 Gresham Street and Charles John Mander of 9 New Square, Lincolns Inn was wound up by court order 5/3/1892.

James Barrows Lockington late of Sutton Lodge, Sutton at Hone, Kent, died aged 62 on 18/10/1894.

"Contract Journal", 14/10/1896 – Fuller Horsey to sell 20/10/1896 at Globe Cement Works, Grays, plant including 0-4-0 loco.

Gauge : 4ft 8½in

    "JUMBO"   0-4-0ST OC HH   1883   New?(a) (1)

(a) possibly new, a copy of a catalogue for The Brush Electrical Engineering Co Ltd dated 1904 lists the Globe Brick Works as a customer to which they had supplied a locomotive.

(1) to Charles Wall Ltd with site.

E.J. & W. GOLDSMITH LTD
GRAYS RUBBISH SHOOT   M73

E.J.& W. Goldsmith until 8/2/1901

Business established in 1848; the company was listed in 1930 with an office address at 110 Fenchurch Street EC3, as Lightermen Contractors & Sailing Barge owners, coal, coke & breeze merchants, also dealers and shippers of whiting chalk, flint, loam, sand & ballast, at Grays, Essex, Rochester, Dartford and Stone Court, Kent; E.J. & W. Goldsmith are listed in rates records for 1910 as having workshops & offices at Town Wharf, Grays. The Rubbish Shoot was located near to the Sewage Works of the Thurrock, Grays & Tilbury Joint Sewage Board, tipping commenced in 1910 using a locomotive and wagons, shoot closed. Possibly that later worked by Charles Wall Ltd. Edward John Goldsmith of Hill Crest, Grays, died on 8/12/1910; Walter James Goldsmith of Milton Grange, Gravesend, Kent, died on 9/4/1927.

Gauge : 4ft 8½in

    PARANA   0-4-0ST OC MW   1141 1889 (a)   (1)

(a) ex Charles Wall Ltd, Grays Plant Depot, earlier Pauling & Co Ltd, High Wycombe (GW/GC Jt) (1901-1907) contract, Buckinghamshire, No.17 PARANA.

(1) to Dartford Coal Wharf, Kent, /1925.

GRAYS CHALK QUARRIES CO LTD
GRAYS WORKS
Richard Meeson & Co until 1862
Meeson & Co until 1858

N74
TQ 608787 & 612787

John Meeson established chalk quarries, brick and tile kilns at Grays at the beginning of the nineteenth century, about 1850 he merged with a local rival George Henry Errington to trade along with Richard Stock Meeson, as Meeson & Co. This partnership was dissolved on 8/3/1858 when George Henry Errington retired, the firm continued trading as Richard Meeson & Co and in 1862 became Grays Chalk Quarries Co Ltd who in their prospectus stated that for a small outlay locomotive power could be introduced and a new pier built. At that time there were two quarries situated either side of Hogg Lane, those to the east known as Titan Pit later became the site of Thos. W. Ward Ltd's Titan Works. Quarries closed /1951. Property and locos taken over by BPCM, by 3/1952 in connection with the Grays works of Light Expanded Clay Aggregates Ltd; BPCM had purchased the majority shareholding of the Grays Chalk Quarries Co in 1927.

"Railway Times" 10/8/1861 - Wanted to hire with option to purchase at end of a few months - two or three small locos, narrow gauge for slow mineral traffic - Meeson & Co, Grays.

"The Engineer" 25/4/1862 - For sale second-hand Loco Steam Engine with tubular boiler, three cylinders, 140lbs wp, 6t 17cwt, to be seen on application to Mr Tilbury, No.25, 181 at the Pantechnicon, Belgrave Square, this engine was first used as a Traction Engine on a tramroad, (this advert is thought to refer to the sale of OPHIR from the Edenham Branch Railway, listed below), still for sale in the "Times" 20/5/1862.

"The Engineer" 16/2/1866 Fuller Horsey to auction 14/3/1866 (later altered to 27/3/1866) re Grays Chalk Quarries Co Ltd, plant inc two locos, one 4w coupled 8x14 by Slaughter Gruning 1863, other 7in by Coley & Co, West London Iron Works, also one new loco boiler & engine, partly finished.

Gauge : 4ft 8½in

-		0-4-0VBT	VCG	Chaplin	140	1860	New	s/s
-		0-4-0VBT	VCG	Chaplin	188	1861	New	s/s
	BRISTOL	0-4-0ST	OC	SG		1863	(a)	(1)
		0-4-0ST	ICG				(b)	(1)
	OPHIR	2-2-0WT	IC	Coley & Co		1854	(c)	(1)
-		4wWT	G	AP	151	1865	New	(2)
-		4wWT	G	AP	167	1865	New	(2)
-		4wWT	G	AP	212	1866	New	(2)
-		4wWT	G	AP	508	1869	New	(2)
	MEESON	0-4-0ST	OC	BE	276	1898	New (e)	(4)
	DUVALS	0-4-0ST	OC	WB	1583	1899	New	(4)
	AMOS	0-4-0ST	OC	WB	1637	1900	New	(5)
	CHARLETON	0-4-0ST	OC	BE	306	1904	New	Scr /1937
6	ALLAN BRUCE	0-4-0ST	OC	AB	680	1890	(d)	(3)

(a) origin unknown.
(b) origin and identity unknown.
(c) earlier Edenham Branch Railway, Lincolnshire, /1862, OPHIR, (converted from a road engine by Daniel Gooch, at Swindon Works, GWR, /1855, delivered to Little Bytham, GNR, 7/11/1855).
(d) ex APCM, Brooks Works, /1921.
(e) loaned to Light Expanded Clay Aggregates Ltd, Grays Works, and returned.

(1) to Isaac Watt Boulton, dealer, Ashton-under-Lyne, 28/3/1866.
(2) two of these four locos sold for scrap by /1904; remaining two advertised for sale, 8/1905 and s/s.
(3) to Thos. W. Ward Ltd, Columbia Wharf, Grays, /1939.
(4) to APCM, Bevan's Works, Kent, by 23/2/1951.
(5) scrapped 1/1954 by a firm from Bexleyheath, Kent.

GRAYS CO-OPERATIVE SOCIETY LTD
GRAYS COAL DEPOT

M75
TQ 621775

The first barge arrived at Grays Co-op Wharfe with wheat in /1875, the adjacent jetty had a narrow gauge tramway which ran to the Bakery via an overhead gantry. The jetty had a grab crane with narrow tracks either side of the crane, in the 1960's there were two 'Granby' wagons built by J.F. Howard of Bedford each with a single dropside door which transported coal from the jetty to the bakery boilers, the line is thought to have been cable operated. Grays Co-op advertised for a driver for a Grafton steam crane with grab, 2/1/1909. Latterly the yard of Seabrooke's Brewery. Closed and demolished in 1969.

Gauge : 4ft 8½in

DAGENHAM   0-4-0ST OC HC   1564 1925 (a)   Scr 10/1955

(a) ex Chas. Wall Ltd, contractor, Grays, /1940.

GREATER LONDON COUNCIL
GASCOIGNE ROAD SEWAGE PUMPING STATION, Barking

P76
TQ 446830

Barking & Ilford Joint Sewage Board until 1962
Works located in Alfred's Way, Barking.

Gauge : 2ft 0in

-    4wDM   RH 354028 1953 (a)   (1)

(a) earlier S.S. Stott & Sons Ltd, Ilford Sewage Works contract.

(1) to A.M.Keef, Bampton, Oxfordshire, 6/1974;
thence to Dowty Railway Preservation Society, Ashchurch, Gloucestershire.

GREAT WAKERING BRICK CO LTD
MILLHEAD BRICKWORKS, near Southend-on-Sea

J77
TQ 952888

D. & C. Rutter Ltd until 1923
D. & C. Rutter 1865 until c1909
Company registered 6/2/1923 to acquire and take over the business of brick and tile makers carried on by D. & C. Rutter Ltd. Company went into voluntary liquidation 11/12/1961. Works closed in 1962.

Gauge : 2ft 0in

-	4wPM	KC			(a)	s/s c/1963
-	4wPM	MR	5027	1929	New	s/s /1963
-	4wPM	MR	21520	1955	New	(1)

(a) origin and identity unknown.

(1) to Milton Hall (Southend) Brick Co Ltd, Star Lane Brickworks, c/1963.

GRIFFITHS WHARF CO
NORTH WOOLWICH WHARF
A78
TQ 427798

On 11/6/1889 Mr W. Griffiths of 283 Kingsland Road, East London asked Woolwich Council for permission to lay a siding from the GER across Green Lane to part of the former works of W.T. Henley & Co (submarine cable makers), Griffiths had taken over the site to use as a wharf for handling stone. The partnership between William Griffiths and Fred Griffiths carrying on business under the style or firm of W. and F. Griffiths and Griffiths Wharf Co was dissolved 8/6/1916. Directories show occupants of the wharf as William Griffiths & Co Ltd (incorporated 4/10/1900), contractors & quarry owners, works closed c1939.
Gauge : 4ft 8½in

	IDA	0-4-0ST	OC	TG	227	1900	New	s/s
	JOE	0-4-0ST	OC	HE	620	1895	(a)	(1)
	-	0-4-0PM		MW	1951	1918	(b)	(2)

(a) ex S. Pearson & Sons Ltd, Royal Albert Dock Extension (1912-1918) contract, on hire by 12/1918.
(b) ex MoM, Sandwich, Kent, 1697, c/1922.
(1) s/s after 5/1919.
(2) s/s after 3/1937.

HALL & CO LTD
ROMFORD GRAVEL PITS
L79
TQ 512876

Company registered in 6/6/1918 with pits in Oldchurch Road, Romford opened in 1927, and listed in HMSO list of quarries for 1928 until at least 1939, pits closed /1940. Later, by 1948, there were pits at Sandy Lane, Aveley until about 1959-1960. The Quarry directories for 1953 also list pits at Waltham Road, Boreham until at least 1965-1966.
Gauge : 2ft 0in

		4wPM		MR	4032	1926	New	s/s
		4wPM		MR	4048	1927	New	s/s
		4wPM		MR	4502	1927	New	(3)
		4wPM		H	940	1928	New	s/s
		4wPM		MR	4510	1928	New	s/s
		4wPM		MR	4516	1928	New	s/s
		4wPM		MR	4521	1928	New	s/s
		4wPM		MR	5061	1930	New	(1)
		4wPM		MR	5204	1930	New	s/s
18		4wPM		MR	5301	1931	New	(2)
		4wDM		MR	5645	1933	New	s/s
23		4wDM		MR	5646	1933	New	(4)

(1) to MR, 19/5/1933, resold to Greenham Plant Hiring Co, consigned to Paterson & Dickinson Ltd, Southall, Middlesex, 31/1/1934.
(2) to Langney Point Pits, Sussex.
(3) returned to MR, 26/5/1936.
(4) to Washington Pits, Sussex, 1/1937.

HALL & HAM RIVER LTD

The Ham River Grit Co Ltd (registered 3/9/1923) until 15/3/1961
The company also operated pits at Fishers Green, Waltham Abbey (listed in quarry directories for 1948 until at least 1965-1966), and pits on the arterial road at Stifford, West Thurrock, listed from 1948 until at least 1957-1958.

BRIGHTLINGSEA PITS   H80
TM 094174

Pits in Robinson Road, Brightlingsea listed in Quarry directories for 1963/64 and 1965/66. Rail traffic ceased and track lifted in 1964.

Gauge : 2ft 0in

LO 29	(122)	4wDM	RH 338440	1955	New	(1)
LO 30	123	4wDM	RH 393327	1956	New	(4)
LO 16		4wDM	RH 187064	1937	(a)	(3)
LO 13		4wDM	RH 211686	1942	(b)	(2)

(a)   ex Feltham Sand & Gravel Co Ltd, Heston Airport Pits, Middlesex, c/1958.
(b)   ex South Ockendon Pits, c/1961.

(1)   to Chertsey Shops, Surrey, 8/1962.
(2)   returned to South Ockendon Pits, 3/1964.
(3)   to South Ockendon Pits, 3/1965.
(4)   to S.B. Wheeler & Sons, Hythe Quay, Colchester, for scrap, c4/1967.

EASTBROOK FARM PITS, Rush Green, near Romford   L81
TQ 510862

Pits at Eastbrook Farm, Dagenham Road, Rush Green, Romford, listed in quarry directories for 1959-1960 until at least 1965-1966. Rail traffic ceased c1961; pits worked out and later flooded.

Gauge : 2ft 0in

LO 31	(No.124)	4wDM	RH 398099	1956	(a)	(2)
No.125		4wDM	RH 398100	1956	(a)	(1)

(a)   ex Chertsey Depot, Surrey, by 7/1958.

(1)   to Land End Gravel Pits, Berkshire, 7/1962.
(2)   to Chertsey Depot, Surrey, after 9/1962.

SOUTH OCKENDON PITS   L82
TQ 580824

Pits on land at Belhus Farm, South Ockendon, listed in quarry directories for 1939 until 1965-1966 at least; pits worked out and later flooded.

Gauge : 2ft 0in

LO 10	(112)	4wDM	RH 192876	1938	New	s/s c/1967
	113	4wDM	RH 193966	1939	New	(1)
LO 14	(117)	4wDM	RH 200771	1940	New(c)	(2)
LO 12	(114)	4wDM	RH 211685	1942	(a)	s/s c/1967
LO 13	(115)	4wDM	RH 211686	1942	(b)	s/s c/1967

LO 15	R 132		4wDM	RH	277278	1949	(d)	s/s c/1967
LO 17			4wDM	RH	186327	1937	(d)	s/s c/1967
LO 16			4wDM	RH	187064	1937	(g)	s/s c/1967
LO 29			4wDM	RH	338440	1955	(e)	s/s c/1967
LO 32			4wDM	RH	398100	1956	(f)	s/s c/1967

(a) ex Harlington Pits, Middlesex, by 9/1946.
(b) ex Harlington Pits, Middlesex, by 4/1947,
to Brightlingsea Pits, c/1961; returned 3/1964.
(c) to Chertsey Depot, Surrey, c/1957; returned.
(d) ex Feltham Sand & Gravel Co Ltd, Heston Airport Pits, Middlesex, c/1958, (after 24/7/1957).
(e) ex Chertsey Depot, Surrey, 3/1963;
to Chertsey Depot, Surrey, 6/1964; returned 9/1965.
(f) ex Chertsey Depot, Surrey, 3/1964.
(g) ex Brightlingsea Pits, 3/1965.

(1) to Hall & Co Ltd, Washington Depot, Sussex, c/1961.
(2) to Chertsey Depot, Surrey, c6/1967.

HARRISONS (LONDON) LTD
PURFLEET WHARF
N83
TQ 551781

Company registered 2/8/1913 at 66 Mark Lane, London EC as importers & exporters and dealers in coal and coke, listed as coal contractors at Cornwall Building, London Road, Purfleet, established on land purchased from the Whitbread Estate on 24/6/1920. Wharf leased to Esso Petroleum Co Ltd, the company paid sums of money to the Petroleum Board during 1940-1941and in 1945-1946; also during the latter period to William Cory. As these charges were made at an hourly rate they would appear to be for the services of those two companies locos for shunting Harrisons Wharf; locomotive use appears to have been in connection with coal storage on behalf of the government. On 23/12/1943 the Ministry of War Transport (Coal Division) requested from the Railway Executive Committee the loan of a loco and tip wagons for their agents Harrisons (London) Ltd. On 31/12/1943 the LNER offered to loan Sentinel loco 87; on 14/1/1944 this offer was accepted and the loco dispatched from Ardsley to Purfleet. On 30/9/1945, Sentinel 64 was here on loan.
Gauge : 4ft 8½in

-		0-4-2ST	OC	KS	3129	1918	(a)	(1)
TW 317		0-6-0ST	IC	MW	2005	1921	(b)	(2)

(a) ex Mitchell Engineering Ltd, 5/1941;
earlier Wm. Jones Ltd, dealers, Greenwich, London.
(b) ex Thos. W. Ward Ltd Grays, on hire 12/1946-1/1947.

(1) to Thos. W. Ward Ltd, Grays, 4/1945; scrapped /1953.
(2) returned to Thos. W. Ward Ltd, off hire, after 1/1947, by /1949.

G.J. HAWKES & SONS
BOREHAM SAND PITS, Chelmsford   G84
                                                                TL 755114

Company with an address given as Rainsford Works, Rainsford End, Chelmsford and listed in HMSO list of quarries for 1931 with pits at Boreham until at least 1955-1956. G.J. Hawkes & Sons Ltd went into voluntary liquidation on 20/8/1958.

Gauge : 2ft 0in

| | 4wPM | L | 3270 | 1930 | New | s/s |

HAWKWELL BRICKFIELDS LTD
HAWKWELL BRICKWORKS   J85
                                                                TQ 861922

Company registered 8/6/1937 with an address at 59 High Street, Grays; went into voluntary liquidation 23/11/1962. Locomotive haulage dispensed with, works taken over by Eastwoods Ltd, and operated from c1955/56 – c1957/58 and later closed.

Gauge : 2ft 0in

| | 4wDM | RH | 193972 | 1938 | (a) | s/s |

(a) ex M.E. Engineering Ltd, Cricklewood, Middlesex, 11/1954.

HERTS PLANT HIRE LTD
NEWMARKET ROAD SAND & GRAVEL PITS, Great Chesterford   F86
                                                                TL 500433

Company registered 27/8/1942 at 8 Windermere Avenue, St Albans as hirers of excavating equipment; had sand and gravel pits at Great Chesterford.

Gauge : 2ft 0in

| | 4wDM | MR | 8602 | 1940 | (a) | (1) |

(a) ex Buckland Sand & Silica Ltd, Surrey, 10/1945.

(1) to Savages Ltd, Kings Lynn, Norfolk, later George W. Bungey Ltd, Hayes, Middlesex, by 4/1951.

HIGHWAYS CONSTRUCTION LTD
Beckton   A87

In 1915 a clinker tip from the retort houses at Beckton Gas Works caught fire by spontaneous combustion and was on fire until 1919. About 1923 Highways Construction Ltd bought the clinker tip and crushed it to make clinker asphalte for roads on the Kent County Council arterial road. The company used a Ruston steam excavator and narrow gauge track with at least one 20hp and one 40hp Simplex loco in use.

Gauge : narrow

| 4wPM | [MR/FH?] | (a) | s/s |
| 4wPM | [MR/FH?] | (a) | s/s |

(a) origin and identity unknown.

IMPERIAL CHEMICAL INDUSTRIES LTD, Alkali Division
SILVERTOWN WORKS B88
ICI (Alkali) Ltd until 30/12/1944   TQ 409798
Brunner Mond & Co Ltd incorporated 24/2/1881until 12/1931
The works (established by 1900) produced TNT during WW1, during which a large explosion occurred here causing extensive damage in the surrounding area. The works closed in 3/1961 and the premises were subsequently acquired by Wood & Plastics Ltd.

Gauge : 4ft 8½in

SOLVAY	0-4-0WT	OC	EB	10	1880		
	reb		KS		1924	(a)	(1)
WALTER	0-4-0ST	OC	RP	13111	1888	(a)	(1)
HASSALL	4wVBT	VCG	S	6893	1927	(b)	(4)
WHEELOCK	4wVBT	VCG	S	7297	1928	(b)	(4)
J.B.GANDY	4wDM		RH	299103	1950	New	(8)
ETTRICK	0-4-0ST	OC	HL	3721	1928	(c)	(5)
BLACK	0-4-0WT	OC	KS	3048	1917	(d)	(6)
CROOKES	0-4-0WT	OC	KS	4199	1920	(e)	(7)
-	0-4-0ST	OC	MW	1135	1891	(f)	(3)
	0-4-0ST	OC				(g)	(2)

(a) ex Winnington Works, Cheshire, /1912.
(b) ex Sandbach Works, Cheshire, /1932.
(c) ex Winnington Works, Cheshire, c/1951.
(d) ex Winnington Works, Cheshire, 1/1953.
(e) ex Winnington Works, Cheshire, 9/1957.
(f) ex BPCM, Wouldham Works.
(g) origin and identity unknown.

(1) to F.J. Church & Sons, Stamford Hill, Middlesex, /1933.
(2) to Thos. W. Ward Ltd, Silvertown.
(3) to British Alizarine Co Ltd, Trafford Park, Manchester.
(4) scrapped by George Cohen, Sons & Co Ltd, /1952.
(5) to Abelson, Sheldon, Birmingham, 3/1953;
    later to Richard Thomas & Baldwins Ltd, Blisworth, Northamptonshire.
(6) to Thos. W. Ward Ltd, Silvertown, for scrap, 11/1957.
(7) scrapped on site by Thos. W. Ward Ltd, 6/1961.
(8) to Tunstead Works, Derbyshire, 4/1961.

Gauge : 2ft 6in

(1)	4wBE	BEV	194	1920	New	(1)
2	4wBE	BEV	393	1922	New	(1)
3	4wBE	BEV	622	1925	New	(1)
4	4wBE	WR	778	1930	New	(1)

(1) to Winnington Works Crystal Plant, Cheshire.

IND COOPE & ALLSOPP LTD
ROMFORD BREWERY   E89
Ind, Coope & Co Ltd until 6/1934   TQ 513886
Ind, Coope & Co (1912) Ltd until 1/1923
Ind, Coope & Co Ltd until 11/10/1912

Brewing was carried out at the Star Inn, Romford by 1750 and this business was purchased by Messrs Ind & Grosvenor in 1799. In 1845 Mr C.E. Coope joined the firm which became known as Ind, Coope & Co. This firm owned the brewery on a site which was adjacent to the north side of Romford Station but on a lower level. Sidings within the brewery together with a wagon hoist were provided by the Eastern Counties Railway in 1853 and by 1862 the sidings had been extended and the hoist replaced by an inclined connection. Between 1897 and 1899 the wagon turntable in the GER goods yard which gave access to the brewery sidings was replaced by a sharp curve, and this enabled steam locomotives to work within the brewery in place of the horses earlier employed. Two single road locomotive sheds were built.
From 1955 motive power for the much reduced rail traffic was provided by a diesel road tractor fitted with buffers, until rail traffic ceased completely in 1963.
References: Brewery Railways, Ian P Peaty, David & Charles, 1985
Industrial Railway Record No.95.
"Railway Bylines", Ian P. Peaty, December 2003.

Gauge : 4ft 8½in

	OSCAR	0-4-0ST	OC			1872	New?	s/s
	ECLIPSE	0-4-0ST	OC	FW	160	1872	New	s/s
3		0-4-0ST	OC	HL	2513	1902	New	s/s
	-	0-4-0ST	OC	P	1142	1908	New	(1)
3		0-4-0ST	OC	HL	3540	1923	New	(2)
2		0-4-0ST	OC	HL	2345	1896		
		reb	TW			1925	(a)	(3)
5		0-4-0ST	OC	AB	2028	1937	New	Scr c/1952
1		0-4-0ST	OC	HL	3539	1923	(b)	(4)
No.2		0-4-0DM		Bg	3227	1951	New	(5)

(a)   ex Burton Brewery, Burton-on-Trent, Staffordshire, /1925.
(b)   ex Burton Brewery, Burton-on-Trent, Staffordshire, c/1948.

(1)   to Grovesend Steel & Tinplate Co Ltd, Gorseinon, Glamorgan.
(2)   to Burton Brewery, Burton-on-Trent, Staffordshire, /1937.
(3)   returned to Burton Brewery, Burton-on-Trent, Staffordshire.
(4)   returned to Burton Brewery, Burton-on-Trent, Staffordshire, by /1951.
(5)   to Burton Brewery, Burton-on-Trent, Staffordshire, /1955.

INDUSTRIAL CHEMICALS LTD
TITAN WORKS, Hogg Lane, Grays   N90
                                                                                            TQ 613783

The company, part of Industrial Chemicals Group Ltd (registered 10/3/1976) supplies chemicals to the detergent, paper, water treatment and chemical industries, the works at Grays was established in 1976 on the site of Thos. W. Ward's Titan Works and is not rail connected, the loco listed below was still stored on site in 2009.

Gauge : 4ft 8½in

-	4wDH	TH	144v	1964	(a)

(a) ex Proctor & Gamble Ltd, West Thurrock, c4/1996 (after 13/2/1996).

INNS & CO LTD

LITTLE MARDYKE PITS, South Hornchurch **L91**
TQ 509838

Company registered 4/3/1935 had pits on land adjacent to Mardyke Farm listed in HMSO list of quarries for 1948 until 1963-1964. The rail system was abandoned in favour of road transport, c10/1964. The company also operated other pits in Essex; HMSO list of quarries for 1934 lists **Wicken No.2** pit at Wicken Bonhunt (TL 491336?). Also listed in 1948 are sites at **Barnston Hall**, Great Dunmow, and **Langridge Farm**, Nazeing (see under Redland-Inns Gravel Ltd), and by 1959-1960 at **Wash Farm**, Fordham near Colchester (TL 917272).

Gauge : 2ft 0in

-	4wPM	MR	7046	1937	(a)	(1)
-	4wDM	MR	7107	1936	(b)	(1)
-	4wDM	MR	7226	1938	(c)	(6)
-	4wDM	MR			(c)	(6)
-	4wDM	MR	8598	1940	(d)	(2)
-	4wDM	MR	7457	1939	(d)	(2)
-	4wDM	MR	7040	1937	(e)	(4)
-	4wDM	MR	8675	1941	(f)	(5)
-	4wDM	MR	5277	1931	(g)	(3)

(a) ex Denham Pits, Harefield, Middlesex by 9/1952.
(b) ex Harpur Lane Pits, Hertfordshire, by 9/1952.
(c) ex Colney Heath Pits, Hertfordshire. c/1955.
(d) ex Waterford Workshops, Hertfordshire, by 10/1957.
(e) ex Moor Mill Pits, Hertfordshire, by 10/1957.
(f) ex Moor Mill Pits, Hertfordshire, c/1958, by 6/1959.
(g) ex Waterford Workshops, Hertfordshire. c/1961, by 4/1963.

(1) to Waterford Workshops, Hertfordshire, c/1955.
(2) to Denham Pits, Harefield, Middlesex, c/1961.
(3) to Broxbourne Pits, Hertfordshire, c/1964.
(4) to Langridge Farm Pits, Nazeing, c/1961.
(5) to Langridge Farm Pits, Nazeing, c/1964, by 5/1965.
(6) s/s after 3/1957.

RODING PITS E92

The HMSO list of quarries for 1948, lists Roding Pit, Kelvedon Common, near Ongar (possibly TL 571996). This was not listed in quarry directory for 1953. Pits closed.
Gauge : 2ft 0in

-	4wPM	MR	5319	1931	(a)	s/s

(a) ex Waltham Cross Pits, Hertfordshire, by 5/4/1938.

WHITE COLNE PITS F93
TL 875286?
Pits on land adjacent to Fox & Pheasant Farm, listed in quarry directories for 1939 to 1957-1958 but not 1959-1960.
Gauge : 2ft 0in

    -                   4wDM (ex PM) MR   5258 1931   (a)   (1)

(a)   ex MR 30/4/1935, reconstructed loco converted to diesel 11/4/1935.
(1)   to Waterford Workshops, Hertfordshire.

INTRADE LTD
FRESH WHARF, Barking P94
TQ 439835 approx
This company was registered on 21/7/1927 with an address at 185a Wardour Street, W1. Listed in directories as Wharfingers at Hewetts Quay, Abbey Road, Barking also with a depot in Highbridge Road, (Fresh Wharf) Barking as producers of concrete products; also traded as contractors and is thought to have operated a rubbish tip at Warley.
"Contract Journal" 13/7/1949, for sale -- plant including R&R 2ft gauge diesel locos – Intrade Ltd, Warley, Essex.
"Contract Journal" 1/4/1953, H Butcher to auction 5/5/1953 re Intrade Ltd, Barking – plant including R&R 2ft gauge diesel loco.
Gauge : 2ft 0in

                      4wDM   R&R   1938   (a)   s/s

(a)   origin and identity unknown.

KYNOCH LTD
KYNOCHTOWN WORKS K95
G. Kynoch & Co Ltd until 9/3/1897   TQ 742827
Kynoch and Co Ltd 31/7/1873 until 16/7/1884
George Kynoch established his business in 1850, Kynoch & Co were major ammunition manufacturers with a works known as Lion Works at Witton, Birmingham. The development of this site was undertaken to expand production.
Construction of the new explosives works, about two miles east of Corringham, was commenced early in 1897 and the works was in production within a year. The numerous widely dispersed buildings were served by a highly complex network of narrow gauge tramways. It is unlikely that mechanical power was used on these tramways. The works was taken over by Cory Bros. & Co Ltd in 1921, and Kynochtown renamed Corytown later Coryton. The steam locos worked and were owned by the Corringham Light Railway, a separate company but in practice a subsidiary of Kynoch Ltd who had the majority shareholding; providing access to the works from Corringham village and also a link to the LT&SR line at Thames Haven.
The "Times" 11/10/1919 carried an advert by Kynoch Ltd for a dismantling sale at "Shellhaven Works" Kynochtown to be held on 27/10/1919 which included surplus portable buildings, plant, machinery and effects including four miles of 20in gauge tram lines and 1,700 yards of standard gauge railway lines. On 3/7/1920 a further advert by Kynoch Ltd

of Birmingham offered for sale extensive properties which comprised 800 acres of land and included the Kynochtown Factory and village, also included was the Corringham Light Railway. On 30/10/1926 an auction was advertised of the Plant and Machinery of the Explosives Works, Corytown, being the entire equipment of the works; the auction to be held by Fuller Horsey & Co on 2/11/1926 by order of the directors of Cory Bros & Co Ltd.

Reference : The Corringham Light Railway, Peter Kay, 2008.

Gauge : 4ft 8½in

	CORDITE	0-4-0WT	OC	K	T109	1884	(a)	(2)
	KYNITE	0-4-2T	OC	KS	692	1901	New	(3)
2		0-4-0ST	OC	KS	1283	1915	New	(1)
	-	0-6-0ST	OC	AE	1771	1917	New	(3)

(a) ex Linton & Geen, Tredegar Dock Contract, South Wales, c/1900.
(1) to Eley Bros Ltd, Edmonton, Middlesex, c/1919.
(2) possibly to Cory Bros & Co Ltd with site, otherwise s/s after c/1915.
(3) to Cory Bros & Co Ltd with site.

LAFARGE ALUMINOUS CEMENT CO LTD
FONDU WORKS, West Thurrock　　　　　　　　　　　　　　　　　　**N96**
Eldonwall Industrial Estates Ltd (registered 13/5/1966) until /1976　　　　TQ 573779

The Lafarge Cement Co Ltd (registered 4/7/1923) bought seven acres of land from Thurrock Chalk & Whiting Co for a works which started up in 1926, chalk being supplied by the latter company. Due to problems of flints in the chalk, limestone was later used and bauxite imported. In 1948 it was stated that the Thurrock Chalk & Whiting Co provided locomotive power for traffic of Alpha Cement Co and Lafarge Aluminous Cement Co Ltd. In 1975 the Eldonwall company purchased more land which included the site of the then existing loco shed and rail and transport operations were taken over by Lafarge but operated by a new subsidiary company, Fairshare Transport Ltd. In 1981 a rail connection was made with the former Tunnel Portland Cement railway and use was made of their jetty. Production of clinker ceased in 1985 and clinker was subsequently imported from Dunkirk and moved by road. Track removed in 1989.

Gauge : 4ft 8½in

1	4wDH	RR	10247	1966	New	(1)
2	4wDH	RR	10248	1966	New	(1)
3	4wDH	RR	10249	1966	New	(1)

(1) to Blue Circle Industries Ltd, Dunbar Works, Lothians, c/1988.

LAKE & ELLIOT LTD
ALBION WORKS, Braintree　　　　　　　　　　　　　　　　　　**F97**
Lake & Co until 8/6/1910　　　　　　　　　　　　　　　　　　　TL 766227

Firm established c1894 by W.B. Lake and joined three years later by Mr E.F. Elliot as manufacturers of cycle and motor tools & accessories who also made a range of jacks under the brand name "Millennium". In 1910 the company built a new factory which opened 31/12/1910 with a new electric furnace installed, powered by a 450hp gas engine. The furnace was capable of melting 2½tons of metal every four hours for the production of steel castings. In 1919 The Unit Construction Company Ltd built a new stores building. The company made petrol locomotives according to their advertising material. These locomotives were advertised as capable of hauling 100tons and weighed 5¼ tons; they were advertised for sale at a price of £300 without power unit which was a Fordson Tractor. The brochure stated that the power unit could easily be placed in position should the purchaser wish, and would be fitted free of cost if consigned, carriage paid to the company's railway sidings, the cost of the tractor, minus wheels quoted as approximately £125 delivered; perhaps they did not in fact manage to sell many. The company was taken over by the Suter engineering group in 1985.

Gauge : 4ft 8½in

		tank loco			(a)	s/s
-		4wPM	Lake & Elliot	c1923	New	(1)
-		4wPM	Lake & Elliot	c1924	New	(2)

(a) origin and identity unknown.
(1) to Hoffmann Manufacturing Co Ltd, Chelmsford, c/1959.
(2) to Colne Valley Railway, Castle Hedingham, 12/2/1977.

W.T. LAMB & SONS LTD
HAMBRO HILL BRICKWORKS, Rayleigh J98
W.T. Lamb & Sons until 20/12/1935 TQ 809918
Pressed Brick and Tile Company Ltd until 12/1926
Brick & Tile Manufacturers Corporation Ltd until 7/7/1923
Rayleigh Brick & Tile Company Ltd until 1921
Rayleigh Brick & Tile Co trading as **Rayleigh Brick & Tile Works**, until 21/7/1919
Rayleigh Brick & Tile Company Ltd until 1907
Theophilus Sneeds Eli Plowman, trading as **W. Clover & Co** until 23/4/1903
William Clover jnr, trading as **W. Clover & Co**
William Clover until 1900

Brickworks established by William Clover in the 1890s and traded as W. Clover & Co from 1900 to 1903. Theophilus Sneeds Eli Plowman brick manufacturer of Shefford, Bedfordshire trading as W. Clover & Co had acquired the works which was sold in 1903 to the Rayleigh Brick & Tile Company Ltd of which Eli Plowman was a director and share-holder, the company went into voluntary liquidation on 19/3/1906. A new company of the same name was formed in 1919 to take over as a going concern the business of the Rayleigh Brick & Tile Co under the style or firm of The Rayleigh Brick & Tile Works; still in the ownership of Eli Plowman. The site covered 25 acres had a 16 chamber continuous kiln with a capacity of 400,000 bricks. The works also had a large machine house, engine house and a six track tramway tunnel dryer; an engine and gas producer plant by Tangye. A siding from the GER ran into the works alongside the kilns and a tramway ran throughout the yard and into the drying chambers; a receiver was appointed 20/12/1920. The works was taken over in 1921 by the Brick and Tile Manufacturers Corporation Ltd and passed to the Pressed Brick & Tile Co Ltd in 1923 who in turn sold it to W.T. Lamb & Sons in 1926.

Gauge : 2ft 0in

		4wDM	RH 174528 1935	(a)	(1)
-					

(a) ex Canewdon Brick, Sand & Ballast Co, Canewdon Brickworks, Ballards Gore, after 10/2/1939 by 14/7/1944.
(1) to W.T.Lamb & Sons Ltd, Wickford Brickworks, after 29/7/1953 by 3/3/1955.

WICKFORD BRICKWORKS　　　　　　　　　　　　　　　　　　　G99
W.T. Lamb & Sons until 20/12/1935　　　　　　　　　　TQ 746929
Harry Burton Haylock until 1925
Cornish & Haylock until 1921
Shenfield & Cranham Brick & Tile Co Ltd until 1914
Daniel Cornish & Co until 1906
Orbell Cornish until 1895

Brickworks located in Nevendon Road, Wickford established by Orbell Cornish in 1890 and passed to Daniel Cornish in 1895 trading as Daniel Cornish & Co. In 1906 it became the works of the Shenfield & Cranham Brick & Tile Co Ltd operated by Daniel Cornish. In 1914 the works passed to Cornish & Haylock a partnership between John Cornish (a brother of Daniel) and Harry Burton Haylock until the partnership was dissolved in 1921 when Harry Burton Haylock continued to trade alone. In 1925 the works was taken over by W.T. Lamb & Sons and continued to operate until c/1962 as W.T. Lamb & Sons Ltd.

Gauge : 2ft 0in

	4wDM	RH	174528	1935	(a)	s/s

(a)　ex Rayleigh Brickworks, after 29/7/1953 by 3/3/1955.

LAND RECLAMATION CO LTD
PITSEA RUBBISH TIPS　　　　　　　　　　　　　　　　　　　K100
TQ 741853 & TQ 745842

Company registered 15/3/1935 at 134 Fenchurch Street EC3 whose directors were listed as R. Wallace and A.C. Hutchins. The company was formed to acquire Little Mussels Farm, Bowers Gifford from **Flower & Everett Ltd**, complete with wharf, gantry and erected plant to carry on the business of removers & depositors of refuse for public authorities. The company was associated with **Inns & Co Ltd**, and later with an office address at 286 Pentonville Road, London N1.

It had a landfill operation on Bowers Gifford Marshes near Pitsea where it owned about 1000 acres of low lying land, operations commenced in 1937 when rubbish was transported from the boroughs of Camberwell, Poplar and Westminster by barge to a jetty on Vange Creek (TQ 741853); here the barges were unloaded by three electric grab cranes, the rubbish being transported to the tipping area by a narrow gauge railway using jubilee skips. In 1946 145,000 tons of rubbish was disposed of. In 1947 a new jetty was constructed at (TQ 745842) and a further two cranes installed due to the increase in barge traffic; by the end of 1949 this had increased to an estimated 291,000 tons with rubbish being transported from the boroughs of Bethnal Green, Holborn and Stepney and later from Paddington and St. Marylebone via a wharf at Westminster.

"Contract Journal" 24/11/1955 for sale – plant including six diesel locos – Land Reclamation Ltd, London.

"Contract Journal" 26/4/1956 for sale – plant including four 40hp diesel locos & approx 1mile of track, seen Pitsea area – Land Reclamation Ltd, London N1.

Gauge : 2ft 0in

	4wDM	RH	174945	1935	New	(1)
	4wDM	RH	174946	1935	New	(5)
	4wDM	RH	179884	1936	New	(4)
	4wDM	RH	183073	1937	New	(2)
	4wDM	RH	183757	1937	(e)	(3)

-	4wDM	MR	7128 1937	(a)	(6)
-	4wDM	MR	7214 1938	New(b)	(6)
-	4wDM	MR	7928 1941	New(b)	(7)
-	4wDM	MR	7403 1939	(c)	(6)
-	4wDM	MR	10030 1948	New(d)	(7)
-	4wDM	MR	10114 1949	New(d)	(7)
-	4wDM	MR	10154 1950	New(d)	(7)
-	4wDM	MR	10155 1950	New(d)	(7)
-	4wDM	MR	10342 1952	New(d)	(7)

(a) ex Diesel Loco Hirers Ltd, Bedford, on hire 9/6/1938.
(b) consigned to Wakeley Bros & Co Ltd, Honduras Wharf, Bankside, London, wharfingers and lightermen, possibly delivered to site by barge.
(c) ex Diesel Loco Hirers Ltd on hire, purchased 7/7/1948.
(d) consigned to Joseph Westwood & Co Ltd, Napier Yard, Millwall, London, possibly delivered to site by barge.
(e) ex hire to Bradgate Granite Quarries Ltd, Aylesford, Leicestershire, by 12/2/1938.
(1) s/s after 13/6/1950.
(2) s/s after 12/2/1951.
(3) s/s after 25/10/1951.
(4) s/s after 18/11/1952.
(5) s/s after 1/6/1954.
(6) to Joseph Arnold & Sons Ltd, Leighton Buzzard, Bedfordshire, 2/1956.
(7) noted derelict in a field at Basildon c/1957, (possibly the site of M.C. Plant & Equipment Ltd, Archers Fields, Burnt Mills Road, Basildon), s/s

LIGHT EXPANDED CLAY AGGREGATES LTD
GRAYS WORKS   N101
Subsidiary of **APCM**   TQ 609775
Company registered 2/6/1948 with an office at Aldermans House, Bishopsgate, EC2, opened a works at Grays in 1951. However the company got into difficulties and production ceased 31/12/1954; Rail traffic ceased and the company went into voluntary liquidation 13/6/1962. Rail traffic was operated by locos loaned from BPCM Wouldham Works and the Grays Chalk Quarries Co Ltd, (property and locos of the latter company taken over by BPCM Ltd by 3/1952). Locos were stabled at the shed of the Grays Chalk Quarries Co Ltd until 1954. Site acquired by Lightweight Precast Concrete Co Ltd (registered 28/7/1950) of 14 Dartmouth Street, London SW1.
Gauge : 4ft 8½in

MEESON	0-4-0ST	OC	BE	276	1898	(a)	(1)
(MARGAM)	0-4-0ST	OC	MW	1306	1895	(b)	(2)

(a) ex Grays Chalk Quarries Co Ltd, /1951.
(b) ex BPCM, Wouldham Works, on loan, c/1952, by 3/1952.
(1) returned to Grays Chalk Quarries Co Ltd.
(2) out of use at Light Expanded Clay Aggregates, Grays Works, by 9/1954 scrapped by Norrington, of Grays, 8/1957.

LONDON COUNTY COUNCIL
BARKING SEWAGE WORKS and NORTHERN OUTFALL, Beckton　　A102
TQ 450818

Works built by **George Furness** 1860-1865, later a sewage treatment plant was built by John Mowlem 1887-1890, the works served the central parts of London north of the Thames. The Northern Outfall sewer ran from Barking Creek to Abbey Mills pumping station TQ 387831 where the flow from the three main intercepting sewers north of the Thames converge. Rail system abandoned, most of track removed by 4/1964.

"Machinery Market" 13/2/1914 for sale – plant including tank loco at Beckton Outfall Works – London County Council.

Reference:　Industrial Locomotive No.76.

Gauge : 3ft 6in

No.2	18		0-4-0IST	OC	WB	1424	1894	New	Scr c6/1957
	-		0-4-0ST?		DK		c1887	New(a)	s/s
	-		0-4-0T	OC	AE	1668	1913	New	(1)
	-		4wPM		Bg/DC	2081	1934	New	Scr /1946
No.3			4wDM		RH	235743	1945	New	(2)

(a)　identity and origin unknown.

(1)　s/s after 6/1925.

(2)　to Ronald L. Baker (Shipbreakers), Pitsea Wharf, c8/1965.

LONDON FERRO-CONCRETE CO LTD
Unknown Location　　X103

Company listed as reinforced concrete engineers with an address at 9 Tufton Street, SW1, and a yard at Pleasant Place, SE5. Loco listed as delivered here new c/o J.T. Luton & Sons Ltd of Ilford, this company was an established firm of road & sewage contractors in East London with an address at 14 Forest Lane, E15.

Gauge : 2ft 0in

-	4wPM		L	13258	1940	New	s/s

LONDON & GRAYS SEABORNE COAL CO LTD
GRAYS COAL DEPOT　　N104

Read & Company until 26/6/1894

Coal Merchants occupying part of the site owned by Clark and Standfield. The company ceased operations about 4/1896 when a receiver was appointed.

Gauge : 4ft 8½in

AJAX	0-4-0ST	OC	MW	983	1886	(a)	(1)

(a)　hired by Chas H. Read from Standfield & Sons, Grays, by agreement dated 29/9/1893, earlier Foster & Barry, Finsbury Park-Hornsey (GNR) (1886-1888) contract, Middlesex, AJAX.

(1)　returned to Standfield & Sons, Grays, later Thos Docwra & Sons, [probably West Molesey Reservoirs (1898-1904) contract, Surrey], c/1900.

LONDON & INDIA DOCKS CO

London & West India Docks Joint Committee until 1/1/1901

Formed by the amalgamation of the **London & St Katharine Dock Co** and the **East & West India Dock Co** on 1/1/1889, this entire undertaking was taken over in 1909 by **Port of London Authority**.

ROYAL ALBERT & ROYAL VICTORIA DOCKS   A105

London & St Katharine Dock Co until 1/1/1889

The Victoria Dock was constructed by the **Victoria Dock Co** and opened on 26/11/1855 by Prince Albert. It was purchased by the London & St Katharine Dock Co in 1864. The Royal Albert Dock was built by the London & St Katharine Dock Co and opened 24/6/1880.

"The Engineer" 14/8/1896 for sale three tank locos with vacuum & hand brakes, also two sidetank tipping engines, London & India Docks Committee, London.

Gauge : 4ft 8½in

No.		Type		Builder	No.	Year		
No.16	(No.9 until/84)							
	VICTORIA	0-4-0ST	OC	Shanks		1870	(a)	(6)
No.17	(No.10 until /84)							
	ALBERT	0-4-0ST	OC	Shanks		1872	(a)	(7)
	CHELSEA	[0-6-0ST	IC	Bton ?		c1865?]	(b)	Scr
No.1		0-6-0ST	OC	D	1438	1881	New	(9)
No.2		0-6-0ST	OC	D	1439	1881	New	(9)
No.3		0-6-0ST	OC	D	1440	1881	New	(9)
No.4		0-6-0ST	OC	D	1441	1881	New	(9)
No.5		2-4-0T	OC	Crewe		1849	(c)	(1)
No.6		2-4-0T	OC	Crewe		1849	(d)	(1)
No.7		2-4-0T	OC	Crewe		1847	(e)	(1)
No.8	("LONG WIND")	0-6-0ST	IC	Longridge		1847		
		Reb		Bton 1860,		1865	(f)	(2)
No.11A	(No.11 until /02)	0-6-0ST	OC	FW	149	1872	(g)	(8)
No.12A	(No.12 until /02)	0-6-0ST	OC	FW	288	1875	(h)	(7)
No.13		0-6-0ST	OC	FW	263	1875	(h)	(9)
No.14		0-4-0ST	OC	YE	284	1876	(j)	(9)
No.15		0-4-0ST	OC	FE		1884	(k)	(9)
No.16		0-4-0ST	OC	MW	893	1884	(k)	(3)
No.16	(No.9 until /00)	0-4-0ST	OC	MW	905	1884	(k)	(9)
No.18		0-4-0ST	OC	RP		c1870	(k)	(9)
No.19	(No.10 until /01)	0-4-0ST	OC	HE	343	1884	(k)	(9)
No.20		0-4-0Tram		FE			(k)	(4)
No.5		0-6-0ST	OC	RS	2844	1896	New	(9)
No.6		0-6-0ST	OC	RS	2845	1896	New	(9)
No.7		0-6-0ST	OC	RS	2981	1900	New	(9)
No.8		0-6-0ST	OC	RS	2982	1900	New	(5)
No.9		0-6-0ST	OC	RS	2983	1900	New	(9)
No.10		0-6-0ST	OC	RS	3070	1901	New	(9)
No.11	LOOE	0-6-0ST	OC	RS	3050	1901	(l)	(9)
No.12		0-6-0ST	OC	RS	3094	1902	New	(9)

(a) ex Alfred Giles, Southampton Dock Co, Hampshire, 12/1878; one of these locos ex Royal Albert Dock extension contract after 10/1887.

(b) ex Lucas & Aird on hire, c/1880.

(c)   ex LNWR, 1819, 4/1881.
(d)   ex LNWR, 1927, 4/1881.
(e)   ex LNWR, 1911, 4/1881.
(f)   ex Lucas & Aird, contractors, 5/1881, earlier LBSCR 104, originally 75.
(g)   ex J. Dickson Jnr, contractor, Liverpool, 6/1882.
(h)   ex J. Dickson Jnr, contractor, Liverpool, 10/1883.
(j)   ex Benton & Woodiwiss, contractors, RUTLAND.
(k)   ex Royal Albert Dock extension contract, after 10/1887.
(l)   ex Liskeard & Looe Railway, Cornwall, /1901.

(1)   to George Cohen, Sons & Co Ltd, Canning Town, for scrap, 8/1896.
(2)   sold for scrap, 9/1899.
(3)   to MacKay & Davies, Gloucester, /1900.
(4)   to Clark & Standfield, Grays Engineering Works, Grays, c/1900.
(5)   to Tilbury Docks.
(6)   to Stone Court Chalk Land & Pier Co Ltd, Kent, by/1896.
(7)   to East & West India Dock by 1901.
(8)   to East & West India Dock by 1902.
(9)   to PLA, with site, /1909.

Note: The information in the above list is based on surviving PLA records, which are, however, incomplete; other sources have given conflicting numbers for some of the locos, e.g., VICTORIA was 16 from 1884 : MW 893 was 19 until c/1901.

TILBURY DOCKS   M106
East & West India Dock Co until 1/1/1889
The opening of the Albert Dock in 1880 by the London & St. Katharine Dock Co reduced trade at the East and West India Docks. Consequently a major new dock down-river at Tilbury was planned. Construction was undertaken by Kirk & Randall from 1882 to 1884 and completed by Lucas & Aird from 1884 to 1886 (which both see). Opening was on 17/4/1886, although at first traffic levels were low. Private siding agreement between the East & West India Dock Co & the LT&SR dated 8/11/1883.
Gauge : 4ft 8½in

No.3	WASP		0-6-0ST	IC	MW	872	1883	(a)	(2)
No.4	BAT [GNAT?]	0-6-0ST	IC	MW	694	1878	(a)	(2)
No.6	MOTH		0-6-0ST	IC	HE	234	1880	(a)	(2)
No.7	FLY		0-6-0ST	IC	MW	873	1883	(a)	(2)
No.8	AGENOR	0-6-0ST	OC	RS	2982	1900	(b)	(2)
No.2	AJAX		0-6-0ST	OC	RS	3053	1901	New	(2)
No.1	ANT		0-6-0ST	IC	MW	663	1877	(c)	(1)
	HECTOR		0-6-0ST	OC	RS	3120	1904	New	(2)
	JASON		0-6-0ST	OC	RS	3170	1905	New	(2)
	NESTOR		0-6-0ST	OC	RS	3296	1907	New	(2)

(a)   ex Lucas & Aird, contract on site, /1886.
(b)   ex Royal Docks.
(c)   ex East & West India Docks, by 30/6/1902.

(1)   to Hargreaves Colls Ltd, Bank Hall Colliery, Lancashire, /1907.
(2)   to PLA, with site, /1909.

LONDON & THAMES HAVEN OIL WHARVES LTD
REEDHAM WHARVES, Thames Haven　　　　　　　　　　　　　　　　K107
Loco Shed TQ 728815

Wharves originally developed for the import of fish for despatch by rail to London and later live cattle in 1855; Thames Haven handled over a third of live cattle imported into Britain, this trade was withdrawn in 1882 with the building of Tilbury Docks and introduction of new legislation affecting cattle slaughter. With the advantages of its geographical location Thames Haven became the site for the import of petroleum spirit the original petroleum wharf being built in 1875; in 1876 the first cargo of petroleum was landed at Thames Haven and for the next forty years every ton of petroleum spirit imported into London passed through the wharves. **London & Thames Haven Oil Wharves Ltd** (LATHOL) was registered 24/5/1898 to take over the **London and Thames Haven Petroleum Wharf Ltd** and to develop oil storage facilities, which at the time were modest – comprising fifteen storage tanks with a capacity of 8,000 tons and one discharge jetty. As the size of tankers grew so did the requirement for larger storage tanks and more deep water jetties, by 1914 tankage capacity had grown to 300,000 tons consisting of 75 storage tanks served by three jetties. Between the wars continuous expansion took place and the first oil refineries came into operation; by 1939 total capacity was nearly one million tons and five deep water jetties had been constructed. The war years found Thames Haven serving the Petroleum Board and handling stocks of whale oil, molasses and vegetable oils. In 1948 an agreement was entered into with the Shell Refining Co to store crude oil and feedstocks for the Shell Haven refinery as well as its refined product. Between 1955 and 1957 storage capacity for crude oils was increased with the erection of tanks each holding nearly 17,000 tons (over 3¾ million gallons); between 1957 and 1959 two further jetties were constructed capable of accommodating the largest tankers in service the contract let to Peter Lind & Co Ltd; each jetty was 345ft long and 40ft wide. Expansion continued and the premises were taken over by Shell UK Ltd in 1972.

Reference : The Thames Haven Railway, Peter Kay, author, 1999.

Gauge : 4ft 8½in

	0-4-0PM		McP				
-			BgC	566	1916	New	Scr
(2)	0-6-0F	OC	AB	1553	1917	(a)	(3)
(No.4)	0-6-0F	OC	AB	1551	1917	(a)	(3)
(3)	0-4-0F	OC	AB	1472	1916	(b)	(2)
-	0-6-0DM		HE	1697	1932	(c)	(1)
-	0-4-0DM		HE	4250	1951	New	(5)
-	0-4-0DM		HE	4525	1953	New	(5)
-	0-6-0DH		JF	4240016	1964	New	(4)
-	4wDH		TH	187V	1967	New	(4)

(a)　ex MoM, Gretna, Dumfriesshire, by 10/1922.
(b)　ex MoM, Morecambe, Lancashire, possibly 11/1919, by 10/1922.
(c)　ex HE, 10/1949. (earlier WD 70027; originally LMS 7051).

(1)　returned to HE, by /1951.
(2)　to Thos. Hedley & Co Ltd, West Thurrock, c2/1955.
(3)　scrapped on site by Gomm & Searle Ltd, Bow, East London, 8/1966.
(4)　to Shell UK Ltd, Thames Haven, /1972.
(5)　to Southern Counties Trading & Demolition Co Ltd, Bedhampton, Hampshire, 9/1972.

Gauge : 600mm.

		4wDM	FH	2416	1941	(a)	s/s
		4wDM	HE	2591	1942	(b)	s/s

(a) ex FH, earlier WD.
(b) ex MoS by 23/4/1945.

ABRAM LYLE & SONS LTD
PLAISTOW WHARF SUGAR REFINERY   **B108**
Abram Lyle & Sons 1882 until 2/4/1890   TQ 401799

Abram Lyle of Greenock, a sugar refiner and ship-owner went into business with John Kerr at the Glebe Refinery on the Clyde. After Kerr's death, the Lyle's sold their interests in the partnership. In 1881 Abram Lyle bought two adjacent sites on the north bank of the Thames to construct a refinery at Plaistow Wharf, about 1½ miles upstream of Henry Tate's refinery at Silvertown. Work started on a new refinery in October 1881 and production began in 1883; Lyle's speciality was the production of "Golden Syrup". The company merged in 1921 with Henry Tate & Sons Ltd to form Tate & Lyle Ltd.
Reference : Sugar And All That, A History of Tate & Lyle, Antony Hugill, 1978.

"Mechanical World" 24/3/1883 wanted – saddletank loco to haul 20tons up 1in40 – Abram Lyle & Sons, 37 Mincing Lane, London.

MARLEY TILE (AVELEY) CO LTD
Stifford Road, North Stifford   **L109**
A subsidiary of **The Marley Tile (Holding) Co Ltd**   TQ 580803

Company registered 11/9/1934 to make concrete roofing tiles and later concrete blocks, production commencing in 1935. It had a sand and gravel pit on land leased from Thames Land Co Ltd; the site was closed in 1970.

Gauge : 2ft 0in

		4wDM	RH	173394	1934	New	(1)
		4wDM	RH	187083	1937	(a) s/s after 18/7/1956	

(a) ex Thos. W. Ward Ltd, Titan Works, Grays, after 2/9/1947, by 17/3/1948.
(1) to William Jones Ltd, dealer, Greenwich, London by 15/3/1958.

MARSH & SONS LTD
Westcliffe   **J110**

Company registered 31/7/1909 as carmen and contractors, at 33 Leigh Road, Southend-on-Sea, with Mr G. Marsh as sole governing director; also listed in quarry directories for 1929 and 1933 as sand & ballast merchants with an address at 253 London Road, Westcliffe, Southend-on-Sea. George Marsh of 8 St. Benet's Road, Southend is listed in 1937 as a gravel merchant.

Gauge : 2ft 0in

	4wDM	RH	166010	1932	(a)	(1)

(a) ex RH, Lincoln 3/9/1934, on hire to George Marsh, 30 Sutton Street, Southend-on-Sea
(1) to RH, Lincoln, off hire after 22/11/1935; to Kinnear Moodie, 299 Hither Green Lane, Lewisham, London, on hire, per J.J. Johnston & Co Ltd, by 22/12/1936.

METROPOLITAN WATER BOARD

LEA BRIDGE WORKS   C111
East London Waterworks Co from 1807 until 18/12/1902   TQ 360867

Pumping station and water treatment plant with a number of steam pumping engines. A siding was laid from the main line in 1890 to supply coal and bring in supplies, track lifted in 1950.
Gauge : 4ft 8½in

-	0-4-0ST	OC	MW	786	1881	(a)	
			reb		1919		(1)
ITCHEN	0-4-0ST	OC	AE	2037	1931	(b)	(2)
-	4wDM		FH	1844	1933	New	(3)

(a) ex Battersea Works, London, after 10/1919, by 4/8/1923.
(b) ex John Mowlem & Co Ltd, Chingford Reservoir (1935-1951) contract, /1944.

(1) scrapped by George Cohen, Sons & Co Ltd, c1/1934.
(2) returned to John Mowlem & Co Ltd, Chingford Reservoir (1935-1951) contract, /1949.
(3) to Kempton Park Works, Middlesex, 6/1951.

MAYNARD RESERVOIR   D112
earlier **East London Waterworks Co**

Reservoir built in 1869, details of work unknown, possibly maintenance works.
Gauge : 2ft 0in

-	4wDM	FH	[3317? 1948]	(a)	(1)

(a) earlier Charlton Road Depot, Shepperton, Middlesex, here by 13/10/1956.

(1) later Crouch End Reservoir contract, Middlesex, by 16/2/1964.

MINISTRY OF DEFENCE, ARMY DEPARTMENT
SHOEBURYNESS DEPOT and TRAMWAY   J113
War Department until 1/4/1964   Loco Sheds TQ 946856

The first land on the Ness in the Parish of Shoebury was purchased by the Board of Ordnance in 1849, by the 1860s a standard gauge tramway had been laid using horse drawn power and from the 1870s a steam traction engine was in use. In the 1880s more land was purchased north east of the village and 'New' Ranges were established over a few years from 1889, these ran north east to Havengore Creek. A rail link from the Old Ranges was constructed about 1890. By 1914 the rail system had expanded as new trial batteries and other facilities were built, these eventually reached the mouth of Havengore Creek.

Havengore Island was purchased by the War Office in 1902 followed in 1914-15 by New England and Foulness Islands, a lifting bridge across Havengore Creek and a road across the islands were completed by the mid 1920s. As part of the road construction project a standard gauge line was laid from the mainland over the bridge and alongside the road to about half way across Foulness with a narrow gauge line continuing thereafter. All this line from Havengore Island was lifted once the project was completed in 1925. By this time a standard gauge line had been laid from Havengore Bridge to a trials battery near Havengore Head and in 1936 a short spur was laid from here on to New England. All this was lifted about a year before a new bridge was built without rail facility in 1987-88. A short 2ft 6in gauge system was laid down on Havengore Island in 1995 and is still in use.

On 1/4/1964 the War Department was retitled Ministry of Defence, Army Department, who continued to be responsible for the railway installations at the Shoeburyness Depot and the ranges of the Proof & Experimental Establishment. These were operated by standard Army Department locomotives with frequent transfers. Regular rail traffic ceased 27/11/1991 but the rail system remains and is used from time to time for special traffic.

References :   Guns and Gunners at Shoeburyness, Tony Hill, Baron Books, 1999.
                Railway Observer, January 1955.
                Industrial Railway Record No.128, 153, 155, 160.

Gauge : 4ft 8½in

           NICHOLSON   0-6-0ST   IC   MW   847 1882
earlier Lucas & Aird, Cudworth, Yorkshire, c2-3/1887.
to Royal Arsenal, Woolwich, London.

1   WD 3856   0-6-0ST   IC   MW   889 1883
ex Lucas & Aird contractors, c1888-1889.
for disposal 5/5/1921, s/s.

                                0-6-0ST   IC   MW   944 1885
ex Lucas & Aird contractors, c1888-1889.
s/s after /1921.

                                0-6-0ST   IC   MW   958 1885
ex Lucas & Aird contractors, c1888-1889.
s/s after /1921.

                                0-6-0ST   IC   MW   969 1885
ex Lucas & Aird contractors, c1888-1889.
s/s

           SHOEBURYNESS   0-6-0CT   OC   VF   1436 1895   New
to Admiralty, Chatham Dockyard, Kent, c/1904.

2   2-4-2ST   OC   NBH   16032 1903   New
s/s after /1919.

3   2-4-2ST   OC   NBH   16601 1905   New
s/s by /1946.

4   A904   2-4-2ST   OC   NBH   17223 1906   New
to Thos. W. Ward Ltd, Silvertown, /1933.

14 (6)   A903   2-4-2ST   OC   NBH   19019 1909   New
to Kings Newton Depot, Derbyshire, 6/5/1948.

5   WD 4189   BORDON   0-6-0ST   OC   AE   1505   1906
ex Woolmer Instructional Military Railway, /1910, for disposal 5/5/1921, (carried J.F. Wake plate on cab).
to John F. Wake & Co Ltd, dealer, Darlington, Co. Durham;
thence to Clay Cross Co Ltd, Derbyshire, after 4/1921, by 1/1924.

7   0-6-0ST   OC   AE   1672   1914   New
to William Jones Ltd, dealer, Greenwich, London, after 9/1926.

8   0-6-0T   OC   HC   1164   1916
built for BPGVR, but requisitioned by Government, despatched here 24/2/1916.
to BPGVR, 15, 5/1920; later GWR 2168.

2791   [0-6-0   IC   Derby   1872]
a Royal Engineers letter book of 30/4/1916 records re tubing of "WD loco No. 2791", if identified correctly this loco was possibly Midland Railway 2791, a Kirtley 0-6-0 goods engine, one of 87 engines in the batch 2707-2794 which were loaned to the ROD. According to the book 'Midland Railway Locomotive History 1844-1966' Stephen Summerson, Irwell Press, the batch of locos 2789-2794 was destined to go to France with the ROD but in fact was never sent, neither was the order cancelled; given this it seems likely that 2791 may have found its way to Shoeburyness.

                              2-2-0T   OC   9E   781   1906
ex MoM, Crossgates, Leeds, Yorkshire, /1917; originally LSWR 736.
to James Brown Ltd, Sittingbourne, Kent, for scrap 10/1921

12 EARL OF CARYSFORT   2-4-2T   IC   Crewe 2726   1884
    (WD 4190)
ex Dublin & South Eastern Railway, Ireland, /1917, 63, rebuilt from 5ft 3in gauge at Crewe.
for disposal 6/5/1921, for sale in Surplus 2/10/1922,
to John F. Wake & Co Ltd, dealer, Darlington, Co. Durham.

                              2-2-0T   OC   9E   795   1906
ex IWD, Richborough Port, Kent, via repairs at SECR Ashford Works, 10/1917 - 1/1918; originally LSWR 740. returned to Richborough by 10/1919

11   (WD 4222)   2-2-0T   OC   9E   798   1906
ex LSWR 742, 3/1917, for disposal 6/5/1921.
to James Brown Ltd, Sittingbourne, Kent, for scrap 10/1921.

7   (WD 4147)   0-6-0ST   OC   P   697   1897
ex John Best & Sons Ltd, contractors, Edinburgh, /1917.
for disposal 5/5/1921, s/s.

WD 95 (WD 4098)   2-4-0T   IC   BP   2467   1885
ex Woolmer Instructional Military Railway, /1918-/1919, (overhauled Darlington /1918)
for disposal 10/5/1921, for sale in Surplus 2/10/1922, s/s.

       THE BEAR   0-4-0ST   OC   MW
origin unknown, reported as exploded in /1919.
possibly for sale in Surplus in 6/1921, s/s.

5A   0-6-0PT   OC   BLW 46407 1917
ex ROD 654.
to Kings Newton Depot, Derbyshire, 6/5/1948.

7   ROYAL ENGINEER   0-6-0PT   OC   BLW 46489 1917
ex ROD 660.
to Chilwell Depot, Nottinghamshire.

10   (WD 4919)   0-6-0PT   OC   BLW 46956 1917
ex ROD 690, fitted with new boiler 50280/1936, plated Ruston, Lincoln, England.
to Easkmeals Depot, Cumberland, 16/4/1941.

SCHOOL OF GUNNERY   0-4-0ST   OC   NR   5936   1902
ex Royal Arsenal, Woolwich, London, /1920.
s/s by /1946.

                          0-6-0T   [Crewe]
origin and identity unknown, thought to be ex North Staffordshire Railway, possibly one of Crewe 431, 513, 569, 1261, purchased by the NSR from the LNWR in /1900 and withdrawn between 1913-1916, here by 6/9/1919.
scrapped by 7/1949.

7   4wBE   1924
origin and identity unknown.
s/s.

WD 884 (813) (7) (6)   4wBE   EE   687   1925 New
rebuilt as 4wDE. Became WD 813, later 884.
to Standard Brick & Sand Co Ltd, Redhill, Surrey, 12/1959.

3E, 7E   0-4-0BE   Electromobile
                                           W247   1928 New
to Thos. W. Ward Ltd, Titan Works, Grays, 6/1948.

8E   Bo-BoBE   EE   785   1930 New
to Royal Arsenal, Woolwich, London, 2/1949.

12 (1)   (71660)   0-6-0ST   OC   AE   1819   1919
ex WD, Hilsea, Hampshire, /1931.
to Thos. W. Ward Ltd, Titan Works, Grays, 7/1947,
thence to NCB, North East Division, Area 3, Cortonwood Colliery, Yorks (WR), /1948.

                          0-4-0DE   AW   D23   1933 New
ex AW on demonstration.
to North Thames Gas Board, Beckton By-Products Works, on demonstration.

WD 011 (71677) 9   0-6-0ST   IC   HC   1510   1923
ex McAlpine, No.48, per Abelson, c12/1939, by 4/1940.
to Thos. W. Ward Ltd, Columbia Wharf, Grays, by 11/3/1959.

11   (71678)   0-6-0ST   OC   AB   1297 1915
ex George Cohen, Sons & Co Ltd, 28/12/1939, renumbered WD 71678.
to Abelson, Sheldon, Birmingham, 8/1950.

8   (71687)   0-4-0DM   JF   22890 1939
ex Corsham Depot, Wiltshire, /1940, renumbered WD 71687, later 854.
to West Hallam Depot, Derbyshire.

No.1   0-6-0DM   HE   1846 1936
ex Corsham Depot, Wiltshire, by 4/1938.
to Corsham Depot, Wiltshire, by 9/1943.

WD 229 (70229, WD 4098)   0-6-0ST   IC   HC   845   1908
ex South Wales Coalite Co Ltd, Wern Tarw Colliery, Glamorgan, by 7/1941.
to Steel Breaking & Dismantling Co Ltd, Chesterfield, Derbyshire, for scrap, 11/11/1946.

WD 68   0-6-0ST   IC   HC   1529 1924
ex Longtown Depot, Cumberland, 15/6/1943.
to GWR Slough, Buckinghamshire, 13/7/1944.

WD 125 (75063) 22   0-6-0ST   IC   RSHN 7099 1943 New
to Tidworth Depot, Hampshire, 18/4/1947.

       75062   23   0-6-0ST   IC   RSHN 7098 1943
ex Kings Newton Depot, Derbyshire, 1/1944.
to Kings Newton Depot, Derbyshire, 1/6/1944.

       75060   0-6-0ST   IC   RSHN 7096 1943
ex WD Kings Newton, Derbyshire, 1/1944.
to Ministry of Fuel & Power, 6/1946.

WD 168 (75019) 25   0-6-0ST   IC   HE   2868 1943
ex Donnington Depot, Shropshire, 1/1944, to Coalville Depot, Leicestershire, 4/1945, ex ?
here by 11-12/1952, awaiting overhaul 11-12/1954, to Bicester Depot Oxfordshire for
overhaul 3/1955, ex Bicester Depot, 11/1955, OOU by 27/9/1960, to HE 9/1961.

WD 132 (75113)   0-6-0ST   IC   HE   3163 1944 New
to Bicester Depot, Oxfordshire, 7/7/1944.

WD 130 (75107) 23   0-6-0ST   IC   HE   3157 1944 New
to Tidworth Depot, Hampshire, 18/4/1947.

WD 126 (75099) 24   0-6-0ST   IC   HC   1762 1944 New
to Bicester Depot, Oxfordshire, 21/6/1956.

WD 801 (70232) 1   4wDM   RH   221648 1944
ex Hilsea Depot 9/8/1956,
to Bicester Depot, Oxfordshire, 11/11/1958.

WD 804 (72211)   2   4wDM   RH 224342 1944
ex Feltham Depot, Middlesex, by 2/12/1952,
to Bicester Depot, Oxfordshire, after 4/1958, by 7/1959.

WD 805 (72212)   3   4wDM   RH 224343 1944   New
to Bicester Depot, Oxfordshire, after 13/8/1951, by 1/1959.

WD 806 (72213)   4   4wDM   RH 224344 1944   New
to Bicester Depot, Oxfordshire, by 10/2/1949.

           71450 26   0-6-0ST IC HE   3214 1945
ex Barby Depot, Northamptonshire, 16/7/1945,
to Bicester Depot, Oxfordshire, 8/1958, ex Bicester Depot, 3/1960,
to Thos. W. Ward Ltd 1/1961.

WD 180 (75280) 28   0-6-0ST IC RSHN 7210 1945
ex Barby Depot, Northamptonshire, 16/7/1945.
to Bicester Depot, Oxfordshire, 11/11/1958.

WD 183 (75284) 29   0-6-0ST IC VF   5274 1945
ex Barby Depot, Northamptonshire, 16/7/1945.
to Bicester Depot, Oxfordshire, 13/1/1958.

WD 181 (75282)   0-6-0ST IC VF   5272 1945
ex Longmoor Depot, Hampshire, 21/7/1945.
to Longmoor Depot, Hampshire, 10/1958.

WD 164 (71477)   5   0-6-0ST IC RSHN 7286 1945
ex Long Marston Depot, Warwickshire, 8/1945.
to Bicester Depot, Oxfordshire, 18/6/1946.

WD 143 (75152) 21   0-6-0ST IC WB   2740 1944
ex Long Marston Depot, Warwickshire, 8/1945.
to Cairnryan Depot, Galloway, 18/6/1946.

WD 161 (71448) 14   0-6-0ST IC HE   3212 1945
ex Long Marston Depot, Warwickshire, 8/1945, to Yantlet Depot, Kent, 13/2/1946,
to Hunslet Engine Co, Leeds, Yorkshire, for overhaul 8/4/1952, ex Bicester Depot,
Oxfordshire, 8/7/1960, to NCB North Gawber Colliery, Yorkshire, 4/1964.

WD 153 (75191) 30   0-6-0ST IC RSHN 7141 1944
ex Long Marston Depot, Warwickshire, 8/1945.
to Cairnryan Depot, Galloway, 18/6/1946.

WD 173 (75252) 27   0-6-0ST IC WB   2775 1945
ex Barby Depot, Northamptonshire, 2/1946, to Bicester Depot, Oxfordshire, 27/9/1954, ex
Bicester Depot, 5/1955.
to Thos. W. Ward Ltd, Columbia Wharf, Grays, 12/1960.

103   0-6-0ST   IC   MW   212   1866
ex Air Ministry, Halton, Buckinghamshire, by 14/2/1948.
to Thos. W. Ward Ltd, Columbia Wharf, Grays, 23/6/1949.

WD 852 (70028)   0-4-0DM   JF   22889 1939
ex Bicester Depot, Oxfordshire, 1/1949.
to Queensferry Depot, Flintshire, by 9/1952.

WD 853 (70238) 7   0-4-0DM   JF   22976 1942
ex Bicester Depot, Oxfordshire, 2/1949.
to Bicester Depot, Oxfordshire, 30/6/1953.

WD 162 (71449) 12   0-6-0ST   IC   HE   3213 1945
ex WD Long Marston, Warwickshire, 4/1949.
to Thos. W. Ward Ltd, Columbia Wharf, Grays, 12/1960.

WD 169 (75176) 23   0-6-0ST   IC   WB   2764 1944
ex Queensferry Depot, Flintshire, 7/1952.
to Bicester Depot, Oxfordshire, 27/8/1957.

WD 184 (75285) 22 30   0-6-0ST   IC   VF   5275 1945
ex Hunslet Engine Co, Leeds, after overhaul 16/9/1952.
to Hunslet Engine Co, Leeds, for overhaul 7/12/1956.

WD 102 (75035) 27   0-6-0ST   IC   HE   2884 1943
ex Bicester Depot, Oxfordshire, 9/1954.
to Bicester Depot, Oxfordshire, 17/11/1955.

WD 109 (75044) 24   0-6-0ST   IC   HE   2893 1943
ex Bicester Depot, Oxfordshire, 6/1956.
to Hunslet Engine Co, Leeds, Yorkshire, 9/1960.

WD 808 (72215) 5   4wDM   RH   224347 1945
ex Bicester Depot, Oxfordshire, 7/1956.
to Bicester Depot, Oxfordshire, 23/3/1960.

WD 120 (71528) 11   0-6-0ST   IC   AB   2182 1944
ex Bicester Depot, Oxfordshire, 12/7/1956,
to Steventon Depot, Berkshire, 6/1958, ex Steventon Depot, by 1/1959.
to Thos. W. Ward Ltd, Columbia Wharf, Grays, 8/1963.

WD 124 (75049) 22   0-6-0ST   IC   HE   2898 1943
ex Bicester Depot, Oxfordshire, 7/1956.
scrapped on site 3/1959.

120   (8323, WD1000)   0-4-0DM   AB   419 1957 New
to Bicester Depot, Oxfordshire, 10/1/1966, returned 26/9/1966.
to Bridge Metals, Basildon, 12/1978.

WD 166 (71530) 23   0-6-0ST IC AB   2184 1945
ex Long Marston Depot, Warwickshire, 20/12/1957.
to NCB Ashington Colliery, Northumberland, 6/1963.

WD 146 (75165) 26   0-6-0ST IC WB   2753 1944
ex Bicester Depot, Oxfordshire, 22/9/1958.
to Hunslet Engine Co, Leeds, Yorkshire, 12/10/1965.

WD 185 (75286) 28   0-6-0ST IC VF   5276 1945
ex Bicester Depot, Oxfordshire, 22/9/1958.
to NCB Prince of Wales Colliery, West Yorkshire, 7/1965.

WD 844 (72238)   0-4-0DM   AB   371   1945
ex Bicester Depot, Oxfordshire, 12/11/1958, to Bicester Depot, 12/1959,
ex Thatcham Depot, Berkshire, 11/1960.
to Bicester Depot, Oxfordshire, 14/12/1964.

   236 (WD 845, 72239)   0-4-0DM   AB   372   1945
ex Bicester Depot, Oxfordshire, 9/2/1959, to Bicester Depot, 20/4/1965,
ex Bicester Depot, 28/6/1966.
to Bicester Depot, 14/10/1971.

WD 841 (72235)   0-4-0DM   AB   368   1945
ex Bicester Depot, Oxfordshire, 26/4/1960.
to Bicester Depot, Oxfordshire, by 16/6/1966.

WD 112 (75076)   0-6-0ST IC RSHN 7112 1943
ex Bicester Depot, Oxfordshire, 4/1960.
possibly to South Wales for scrap, 11/1964, (possibly transported by Wynn & Co of Newport?)

WD 171 (75250)   0-6-0ST IC WB   2773 1945
ex Bicester Depot, Oxfordshire, 20/9/1960.
to E.C. Steele & Co Ltd, Hamilton, Lanarkshire, 3/1969, scrapped on site.

WD 170 (75179)   0-6-0ST IC WB   2767 1945
ex Bicester Depot, Oxfordshire, by 26/10/1961, to Old Dalby Depot, Leicestershire, 1/1967,
ex Old Dalby Depot, /1968?
to E.C. Steele & Co Ltd, Hamilton, Lanarkshire, 3/1969, scrapped on site.

WD 148 (75180)   0-6-0ST IC RSHN 7130 1944
ex Bicester Depot, Oxfordshire, 18/1/1962.
to J. Cashmore Ltd, Great Bridge, Staffordshire, 31/8/1967.

WD 188 (75294)   0-6-0ST IC VF   5284 1945
ex Bicester Depot, Oxfordshire, 6/9/1962.
to J. Cashmore Ltd, Great Bridge, Staffordshire, 31/8/1967, scrapped on site.

96   (WD 201)   0-6-0ST IC HE   3801 1953
ex Bicester Depot, Oxfordshire, 25/7/1963.
to Steel Breaking & Dismantling Co Ltd, Chesterfield, Derbyshire, scrapped on site 10/1970.

94   (WD 194)   0-6-0ST   IC   HE   3794 1953
ex Bicester Depot, Oxfordshire, 18/2/1964.
to Lakeside & Haverthwaite Railway Co Ltd, Lancashire, 12/9/1973.

411   (WD 8211)   0-4-0DH   NBQ   27646 1959
ex Bicester Depot, Oxfordshire, 6/11/1964.
to Bicester Depot, Oxfordshire, 11/1/1967.

WD 825 (70039)   0-4-0DM   AB   354   1941
ex Bicester Depot, Oxfordshire, 11/12/1964.
to Bicester Depot, Oxfordshire, 7/9/1966.

                           [0-6-0DH   EEV   D911 1964]
a successful trial was carried out in the mid 1960s with a 400hp diesel loco believed to have been built by English Electric Co Ltd, this may have been the loco listed above which was the demonstration loco at the time, no further details available.

222   (WD 829, 72220)   0-4-0DM   VF   5256 1945
                                           DC   2175 1945
ex Bicester Depot, Oxfordshire, 5/9/1966, to Bicester Depot, 28/1/1971,
ex Bicester Depot, 13/10/1971.
to Bicester Depot, Oxfordshire, 29/4/1981.

95   (WD 200)   0-6-0ST   IC   HE   3800 1953
ex Bicester Depot, Oxfordshire, 11/10/1966.
to Kent & East Sussex Railway, Rolvenden, Kent, 31/12/1970.

97   (WD 202)   0-6-0ST   IC   HE   3802 1953
ex Bicester Depot, Oxfordshire, 11/10/1966.
to Steel Breaking & Dismantling Co Ltd, Chesterfield, 11/1970, scrapped on site.

90   (WD 190)   0-6-0ST   IC   HE   3790 1952
ex Long Marston Depot, Warwickshire, 3/12/1967.
to Stour Valley Railway Preservation Society, 22/6/1971.

91   (WD 191)   0-6-0ST   IC   HE   3791 1952
ex Long Marston Depot, Warwickshire, 3/12/1967.
to Kent & East Sussex Railway, Rolvenden, Kent, 4/2/1972.

110   (WD 813)   4wDM   RH   411319 1957
ex Bicester Depot, Oxfordshire, 4/9/1968.
to Inchterf Depot, Strathclyde, 5/11/1975.

92   (WD 192)   0-6-0ST   IC   HE   3792 1953
ex Long Marston Depot, Warwickshire, 9/4/1969, to Long Marston Depot, 21/6/1972, ex Marchwood Depot, Hampshire, 29/3/1979, to Marchwood Depot, 6/1980, ex Marchwood Depot, 11/1982.
to Museum of Army Transport, Beverley, Humberside, 12/7/1984.

WD 196 0-6-0ST IC HE 3796 1953
ex Longmoor Depot, Hampshire, 1/1970.
to Kent & East Sussex Railway, Rolvenden, Kent, 20/6/1970.

601 (WD 878, 70272) 0-6-0DE Derby 1945
ex Bicester Depot, Oxfordshire, 5/5/1970, to USAF Welford Park Depot, Berkshire, 6/7/1971, ex Thatcham Depot, Berkshire, 31/10/1972.
to Bicester Depot, Oxfordshire, 2/2/1973.

610 (WD 890) 0-8-0DH S 10143 1963
GENERAL LORD ROBERTSON
ex Longmoor Depot, Hampshire, 22/6/1970,
to Thomas Hill Ltd, Kilnhurst, South Yorkshire, for overhaul 19/7/1977, returned 3/1/1978.
to Mid-Hants Railway, Ropley, Hampshire, 7/5/1985.

234 (WD 843, 72237) 0-4-0DM AB 370 1945
ex Bicester Depot, Oxfordshire, 27/1/1971, to Bicester Depot, 22/9/1980,
ex Bicester Depot, 28/4/1981.
to Ashchurch Depot, Gloucestershire, 14/5/1985.

424 (WD 8218) 0-6-0DH RH 459520 1961
ex Bicester Depot, Oxfordshire, 9/8/1977.
to Bicester Depot, Oxfordshire, 13/6/1996.

431 (WD 8255) 0-6-0DH RH 466622 1962
ex Kineton Depot, Warwickshire, 20/1/1981.
to Thos Hill Ltd, Kilnhurst, South Yorkshire, 4/8/1981.

430 (WD 8224) 0-6-0DH RH 466621 1961
ex Baguley-Drewry, Burton on Trent, Staffordshire, 15/10/1982.
to Moreton-on-Lugg Depot, Worcestershire, 18/12/1990.

423 (WD 8217) 0-6-0DH RH 459518 1961
ex Marchwood Depot, Hampshire, 15/3/1985.
to RNAD Dean Hill, Wiltshire, 21/8/1990.

425 (WD 8219) 0-6-0DH RH 459519 1961
ex Longtown Depot, Cumbria, 4/5/1985.
to Ludgershall Depot, Wiltshire, c8/1996.

433 (WD 8228) 0-6-0DH RH 468043 1963
ex Longtown Depot, Cumbria, 23/4/1992.
to Ludgershall Depot, Wiltshire, c8/1996.

259 4wDH TH 299v 1981
ex Bicester Depot, Oxfordshire, 13/6/1996.
to LH Group Services Ltd, Barton-Under-Needwood, Staffordshire, for overhaul 22/7/2004.

01524 (261)   4wDH   TH   301v 1982
ex Ludgershall Depot, Wiltshire, 20/8/1996.
to LH Group Services Ltd, Barton-Under-Needwood, Staffordshire, for overhaul 28/10/2004.

01525 DRAPER (264)   4wDH   TH   306v 1983
ex Ludgershall Depot, Wiltshire, 20/8/1996.
to LH Group Services Ltd, Barton-Under-Needwood, Staffordshire, for overhaul 3/3/2004, ex LH Group Services Ltd, 22/7/2004.

01522 (254)   4wDH   TH   272v 1977
ex LH Group Services Ltd, Barton-Under-Needwood, Staffordshire, 3/3/2004.
to Bicester Depot, Oxfordshire, 16/11/2006.

01528 (267)   4wDH   TH   309v 1983
ex LH Group Services Ltd, Barton-Under-Needwood, Staffordshire, 28/10/2004.

3/2203 (No.1)   4wPM   Caledon
ex MoM France, /1920.
one of these to Bicester Depot, Oxfordshire, for scrap; the other converted at Shoeburyness to a mobile boiler testing unit, s/s c10/1952.

3/2202 (No.2)   4wPM   Caledon
ex MoM France, /1920.
one of these to Bicester Depot, Oxfordshire, for scrap; the other converted at Shoeburyness to a mobile boiler testing unit, s/s c10/1952.

3/2178 (4)   2w-2PMR   BgC   1023 1918
ex MoM France, /1920, to Eskmeals Depot, Cumberland, 26/6/1940, returned 26/6/1942, converted to trailer 16/1/1945.
s/s.

3/2201 (15) (LR 3034)   4wPM   MR   1313 1918
converted from 60cm gauge, to Yantlett Depot, Kent, by 10/1944.
to David S. Burleigh, Wembley, Middlesex, 5/1/1951.

                        4wBER
origin and identity unknown.
to E. Boydell & Co Ltd, Manchester, 14/10/1949.

17   4wPER (ex BER) Electromobile 1927 New
converted to petrol, fitted with 28hp Parsons engine /1935.
to Bicester Depot, Oxfordshire, 24/4/1948.

9114 3/2228 (10) (3) 2w-2DMR   Bg/DC 1837 1937 New
to Bicester Depot, Oxfordshire, 25/7/1961.

3/2371 3/337 (16)   2w-2PMR   Wkm   3096 1942 New
s/s c/1959.

3/987 (19A)   2w-2PMR   Wkm   3227 1943 New
s/s 20/7/1959.

Essex Page 125

3/2145 (20)   4wPM   Bg/DC 2201 1943   New
to Yantlet Depot, Kent, c/1946-1947 by 7/1948.

3/3224 (18)   2w-2PMR   Wkm 3860 1945   New
s/s c/1957.

3/984   2w-2PMR   Wkm
origin and identity unknown
s/s 14/3/1947.

WD 9040 (7)   2w-2PMR   Wkm 6963 1955
ex Eskmeals Depot, Cumberland.
to Bicester Depot, Oxfordshire, 1/11/1962.

WD 9041 (8)   2w-2PMR   Wkm 7390 1956   New
to Bicester Depot, Oxfordshire, 1/11/1962.

WD 9042 (9)   2w-2PMR   Wkm 7391 1956   New
to Bicester Depot, Oxfordshire, 28/11/1962.

       17   2w-2PMR   Wkm
origin and identity unknown, here by 9/1961.
s/s by 1969.

WD 9031   2w-2PMR   Wkm 8089 1958
ex Bicester Depot, Oxfordshire, 26/9/1962,
to Longmoor Depot, Hampshire, 6/7/1964; ex Bicester Depot, 17/1/1973,
to Bicester Depot, 9/1/1976, ex Bicester Depot, 7/12/1976;
to RAF Caerwent, Gwent, 10/5/1978.

WD 9022   2w-2PMR   Wkm 8086 1958
ex Bicester Depot, Oxfordshire, 4/10/1962.
to Bicester Depot, Oxfordshire, 12/10/1966.

WD 9037   2w-2PMR   Wkm 8197 1958
ex Bicester Depot, Oxfordshire, by 30/4/1964.
to Bicester Depot, Oxfordshire, 4/8/1970.

WD 9101   2w-2PMR   Bg/DC 1895 1950
ex Bicester Depot, Oxfordshire, 23/9/1966,
to Bicester Depot, 12/11/1970, ex Bicester Depot, 10/2/1971.
dismantled, engine & gearbox sold 4/1977, frame & body retained as trolley No 9248, to
East Anglian Railway Museum, Chappel & Wakes Colne Station, c/2003

WD 9104 (WD 9034)   2w-2DMR   Wkm 7397 1957
ex Longmoor Depot, Hampshire, 15/6/1970.
to Yardley Chase Depot, Northamptonshire, 16/4/1979.

WD 9129   4wDHR   BD   3745 1976   New
to Bicester Depot, Oxfordshire, /1979, returned by 11/1979,
to Eskmeals Depot, Cumbria, 2/9/1987, returned 5/10/1987.
to Kineton Depot, Warwickshire, 25/3/1991.

WD 9117   4wDHR   BD   3706 1975
ex Long Marston Depot, Warwickshire, 8/8/1979.
to Kineton Depot, Warwickshire, 25/3/1991.

RBGT 1 (TNS 108)   2w-2DMR   Robel   1983
                                                               56.27-10-AG39
ex Long Marston Depot, Warwickshire, 5/10/1998.
to Luggershall Depot, Wiltshire, 5/2004.

Gauge : 2ft 0in
                            0-4-0T   [OK]
origin and identity unknown advertised for sale in "Surplus" 15/6/1921 as lying at Yantlett
again advertised for sale on 2/10/1922 at RE Shoeburyness, loco had 6x10 cylinders.
s/s.

      LR 234   4wPM   MR   234 1916
originally MoM France. Here by 13/4/1922.
s/s.

      LR 1721   4wPM   MR   320 1917
originally MoM Longmoor, Hampshire. Here by 13/4/1922.
s/s.

      LR 3034   4wPM   MR   1313 1918
originally MoM France. Here by 13/11/1924.
converted to standard gauge.

      LR 2364   4wPM   MR   1643 1918
originally MoM France.
s/s.

      LR 2465   4wPM   MR   1744 1918
originally MoM France.
s/s.

                            4wDM   HE   1974 1939   New
to Royal Engineers, COD, Old Dalby, Melton Mowbray, Leicestershire, by11/1953; later with
J. Flavell Ltd, Wakefield, Yorkshire (WR).

                            4wDM   HE   1975 1939   New
later Royal Engineers, Pitsea, by 18/8/1943.

                            4wDM   RH   202967 1940   New
later with A. Waddington & Son Ltd, contractors, Middlesex, by 30/6/1954.

                            4wDM   RH 202998 1941   New
s/s.

        MP17 EC377   4wDM   RH 217967 1942
New to MoS, origin unknown, here by 12/8/1943.
to Ministry of Public Building & Works, Shoeburyness.

        MP18 EC480   4wDM   RH 202969 1940
New to MoS, origin unknown, here by /1946.
to Ministry of Public Building & Works, Shoeburyness.

        MP19 EC481   4wDM   RH 202970 1940
New to MoS, origin unknown, here by /1946.
s/s.

        758003   4wDM   MR   8615 1941
ex BAOR Arsbeck, Germany, 1/1980.
scrapped on site.

        758008   4wDM   MR   8640 1941
ex BAOR Arsbeck, Germany, 1/1980, converted to Brake Wagon;
to Ludgershall Depot, Wiltshire, 21/9/1981, returned from Aldershot Show after 7/1982,
to Wickford Narrow Gauge Railway Group, Wickford, 4/11/1985.

        758009   4wDM   MR   8641 1941
ex BAOR Arsbeck, Germany, 1/1980.
to ROF East Riggs, Dumfries, 3/4/1981.

        758019   4wDM   MR   8820 1943
ex BAOR Arsbeck, Germany, 1/1980.
to Lydd ranges, Kent, 7/1982.

        758220   4wDM   MR   8745 1942
ex BAOR Arsbeck, Germany, 1/1980.
to ROF East Riggs, Dumfries, 3/4/1981.

        758035   4wDM   MR   8903 1944
ex BAOR Arsbeck, Germany, 1/1980.
to Ludgershall Depot, Wiltshire, 21/9/1981, returned from Aldershot Show after 7/1982;
to Lydd ranges, Kent, 9/1984.

        758036   4wDM   MR   8857 1944
ex BAOR Arsbeck, Germany, 1/1980.
s/s c1988.

        758039   4wDM   MR   8887 1944
ex BAOR Arsbeck, Germany, 1/1980.
to Ludgershall Depot, Wiltshire, 21/9/1981, returned from Aldershot Show after 7/1982;
to Lydd ranges, Kent, 9/1984.

758221   4wDM   MR   8886 1944
ex BAOR Arsbeck, Germany, 1/1980.
to Lydd ranges, Kent, by 7/1982.

758227   4wDM   MR   8813 1943
ex BAOR Arsbeck, Germany, 1/1980.
to Lydd ranges, Kent, 16/7/1980.

758138   4wDM   RH 223696 1944
ex Lydd ranges, Kent, c/1984.
s/s c/1987.

758366   4wDM   RH 202000 1940
ex Lydd ranges, Kent, here by 16/3/1984.
to Lydd ranges, Kent, after 4/11/1985, by 17/7/1986.

758263   4wDM   RH 191646 1938
ex Lydd ranges, Kent, c/1984.
to Lydd ranges, Kent, by 17/7/1986.

758365   4wDM   RH 201999 1940
ex Lydd ranges, Kent, by 9/1984.
s/s after 4/11/1985.

RTT/767091   2w-2PMR   Wkm 3030 1941 New
to Southern Counties Demolition & Trading Co, Bedhampton, Hampshire, by 8/1969.

RTT/767093   2w-2PMR   Wkm 3031 1941 New
to Southern Counties Demolition & Trading Co, Bedhampton, Hampshire, by 8/1969.
thence to Wey Valley Light Railway, Farnham Surrey, 12/2/1972.

RTT/767092   2w-2PMR   Wkm 3032 1941 New
to Southern Counties Demolition & Trading Co, Bedhampton, Hampshire, by 8/1969.

RTT/767094   2w-2PMR   Wkm 3033 1941 New
to M.E. Engineering Ltd, Neasden, Greater London, c/10/1983.

                2w-2PMR   Wkm 3034 1941 New
to Southern Counties Demolition & Trading Co, Bedhampton, Hampshire, by 8/1969.

                2w-2PMR   Wkm 3287 1943 New
to Southern Counties Demolition & Trading Co, Bedhampton, Hampshire by 8/1969.

RTT/767095   2w-2PMR   Wkm 3414 1943 New
to M.E. Engineering Ltd, Neasden, Greater London, c10/1983.

Gauge : 2ft 6in
Yard No.1066   RAMBO   4wBE   CE   B0483   1976
                         Reb   CE   B4408   2005
ex RNAD Crombie, Fife, 3-4/1995, to CE for overhaul c/1998, returned, to CE for overhaul by 21/4/2005, returned 4/7/2005.

Yard No.1067   PREDATOR   4wBE   CE   B0483   1976
ex RNAD Crombie, Fife, 3-4/1995, to CE for overhaul by 31/3/1998, returned by 30/6/1998.
to CE for overhaul by 21/4/2005, work cancelled, sold to McGregor Railway Services, Bicester, Oxfordshire, /2005, resold to CE 9/2005, (still at CE /2009).

Yard No.1068   TERMINATOR 4wBE   CE   B0483   1976
                         Reb   CE   B4253   1998
                         Reb   CE   B4408   2004
ex RNAD Crombie, Fife, 3-4/1995, to CE for overhaul c12/1997, returned by 31/3/1998, to CE for overhaul after 5/3/2004 by 31/8/2004, returned c11/2004.

The following locomotives were used in "Train Busting" exercises during WW2 and were scrapped at Shoeburyness after being damaged beyond repair. On 25/2/1943 a 4-4-0 with outside cylinders was used in weapons trials; two Hawker Hurricane fighter aircraft from A&AEE Boscombe Down fitted with various types of weapons including 20 & 40mm cannons carried out attacks on the locomotive (Dübs 3404). A Rocket with a 60pr high explosive shell was the most effective, it struck the boiler ahead of the throat plate and blew a 4ft by 2ft hole in the boiler barrel, the throat plate being forced back into the firebox and the tube plate forced away from the crown sheet rendering the boiler beyond repair, the loco was not steamed during the trials.

Reference: National Archives, AIR37/229.

Gauge : 4ft 8½in

14391	LOCH SHIN	4-4-0	OC	D	3404	1896	(a)	Scr
1239		4-4-0	IC	Dar	548	1908	(b)	Scr
14405	BEN RINNES	4-4-0	IC	HR		1899	(c)	Scr

(a)   ex LMSR /1941, withdrawn 8/1941.
(b)   ex LNER /1942, withdrawn 12/1942.
(c)   ex LMSR /1944, withdrawn 9/1944.

MINISTRY OF DEFENCE
ROYAL ARMAMENT RESEARCH & DEVELOPMENT ESTABLISHMENT, ROYAL GUNPOWDER FACTORY, Waltham Abbey (R.O.F. 2)   D114
Loco Shed TL 377012

An extensive explosives factory which comprised two separated areas. The North Site lay between the Old River Lea and the River Lea Navigation north of Waltham Abbey whilst the South Site extended from the Old River Lea east to Quinton Hill (and the A112 road) south of Waltham Abbey. Hand worked tramways, thought to be of 2ft 3in gauge, were installed on the North Site in 1857 while by 1894 extensive tramways, probably of 18in gauge, were in use at the South Site. In 1916 a new 18in gauge locomotive worked system about two miles long was built from the North Site, with a loco shed at TL 377012; this passed under the Waltham Cross - Waltham Abbey (later A121) road, to serve the South Site and terminated at an interchange siding with the standard gauge lines of the Royal Small Arms Factory (Enfield, Middlesex) (at approximately TQ 376992). Rail traffic continued until explosive manufacture ceased c10/1943. The line between North and South sites was by 1952 dismantled although some locomotives remained in limited use in 1954, with three still in use in 1958 and at least one remained until 1962.

Reference: Industrial Railway Record No.117

Gauge : 1ft 6in

-	4wBE	BEV	59	1918	New	s/s
-	4wPM+	RP	51697	1917	New	s/s
-	4wPM+	RP	51707	1917	New	s/s
-	4wPM+	RP	51901	1917	New	(1)
-	4wPM+	RP	51927	1917	New	s/s
-	4wBE	GB	1668	1940	New	(2)
-	4wBE	GB	1669	1940	New	(2)
-	4wBE	GB	1670	1940	New	(2)
-	4wBE	GB	1671	1940	New	(2)
-	4wBE	GB	1672	1940	New	(2)
-	4wBE	GB	1673	1940	New	(2)
-	4wBE	GB	1851	1942	New	(2)
-	4wBE	GB	1852	1942	New	(2)
-	4wBE	GB	1861	1942	New	(2)
-	4wBE	GB	1862	1942	New	(2)

+ 10hp ZLH type locomotive, started on petrol then run on paraffin

(1) rebuilt to 2ft 0in gauge by /1933;
to Oakeley Slate Quarries, Blaenau Ffestiniog, Merioneth, via H. Gardam & Co Ltd, Staines, Middlesex, /1934.

(2) three locomotive still here in /1958, at least one until /1962; all s/s.

MINISTRY OF PUBLIC BUILDING & WORKS
SHOEBURYNESS DEPOT   J115

Narrow gauge equipment within the Ministry of Defence establishment. Tramway closed.

Gauge : 2ft 0in

MP 17	4wDM	RH	217967	1942	(a)	(1)
MP 18	4wDM	RH	202969	1940	(a)	(1)

(a) ex WD, Shoeburyness.

(1) to Track Supplies & Services Ltd, Wolverton, Buckinghamshire, 2/1973.

MINISTRY OF MUNITIONS
DAGENHAM DOCK
P116
TQ 487819

A National Cartridge and Box Repair Factory No.98 was established on Samuel Williams, Dagenham Dock estate in 1916, partly using an existing factory unit built by the company. The MoM erected a small jetty serving their main factory and utilised the company's internal user wagons and tarpaulins for Government traffic. The Government also commandeered several of the companys tugs and barges. On 9/11/1917 authorisation was given to spend £650 for the purchase of a locomotive.

Reference: Industrial Locomotive No.93

Gauge : 4ft 8½in

        -        0-4-0ST   OC   AB   193 1878   (a)   (1)

(a)   ex Clippens Oil Co Ltd, Clippens Works, Loanhead, Midlothian, hire, by 2/1918.

(1)   returned to Clippens Oil Co Ltd, Clippens Works, Loanhead, Midlothian, after 4/1918.

MUREX LTD
RAINHAM FERRY WORKS
L117
Murex Co Ltd until 10/6/1920
TQ 517805

Murex Company Ltd was incorporated 9/7/1913 with an address at 1 London Wall Buildings, London, EC. The company was formed to acquire the assets of **Murex Magnetic Company Ltd** and to carry on the business of smelting and refining non-ferrous metals including gold, silver, zinc & copper and also as Iron founders. The works was on the north bank of the River Thames about a mile south-west of Rainham with an internal narrow gauge system to a wharf. Rail traffic ceased c/1961 although the locomotives remained on site for some years before disposal.

Gauge : 2ft 0in

No.1		4wDM	MR	7902	1939	New	(3)
No.2	No.101	4wDM	MR	7927	1941	New	(3)
-		4wDM	MR	7981	1946	New	(1)
No.3		4wDM	MR	10130	1949	New	(2)

(1)   to Pilkington Bros Ltd, Bickerstaffe Sand Pits, Lancashire.

(2)   to Taylor Woodrow Construction Ltd, Holyhead Aluminium Smelting Works Contract, Caernarvonshire, c/1968.

(3)   to Llanberis Lake Railway Co Ltd, Llanberis, Caernarvonshire, 28/4/1971.

W.C. MURRELL
Rainham & Barking
P118
Barking shoot TQ 461820

William Charles Murrell was described as a coal factor, merchant, and forage contractor of 16 Dockhead, and Murrell's Wharf, Bermondsay (TQ 341798) who established his business about 1859. He also had premises at the Regent's Canal Basin, Limehouse, 1 Angel-Park-Gardens, Brixton and 11 Kennington Park Road. His business went into receivership in 1889 with debts of £50,111. Charles Murrell operated rubbish shoots at Rainham Creek by 1913 and at Barking by 1915. The locos listed below may have worked at either of these locations. (also addresses in Hunslet, spares records given as 5 Albion Place, Blackfriars Bridge (1915-17) & Wharf, Upper Ground Street Blackfriars (1917) and Exors 31 Stamford Street, Blackfriars (1919).

Gauge : 4ft 8½in

		0-4-0ST	OC	MW	559	1875	(a)	s/s
		0-4-0ST	OC	HE	103	1879	(b)	(1)
CALDEW		0-4-0ST	OC	MW	225	1867	(c)	s/s

(a) ex John Aird & Sons, Nottingham – Melton Mowbray (Mid Rly) (1874-1879) contract, Notts.
(b) ex Buxton Lime Firms Co Ltd, Great Rocks Quarry, Derbyshire, after 4/1902, by 6/1915.
(c) ex Charles Wall Ltd, Chingford reservoir contract.

(1) s/s after10/1919.

NORTH THAMES GAS BOARD

BECKTON GAS WORKS A119
Gas Light & Coke Co. until 1/5/1949 ST 445815

A long established and major town gas producing works, founded in 1868 by the Gas Light & Coke Co and built by John Aird. The rail system extended to about 70 miles of track and operated 1100 wagons (including the By-Products Works), the works had two loco sheds one of which was a roundhouse. Coal was received on river jetties and taken on an elevated railway to retort house bunkers; after closure of retorts a gas pipeline was laid to Canvey Island gas terminal for gas imported by ship from overseas. Rail traffic ceased c8/1970. Works closed for gas production. Site administered by British Gas Corporation, North Thames Region after 1/1/1973.

Reference : "Gas Light and Steam", Malcolm Millichip, 1994.

Gauge : 4ft 8½in

No.1	(LORD MAYOR)	0-4-0WT	OC	N	1561	1870	New	
			reb	Beckton		1929		(14)
No.2	(R.A.GRAY)	0-4-0WT	OC	N	1562	1870	New	(4)
No.3	(HON. HOWE BROWN)	0-4-0WT	OC	N	1659	1872	New	Scr /1934
1		0-4-0VBT	VCG	Chaplin	1675	1874	New	Scr
2		0-4-0VBT	VCG	Chaplin	1756	1874	New	Scr
3		0-4-0VBT	VCG	Chaplin	1757	1874	New	Scr
No.4	(SIMON ADAM BECK)	0-4-0ST	OC	MW	457	1875	New	Scr c/1931
No.5	(QUEEN'S COUNSEL)	0-4-0WT	OC	N	2151	1876	New	Scr /1927
No.6		0-4-0WT	OC	N	2227	1877	New	Scr /1934
No.7		0-4-0WT	OC	N	2228	1877	New	Scr /1934
No.8		0-4-0WT	OC	N	2380	1878	New	Scr /1934
No.9		0-4-0T	OC	N	2382	1878	New	Scr /1938
No.10		0-4-0T	OC	N	2465	1879	New	
			reb	Beckton		1929		Scr /1962
No.11		0-4-0T	OC	N	2466	1879	New	
			reb	Beckton		1928		Scr /1961

No.12			0-4-0T	OC	N	2597	1880	New	
			reb		Beckton		1931		(10)
(No.13)			0-4-0T	OC	N	2598	1880	New	
			reb		Beckton		1929		Scr c2/1967
No.14			0-4-0ST	OC	N	3097	1883	New	Scr /1935
No.15			0-4-0ST	OC	N	3345	1884	New	Scr /1930
No.16			0-4-0ST	OC	N	3451	1885	New	
			reb		Beckton		1936		(7)
No.17			0-4-0ST	OC	BH	865	1886	New	
			reb		Beckton		1936		(7)
No.18			0-4-0T	OC	BH	864	1886	New (h)	(9)
No.19			0-4-0T	OC	N	3789	1888	New	
			reb		Beckton		1931		Scr /1962
No.20			0-4-0T	OC	N	4249	1890	New	
			reb		Beckton		1929		(10)
No.21			0-4-0ST	OC	N	4250	1890	New	
			reb		Beckton		1938		(9)
No.22			0-4-0T	OC	N	4414	1891	New	
			reb		Beckton		1931		(11)
No.23			0-4-0T	OC	N	4408	1892	New	
			reb		Beckton		1929		Scr c/1967
No.24			0-4-0T	OC	N	5086	1896	New (h)	
			reb		Beckton		1930		(7)
No.25			0-4-0ST	OC	N	5087	1896	New	
			reb		Beckton		1938		(12)
No.26			0-4-0T	OC	N	5228	1897	New	
			reb		Beckton		1929		(10)
No.27			0-4-0T	OC	N	5229	1897	New	Scr /1962
No.28			0-4-0T	OC	N	5230	1897	New	
			reb		Beckton		1930		Scr /1962
No.29			0-4-0T	OC	N	5231	1897	New	
			reb		Beckton		1930		(13)
No.30			0-4-0T	OC	Beckton	1	1902	New	(10)
No.31			0-4-0T	OC	Beckton	2	1902	New	(9)
No.32			0-4-0ST	OC	MW	901	1885	(a)	(1)
No.32			0-4-0ST	OC	MW	1832	1913	New	Scr 6/1955
No.33			0-4-0ST	OC	AB	636	1889	(b)	Scr c/1949
No.34			0-4-0ST	OC	AB	262	1883	(c)	Scr /1935
No.35			0-4-0ST	OC	AB	957	1902	(d)	Scr c/1949
No.36			0-4-0ST	OC				(e)	(2)
No.37			0-4-0ST	OC	HC	287	1887	(g)	Scr /1930
No.38			0-4-0ST	OC	HC	657	1903	(f)	Scr /1938
No.39			0-4-0ST	OC				(e)	(3)
No.40			0-4-0ST	OC	HE	1335	1919	New	(9)
			2w-2PMR		BgC	790	1920	New	s/s
No.4	(No.41)		0-4-0T	OC	AB	1720	1921	New	Scr /1961
No.36	(No.42)		0-4-0T	OC	AB	1721	1921	New	Scr /1961
No.39	(No.43)		0-4-0T	OC	AB	1722	1921	New (h)	(11)
			2w-2PMR		BgC	1335	1923	New	s/s
No.2			0-4-0VBT	VCG	S	6951	1927	New	Scr /1938
		Reb of	0-4-0WT	OC	N	1562	1870		

Essex Page 134

No.5	0-4-0ST	OC	HL	3742	1929	New	(5)
No.15	0-4-0ST	OC	P	1811	1930	New	(8)
No.37	0-4-0ST	OC	P	1837	1931	New	Scr /1962
No.2	0-4-0ST	OC	P	1966	1939	New	Scr c2/1967
No.3	0-4-0ST	OC	P	1932	1937	New	(8)
No.5	0-4-0ST	OC	HL	3794	1931	New	(9)
No.6	0-4-0ST	OC	P	1933	1937	New	(10)
No.7	0-4-0ST	OC	WB	2657	1942	New	Scr c2/1967
No.8	0-4-0ST	OC	WB	2658	1942	New	Scr /1961
No.9	0-4-0ST	OC	MW	1427	1899		
	reb		MW		1916		
	reb		HC		1945	(j)	(9)
No.38	0-4-0ST	OC	HC	522	1899	(k)	(7)
No.14	0-4-0ST	OC	RSHN	7474	1949	New	Scr /1962
	4wVBT	VCG	S	9398	1950	(l)	(6)
No.33	0-4-0ST	OC	P	2123	1951	New	Scr /1962
No.34	0-4-0F	OC	RSHN	7665	1951	New	(25)
No.35	0-4-0F	OC	RSHN	7803	1954	New	(27)
	4wDM		RH	421416	1958	(m)	(28)
No.9	0-4-0ST	OC	P	1574	1920	(n)	(15)
No.3	0-4-0T	OC	N	4571	1892	(o)	s/s c/1968
41	4wDM		FH	3885	1958	New	(18)
42	4wDM		FH	3889	1958	New	(17)
43	4wDM		FH	3907	1959	New	(16)
44	4wDM		FH	3908	1959	New	(20)
45	4wDM		FH	3909	1959	New	(24)
46	4wDM		FH	3910	1959	New	(18)
47	4wDM		FH	3911	1959	New	(19)
48	4wDM		FH	3912	1959	New	(21)
49	4wDM		FH	3959	1961	New	(22)
50	4wDM		FH	3960	1961	New	(23)
51	4wDM		FH	3961	1961	New	(26)

Gauge : 4ft 8½in. Locomotives used at New Coke Ovens.

0-4-0WE		WSO	1408	1929	New	s/s c/1973
0-4-0WE		WSO	4191	1945	New	s/s c/1973
0-4-0WE		WSO	5785	1953	New	s/s c/1973

LNER Sentinel locos 78 & 87 were here on loan in /1940

(a) ex Thos. W. Ward Ltd, Silvertown, /1909, (or George Cohen Sons & Co Ltd, 9/1909) earlier J.T. Firbank, Acton-Northolt (GWR) (1899-1905) contract, Middlesex.
(b) ex Morrison & Mason, Portsmouth, Hampshire, via B. Goodman, dealers, 6/1916, damaged by enemy action 1/1941.
(c) ex Steel & Turner, Barrasford, Northumberland, via B. Goodman, dealers, 8/1916.
(d) earlier Morrison & Mason, Portsmouth Dockyard (Admiralty) (1908-1912) contract; via B. Goodman, dealers 3/1917.

(e)   earlier Morrison & Mason, Portsmouth Dockyard (Admiralty) (1908-1912) contract; here by c/1918,
(one of these two locos appears to be AB 266, this loco carried 295 on the motion but was renumbered 266 prior to despatch from AB (AB 266 was at WD Catterick by 6/1917 then via J. F. Wake to here, /1917 ?)
(f)   ex New Explosives Co Ltd, Stowmarket, Suffolk, /1918.
(g)   ex MoM, Gretna, Dumfriesshire, c/1919.
(h)   damaged by enemy action 1/1941.
(j)   ex Mansfield Standard Sand Co Ltd, Nottinghamshire, /1946.
(k)   ex Wandsworth District Gas Co, Kingston, Surrey, by 3/1947.
(l)   ex S on demonstration 10/1951.
(m) ex RH, Lincoln, on trial by 12/4/1958.
(n)   ex Beckton By-Products Works, /1965.
(o)   ex Beckton By-Products Works, /1966.

(1)   to Bromley-by-Bow Works, /1912.
(2)   to Bromley, c9/1920.
(3)   to Bromley, c/10/1920.
(4)   rebuilt as 0-4-0 VBT S6951, /1927.
(5)   to Southall Works, Middlesex, /1932.
(6)   returned to S off demonstration, after 10/1951.
(7)   scrapped by George Cohen, Sons & Co Ltd, 8/1958.
(8)   to Lacmots Ltd, (for Italy), via Drew & Sawyer, Kent, /1959.
(9)   scrapped on site by Drew & Sawyer, Belvedere, Kent, 6/1959.
(10) scrapped on site by George Cohen, Sons & Co Ltd, 5/1960.
(11) scrapped on site by H.F.A. Dolman Ltd, Southend, 11/1960.
(12) to Southall Gas Works, Middlesex, /1961.
(13) to Beckton By-Products Works, 6/1962.
(14) to National Trust, Penrhyn Castle Museum, Caernarvonshire, 12/12/1963.
(15) returned to Beckton By-Products Works, c10/1965.
(16) to Kensal Green Gas Works, London, 2/1968.
(17) to British Industrial Sand Ltd, Redhill, Surrey, 11/1968.
(18) to W.R. Cunis Ltd, Great Coldharbour Rubbish Shoot, /1968.
(19) to British Industrial Sand Ltd, Middletown Towers, Norfolk, c2/1969.
(20) to British Industrial Sand Ltd, Redhill, Surrey, c2/1969.
(21) to Thames Metal Co Ltd, Greenwich, London, 9/1970.
(22) to Edward Ash (Plant) Ltd, Deptford, London, 6/1970;
      thence to Powell Duffryn Wagon Co Ltd, Maindy Works, Glamorgan,
      per Thos. W. Ward Ltd, 1/1972.
(23) to Wagon Repairs Ltd, Stoke-on-Trent, Staffordshire, c12/1970.
(24) to Wagon Repairs Ltd, Long Eaton, Derbyshire, c12/1970.
(25) scrapped on site by Dismantling Contractors Ltd, of Ewell, Surrey, /1971.
(26) to C. & K. Metals Ltd, Barking, /1971;
      thence to British Industrial Sand Ltd, Redhill, Surrey, c11/1972.
(27) to Carless, Cape & Leonard Ltd, Harwich Refinery, Parkeston,
      per C. & K. Metals Ltd, Barking, /1972.
(28) to Lea Bridge Gas Works after trial.

BECKTON BY-PRODUCTS WORKS   A120
Gas Light & Coke Co. until 1/5/1949   TQ 435820

Extensive chemical works adjacent to the gas works with its own loco fleet and loco shed, also processed by-products from other works.
Gauge : 4ft 8½in

No.1		0-4-0T	OC	N	4444	1892	New	(6)
No.2		0-4-0T	OC	N	4445	1892	New	Scr c7/1967
No.3		0-4-0T	OC	N	4571	1892	New	(5)
No.4		0-4-0T	OC	N	4572	1892	New	
		reb		Beckton		1930		(4)
No.5		0-4-0T	OC	NR	5348	1898	New	
		reb		Beckton		1925		(4)
No.6		0-4-0T	OC	NR	6302	1902	New	(3)
No.7		0-4-0ST	OC	AB	757	1895	(a)	(1)
No.9		0-4-0ST	OC	P	1574	1920	New(g)	Scr c7/1967
No.10		0-4-0ST	OC	P	1575	1920	New	Scr 7/1962
No.8		0-4-0ST	OC	AB	730	1893		
		reb		Wake		1921	(b)	(4)
No.11		0-4-0ST	OC	P	1576	1921	New	(4)
		0-4-0DE		AW	D23	1933	New(c)	(2)
No.13		0-4-0ST	OC	HL	3308	1918	(d)	Scr c7/1967
No.12		0-4-0F	OC	HL	3595	1924	(e)	(7)
No.14		0-4-0ST	OC	P	2083	1947	New	Scr c7/1967
No.15		0-4-0ST	OC	P	2099	1948	New	Scr 7/1962
No.29		0-4-0T	OC	N	5231	1897		
		reb		Beckton		1930	(f)	Scr c7/1967

(a) ex B. Goodman 7/1916 via/per J. Wardell & Co, dealers, London,
      earlier Boyd & Forrest, contractors, Kilmarnock.
(b) ex John F. Wake & Co Ltd, dealer, Darlington, Co. Durham, /1921;
      earlier Mawson Clarke & Co Ltd, Dunston, Co. Durham.
(c) ex AW on demonstration.
(d) ex British Industrial Plastics Ltd, Oldbury, Staffordshire, 7/1946.
(e) ex County of London Electric Supply Co Ltd, Barking, by 3/1947.
(f) ex Beckton Gas Works, 15/6/1962.
(g) to Beckton Gas Works, /1965; returned c10/1965.

(1) to Trent Concrete Ltd, Colwick Works, Nottingham, after 12/1920, by 8/1925.
(2) to Admiralty, Chatham Dockyard, Kent, /1934.
(3) dismantled /1957; scrapped by H.F.A. Dolman Ltd, Southend, 1/1961.
(4) scrapped on site by Thos. W. Ward Ltd, 6/1961.
(5) to Beckton Gas Works, /1966.
(6) to Alan Bloom, Bressingham Hall, Norfolk, 10/1968.
(7) to ?, /1969.

BROMLEY-BY-BOW GAS WORKS   C121
Gas Light & Coke Co until 1/5/1949   TQ 388824
Imperial Gas Light and Coke Co until 3/1876

Constructed by the Imperial Gas Light and Coke Company, the works was obsolescent in design from the start and was still not completed when the Gas Light & Coke Co took over in 1876, production commenced in 12/1873. The works had its own dock where coal was delivered by barge via the River Lea having been off-loaded from colliers at Beckton. Situated on the south side of the ex-LT&SR line east of Bromley-by-Bow Station and west of the ex-GER Stratford - North Woolwich line the works was later served by standard gauge sidings from 1912 which connected with the west to south curve linking these two lines. A narrow gauge system was also used in the works. Production ceased in 1970 and works closed, today the works offices house the London Gas Museum.

Reference: Gas Light and Steam, Malcolm Millichip 1994.

Gauge : 4ft 8½in

	ROSE	0-4-0ST	OC	MW	901	1885	(a)	Scr c/1927
	PEGGY	0-4-0ST	OC	AB	266	1894		
				reb	Wake		(b)	s/s
No.6		0-4-0ST	OC	BH	774	1884	(c)	Scr c12/1950
No.4		0-4-0ST	OC	AB	1666	1920	New	(1)
No.5		0-4-0ST	OC	AB	1674	1920	New	s/s /1966
No.1		0-4-0T	OC	N	4397	1891	(d)	(1)
No.3		0-4-0ST	OC	RSHN	7309	1946	New	(1)
No.2		0-4-0ST	OC	P	2135	1953	New	s/s /1966
6		4wDM		FH	3994	1962	New	(2)

BR 51207 was here on loan in /1950 & 6/1951, 51253 here on loan 6/1951.

(a)   ex Beckton Gas Works, No.32, /1912.
(b)   ex Beckton Gas Works, /1918.
(c)   ex John F. Wake & Co Ltd, dealer, Darlington, Co. Durham, 5/1919;
      earlier Sir W.G. Armstrong, Whitworth & Co Ltd, Elswick, Newcastle upon Tyne.
(d)   ex Kensal Green Gas Works, London, by 3/1934.
(1)   to Mayer-Newman & Co Ltd, Canning Town, for scrap, /1963.
(2)   to H.F.A. Dolman Ltd, Southend, 8/1970;
      thence to W.R. Cunis Ltd, Great Coldharbour Rubbish Shoot, c/1971.

Gauge : 2ft 9in

	MABEL	0-4-0ST	OC	WB	1444	1894	(a)	Scr
	IRIS	0-4-0ST	OC	WB	1553	1898	New	Scr
	LYDIA	0-4-0ST	OC	WB	1734	1904	New	Scr
	BEATRICE	0-4-0ST	OC	WB	1855	1909	New	Scr /1929
	RUTH	0-4-0ST	OC	WB	1964	1914	New	Scr

(a)   ex WB, /1898.

LEA BRIDGE GAS WORKS   C122
South Eastern Gas Corporation Ltd until 1949   TQ 363870
Lea Bridge District Gas Company until 1939
Lea Bridge District Gas Light & Coke Co Ltd from 17/7/1868
County and General Consumers Company until 1868
South Essex Gas Light and Coke Co established 25/8/1852

Small gas works (which commenced production in 1853) with standard gauge sidings connecting with the ex-GER Lea Valley line south of Lea Bridge station. Production ceased in 1968, railway and works dismantled.

Gauge : 4ft 8½in

| | 4wDM | RH | 421416 | 1958 | New | (1) |

(1)   to T. & M. Beaton Bros, Chiswick, /1970.

F.A. NORTON
SANDON HALL GRAVEL PIT, Sandon   G123
                                                          TL 743040

F.A. Norton Esq of Ramsden Heath, Billericay operated sand & gravel pits at Sandon near Chelmsford listed in quarry directories for 1943, later by 1948 pits operated by F. A. Norton & Sons at Bradwell on Sea until at least 1959/1960.

Gauge : 2ft 0in

| | 4wPM | MR | 3983 | | (a) | (1) |
| | 4wPM | MR | 3790 | 1926 | (b) | s/s |

(a)   unidentified loco sent to MR for clutch repairs 24/1/1936 and renumbered as 3983 for record purposes.
(b)   ex Petrol Loco Hirers 27/5/1937, earlier Dunlop & Co, Glasgow, rebuild of MR 1387.
(1)   sold to MR, thence to W.H. Jones Ltd, Coventry, 4/4/1938.

R.G. ODELL LTD
WESTWICK RUBBISH SHOOT, Canvey Island   J124
                                                          TQ 752833

Company registered 21/9/1921 as S. F. Morris & Co Ltd with S. F. Morris and R. G. Odell as directors. R. G. Odell Ltd was listed in 1938 as Lightermen of 68 Horseferry Road, London, SW1. Planning permission was granted in 1936 for a rubbish shoot; silt dredged from the River Thames was also deposited here on 110 acres of marsh land; rubbish was transported by barge from Albion Wharf, Lombard Road, Battersea. A narrow gauge tramway was used to convey landfill material from a wharf on Holehaven Creek. Tramway closed, and track lifted in 1972. Company went into voluntary liquidation on 30/10/1972.

Gauge : 2ft 0in

	4wDM	RH	179004	1936	New	(2)
	4wDM	RH	183741	1937	New	(1)
	4wDM	RH	218032	1943	New	s/s c/1971
	4wDM	FH	3307	1948	(a)	(3)

(a) ex George Cohen, Sons & Co Ltd, Wood Lane Depot, London, 23/6/1953; earlier Ashover Light Railway, Derbyshire.
(1) to Walton-on-Thames Works, Surrey, after 8/3/1951.
(2) to Walton-on-Thames Works, Surrey, after 30/10/1952.
(3) to East Anglian Transport Museum, Carlton Colville, Suffolk, c10/1972.

CHAS.T. OLLEY & SONS
GROVES FARM SAND & GRAVEL PITS, South Ockendon L125
TQ 596842

Pits located in North Road, South Ockendon, listed in HMSO list of quarries for 1948 and quarry directories until at least 1959/60. Rail system abandoned and replaced by dumpers.
Gauge : 2ft 0in

-	4wPM	FH			#	(3)
-	4wPM	FH			#	(3)
-	4wDM	MR	8588	1941	(a)	(1)
-	4wDM	MR	7331	1938	(b)	(2)

\# Per Motor Rail spares records, spares for locos were sent here on dates between 6/9/1946 and 21/7/1947, so presumably at least one of these locos was here by 9/1946.

(a) ex Motor Rail Ltd, 3/3/1947.
(b) ex Westfield Transport Co Ltd, Mansfield, Nottinghamshire, 6/1947.
(1) to Motor Rail Ltd, 12/1947;
thence to Bedford Silica Sand Mines Ltd, Leighton Buzzard, Bedfordshire, 3/1948.
(2) to Motor Rail Ltd, 12/1947.
(3) to East London Transport Co Ltd, South Ockendon, c/1948.

PETERS & BARHAM LTD
Peters & Barham until 23/3/1932
Sand & Gravel pit owners listed in quarry directories for 1933 with pits at **Romford Field**, Hornchurch Road, Romford, and in 1934 with pits at **Princes Road**, Chelmsford. By 1939 also at Sandy Lane, Aveley, **Gerpins Pit**, Gerpens Lane, Upminster, and **Kemps Farm**, South Ockendon, and in 1948 additional pits at Great Waltham, St Osyth, **Nine Acre Shaw**, West Thurrock (TQ 588790), & **Warren Lane**, Grays. In 1953 pits had opened at Asheldham, Birch, Thorpe-le-Soken, Roxwell & Fingringhoe (see also Barham & Tait Ltd), by 1966 pits at **Eastbrook End**, Dagenham (TQ 510862) & **Brook Farm**, Alresford, near Wivenhoe, also **Sandfords Farm**, Hatfield Peverel (TL 811119). J.F. Barham & H.J. Peters were also directors of **Clacton Sand and Ballast Ltd** (registered 12/4/1937).

The company stripped and processed overburden at the north quarry of Wouldham Cement Works. A narrow gauge tramway using diesel locos and skips was used to serve the screening plant. This is the site referred to in quarry directories as being at Warren Lane, Grays.

"Contract Journal", 26/12/1945, H. Butcher to auction 22/1/1946 re Peters & Barham Ltd, Romford – plant including Simplex locos.

"Contract Journal", 5/3/1947 H. Butcher to auction 22/4/1947 re Peters & Barham Ltd, Romford – plant including Simplex locos.

"Contract Journal", 15/11/1950 for sale – plant inc, three 2ft gauge RH diesel locos, 10hp, 16hp & 20hp - Peters & Barham Ltd, Romford.

ROMFORD FIELD PITS, Hornchurch Road, Romford　　　　　　　　　　L126

Gauge : 2ft 0in

-	4wPM	MR	1740	1918	(a)	s/s

(a)　originally MoM France, here by 28/3/1934.

PETROPLUS HOLDINGS AG
CORYTON REFINERY, Stanford-le-Hope
For history of site and details of locomotives see entry for **B P Oil Co Ltd.**

PHOENIX TIMBER CO LTD
FROG ISLAND TIMBER YARD, Rainham　　　　　　　　　　　　P127
TQ 511810

Company of timber importers registered on 9/11/1925 with standard gauge sidings serving a timber wharf on the River Thames about a mile south-west of Rainham village. In 1948 the company purchased and installed two pontoons which originally formed part of Mulberry Harbour, these pontoons formed a floating jetty 400ft long linked to the mainland by a bridge 240ft long. Rail system closed and dismantled.

"Machinery Market" 28/10/1955 for sale – 20hp std gauge diesel loco FCH 3264 – Phoenix Timber Co Ltd, Rainham

Gauge : 4ft 8½in

-	4wDM	FH	2914	1944	(a)	(1)
-	4wDM	FH	3264	1947	New	(2)

(a)　ex Weston-Super-Mare & District Gas Co, Somerset.

(1)　to Thos. W. Ward Ltd, Titan Works, Grays, /1951;
　　　thence to Pittrail Ltd, Aldridge, Staffordshire, /1957.
(2)　to Broom Wade Ltd, Desborough Works, High Wycombe, Bucks,
　　　per M.E. Engineering Ltd, Cricklewood, Middlesex, /1957.

PLESSEY CO LTD
ILFORD FACTORY　　　　　　　　　　　　　　　　　　　　　　D128
Company registered 12/12/1917

During WW2 work was suspended on the eastern extensions of the LT Central Line underground railway. The partly completed deep level tunnels under Eastern Avenue between Gants Hill and Wanstead stations were utilised to become a munitions factory, which was served by a narrow gauge railway system. At the end of the war, the factory was closed and dismantled and the 'tube' extensions completed and put into service.

Reference : The Narrow Gauge No.112

Gauge : 1ft 6in

-	0-4-0BE	WR	2063	1941	New	(1)
-	0-4-0BE	WR	2064	1941	New	s/s
-	0-4-0BE	WR	2065	1941	New	(1)
-	0-4-0BE	WR	2066	1941	New	s/s
-	0-4-0BE	WR	2067	1941	New	s/s
-	0-4-0BE	WR	2068	1941	New	s/s
	0-4-0BE	WR	2352	1942	New	s/s
	0-4-0BE	WR	2353	1942	New	s/s
	0-4-0BE	WR	2354	1942	New	s/s
	0-4-0BE	WR	2355	1942	New	s/s
	0-4-0BE	WR	2356	1942	New	s/s
	0-4-0BE	WR	2357	1942	New	s/s

(1) later with Sir Robert McAlpine & Sons Ltd, contractors; rebuilt to 2ft 0in gauge by 7/1962.

PORT OF LONDON AUTHORITY

This Authority was formed 31/3/1909 to take over responsibility for the London Docks. The oldest of these and nearest to the City of London were the London & St Katharine Docks, on the north bank of the Thames just east of Tower Bridge; these were not rail served.

To the east, the Surrey Commercial Docks lay on the south bank of the Thames and then on the north side again, on the Isle of Dogs, the Millwall and East and West India Docks in the County of London. Continuing eastwards the Royal group of docks were located in the area of Silvertown; further down river were the Tilbury docks which dealt with more modern, larger ships. All except Tilbury Docks have now closed. Part of the area of the Royal docks has been re-developed and incorporates the London City Airport. Locomotives owned by PLA were often repaired and overhauled by Harland & Wolff Ltd who had a dry dock and works at No.9 site, King George V Dock.

Reference : "P.L.A. Railways", Thomas B. Peacock, The Locomotive Publishing Company Ltd, 1952

"Railway Bylines", M. Smith, February-March & April-May 1998

ROYAL DOCKS   A129

Loco Shed TQ 412808

Locomotives were kept at **Custom House**, serving the **Royal Group of Docks** and at **Tilbury** which had its own engineering workshops.

Gauge : 4ft 8½in

No.1		0-6-0ST	OC	D	1438	1881	(a)	sold /1917
No.2		0-6-0ST	OC	D	1439	1881	(a)	sold /1920
No.3		0-6-0ST	OC	D	1440	1881	(a)	sold /1917
No.4		0-6-0ST	OC	D	1441	1881	(a)	sold /1920
No.5		0-6-0ST	OC	RS	2844	1896		
		reb		R		1915	(a)	(5)
No.6		0-6-0ST	OC	RS	2845	1896		
		reb		HW *		1915	(a)	Scr /1926
No.7		0-6-0ST	OC	RS	2981	1900		
		reb				1917	(a)	Scr /1930

No.9		0-6-0ST	OC	RS	2983	1900		
			reb	HW *		1915	(a)	(2)
No.10		0-6-0ST	OC	RS	3070	1901		
			reb		1914,	1921	(a)	(22)
No.11	(LOOE)	0-6-0ST	OC	RS	3050	1901		
	(name removed /1911)		reb			1914		
			reb	HW *		1923	(a)	(16)
No.12		0-6-0ST	OC	RS	3094	1902		
			reb			1919	(a)	(4)
No.13		0-6-0ST	OC	FW	263	1875	(a)	sold /1917
No.14		0-4-0ST	OC	YE	284	1876	(a)	sold /1917
(No.15)		0-4-0ST	OC	FE		1884	(a)	sold /1914
No.15		0-4-0ST	OC	MW	905	1884	(a)	sold /1920
		0-4-0ST	OC	RP		c1870	(a)	sold /1912
No.34		0-4-0ST	OC	MW	1008	1887	(b)	(3)
No.35		0-4-0ST	OC	MW	1106	1888	(c)	(1)
No.38		0-6-0T	OC	AB	1236	1911	New	(15)
No.39		0-6-0T	OC	AB	1237	1911	New	(11)
No.41		0-6-0T	OC	AB	1294	1912	New	Scr 2/1946
No.42		0-6-0T	OC	AB	1300	1912	New	(8)
No.43		0-6-0T	OC	AB	1301	1912	New	(23)
No.44		0-6-0T	OC	AB	1302	1913	New	(17)
No.45		0-6-0T	OC	HC	1101	1915	(u)	(28)
No.49		0-6-0T	OC	HC	1103	1915	(x)	Scr c/1961
No.50		0-6-0T	OC	HC	1153	1915		
			reb			1925	New	(7)
No.51		0-6-0T	OC	HC	1154	1915	New	(10)
No.52		0-6-0T	OC	HC	1155	1915	New	(12)
No.58		0-6-0T	OC	HC	1244	1917		
			reb			1927	New	(21)
No.59		0-6-0T	OC	HC	1245	1917	New	(20)
No.60		0-6-0T	OC	HC	1254	1917	New	(6)
No.61		0-6-0T	OC	HC	1255	1917		
			reb			1927	(d)	(14)
No.62		0-6-0T	OC	HC	1323	1918	New	(13)
No.63		0-6-0T	OC	HC	1324	1918		
			reb			1926	New	(9)
No.64		0-6-0T	OC	HC	1414	1920		
			reb			1927	New	(19)
No.65		0-6-0T	OC	HC	1415	1920		
			reb			1926	New	(29)
No.66		0-6-0T	OC	HC	1453	1921	New	(29)
No.68		0-6-0T	OC	HC	1455	1921	New	(29)
No.69		0-6-0T	OC	HL	3529	1922	(w)	Scr /1960
No.70		0-6-0T	OC	HL	3530	1922	(p)	(25)
No.71		0-6-0T	OC	HC	1596	1927	New	(26)
No.72		0-6-0T	OC	HC	1597	1927	New	Scr c/1961
No.73		0-6-0T	OC	HC	1454	1921		
	(No.67 until 1960)		reb			1928	New	(18)

No.73	0-6-0T	OC	HC	1598	1927	New	Scr /1960
No.76	0-6-0T	OC	HC	1719	1943	(f)	Scr c/1961
No.77	0-6-0T	OC	HC	1720	1943	(e)	Scr /1960
No.78	0-6-0ST	IC	HC	1748	1943	(j)	(29)
No.79	0-6-0ST	IC	HE	2414	1941	(g)	(32)
No.80	0-6-0ST	IC	HE	2876	1943	(k)	(33)
No.81	0-6-0ST	IC	HE	2881	1943	(l)	(27)
No.82	0-6-0ST	IC	RSHN	7103	1943	(i)	(34)
No.83	0-6-0ST	IC	RSHN	7104	1943	(m)	(31)
No.84	0-6-0ST	IC	RSHN	7105	1943	(h)	(27)
No.85	0-6-0ST	IC	RSHN	7107	1943	(n)	(27)
No.86	0-6-0ST	IC	RSHN	7113	1943	(o)	(31)
No.87	0-6-0ST	IC	HE	2878	1943	(r)	(31)
No.88	0-6-0ST	IC	HE	3166	1944	(s)	(31)
No.90	0-6-0T	OC	HC	1873	1954	(t)	(35)
No.91	0-6-0T	OC	HC	1874	1954	(v)	(30)
-	0-4-0ST	OC	P	2025	1942	(q)	(24)
No.200	0-6-0DE		YE	2619	1956	(y)	(47)
No.201	0-6-0DE		YE	2620	1956	(y)	(45)
No.202	0-6-0DE		YE	2630	1956	(z)	(47)
No.203	0-6-0DE		YE	2633	1957	(y)	(46)
No.204	0-6-0DE		YE	2640	1957	(y)	(45)
No.206	0-6-0DE		YE	2690	1959	New	(38)
No.207	0-6-0DE		YE	2691	1959	New	(37)
No.208	0-6-0DE		YE	2739	1959	New	(43)
No.209	0-6-0DE		YE	2740	1959	New	(41)
233	0-6-0DE		YE	2758	1959	New	(39)
234	0-6-0DE		YE	2759	1959	New	(44)
235	0-6-0DE		YE	2760	1959	New	(40)
236	0-6-0DE		YE	2761	1959	New	(48)
237	0-6-0DE		YE	2762	1959	New	(42)
238	0-6-0DE		YE	2763	1959	New	(36)
239	0-6-0DE		YE	2769	1959	New	(48)
240	0-6-0DE		YE	2770	1959	New	(45)

* HW = Harland & Wolff.

(a) ex London & India Docks Co, /1909.
(b) ex Millwall Dock Co, /1917.
(c) ex Millwall Dock Co, c/1923.
(d) ex West India Docks.
(e) ex WD 70071, loan, 7/1943; purchased 5/1946, originally WD Bicester, Oxfordshire.
(f) ex WD 70070, loan, 8/1943; purchased 5/1946,
originally COD, Old Dalby, Leicestershire.
(g) ex WD 70066, loan, 8/1943; purchased 5/1946.
(h) ex WD 75069, loan, 1/1944; purchased 5/1946.
(i) ex WD 75067, loan, 2/1944; purchased 5/1946, boiler to HE for overhaul by 11/7/1955.
(j) ex WD 75089, loan, 5/1944; purchased 5/1946, originally WD Kineton, Warwickshire.

(k) ex WD 75027, loan, 5/1944; purchased 5/1946, to Harland & Wolff, Tilbury for repairs 20/6/1945, to HE for overhaul and replacement boiler by 1/7/1957 and returned.
(l) ex WD 75032, loan, 5/1944; purchased 5/1946.
(m) ex WD 75068, loan, 5/1944; purchased 5/1946.
(n) ex WD 75071, loan, 5/1944; purchased 5/1946.
(o) ex WD 75077, loan, 5/1944; purchased 5/1946.
(p) ex Tilbury Docks 9/1944.
(q) ex Abelson, (Engineers) Ltd, Sheldon, Birmingham, hire, 12/1953.
(r) ex Tilbury Docks 7/1954, to HE for overhaul and new boiler by 1/5/1957 and returned.
(s) ex Tilbury Docks 11/1954.
(t) ex Tilbury Docks 1/1957.
(u) ex Tilbury Docks, 6/1957.
(v) ex Tilbury Docks 7/1957.
(w) ex Tilbury Docks, /1958.
(x) ex Tilbury Docks, /1959.
(y) ex Tilbury Docks, 10/1959.
(z) ex Tilbury Docks c/1968.

(1) to G. Shellabear & Sons Ltd, contractors, c/1926.
(2) sold for scrap, 11/1926.
(3) to A.R.Adams & Son, dealers, Newport, Gwent, /1927;
     thence to Thurrock Chalk & Whiting Co Ltd.
(4) to George Cohen, Sons & Co Ltd, /1931;
     thence to Fountain & Burnley Ltd, North Gawber Colliery, West Yorkshire.
(5) to George Cohen, Sons & Co Ltd, Canning Town, /1937 after 10/7/1937.
     thence to Colvilles Ltd, Rutherglen, Lanarkshire.
(6) to Tilbury Docks; returned 4/1/1942; to Tilbury 24/3/1944; returned 6/1956; scrapped, /1960.
(7) to MSC, loan, 4/1940; returned 3/1946; scrapped /1948.
(8) to MSC, loan, 7/1940; returned 2/1946;
     to Aberthaw & Bristol Channel Portland Cement Co Ltd, Rhoose, South Glamorgan, /1948.
(9) to MSC, loan, 7/1940; returned 4/1946; to Millwall Docks 10/1947; returned 11/1949, Wdn11/57, scrapped 10/1959, by George Cohen Sons & Co Ltd.
(10) to MSC, loan, 8/1940; returned /1945; to Tilbury Docks 10/1947; returned 5/1955, scrapped /1960.
(11) to MSC, loan, 8/1940; returned 1/1946; to Tilbury Docks, 7/1946.
(12) to MSC, loan, 11/1940; returned 12/1945; scrapped /1956.
(13) to Port of Bristol Authority, Avonmouth, Bristol, 26/11/1940; returned 1/1946, Scr /1960.
(14) to MDHB, loan, 1/1941; returned 11/1945; to HC, 4/1947, per G.E. Simm (Machinery) Ltd, thence to NCB, Bowes Railway, Co Durham, 11/1947.
(15) to MDHB, loan, 1/1941; returned 11/1945, to Tilbury Docks, 7/1946.
(16) to Millwall Docks, 9/1943.
(17) to Tilbury Docks, 5/1944.
(18) to Tilbury 5/1944 returned 1/1946, to Cox & Danks, for scrap, c4/1963.
(19) to Tilbury Docks 6/1944 returned 1/1946, scrapped by George Cohen, Sons & Co Ltd, /1953.
(20) to Tilbury Docks, returned, 1/1956, scrapped /1960.
(21) to Tilbury Docks 6/1944, returned /1958, Scr /1960.
(22) to Millwall Docks, 8/1944.

(23) to T. Hall & Sons (Llansamlet) Ltd, Llansamlet, West Glamorgan, /1948; thence to Dalton Main Colls Ltd, Silverwood Colliery, Yorks.
(24) returned to Abelson (Engineers) Ltd, Sheldon, Birmingham, 2/1954.
(25) scrapped on site by George Cohen, Sons & Co Ltd, 20/2/1956 – 27/2/1956.
(26) Wdn 9/1956, scrapped on site by George Cohen Sons & Co Ltd, 10/1959.
(27) Wdn /1958, scrapped on site by George Cohen Sons & Co Ltd, 10/1959.
(28) scrapped on site by George Cohen, Sons & Co Ltd, 10/1959 - 11/1959.
(29) to Cox & Danks Ltd, Bromley-by-Bow, for scrap, 30/4/1960; scrapped 5/1960.
(30) to Millwall Docks, 4/1960.
(31) to NCB, Ashington Colliery, Northumberland, 10/1960.
(32) to NCB, Ackton Hall Colliery, West Yorkshire, 10/1960.
(33) to NCB, Prince of Wales Colliery, North Yorkshire, 10/1960.
(34) to NCB, Hafod Colliery, Denbighshire 11/1960.
(35) to Cox & Danks, for scrap, c4/1963.
(36) to Millwall Docks, /1965.
(37) to Workington Iron & Steel Co, Moss Bay Works, Cumberland, 6/1970.
(38) to RTB, Redbourn Works, Scunthorpe, Lincolnshire, 17/6/1970.
(39) to GKN Ltd, Tremorfa Works, South Glamorgan, c/1971.
(40) to GKN, Tremorfa Works, South Glamorgan, 3/1971.
(41) to Shell Mex & BP Ltd, Killingholme, Lincolnshire, c3/1971.
(42) to GKN, Tremorfa Works, South Glamorgan, 4/1971.
(43) to British Steel Corporation, Tinsley Park Works, Sheffield, 5/1971.
(44) to Sheerness Iron & Steel Co Ltd, Sheerness, Kent, c5/1972.
(45) to GKN Ltd, Tremorfa Works, South Glamorgan, c5/1972.
(46) to GKN Ltd, Temorfa Works, South Glamorgan, 8/7/1972.
(47) to GKN Ltd, Tremorfa Works, South Glamorgan, 1/1973.
(48) to GKN Ltd, Tremorfa Works, South Glamorgan, /1973.

Loan & Hire locos

4385	0-6-0T	OC	Dav	2534	1943	(a)	(1)
4398	0-6-0T	OC	Dav	2547	1943	(a)	(1)
W.D.136	0-6-0ST	IC	HE	3172	1944	(b)	(2)
68574	0-6-0T	IC	Str		1896	(c)	(3)
68578	0-6-0T	IC	Str		1896	(c)	(3)
68644	0-6-0T	IC	Str		1912	(d)	(4)

(a) ex GWR South Wales, loan, 6/7/1944, on loan from USA/TC.
(b) ex WD Bicester, 8/1955.
(c) ex BR Stratford Shed, 5/3/1956.
(d) ex BR Stratford Shed, 16/7/1956.

(1) returned to USA/TC, shipped to France, 11/1944.
(2) returned to WD Bicester for overhaul 10/1956.
(3) returned to BR Stratford, 10/12/1956.
(4) returned to BR Stratford, after 30/3/1957.

TILBURY DOCKS M130

Engineering Workshops TQ 634763

No.	Name	Type	Cyl	Maker	Works No.	Year	Disp	Notes
No.17	BAT	0-6-0ST	IC	MW	694	1878	(a)	
								sold for scrap 10/1919
No.18	WASP	0-6-0ST	IC	MW	872	1883	(a)	sold /1919
No.19	FLY	0-6-0ST	IC	MW	873	1883	(a)	Scr /1919
No.20	AGENOR (No.8)	0-6-0ST	OC	RS	2982	1900		
	(name removed by /1911)	reb		Tilbury		1924	(a)	(3)
No.21	AJAX (name removed by /1911)							
		0-6-0ST	OC	RS	3053	1901	(a)	(1)
No.22	HECTOR (name removed by /1911)							
		0-6-0ST	OC	RS	3120	1904	(a)	(2)
No.23	JASON	0-6-0ST	OC	RS	3170	1905	(a)	(4)
No.24	NESTOR	0-6-0ST	OC	RS	3296	1907	(a)	sold c/1938
No.27	MOTH	0-6-0ST	IC	HE	234	1880	(a)	sold /1912
No.38		0-6-0T	OC	AB	1236	1911	(g)	Scr c/1951
No.39		0-6-0T	OC	AB	1237	1911	(g)	Scr c/1951
No.40		0-6-0T	OC	AB	1238	1911	New	(5)
No.44		0-6-0T	OC	AB	1302	1913	(c)	(10)
No.45		0-6-0T	OC	HC	1101	1915	New	(17)
No.46		0-6-0T	OC	HC	1102	1915	New	(7)
No.49		0-6-0T	OC	HC	1103	1915	New(b)	(20)
No.51		0-6-0T	OC	HC	1154	1915	(h)	(13)
No.57		0-4-0ST	OC	HL	3177	1916		
		reb				1926	(i)	(16)
No.58		0-6-0T	OC	HC	1244	1917		
		reb				1927	(f)	(19)
No.59		0-6-0T	OC	HC	1245	1917	(k)	(14)
No.60		0-6-0T	OC	HC	1254	1917	(k)	(6)
No.64		0-6-0T	OC	HC	1414	1920		
		reb				1927	(f)	(9)
No.73		0-6-0T	OC	HC	1454	1921		
	(No.67 until /1960)	reb				1928	(c)	(9)
No.69		0-6-0T	OC	HL	3529	1922	New	(19)
No.70		0-6-0T	OC	HL	3530	1922	New	(8)
No.87		0-6-0ST	IC	HE	2878	1943	(d)	(11)
No.88		0-6-0ST	IC	HE	3166	1944	(e)	(12)
No.90		0-6-0T	OC	HC	1873	1954	New	(15)
No.91		0-6-0T	OC	HC	1874	1954	New	(18)
No.200		0-6-0DE		YE	2619	1956	New	(21)
No.201		0-6-0DE		YE	2620	1956	New	(21)
No.202		0-6-0DE		YE	2630	1956	New	(23)
No.203		0-6-0DE		YE	2633	1957	New	(21)
No.204		0-6-0DE		YE	2640	1957	New	(21)
No.205		0-6-0DE		YE	2641	1957	New	(24)
210		0-4-0DE		YE	2853	1961	(j)	(26)
211		0-4-0DE		YE	2854	1961	(j)	(25)
230		0-6-0DE		YE	2755	1959	New	(22)
231		0-6-0DE		YE	2756	1959	New	(27)
232		0-6-0DE		YE	2757	1959	New	(27)

(a)   ex London & India Docks Co, /1909.
(b)   to MSC, loan, 9/1940; returned 1/1946.
(c)   ex Royal Docks, 5/1944.
(d)   ex WD 75029, loan, 5/1944; purchased 5/1946.
(e)   ex WD 75116, loan, 5/1944; purchased 5/1946.
(f)   ex Royal Docks, 6/1944.
(g)   ex Royal Docks, 7/1946.
(h)   ex Royal Docks, 10/1947.
(i)   ex Millwall Docks, 1/1956.
(j)   ex Millwall Docks, /1965.
(k)   ex Royal Docks.

(1)   scrapped by George Cohen, Sons & Co Ltd, /1932.
(2)   to George Cohen, Sons & Co Ltd, /1932;
      thence to Madeley Collieries Ltd, Staffordshire, /1933, after rebuild by YE.
(3)   scrapped by Thos. W. Ward Ltd, /1934.
(4)   to Stanton Ironworks Co Ltd, Market Overton Mines, Rutland,
      via George Cohen, Sons & Co Ltd and RSH, 6/1937.
(5)   to MSC, loan, 9/1940; returned 1/1946; scrapped by Thos. W. Ward Ltd /1952.
(6)   to Royal Docks, 4/1/1942, ex Royal Docks, 24/3/1944, to Royal Docks, 6/1956.
(7)   to HC, 4/1942; thence WD Chillwell, Nottinghamshire, WD 211, later 70211.
(8)   to Royal Docks, 9/1944.
(9)   to Royal Docks, 1/1946.
(10) scrapped by George Cohen, Sons & Co Ltd, 2/1953.
(11) to Royal Docks, 7/1954.
(12) to Royal Docks, 11/1954.
(13) to Royal Docks, 5/1955.
(14) to Royal Docks, 1/1956.
(15) to Royal Docks, 1/1957.
(16) to Millwall Docks, 1/1957.
(17) to Royal Docks, 6/1957.
(18) to Royal Docks, 7/1957.
(19) to Royal Docks, /1958.
(20) to Royal Docks, /1959.
(21) to Royal Docks, 10/1959.
(22) to Millwall Docks, /1965.
(23) to Royal Docks, c/1968.
(24) to Shell-Mex & BP Ltd, Purfleet, c/1970.
(25) to Tunnel Portland Cement Co Ltd, West Thurrock, c3/1970.
(26) to British Steel Corporation, Brymbo Works, Denbighshire, 15/12/1971.
(27) to GKN Ltd, Tremorfa Works, South Glamorgan, c4/1973.

Locomotives have also been borrowed from BR and its predecessors from time to time: also 0-6-0T from USA/TC as follows: 1305, 4375 (6/1944 - 9/1944), 4385/98, 6165 (7/1944 - 9/1944) : and 0-6-0ST from WD : 108 (12/1952 - 9/1954), 106 (2/1954 - 7/1954), 136 (7/1955 - /1956).

PRINTAR INDUSTRIES LTD
SILVERTOWN WORKS, Prince Regent's Wharf, Silvertown B131
Subsidiary of **Burt Boulton & Haywood Ltd** TQ 413800
Prince Regent Tar Co Ltd from 4/7/1934 until c/1959
Company registered in 1934 as tar distillers also with a works in Hertford. The works in existence by 1882, but was destroyed during WW2; it was rebuilt in 1947, closed /1969 and demolished by 12/1971.

Gauge : 60cm

		4wPM	FH	1738	1931	New	(1)

(1) destroyed in air raid.

Gauge : 4ft 8½in

		0-4-0F	OC	WB	2851	1947	New	Scr c11/1969

PROCTOR & GAMBLE LTD
WEST THURROCK WORKS N132
Thomas Hedley & Co Ltd until /1962 TQ 595773
Chemical works built in 1938, for this company of soap manufacturers. Located on the south side of the ex-LT&SR line about a mile west of Grays Station and served by standard gauge sidings connecting with this line. Earlier shunted by Thames Land Co Ltd, rail traffic ceased 1994 (after18/5/1994 by 17/8/1994).

Gauge : 4ft 8½in

		0-4-0F	OC	AB	1472	1916	(a)	(1)
		0-6-0F	OC	WB	2370	1928	(b)	(2)
		4wDH		TH	144v	1964	(c)	(3)

(a) ex London & Thames Haven Oil Wharves Ltd, c2/1955.
(b) ex BP Chemicals (UK) Ltd, Salt End, Hull, East Yorkshire, 12/6/1972.
(c) ex TH, 20/12/1978; originally British Steel Corporation, Port Clarence, Co. Durham.

(1) to Alan Bloom, Bressingham Steam Museum, Norfolk, c12/1978.
(2) to North Norfolk Railway Co Ltd, Norfolk, 12/1979.
(3) to Industrial Chemicals Ltd, Grays, c 4/1996 (after 13/2/1996).

PURFLEET CHALK QUARRIES LTD
PURFLEET CHALK QUARRY N133
TQ 557782

Very old quarry at one time owned by W.H. Whitbread but by 1826 being worked by Meeson, Hinton & Co. Horse worked tramways installed about 1807 and the workings were extensive in 1848. Closed some years before 1863; possibly later worked by Gibbs & Co and certainly then by Purfleet Chalk Quarries Ltd (incorporated 10/4/1902). The latter company was Danish controlled with the main shareholders listed as Charles Otto Trechmann and Otto Kramer Trechmann, listed as Cement Manufacturers of West Hartlepool. The company was formed to acquire and take over a lease dated 18/4/1901 between Trechmann and Samuel Whitbread for land at Purfleet. The quarries supplied chalk to the Trechmann Weekes Works in Kent. Plant, including the two locomotives, 100 tip wagons and two miles of track, was auctioned on 28/5/1920. The company went into voluntary liquidation on 1/6/1920; the site was acquired by Harrisons (London) Ltd.

Reference: Industrial locomotive No.25 (Spring 1982) - Thameside Industrials, Frank Jux. Railway Magazine, April 1950

Gauge : 3ft 6in

Two 10hp petrol locomotives of this gauge were included in the auction of 28/5/1920. (believed to be of German origin with 2cyl engines). Dealer J.F.Wake of Darlington listed two 3ft 6in gauge petrol locomotives in his stock shortly after this date and advertised for sale 16/7/1920, so may have been the purchaser.

PURFLEET THAMES TERMINAL LTD
PURFLEET WHARVES   N134

Purfleet Deep Wharf & Storage Co Ltd until 15/2/1991   TQ 565776
Purfleet Wharf & Sawmills Ltd until 1941
Millar's Karri & Jarrah Forests Ltd registered 24/4/1897 until 24/12/1901

Wharves operated from about 1898 by **Millar's Karri & Jarrah Forests Ltd** who owned forest concession in Australia. A new company **Purfleet Wharf & Sawmills Ltd** (a subsidiary of Millar's was registered 24/12/1901) and was formed to take over the Purfleet operation of Millar's Karri & Jarrah Forests Ltd and to carry on the business of importing timber and operators of sawmills. The wharves were served by standard gauge rail system on the south side of the ex-LT&SR line east of Purfleet Station. The site taken over as WD Purfleet 1914-19. Rail traffic ceased, c/1992.

Gauge : 4ft 8½in

	Name	Type						
	DENMARK	0-4-0ST	OC	MW	1442	1899		
		reb		MW		1913	New(a)	Scr /1934
No.4	DENMARK	0-4-0ST	OC	MW	1619	1904	New(a)	
	(MORNINGTON until /40)	reb		Purfleet $		1940		(5)
No.1	(PURFLEET)	0-4-0ST	OC	TW	420	1876	(b)	
		reb		HE *		1898		
		reb		HE		1928		Scr /1957
	LORD MAYOR	0-4-0ST	OC	HC	402	1903	(c)	(3)
No.5	(MORNINGTON)							
	(ROYAL)	0-6-0ST	OC	AB	1577	1918	(d)	Wdn /1960, Scr
	DARENTH	0-4-0ST	OC	P	1741	1937	(e)	(1)
No.1	(No.2)	0-6-0ST	OC	HC	823	1908	(f)	
	CHURCHILL	reb		Purfleet #		1952		(2)
No.2	(PURFLEET)	0-4-0ST	OC	P	1735	1927	(g)	Wdn /1956, Scr 3/1961
No.3		0-4-0ST	OC	P	1571	1921	(h)	Wdn 3/1961, Scr /1963
	STAR	0-4-0ST	OC	AB	1940	1927	(j)	(4)
	M F P No.3	0-4-0DM		JF	4210140	1958	(k)	(7)
	M O P No.6	0-4-0DM		JF	4210143	1958	(l)	(6)
1		0-4-0DH		RH	437362	1960	New	(9)
2		0-6-0DH		RH	457303	1963	New	(9)
3		0-4-0DH		RH	512463	1965	New	(9)
4		0-4-0DH		RH	512464	1965	New	(9)

5		0-4-0DM	DC	2589	1957		
			RSHD	7922	1957	(m)	(11)
2		0-4-0DE	RH	412716	1957	(n)	(8)
6		0-4-0DM	VF	D297	1956		
			DC	2583	1956	(o)	(10)

*   rebuilt from 0-4-0WT.
\#   rebuilt with boiler and tank of No.1 (TW /76).
$ rebuilt with parts of MW 1442/99, and subsequently carrying plates "reb MW 1442/13".

(a)   to Royal Engineers, Purfleet on loan /1914, returned /1919.
(b)   ex WD, Purfleet, /1919.
(c)   ex Chas. Wall Ltd, contractor, Grays, loan, /1934.
(d)   ex Thos. W. Ward Ltd, Grays, /1936, TW494. earlier National Filling Factory, Pembrey, Dyfed.
(e)   ex BPCM, Johnson's Works, Greenhithe, Kent, loan, /1940.
(f)   ex Thos. W. Ward Ltd, Grays, 12/1945; earlier East Kent Railway, Kent, 2.
(g)   ex Anglo-American Oil Co Ltd, Purfleet, by 23/6/1949.
(h)   ex Anglo-American Oil Co Ltd, Purfleet, /1951.
(j)   ex Thurrock Chalk & Whiting Co Ltd, loan, 4/1959.
(k)   ex Ministry of Power, Goostrey, Cheshire, loan, 5/1959.
(l)   ex Ministry of Power, loan, 6/1959.
(m)   ex South Eastern Gas Board, Dover Gas Works, Kent, 8/1970.
(n)   ex Tunnel Cement Ltd, West Thurrock, c9/1977.
(o)   ex Berry Wiggins & Co Ltd, Kingsnorth, Kent, 19/2/1979.

(1)   returned to BPCM, Johnson's Works, Greenhithe, Kent, off loan, /1945.
(2)   scrapped 7/1957 by George Cohen, Sons & Co Ltd.
(3)   returned to Chas. Wall Ltd, Grays.
(4)   returned to Thurrock Chalk & Whiting Co Ltd, 5/1959.
(5)   scrapped on site, 3/1965.
(6)   to Ministry of Power, Portishead, Somerset, c5/1965.
(7)   to Shell Refining Co Ltd, Ardrossan, Ayrshire, /1967.
(8)   to P. W. Leeman Ltd, East Tilbury, /1978, by 12/11/1978.
(9)   to Essex Iron & Steel Co Ltd, Mountnessing , c6/1992.
(10)   to East Kent Light Railway, Shepherdswell, Kent, c9/1992 by 20/9/1992.
(11)   to Mid Norfolk Railway, Dereham, Norfolk, 1/10/1992.

JOSEPH RANK LTD
PREMIER MILLS, Royal Victoria Dock   B135
TQ 411804

In 1916 Hugh Wood & Co of Gateshead, offered the Port of London Authority one months free trial of a battery electric loco manufactured by the Jeffrey Manufacturing Co of America, the cost of the loco quoted as £2,200. However due to the war the trial was delayed, it finally took place at the Royal Victoria Dock in 1919 on the premises of Joseph Rank Ltd, Millers, where haulage was done with horses, however the trial proved unsatisfactory for the heavier traffic in the docks.
Gauge : 4ft 8½in

-		4wBE	Jeffrey	5136	1919	New(a) (1)

(a) supplied by Hugh Wood & Co, Gateshead, Co. Durham, on trial, /1919.
(1) if identified correctly, later sold to Mather Platt Ltd, Manchester, /1919.

RANSOME HOFFMANN POLLARD LTD
CHELMSFORD WORKS G136
Hoffmann Manufacturing Co Ltd until 1/1/1970   TL 710076

Charles Arthur Barrett was a partner in R. Barrett & Sons, Iron and Brass founders of Beech Street, Barbican. In the early 1890s Mr Barrett commenced manufacture of cycle bearings and carriage hubs under the name of Preston Davis Ball Bearing Co, as an off shoot to the Westminster Engineering Co who at that time were manufacturing presses on a site at Springfield Wharf, Chelmsford. In 1896 Geoffrey F. Barrett joined his cousin at Chelmsford where the Preston Davis Ball Bearing Co, were experiencing technical difficulties. In 1897 Mr G.F. Barrett visited E.G. Hoffmann in New York, as a result the Hoffmann Manufacturing Company was registered 11/1/1898 jointly by E.G. Hoffmann and R. Barrett & Sons Ltd to acquire the patents covering manufacture of steel balls held jointly by E.G. Hoffmann & The American Machine Co. Construction work started on a small factory situated on four acres of land for the production of steel balls mainly for bicycles and was completed in 1899. In 1914 the factory extended considerably and had by then a labour force of 1,800. A further factory was erected in 1917 when the company was purchased by the government and nationalised in 9/1917 as MoM National Ball Bearing Factory No.156; it expanded during the latter part of the First World War to cope with demand from the MoM. The contractor Henry Martin of Northampton brought in a small steam loco to haul materials for construction of the works. About 1920 roller bearing axleboxes were introduced for tramcars and soon afterwards similar boxes supplied for railway service. In 1923 the government sold its interest in the company to the United Steel Co Ltd and the Patent Tyre Co Ltd, five years later in 1928 the share capitol was acquired by Brown Bayley Steel Works Ltd of Sheffield.

The company became Ransome Hoffmann Pollard Ltd on 1/1/1970 (an amalgamation of Ransome & Marles Bearing Co Ltd, Hoffmann Manufacturing Co Ltd & Pollard Bearings Ltd), in 1987 the company became United Precision Industries. Rail traffic ceased and works closed 30/9/1989.

Gauge : 4ft 8½in

- 0-4-0ST OC MW 814 1882 (a) s/s
- 4wDM(ex PM) Lake & Elliot c1923 (b) s/s c/1968

(a) earlier Lucas & Aird, Hull & Barnsley Railway (1880-1886) contract, East Yorkshire.
(b) ex Lake & Elliot Ltd, Braintree, c/1959.

REDLAND-INNS GRAVEL LTD
NAZEING PITS E137
Inns & Co Ltd until 12/1965   TL 377056

Narrow gauge system served gravel pits on land at Langridge Farm in the Lea Valley about a mile south of Broxbourne, listed in HMSO list of quarries for 1948. The tramway was closed 25/10/1972 and replaced by road transport.

Gauge : 2ft 0in

		4wDM	MR	7146	1936	(a)	(2)
	-	4wDM	MR	7306	1938	(b)	(4)
	-	4wDM	MR	7467	1940	(b)	(1)
2		4wDM	MR	7040	1947	(c)	(3)
3		4wDM	MR	8675	1941	(d)	(4)
	-	4wDM	MR	7143	1936	(e)	(4)
	-	4wDM	MR	7073	1941	(f)	s/s c/1973
	-	4wDM	MR	7441	1939	(g)	s/s c/1970
	-	4wDM	MR	7358	1939	(g)	(4)
	-	4wDM	MR	7380	1939	(h)	(4)
	-	4wDM	MR	7398	1939	(j)	s/s c/1973
	-	4wDM	MR	7456	1939	(j)	(4)
	-	4wDM	MR	5933	1935	(j)	(5)

(a) originally Diesel Loco Hirers Ltd, Bedford;
thence to Lennox & Co Ltd (location unknown), by 5/1951, here by 4/1953.
(b) originally Diesel Loco Hirers Ltd, Bedford;
to Inns & Co Ltd, (location unknown), by 6/1943, here by 4/1953.
(c) ex Little Mardyke Pits, c/1961.
(d) ex Little Mardyke Pits, c/1964, by 5/1965.
(e) ex Waterford Workshops, Hertfordshire, by 5/1965;
earlier Bradgate Granite Quarries Ltd, Leicestershire.
(f) ex Waterford Workshops, Hertfordshire, c/1964, by 5/1965;
earlier Sawbridgeworth Pits, Hertfordshire.
(g) ex Sparham Sand Pits, Norfolk, c/1967 (after 9/1966).
(h) ex Broxbourne Pits, Hertfordshire, after 11/1968 by 8/1971.
(j) ex Broxbourne Pits, Hertfordshire, c9/1971.

(1) to Meering Pits, Besthorpe, Nottinghamshire, c/1959.
(2) to Waterford Workshops, Hertfordshire. c/1964.
(3) s/s after 11/9/1971.
(4) exported to Singapore, 1/1973.
(5) to Waterford Workshops, Hertfordshire, /1973.

P.T. READ (FAIRLOP) LTD
FAIRLOP GRAVEL PITS, Barkingside

D138
TQ 455900

Company registered on 28/11/1956 as quarry owners with an address at 24 High Street E11(Wanstead) and listed in quarry directories for 1959/60 but not in 1957/58 with pits in Foxlands Crescent, Dagenham, TQ 508857 and at Fairlop Airfield, Ilford by 1963/64. Loco operation ceased c/1963. The company went into voluntary liquidation on 24/3/1986.

P. T. Read Ltd was registered 5/9/1946 at Downs Works, Downs Road, Clapton and listed as carriers, haulage and road contractors.

Gauge : 2ft 0in

1	JEFF	4wDM	MR	11090	1957	New	(1)
2		4wDM	MR	11110	1961	New	(1)
3		4wDM	MR	11091	1957	New	(1)

(1) scrapped on site by company's own labour, 3/1968.

EDMUND ROCHFORD LTD
STANSTEAD NURSERIES F139
Edmund Rochford until 25/11/1911   TL 510240
Nurseries established by Edmund Rochford in the late nineteenth century and shown on the 25in OS map for 1897. A rope hauled wagon way ran from the nurseries at the Mount down an incline to Stony Common where an engine house with a Davy Paxman stationary steam engine was housed. The wagon way between the extensive glasshouses was probably hand worked until the introduction of loco haulage, remains of track were still in existence in the late 1990s. The company went into voluntary liquidation on 5/7/1973; site cleared and redeveloped for housing by 2009.

Gauge : 2ft 0in

	4wPM	FH	1830	1933	New	(1)
-	4wDM	FH	3075	1945	New	(2)

(1) to M.E. Engineering Ltd, Cricklewood, Middlesex, via Cox & Danks Ltd, 3/1956, later with Boothby Peat Co Ltd, Naworth Castle, Brampton, Cumberland, 12/1957.
(2) to M.E. Engineering Ltd Cricklewood, Middlesex, via Cox & Danks Ltd, 3/1956, later Borough of Cheltenham, Hayden Sewage Works, Cheltenham, Gloucestershire, 3/1958.

ROM RIVER CO LTD
WITHAM WORKS G140
TQ 826146

Company incorporated 25/8/1920 with a registered office at Rainham Road, Romford, went into liquidation in 4/1926. A new company of the same name was registered 4/5/1926 to acquire the assets from the liquidator. Standard gauge sidings served the steel stockholding factory operated by these steel reinforcement engineers.

Gauge : 4ft 8½in

4wDM	FH	3491	1951	(a)	Scr 12/1982
6wDM	KS	4421	1929	(b)	(1)

(a) ex North Thames Gas Board, Kensal Green Gasworks, London, c5/1969.
(b) ex Lichfield Works, Staffordshire, 8/1979.

(1) returned to Lichfield Works, Staffordshire, by 3/1982.

ROM RIVER SAND & GRAVEL PITS
BAKER STREET GRAVEL PITS, Orsett near Grays L141
Alfred Booth & Co Ltd, proprietors   TQ 638810
Pits opened by 1929 and listed in quarry directories for 1933. The company also operated pits in Rainham Road, Romford TQ 521834 approx, listed in 1925, and Poles Wood, Kelvedon Common in 1928 & 1931 but not 1934, TQ 571996 approx. HMSO list of quarries for 1922 also show **Rom River Co Ltd** as operators of Chalk & Clay pits in Saffron Walden. Rail system abandoned, but pits still listed in directories for the 1960s.

Gauge : 2ft 0in

-	4wDM				(a)	s/s /1941
-	4wDM	MR	5909	1934	New	s/s /1941
-	4wDM (ex PM)	MR	476	1917	(b)	s/s

(a) origin and identity unknown.
(b) ex Motor Rail Ltd, Bedford, 2/1934; earlier WDLR 2197.

ROOKERY SAND & GRAVEL PIT
LEYTON
C142
TQ 369867

Pit listed in the quarry directory for 1933 and located adjacent Temple Mills yard on the GE Stratford to Tottenham Hale railway.

"Contract Journal", 15/8/1934 – for sale, plant including Lister 5 ½ hp Auto Rail Truck & track, 2ft gauge – Rookery Gravel Pits, Seymour Road, Leyton.

Gauge : 2ft 0in

-	4wPM	L	(a)	s/s

(a) origin and identity unknown.

ROTAVATORS LTD
WEST HORNDON WORKS
L143

Rotary Hoes Ltd. until 1974
TQ 622880

Rotorvators Ltd was a subsidiary company of Rotary Hoes Ltd (registered 18/7/1938) to carry on the business of manufacturers of Howard Rotary Hoes and cultivators. The company changed its name in 1974 to Howard Machinery Ltd. The works were on the north side of the ex-LT&SR line west of West Horndon Station and were served by standard gauge sidings with a connection via the sidings of Brown & Tawse Ltd. Works closed /1975 and production moved to Ipswich, Suffolk.

Gauge : 4ft 8½in

-	4wDM	Brown & Tawse	(a)	Scr c11/1975

(a) ex Brown & Tawse Ltd, West Horndon; converted from PM c/1963

ROWHEDGE SAND & BALLAST CO LTD

BIRCH BROOK PITS, near Rowhedge
H144
TM 022220

Company registered 29/6/1932, was taken over in 1946 by F.A. Hunnable and Sons of Braintree; this firm was taken over in 1968 by Consolidated Gold Fields Ltd. Pits not listed in directories before 1948. The railway was replaced by road transport c1963. Company went into voluntary liquidation 6/8/1964.

Gauge : 2ft 0in

-	4wDM	RH	175399	1935	(a)	(1)

(a) ex Rowhedge Pits.
(1) to Rowhedge Pits, c/1963.

PLACE FARM PITS, near Old Heath, Rowhedge   H145
TM 022232

Pits located on land adjacent to Place Farm at Old Heath and near to the River Colne. Pits and tramway closed. Site cleared by 9/1967.
Gauge : 2ft 0in

-	4wDM	RH	187101	1937	(a)	(1)

(a) ex Rowhedge Pits, c/1963.

(1) to S. P. Wheeler & Sons, dealers, Colchester, for scrap, c/1966.

ROWHEDGE PITS, Rowhedge   H146
TM 030212

Sand and gravel pits listed in directories for 1934 (but not 1933), served by an extensive narrow gauge rail system. Rail traffic ceased c1963 but much of the track was not lifted until after 1967. By 1967 production had ceased but the site was retained for offices and a plant Depot.
Gauge : 2ft 0in

-	4wPM	HC	P251	1925	(a)	(1)
-	4wDM	RH	172889	1934	(b)	(3)
-	4wDM	RH	178256	1936	New	(3)
-	4wDM	RH	187101	1937	New(c)	(2)
-	4wDM	RH	175399	1935	(d)	(3)

(a) ex Leeds Corporation City Engineers Dept, Leeds, by 8/1933.
(b) ex Winfield Bros (Bourne End) Ltd, Pix Farm Lane Quarries, Bourne End, Hertfordshire, c6/1943 (after 1/12/1942 by 29/6/1943).
(c) ex RH, Lincoln, 24/12/1937; earlier an exhibit at the Public Works, Roads & Transport Exhibition, Royal Agricultural Hall, London, from 15/11/1937 to 20/11/1937.
(d) ex RH, Lincoln, 17/11/1936; earlier on hire to Strakers & Love Ltd, Brancepeth Brickworks, Co. Durham, returned to RH, Lincoln, 13/10/1936; to Birch Brook Pits; returned c/1963.

(1) to George Cohen, Sons & Co Ltd, dealers, /1935;
     thence to Surrey County Council, Highways Dept, Guildford By-Pass contract, Surrey.
(2) to Place Farm Pits, c/1963.
(3) to S.P. Wheeler & Sons, dealers, Colchester, for scrap, c/1966.

ST. ALBANS SAND & GRAVEL CO LTD
MEADGATE FARM PITS, Nazeing   E147
TL 385072

Company registered 25/2/1930. Pits in Nazeing were listed in HMSO list of quarries for 1934 as in Sedge Green Lane. Rail traffic ceased and track lifted c1964. Plant closed and demolished c9/1966.

Gauge : 2ft 0in

-		4wDM	MR	5875 1935	(a)	(2)
R 8		4wDM	MR	5612 1931	(b)	(3)
-		4wPM	MR	7035 1936	(c)	(1)
RB13		4wPM	MR	5289 1931	(d)	(5)
R11	(f. R/10)	4wDM	MR	5345 1931	(e)	(3)
RS4		4wDM	MR	5855 1934	(f)	(3)
R/10		4wDM	MR	8565 1940	(f)	(3)
RB2	(f. RS2)	4wDM	MR	7228 1938	(g)	(6)
RB7		4wDM	MR	5613 1931	(h)	(5)
RB 14	(f. RS 14)	4wDM	MR	8788 1943	(j)	(4)

(a) ex Petrol Loco Hirers Ltd, Bedford, 28/6/1935.
(b) ex Hussey Egan & Pickmere Ltd, contractors, by 6/1937.
(c) originally Petrol Loco Hirers Ltd, Bedford, by 7/1952.
(d) ex A.E. Farr Ltd, contractors, by 5/1953;
    to Smallford Pits, Hertfordshire. /1954 by 12/1954; returned /1955 by 11/1955.
(e) ex Marston Valley Brick Co Ltd, Bedfordshire, by 12/1954.
(f) ex Smallford Pits, Hertfordshire, by 12/1954.
(g) originally Diesel loco Hirers Ltd;
    to St.Albans Sand & Gravel Co Ltd (location unknown), 22/7/1939, here by 12/1954.
(h) ex Smallford Pits, Hertfordshire. c/1957, by 11/1957.
(j) originally Admiralty (intended for Alexandria, Egypt), here by 11/1957.

(1) to Landbeach Sand & Gravel Co Ltd, Cambridgeshire, c/1952.
(2) to Smallford Pits, Hertfordshire, by 5/1953.
(3) to Smallford Pits, Hertfordshire, /1956.
(4) s/s c/1965 after 3/1965.
(5) to Smallford Pits, Hertfordshire. c/1964, by 4/1965.
(6) to Riverside Works, Hertfordshire. c/1965, by 3/1966.

SALVATION ARMY
LAND & INDUSTRIAL COLONY, Hadleigh near Benfleet   J148
TQ 807864

In 1891 the Salvation Army purchased land, which eventually extended to 3200 acres around Hadleigh Castle, to provide rehabilitation work for deprived persons. Extensive narrow gauge tramways were laid to move agricultural produce and bricks. There were three brickworks the first was south west of Home Farm Colony and north of the LT&SR railway at TQ 804859, in existence from 1892 to c1914 or later, the second was south of Home Farm Colony at TQ 807859 in existence during the same period, the third was west of Snipes Lane at TQ 800867 in production from 1898 until 1957, this works had a Hoffmann 24 chamber kiln with a capacity for 80,000 bricks, a tramway ran between the clay pits and the brick works. A surviving photograph shows a tramway in a clay pit to be about 3ft gauge and utilising hand worked wooden wagons. In 1892 a locomotive worked standard gauge line was built to convey bricks made at brickworks north of the LT&SR Pitsea - Southend line, over this railway for shipment at a wharf on Hadleigh Bay. The railway closed in 1914 although the area was still used for farming by the Salvation Army until after 1972.

Reference: Industrial Railway Record No.93, June 1982.

Gauge : 4ft 8½in

	0-4-0ST	OC	P	596 1894	New	(1)

Note : A locomotive named PROSPERITY, details unknown, was reported here from 1893.

(1) to MoM, Woolston, Hampshire, c/1914.

SANDERS & FORSTER LTD
BRIDGWATER ROAD FACTORY, Stratford   C149
TQ 382836

Business established in 1888, the company being registered 17/1/1916 at the Royal Works, 27 Union Road, Rotherhithe; listed in 1938 as Constructional Engineers with an address as Thames Works, Barking; a subsidiary of Chamberlain Group Ltd. An engineering works, with an internal narrow gauge system about 420 yards in length. Built 1954-1955, the line ran between the stockyards, fabricating and paint shops. Located on the south side of the ex-GER line west of Stratford Station. Rail traffic ceased 10/1980. Sanders & Forster (Steel Stockholders) Ltd, went into voluntary liquidation on 23/11/1982.

Gauge : 2ft 0in

-	4wDM	RH	164342 1932	(a)	Scr 1/1963
-	4wDM	OK	6194 1935	(b)	(1)
-	4wDM	FH	4008 1963	New	(6)
-	4wDM	FH	2834 1943	(c)	(2)
-	4wDM	L	25919 1944	(d)	(3)
-	4wDM	RH	226294 1944	(e)	(4)
	4wDM	RH	209429 1942	(f)	(5)
-	4wDM	HE	6285 1968	(g)	(7)

(a) ex L. E. Stubbs & Co Ltd, 2c Well Street, Porthcawl, Mid Glamorgan, South Wales, after 9/1946, by 12/1956.
(b) ex Wm. Jones Ltd, dealers, East Greenwich, hire, /1962, by 6/1962.
(c) ex M.E. Engineering Ltd, Cricklewood, Middlesex, hire, c7/1972, whilst M.E. Engineering repaired FH 4008.
(d) ex M.E. Engineering Ltd, Cricklewood, Middlesex, hire, 10/12/1973.
(e) ex M.E. Engineering Ltd, Cricklewood, Greater London, hire, /1976.
(f) ex M.E. Engineering Ltd, Cricklewood, Greater London, hire, c11/1976.
(g) ex Walker & Partners Ltd, dealers, Staveley, Derbyshire, 12/1976;
originally Millom & Askam Hematite Iron Co Ltd, Florence Mine, Cumberland.

(1) returned to Wm. Jones Ltd, /1963.
(2) returned to M.E. Engineering Ltd, off hire, by 10/1972.
(3) returned to M.E. Engineering Ltd, by 17/1/1974.
(4) returned to M.E. Engineering Ltd, off hire, /1976.
(5) returned to M.E. Engineering Ltd, off hire, c/1977.
(6) to J. Ewing, Surrey, for preservation, 1/2/1981.
(7) to Welsh Highland Light Railway (1964) Ltd, Porthmadog, Gwynedd, 14/2/1981.

SEABROOKE & SONS LTD
THURROCK BREWERY, Grays   M150
TQ 621775

Company incorporated 4/3/1891 to take over existing business carried on by Charles Seabrooke, Thomas William Seabrooke & Jonathan Seabrooke as brewers, coal & coke merchants and general wharfingers. The company went into voluntary liquidation 26/4/1934, the undertaking and assets were acquired by Charrington & Co Ltd. Premises taken over by Grays Co-operative Society Ltd.

Gauge : 4ft 8½in

THOR (originally PETROS)	0-4-0ST	OC	HE	629	1896		(1)
	reb		Seabrooke		1921	(a)	

(a)   ex BPCM Wouldham Works, by 11/1918.
(1)   to H. Covington & Sons Ltd, Wennington, after 3/1927, by /1929.

JOHN SHELBOURNE & CO
RAINHAM RUBBISH SHOOT   P151
TQ 505814 approx

John Shelbourne of Nos.66 & 72 Narrow Street, Limehouse, Middlesex, traded as Lighterman, Steam Dredger, Excavator, and Stone and Ballast contractor, trading as Edward and John Shelbourne. John Shelbourne was adjudicated bankrupt on 6/8/1880 but was discharged on 11/1/1887. John Shelbourne & Co a partnership between Alfred Augustus Taylor and Frank William Bateman at No.70 Fenchurch Street in the City of London, as contractors under the style or firm of John Shelbourne and Co, was dissolved on 31/12/1886. Site acquired by Wm. Cory & Son Ltd by 1909.

Gauge : 3ft 0in

JUBILEE	0-4-0IST	OC	WB	840	1887	(a)	(1)
No.2	0-4-0ST	OC	HC	696	1904	New	(2)

(a)   ex John Fell, Blenheim Palace lake cleaning contract, Woodstock, Oxfordshire, c/1897.
(1)   advertised for sale, 4/1904; s/s.
(2)   to Wm. Cory & Son Ltd, Rainham, with site, by /1909.

SHELL U.K. LTD
SHELL HAVEN REFINERIES, Thames Haven, Stanford-le-Hope   K152
Shell Refining Co Ltd until 1972   East Haven Steam Loco Shed   TQ 737818
Shell Refining & Marketing Co Ltd until 1957   West Haven Diesel Loco Shed TQ 720816

Shell was founded in the 1890s; The Shell Transport & Trading Co Ltd was set up in 1897 (incorporated 18/10/1897); in 1903 a working agreement came into effect with the Royal Dutch Petroleum Co and all Shell companies have since been jointly UK/Dutch owned. In 1903 a joint marketing company The Asiatic Petroleum Co (incorporated 29/6/1903) was formed; in 1907 two new operating companies were set up (one in each country). The UK operating company was The Anglo-Saxon Petroleum Co Ltd (registered 29/6/1907), it was this company which owned the Thames Haven site. In 1911 Anglo-Saxon bought part of Oil Mill Farm between Thames Haven and Shellhaven Creek, a pier was built and in 1914-16 a refinery and tank farm constructed. When commissioned in 1916 the refinery produced fuel oil for the Admiralty. In 1919 it was modified to produce bitumen, further improvements enabled the production of printing ink, luboils and asphalts. In 1946 Asiatic Petroleum Co changed its name to Shell Petroleum Co Ltd. In 1948 Shell decided to build a crude oil refinery on unused marshland to the west of the LATHOL installations, this became known as the 'West Site' with the original Shell installations by Thames Haven Station becoming the 'East Site'. The new refinery with a capacity of 2 million tons per annum was brought into use in 1950; in 1959 a fertilizer complex incorporating an ammonia plant and nitric acid plant was commissioned. In 1969 the Shell Refining Co purchased the LATHOL tankage and jetty facilities at Thames Haven for £21m. By 1992 the plant had a capacity of 4.6 million tons per annum, and covered 2,000 acres, with a 17mile perimeter. It had five jetties, handling tankers of up to 300,000 tons capacity delivering products by road, ship and the UK oil pipeline network; and was served by standard gauge sidings connecting with the ex-LT&SR Thames Haven Branch. The refinery ceased production in 12/1999 and was decommissioned and demolished, apart from the bitumen plant and some storage tanks. In 2006 the site was purchased by DP World, one of the largest marine terminal operators in the world, in 5/2007 outline planning permission was granted for a major new deepwater port and logistics park on the site, known as London Gateway, this to be developed over 10-15 years.

Reference : The Thames Haven Branch, Peter Kay, 1999.

Gauge : 4ft 8½in

STARHAVEN REFINERIES No.1	0-4-0F	OC	AB	1471	1916	(a)	(4)
STARHAVEN REFINERIES No.3	0-4-0F	OC	AB	1437	1916	(b)	(2)
No.19	0-4-0DM		JF	4210005	1949	New	(3)
No.20	0-4-0DM		JF	4210007	1949	New	(1)
No.21	0-4-0DM		JF	4210130	1957	New	(7)
No.22	0-4-0DH		JF	4220031	1964	New	(8)
No.23	0-4-0DH		JF	4220039	1965	New	(9)
	4wDM	R/R	S&H	7509	1968	New	(6)
	0-6-0DH		Sdn		1965	(c)	(5)
19	0-6-0DH		JF	4240016	1964	(d)	(7)
20	4wDH		TH	187v	1967	(d)	(10)
No.24	0-4-0DH		TH	239v	1972	New	(11)
No.25	4wDH		TH	279v	1978	New	(11)
No.26	4wDH		TH	280v	1978	New	(12)
No.27	4wDH		TH	281v	1978	New	(13)
No.28	4wDH		TH	282v	1978	New	(14)

(a)   ex Ministry of Munitions, National Filling Factory, Morecambe, Lancashire, /1920.
(b)   ex Shell Chemicals Ltd, Stanlow Works, Cheshire, by 5/2/1925.
(c)   ex BR, D9538, 4/1970.
(d)   ex London & Thames Haven Oil Wharves Ltd, /1972.

(1)   to Thos. W. Ward Ltd, Templeborough Works, Sheffield, 4/1963;
      thence to C.F.Booth Ltd, Rotherham, South Yorkshire, c/1966.
(2)   to Southern Counties Demolition & Trading Co Ltd, Bedhampton, Hampshire, /1964.
(3)   to Thos. W. Ward Ltd, Templeborough Works, Sheffield, South Yorkshire, /1966.
(4)   to Imperial Paper Mills Ltd, Gravesend, Kent, c8/1968.
(5)   to BR, Swindon, for repairs, /1970;
      thence to British Steel Corporation, Ebbw Vale Works, Gwent, 2/1971.
(6)   possibly to an unidentified freight forwarding company, Southampton, Hampshire, c/1971, otherwise s/s.
(7)   to Dowty Railway Preservation Society, Ashchurch, Gloucestershire, c7/1979.

(8) to Tenterden Railway Co Ltd, Rolvenden, Kent, 1/1980.
(9) to Stour Valley Railway Preservation Society, Chappel & Wakes Colne Station, c4/1980.
(10) to TH, for resale, 17/9/1980;
thence to Norsk Hydro Ltd, Immingham, Humberside, 25/10/1983.
(11) to Mobil Oil Co Ltd, Coryton, c7/1981.
(12) to Shell UK Oil Ltd, Stanlow, Cheshire, 14/7/1993.
(11) to Shell UK Oil Ltd, Stanlow, Cheshire, by 11/9/1993.
(13) to Shell UK Oil Ltd, Stanlow, Cheshire, by c23/9/1995.
(14) to Shell UK Oil Ltd, Stanlow, Cheshire, 23/9/1995.

SILVERTOWN LUBRICANTS LTD
MINOCO WHARF, Silvertown   **B153**
Mineral Oil Corporation until c1901   TQ 410797

Works established in 1896 by the Mineral Oil Corporation which was formed to build a wharf and plant to distil oil imported from Russia. Works taken over by Gulf Oil Corporation in 1929 and later in 1950 became Gulf Oil (GB) Ltd. Company registered on 12/3/1913.

Gauge : 4ft 8½in

       -        0-4-0ST   OC   MW   810   1881   (a)   s/s

(a) originally Lucas & Aird, Hull & Barnsley Railway (1880-1886) contract, East Yorkshire, 236;
[possibly John Aird & Co, Hodbarrow Sea Wall (1899-1905) contract, Cumberland];
here by 4/1930.

SOUTHCHURCH BRICKFIELDS CO LTD
SOUTHCHURCH BRICKWORKS   **J154**
John Ephraim Potter c1883 - 1889   TQ 905849

Brickfield in existence in the 1870's and operated from c1883 by John Ephraim Potter under the name of Southchurch Brickfields Company. From 1883 to 1889 he contracted to make bricks for the Brickfields Estate Company and from 1889 to 1890 was foreman to the company and kept a large fleet of barges to transport bricks; he was adjudged bankrupt in 1890. The company incorporated 21/3/1893, had extensive brickfields with a tramway which ran south to a wharf on the Thames. The plant was auctioned in 1907, and the company went into voluntary liquidation 18/9/1907 and was struck off 10/3/1911. Works later operated by Ernest Roots c1910.

"Contract Journal", 8/3/1899, for sale – not powerful enough for vendors requirements – HE tank loco, new Dec 1895, 6x8, 3ft 4in gauge – Southchurch Brickfields Co Ltd.

Gauge : 3ft 4in

      SOUTHCHURCH   0-4-0ST   IC   HE   640   1895   New   (1)

(1) to Thos. W. Ward Ltd, after 3/1899, by 5/1899, for sale 28/7/1899, thence to Callender's Cable & Construction Co Ltd, Belvedere, Kent, by 11/1899.

SOUTHEND SAND & GRAVEL CO LTD
GREAT WAKERING SAND PITS   J155

Company registered 18/8/1945 with an office address at 16 Warrior Square, Southend-on-Sea. The company was associated with the Milton Hall (Southend) Brick Co Ltd and shared the same office address. The sand pit was located in Star Lane, Great Wakering, location uncertain but possibly on land adjacent to Star Lane Brickworks.

Gauge : 2ft 0in

		4wDM	MR			(a)	s/s after 8/1948

(a)   origin and identity unknown, here by 12/1945.

THE SOUTHEND-ON-SEA ESTATES CO LTD
THORPE BAY BRICKWORKS, Thorpe Bay, near Southend-on-Sea   J156
Company registered 30/8/1900   TQ 922856

The Southend-on-Sea Estates Co Ltd operated brickworks in Barnstable Road, Thorpe Bay and are listed in quarry directories from 1933. The company was associated with the Milton Hall (Southend) Brick Co Ltd, and both shared the same office address at 16 Warrior Square, Southend. In 1971 both companies had the same board of directors. The works produced machine made yellow stock bricks which were fired in a clamp. Works demolished c/1963 and the site returned to agricultural use.

Gauge : 2ft 0in

	4wDM	MR	5853	1934	(a)	(1)
-	4wDM	RH	179880	1936	New	(2)
-	4wDM	RH	179881	1936	New	(2)
-	4wDM	RH	441951	1960	New	(3)

(a)   ex Sand & Shingle Ltd, Hounslow, Middlesex.
(1)   to Baggeridge Brick Co Ltd, Staffordshire.
(2)   to Milton Hall (Southend) Brick Co Ltd, Star Lane Brickworks, c/1963.
(3)   to Milton Hall (Southend) Brick Co Ltd, Cherry Orchard Brickworks, c/1963.

SOUTHERN DEPOT CO
Subsidiary of **British Coal**
and earlier of **National Coal Board**

SOUTHEND COAL CONCENTRATION DEPOT, Southend-on-Sea   J157
operated by **Co-operative Wholesale Society Ltd**   TQ 881865

Coal distribution depot in the sidings at the ex-GER Southend Victoria station. The depot commenced operation in 1967 and was closed c1986.

Gauge : 4ft 8½in

	-	0-4-0DM	JF 4200035	1949	(a)	s/s c9/1986
D2184		0-6-0DM	Sdn	1962	(b)	(1)

(a)   ex Whitwick Granite Co Ltd, Leicestershire, 24/6/1967,
       per L. Sanderson Ltd, dealers, Birtley, Co. Durham.
(b)   ex BR, Worcester, 8/1969.

(1)   to Colne Valley Railway, Castle Hedingham, 17/10/1986.

SOUTH ESSEX WATERWORKS COMPANY
Romford   **E158**
South Essex Waterworks Act of 1921 gave power to the company to drive an adit from an existing well at Romford to a point near to Beam Bridge to the north, given the date of 1921 it seems likely the loco listed here was used on these works.

		4wBE	Electromobile	1921	New	s/s
-						

STENA SEALINK LTD
Company registered 16/11/1992.

HARWICH HARBOUR   H159
British Ferries Ltd   TM 234326
formerly **Sealink Ltd**

Tractive units based on the SPEEDLINK VANGUARD train ferry operating between Harwich and Zeebrugge.

Gauge : 4ft 8½in

PQ 363	4wDH	R/R	NNM 80504	1980	New	s/s c/1988
PQ 364	4wDH	R/R	NNM 80505	1980	New	(1)
PQ 365	4wDH	R/R	NNM 80508	1980	New	s/s c/1988
PQ 380	4wDH	R/R	NNM 81514	1983	New	(2)

(1) to Stena Sealink Ltd, Engineers Yard, Harwich Harbour, after 9/10/1991 by 10/7/1992.
(2) to Sackers (Claydon) for scrap 12/1990.

ENGINEERS YARD, Harwich Harbour   H160
                                                                            TM 234324

Gauge : 4ft 8½in

| PQ 364 | 4wDH | R/R | NNM 80505 | 1980 | (a) | (1) |

(a) ex British Ferries Ltd, Harwich Harbour, after 9/10/1991 by 10/7/1992.
(1) to Banffshire Rolling Stock Ltd, Dufftown, c1/1996, after 25/11/1995 by 5/3/1996.

ROBERT STROUD
MEADS LANE BRICKFIELDS, Seven Kings   D161
                                                                             TQ 457879

Robert Stroud was born in Middlesex in 1856 and moved to Ilford in 1885, directories for 1894 list his address as Ley Street, Ilford. In c1887 he set up as a contractor, building houses for Cameron Corbett an estate developer, he built houses on the Clements Wood, Downshall and Mayfield estates and also undertook work for Ilford Council. He operated brickfields north of Meads Lane, Seven Kings and as a contractor at Barley Lane, Goodmayes. At the turn of the century he was employing over 500 men and lads and kept over 30 horses. Robert Stroud died at his residence Barley Hall, Goodmayes in 12/1925.

"The Times" 25/6/1927, the estate of the late Robert Stroud, Goodmayes, a valuable freehold building estate, an area of about 61 acres with frontages of about 2,900 feet to

Barley Lane and Chadwell Heath Lane, also builders & contractors yard and detached residence known as Barley Hall. Messrs Kemsley will sell by auction at the London Auction Mart on Tuesday 12th July.

Gauge : 2ft 0in

-	0-4-0ST	OC	KS	2469	1916	(a)	(1)

(a) ex "Mundays" c1923-1924, (possibly George Munday & Sons Ltd, Rainham), earlier Wm Muirhead, MacDonald Wilson & Co Ltd, Southend Arterial Road (1921-1925) contract.

(1) to George Cohen, Sons & Co Ltd, dealers,
thence to Tunnel Glucose Refineries Ltd, East Greenwich, c/1937 (by 1/1938).

STUART (THAMESMOUTH) SAND & SHINGLE CO LTD
MUCKING HALL GRAVEL PIT, Mucking, near Stanford-le-Hope   **K162**

Mucking Pits TQ 690813
Golden Gates Pits TQ 684808

Company registered 15/1/1930 by **F.W. Surridge** whose address was listed as Old Crown Cottage, Mucking. Pits listed in HMSO list of quarries for 1931, in existence until after 1967, company dissolved 23/11/1989. In 5/1931 a standard gauge ballast siding was opened on the east side of Thames Haven Junction, connecting to the Thames Haven branch. The narrow gauge was extended north-west from the quarries to an interchange at this siding, locomotive working appears to date from this extension. On 31/5/1947 the Grays & Tilbury Gazette reported on an accident at the Mucking Hall Works when a fifteen year old boy, the driver of a light diesel locomotive hauling four skips, died when the locomotive overturned into a sand hole where he drowned in six feet of water. Rates records show that the gravel pits at Butts Lane (Golden Gates) were out of use in 10/1950 and the railway from Stanford Wharf to Mucking was deleted from the assessment by 2/9/1952, and the wharf deleted from the assessment 3/6/1953. By 1957 the standard gauge siding was disused and the narrow gauge lifted. Company went into voluntary liquidation 8/2/1989.

F.W. Surridge was listed in HMSO list of quarries for 1948 with pits at **Hall Farm**, South Ockendon.

Gauge : 2ft 0in

-	4wDM	FH	1708	1932	New(a)	s/s
-	4wDM	HE	1735	1933	(b)	s/s
JOAN	4wDM	HE				(1)
3	4wDM	OK				(1)
-	0-4-0DM	OK				(1)

(a) ex FH, 21/10/1932;
earlier Demolition & Construction Co Ltd, Barmouth sea defence works (1930-1932) contract, Merioneth.

(b) ex WD, Deptford, London by 31/8/1939, converted from 1ft 6in gauge.

(1) to Thurrock Flint Co Ltd, West Thurrock for scrap, via Thos. W. Ward Ltd, Grays, c/1954.

A visit on 16/8/1950, showed that three 4wD O&K and an 0-4-0D O&K were here, plus 4wDM HE JOAN, all derelict; and pits disused.

CLIFFORD STUBBINGS
MID-ESSEX PITS & BITUMINOUS GRAVEL WORKS,   G163
BROOMFIELD, Chelmsford
TL 719109

Pits at Broomfield listed in the quarry directory for 1929 until at least 1955/56, also by 1934 with pits at Sandon. And by 1948 pits at **Well Farm**, Gosfield and **Ishams Farm & Moor Gardens**, Wickham Bishops.

Gauge : 2ft 0in

		4wDM	RH	166011	1932	(a)	(1)
	-	4wDM	RH	166050	1934	New	(2)

(a)   ex RH, Lincoln, 4/6/1934 on hire; earlier a RH hire loco latterly on loan by RH to Vales Plant Register, 14 Lower Grosvenor Place, London SW1, but no sale or hire was effected.

(2)   to Cubitts & Pauling, ROF Ranskill contract, Retford, Nottinghamshire, after 3/3/1938, by 5/6/1941.

(1)   returned to RH, Lincoln, 5/10/1934; later to Derbyshire Silica Firebrick Co Ltd, Friden, Derbyshire, initially on hire, 15/11/1934.

SANDON PITS, near Chelmsford   G164

Gauge : 2ft 0in

	4wPM	FH	1682	1931	New	s/s

F.W. SURRIDGE

Firm of refuse contractors with an address at Wharf Road, Wandsworth SW18, began operations in 1919 with a contract for Wandsworth Borough Council to collect & dispose of their domestic and trade refuse at a tip on Rainham Marshes. Further contracts were entered into with Hammersmith & Lambeth Councils to dispose of their refuse initially at Rainham and later at Mucking. By 1931 the firm operated two incinerators, ten wharves and operated 100 horse drawn containers & forty motor containers and employed sixty horses & 600 men. In 1930-1931 Surridge acquired an estate of 1300 acres at Mucking and East Tilbury for two new dumps. Frederick Walter Surridge died aged 60 on 21/11/1950 at the Rosery, Roehampton Vale, SW15.

Reference : The Municipal Journal & Public Works Engineer, 1/5/1931.

EAST ESSEX RECLAMATION SCHEME, Mucking   K165

The Grays & Tilbury Gazette of 31/1/1931 reported plans for a super dump at Mucking, two of the contractors Mr F.W. Surridge of Mucking and the Tilbury Contracting & Dredging Co Ltd objected to the by laws proposed by Orsett Rural District Council as being a little too stringent, several amendments were proposed by Mr Surridge including a different system of layering. Hunslet Engine Company records show the locomotives listed below as delivered to the East Essex Reclamation Scheme; records show spares were delivered for these locos to F.W. Surridge, East Essex Reclamation Scheme, Mucking, until 2/1952. (see Mucking

Rubbish Shoot & Stuart (Thamesmouth) Sand & Shingle Co Ltd), one of the Hunslet locos listed below was later fitted with a Fordson engine which proved unsuccessful.

Gauge : 2ft 0in

		4wDM	HE	1710	1932	New	s/s
		4wDM	HE	1711	1932	New	s/s
		4wDM	HE	1712	1932	New	s/s
		4wDM	HE	1713	1932	New	s/s
		4wDM	HE	1714	1932	New	s/s
		4wDM	HE	1715	1932	New	s/s
-		4wDM	HE	2700	1942	New	s/s
-		4wDM	HE	2701	1942	New	s/s
-		4wDM	HE	2702	1942	New	s/s

RAINHAM RUBBISH SHOOT   P166
                                  TQ 515814

Shoot operated from c/1919, had, by 1931, a jetty on the Thames where refuse was handled by steam cranes and conveyed by light railway operated by Fowler locomotives to the dumping site. Refuse was transported from London Boroughs in barges & lighters with capacities of 80-200 tons; 130 men were employed on the dump. (It is thought that the Fowler locos mentioned were in fact those of William Cory and that Surridge may have had an agreement with Cory for the disposal of refuse on his dump.)
Reference : The Municipal Journal & Public Works Engineer, 1/5/1931

MUCKING RUBBISH SHOOT   K167
                                  TQ 690800

Rubbish Shoot located on Mucking Marshes near Mucking Creek in operation from 1931 and shown on the 25in OS map for 1939, the jetty was operated by two electric grab cranes with a tramway to the tip; by 1953 the shoot covered 1036 acres of marsh land; use of locomotives unknown. The batch of Hunslet locos listed above (under East Essex Reclamation Scheme) were delivered to Mucking in 1932 and 1942, it is almost certain that some of these locos worked at this site.

EAST TILBURY RUBBISH SHOOT   M168
                                  TQ 680760

Rubbish Shoot at Bowaters Farm, East Tilbury and located next to that of the Tilbury Contracting & Dredging Co Ltd, in operation from c/1932. In 9/1945 two Stothart & Pitt electric cranes were in use on the jetty; by 1953 the shoot covered 184 acres of marsh land.

Gauge : 2ft 0in

		4wDM	HE			(a)	s/s
	-	4wDM	HE			(a)	s/s
	-	4wDM	HE			(a)	s/s
	JOYCE	4wDM	OK			(c)	s/s
12	GLADYS	4wDM	OK			(c)	s/s
13	JEAN	4wDM	OK			(c)	s/s
14	EILEEN	4wDM	OK	20463	1934	(b)	s/s
15	EILEEN	4wDM	OK			(c)	s/s

16		4wDM	OK	(c)	s/s
17	VERA	0-4-0DM	OK	(c)	s/s

Note, the 4wDM O&K locos above are believed to be 20462 to 20467 of type RL2b delivered to William Jones Ltd, dealers, in 8/1934.

(a) identity unknown; here operational on 28/8/1949.
(b) here operational on 28/8/1949.
(c) here derelict 28/8/1949.

A visit on 28/8/1949 recorded two Hunslet locos fitted with Lister engines both in working order and 14 OK 20463 in working order, 17 VERA was derelict, also present were two dismantled locos derelict, 13 JEAN, 15 EILEEN, and parts of 16, & 12 GLADYS.

TATE & LYLE LTD
THAMES SUGAR REFINERY, Silvertown   B169
Henry Tate & Sons Ltd until 8/1921   TQ 423799
Henry Tate & Sons (1903) Ltd until 2/7/1904
Henry Tate & Sons Ltd until 27/2/1903
Henry Tate & Sons (established 1869) until 20/2/1896

Henry Tate was born in 1819 in Chorley, Lancashire, in 1859 he went into partnership with John Wright a sugar refiner of Manesty Lane, Liverpool and in 1862 opened a small refinery of his own in Earle Street. His partnership with Wright was dissolved on June 30th 1869 and Henry changed the name of the firm to Henry Tate & Sons, taking his two sons Alfred and Edwin into partnership. In 1870 he built a new refinery in Love Lane which opened in 1872 – looking for expansion Henry Tate went to London in 1874 or early 1875 and purchased a derelict shipyard owned by Messrs Campbell & Johnston at Silvertown. The new Thames Refinery was built and was in production by 6/1878 producing cube sugar. In 1921 Henry Tate & Sons Ltd merged with Abram Lyle & Sons Ltd to form Tate & Lyle Ltd, As far as is known Henry Tate and Abram Lyle never met. Private siding agreement dated 6/12/1917.
Reference : Sugar And All That, A History of Tate & Lyle, Antony Hugill, 1978.
"Machinery Market" 17/4/1925 wanted – 12in or over loco for curve 250ft radius and 1in100 – Tate & Lyle Ltd, London EC3.
Gauge : 4ft 8½in

ELSIE	0-4-0ST	OC	MW	1671	1905	(a)	(1)

(a) ex Burry Port Copper Works, Burry Port, West Wales.

(1) to Thos. W. Ward, Silvertown, /1934.

THAMES SAND MARKETING CO LTD
FINGRINGHOE SAND PITS, near Colchester   **H170**
J.J.Prior (Aggregates) Ltd, by c/1948.   TM 042203
Company registered 27/9/1933 to acquire the freehold gravel bearing land on part of Ballast Quay Farm, Fingringhoe, listed in quarry directories for 1934 until at least 1948.
Gauge : 2ft 0in

	-	4wPM	OK	1936	New	(1)

(1)   advertised for sale in "Contract Journal", 11/11/1942; s/s.

THURROCK CHALK & WHITING CO LTD
WEST THURROCK WORKS   **N171**
Subsidiary of **APCM** from 1/10/1958   TQ 573779
Northfleet Coal & Ballast Co Ltd until 12/11/1921

The Northfleet Coal & Ballast Co Ltd bought land in 1902; Thurrock Chalk & Whiting Co Ltd was registered on 12/11/1921 to acquire the undertaking of the Northfleet Coal & Ballast Co Ltd (registered 10/6/1868) and described as shippers of chalk, flint & ballast, whiting manufacturers and wharfingers. Site taken over by Lafarge Aluminous Cement Co Ltd.

Reference : "Railway Bylines", Roger Hateley, June 2003.

Gauge : 4ft 8½in

Name	Type		Builder	No.	Date	Ref	Disposal
NORTHFLEET	0-4-0ST	OC	HH		c1868	(a)	(4)
SWANSCOMBE	0-4-0ST	OC	AB	699	1891	(b)	(6)
KILMARNOCK	0-4-0ST	OC	AB	282	1886	(c)	(1)
LOUGHBOROUGH	0-4-0ST	OC	HH			(c)	Scr /1927
FOX	0-4-0ST	OC	HE	223	1879	(d)	Scr 3/1952
-	0-4-0VBT	VCG	S	6310CH	1926	(e)	(3)
THURWHIT	0-4-0ST	OC	P	1734	1927	New	(7)
-	0-4-0ST	OC	HC	[604	1902?]	(f)	(2)
FELSPAR	0-4-0ST	OC	MW	1846	1914	(g)	(2)
P.H.B.	0-4-0ST	OC	HL	3760	1932	(h)	(7)
ALBERT	0-4-0ST	OC	MW	1008	1887		
	reb		Adams		1916	(j)	Scr 8/1956
ORMSBY	0-4-0DM		JF	22077	1938	New	(5)
GEORGE	0-4-0ST	OC	AB	1281	1912	(k)	(7)
STAR	0-4-0ST	OC	AB	1940	1927	(l)	(7)
COMET	0-4-0ST	OC	WB	2879	1948	New	(9)
PLANET	0-4-0ST	OC	AB	747	1894	(m)	(7)
SOUTHFLEET	0-4-0ST	OC	P	1746	1928	(n)	(8)

(a)   ex Northfleet Wharf, Kent, c/1902-1903.
(b)   ex Northfleet Wharf, Kent, /1912.
(c)   ex Northfleet Wharf, Kent, c/1913 by 6/1913.
(d)   ex J. Oakes & Co Ltd, Pollington Colliery, Derbyshire, by 9/1920.
(e)   rebuild of KILMARNOCK AB 282, /1926, (see above).
(f)   identity unknown, ex George Cohen, Sons & Co Ltd, on hire /1930, possibly HC 604, TERRIER.
(g)   ex George Cohen, Sons & Co Ltd, on hire /1930.
(h)   ex John Mowlem & Co Ltd, Southampton, BEAULIEU, /1933.
(j)   ex A.R. Adams & Son, dealers, Newport, Gwent, /1937; earlier PLA, 34.

(k) ex Thos. W. Ward Ltd, Grays, /1939, TW 965; earlier New Westbury Iron Co Ltd, Wiltshire, to BPCM Wouldham Works on hire /1945 and returned.
(l) ex George Cohen, Sons & Co Ltd, c/1946, by 5/7/1947;
earlier K. & L. Steelfounders & Engineers Ltd, Hertfordshire;
to Purfleet Deep Wharf & Storage Co Ltd, Purfleet, loan, 4/1959; returned 5/1959.
(m) ex James W. Perkins & Co Ltd, Isleworth, Middlesex, (per Ronald Clarke & Co, Brixton), by 30/4/1955; earlier North Thames Gas Board, Fulham, London.
(n) ex APCM, Crown & Quarry Works, Kent, 4/6/1964.

(1) rebuilt as S 6310CH in 1926 (see above).
(2) returned to George Cohen, Sons & Co Ltd.
(3) to Ipswich Docks, Suffolk, c2/1928.
(4) scrapped /1949 (after 9/10/1948).
(5) to Anglo-Iranian Oil Co Ltd, Grain Crossing, Kent, 12/1951.
(6) to London Railway Preservation Society site, Luton, Bedfordshire, 6/2/1966.
(7) to Thos. W. Ward Ltd, Grays, for scrap, 3/1966.
(8) scrapped on site by W. Rice, of Slades Green, Kent, 2-6/6/1967.
(9) scrapped on site by W. Rice, of Slades Green, Kent, w/e 9/9/1967.

THURROCK FLINT CO LTD
FLINT WORKS, West Thurrock   N172

Company registered 31/5/1933 to take over the business of **John Walley** as stone merchants and buyers. The works located between Thurrock Chalk & Whiting Ltd and Tunnel Portland Cement Ltd in existence until at least 1967. The locos listed below were noted here derelict.
Gauge : 2ft 0in

3		4wDM	OK	(a)	Scr
-		4wDM	HE	(a)	Scr
-		0-4-0DM	OK	(a)	Scr
-		0-4-0DM	OK	(a)	Scr
-		0-4-0DM	OK	(a)	Scr
-		0-4-0DM	OK	(a)	Scr

(a) believed ex Stuart (Thamesmouth) Sand & Shingle Co Ltd, Mucking Ballast Pits, Stanford-le-Hope, via Thos. W. Ward Ltd, c/1954 (here by 16/3/1957).

TILBURY CONTRACTING & DREDGING CO LTD
Tilbury Contracting & Dredging Co (1906) Ltd until 13/6/1908
Tilbury Contracting & Dredging Co Ltd until 19/4/1906
London & Tilbury Lighterage Contracting & Dredging Co Ltd until 2/8/1904
London & Tilbury Lighterage Co Ltd until 24/1/1896

Company established in 1884 to carry into effect an agreement dated 15/5/1884 between The East & West India Dock Co and George Edward Wood & John Rowley Jones as the London & Tilbury Lighterage Co Ltd (registered 9/10/1884). The company changed its name to London & Tilbury Lighterage Contracting & Dredging Co Ltd (registered 24/1/1896) – this became the Tilbury Contracting & Dredging Co Ltd 2/8/1904 and was reconstructed in 1906 (registered 19/4/1906). In 1938 it was listed as dredging contractors, lightermen, barge &

tug owners with an address at 63 Petty France, SW1 and Dreadnought Yard, Greenwich, SE10, with a disposal Depot at Tilbury and Lighterage Depot at 50 Mark Lane EC3.

The company operated two rubbish shoots, the first on Little Thurrock Marshes (an area now occupied by Tilbury Docks), the second near Low Street on East Tilbury Marshes where the company had 137 acres of land. This shoot had a wharf with two grab cranes and commenced operations in 2/1924; about 1600 tons of refuse was received each week in barges from Greenwich & Kensington.

An advert appeared in the "The Engineer" for 7/1/1921 offering for sale a loco with 9x11 cyls at Dreadnought Wharf Greenwich, directories for 1921 list the only company there as the Tilbury Contracting & Dredging Co Ltd. Standard gauge track removed c6/1958.

"Machinery Market" 13/5/1958 for sale – 100hp JF diesel loco & two 80hp JF diesel locos, all std gauge and std gauge & 2ft gauge tip wagons – Tilbury Contracting & Dredging Co Ltd, London.

TILBURY RUBBISH SHOOT   M173

Thurrock Marshes Wharf TQ 624755

Rubbish shoot adjacent to that of Covington's received refuse from Chelsea, Kensington & Greenwich Borough Councils, the site covered 20 acres and had a 20ft deep tipping face; shoot closed /1924.

Gauge : 4ft 8½in

THE NIPPER	0-4-0ST	OC	[MW	1383	1898?]		
		reb	Phillips		1909	(a)	(1)
GIPSY	0-4-0ST	OC	HE	387	1886	(b)	(1)

(a) ex Joseph Pugsley & Sons Ltd, dealers, Bristol;
possibly earlier Gramophone Co Ltd, Hayes, Middlesex.
(b) ex S. Pearson & Son Ltd, by 24/4/1912, earlier Blackwall Tunnel construction (1890-1897) contract, London.

(1) to Low Street rubbish shoot, /1924.

LOW STREET RUBBISH SHOOT   M174

TQ 672760

Shoot commenced operations in 1924 with refuse received in barges from Greenwhich Borough Council via Norman Wharf on Deptford Creek, and later from 1/4/1932 refuse from Kensington Borough Council via Chelsea Wharf, Lotts Road, Chelsea. A contract was entered into on 1/4/1934 to transport refuse from Deptford Borough Council; by 1953 the shoot covered 458 acres of marsh land.

Gauge : 4ft 8½in

THE NIPPER	0-4-0ST	OC	[MW	1383	1898?]		
		reb	Phillips		1909	(a)	Scr /1936
GIPSY	0-4-0ST	OC	HE	387	1886	(a)	s/s by 8/1949
No.1	0-4-0DM		JF	19424	1931	New	s/s c6/1958
No.2	0-4-0DM		JF	20550	1934	New	s/s c6/1958
-	0-4-0DM		JF	4160002	1952	New	(1)

(a) ex Tilbury rubbish shoot, /1924.

(1) to NCB, South East Division, Chislet Colliery, Kent, c6/1958.

Narrow gauge locos purchased for intended system but only used for about six months.
Gauge : 2ft 0in

		4wDM	HE	4394	1952	New	(1)
		4wDM	HE	4395	1952	New	(1)
		4wDM	HE	4396	1952	New	(1)

(1) to Twickenham Gravel Co Ltd, Hanworth Workshops, Middlesex, /1957.

TRANSFESA UK LTD
TILBURY RIVERSIDE TERMINAL   M175
TQ 626758

Company registered 18/11/1953 as continental and colonial carriers.
Gauge : 4ft 8½in

003	FLORENCE	0-6-0DE	Hor	1961	(a)

(a) ex RFS (E) Ltd, Doncaster, Yorkshire, 16/8/1997.

F.H. TREVITHICK
Buckhurst Hill   D176

Trials of a patent tramway track by Trevithick who was engineer to the Lisbon Steam Tramways Co Ltd. The track was composed of two parallel timber baulks with a central steel guide rail of the Larmanjat type. The driving wheels of the Sharp Stewart loco (2255 of 1872) had broad flat treads with guidance on the central rail by two small double flanged wheels at either end. A test track of 1,700 feet was laid on a strip of land near Lords Bushes, Buckhurst Hill, in 1872; trials took place on 28/12/1872. The Lisbon Tramway opened on 5/7/1873 but was not successful, the company being wound up by court order on 29/8/1878. The locomotives were offered for sale and some if not all were acquired by Merryweather & Sons Ltd of Greenwich; these were rebuilt as standard gauge conventional locomotives and sold for further use, some to Lucas & Aird contractors for construction of the Tilbury Docks and the Hull & Barnsley Railway.
Reference : The Industrial Railway Record, Vol.3, No.32, pp306-308
The Engineer, 10/1/1873

TUNNEL INDUSTRIAL SERVICES LTD
TUNNEL INDUSTRIAL ESTATE, West Thurrock Works   N177
Company dissolved 19/5/1992
subsidiary of **Tunnel Holdings plc**   TQ 578780
Tunnel Holdings Ltd
Tunnel Cement Ltd until 1/4/1975
Tunnel Portland Cement Co Ltd until 1/10/1968
Tunnel Portland Cement Works Co Ltd until 11/9/1911

In 1874 Tunnel Portland Cement Co (West Thurrock) Ltd purchased property in West Thurrock and commenced production by 1880 – this works closed in 1906. The Works were to be auctioned on 23/7/1908 and were dismantled by Thos. W. Ward in 1913. The site was taken over by a new company in 1911 who built a new works. A standard gauge rail system served this cement works and associated chalk quarries. There was a connection with the ex-LT&SR line midway between Purfleet and Grays Stations together with a line which ran south to a wharf on the River Thames. Works closed for production 30/4/1976 and dismantled shortly afterwards. The site was then utilised as an industrial estate and one loco (YE 2856) was retained to handle possible traffic. Normal rail traffic ceased by 6/1983 and completely by 9/1987. Site of quarry now occupied by Lakeside shopping centre.

"Contract Journal" 6/1/1909 for sale, direct driven cement works single cylinder loco, std gauge 10x12 – T W Ward, Silvertown, (possibly AP 1121 below).

"Contract Journal", 22/5/1912, Fuller Horsey to sell 7/6/1912 re Tunnel Portland Cement Works, West Thurrock – plant including three 12in MW saddle tank locos.

Reference : "Railway Bylines", Roger Hateley, March 2004.

Gauge : 4ft 8½in

Name	Type	Cyl	Builder	Works#	Year	Acquired	Disposal
-	4wWT	G	AP	1121	1875	New	s/s
PORTLAND	0-4-0ST	OC	MW	779	1881	New	s/s
TUNNEL	0-4-0ST	OC	MW	1581	1903	New	(10)
CEMENT	0-4-0ST	OC	MW	336	1871	(a)	(1)
FOLA	0-4-0T	OC	P	1287	1912	New	(2)
ANGLO-DANE	0-4-0ST	OC	P	1318	1913	New(c)	(11)
VIKING	0-4-0ST	OC	P	1319	1913	New	(7)
TUNNEL	0-4-0ST	OC	P	1369	1914	New	(13)
NORSEMAN	0-4-0ST	OC	AB	1711	1921	New	(3)
THOR	0-4-0ST	OC	P	1689	1925	New	Scr 3/1965
FOLA	0-4-0ST	OC	P	1806	1930	New	(11)
"AVONSIDE"	0-4-0ST	OC	AE	1877	1921	(b)	Scr 5/1960
JUBILEE	0-6-0ST	OC	P	1919	1936	New	(12)
CORONATION	0-6-0ST	OC	P	1920	1937	New	(13)
T.P.C. No.1	4wDM		RH	192325	1938	New	Scr 3/1971
WINSTON	0-6-0ST	OC	HC	1725	1941	New	(9)
MONARCH	0-6-0ST	OC	P	1923	1937	(d)	(4)
LAURIE	0-6-0ST	IC	MW	2005	1921	(e)	(5)
BRAMLEY No.6	0-6-0ST	OC	HE	1644	1929		
	reb		HE		1938	(f)	(6)
MARFLEET	0-6-0ST	IC	HE	550	1892	(g)	Scr 5/1960
ODIN	0-4-0ST	OC	P	1821	1931	(h)	Scr 5/1960
POLAND	0-4-0ST	OC	P	1994	1940	(j)	(11)
"WALLASEY" (EN 3)	0-6-0ST	OC	HE	1686	1931	(k)	Scr 4/1965
"SOUTHERN"	0-6-0ST	OC	HE	1688	1931	(l)	Scr 4/1965
	0-4-0DH		JF	4220009	1960	(m)	(8)
No.2	0-4-0DE		RH	412716	1957	(n)	(19)
4	0-4-0DM		RH	252687	1947	(o)	(21)
T.P.C. No.3	0-4-0DE		RH	433676	1960	(p)	(21)
T.P.C. No.5	4wDH		RR	10230	1965	New	(15)
T.P.C. No.6	4wDH		RR	10235	1965	New	(17)
T.P.C. No.7	4wDH		RR	10276	1967	New	(16)
211 (17)	0-4-0DE		YE	2854	1961	(q)	(18)
No.8 (213)	0-4-0DE		YE	2856	1961	(r)	(22)
-	0-4-0DM		RH	310081	1951	(s)	(14)
2	0-6-0DM		HE	4208	1950	(t)	(20)

(a) ex Stockton Iron Furnace Co, Cleveland.
(b) ex Thos. W. Ward Ltd, Grays, TW 330, 11/11/1930;
     earlier Bombay Improvement Trust, India, No.12.
(c) to Pitstone Works, Buckinghamshire, after 10/1937; returned 7/1945.
(d) ex MOFP, Farnley Ironworks Outcrop Site, West Yorkshire, by 7/1945.
(e) ex Thos. W. Ward Ltd, Grays, TW317, LAURIE, hire, 8/1939 and /1950.
(f) ex Thos. W. Ward Ltd, Grays, TW2694, hire, /1953 (by 25/1/1953).
(g) ex Whitehall Security Corporation Ltd, Grays, c2/1955.
(h) ex N.E. Tar Distillers (Sadlers) Ltd, Middlesbrough, Cleveland, c11/1958.
(j) ex Hadfields Ltd, East Hecla Works, Sheffield, 3/1959.
(k) ex Edmund Nuttall, Sons & Co Ltd, Colnbrook Plant Depot, Buckinghamshire, 9/1959.
(l) ex Edmund Nuttall, Sons & Co Ltd, Colnbrook Plant Depot, Buckinghamshire, 11/1959.
(m) ex JF on trial 4/1960.
(n) ex RH, Lincoln, /1961; earlier demonstration loco.
(o) ex Padeswood Hall Works, Buckley, Flintshire, /1961.
(p) ex RH, Lincoln, 6/1961; earlier demonstration loco.
(q) ex PLA, c3/1970.
(r) ex PLA, 213, c5/1971.
(s) ex Pitstone Works, Buckinghamshire, c4/1973.
(t) ex Padeswood Hall Works, Buckley, Flintshire, 11/1976.

(1) to Thos. W. Ward Ltd, Sheffield, South Yorkshire, c/1913.
(2) to P, /1930.
(3) to Pitstone Works, Buckinghamshire, /1937.
(4) to Pitstone Works, Buckinghamshire, by 18/3/1946.
(5) returned to Thos. W. Ward Ltd, Grays, by 10/1950.
(6) returned to Thos. W. Ward Ltd, Grays, /1953.
(7) to P for repair, but scrapped, /1959.
(8) returned to JF after trials, to a site in Warrington, Lancashire, 1/5/1960.
(9) to HC, /1961; scrapped /1961.
(10) to Raven Tinplate Co Ltd, Glanamman, Carmarthenshire, by 7/1914.
(11) to R. Fenwick & Co Ltd, Canning Town, for scrap, c11/1965.
(12) to Ferro Services Ltd, Smallford, Hertfordshire. for scrap, c1/1966.
(13) to P for repair c9/1960 (there 5/10/1960); returned; to George Cohen, Sons & Co Ltd, Cransley, Northamptonshire, for scrap, 3/1968.
(14) to Gartsherrie Works, Lanarkshire, 5/1975.
(15) to Pitstone Works, Buckinghamshire, c9/1976.
(16) to Padeswood Hall Works, Flintshire, c9/1976.
(17) to Padeswood Hall Works, Flintshire, 7/10/1976.
(18) to Pitstone Works, Buckinghamshire, c5/1977.
(19) to Purfleet Deep Wharf & Storage Co Ltd, c9/1977.
(20) to Tenterden Railway Co Ltd, Kent, 21/1/1978.
(21) to P.W. Leeman Ltd, scrap merchants, East Tilbury, c1/1978 by 12/11/1978.
(22) to Chatham Dockyard Trust, Kent, /1988.

Gauge : 2ft 0in

		4wDM	OK	(a)	(1)

(a) ex Aveley Clay Pit, loco stored only.

(1) Scrapped, after 6/1963, by 17/8/1964.

TUNNEL PORTLAND CEMENT CO LTD
AVELEY CLAY PIT
L178
TQ 557803

Pits opened in 1927 with pipeline to West Thurrock works. On 30/7/1938 the "Grays & Tilbury Gazette" reported on an accident which happened in the early hours of Wednesday 20/7/1938 when the driver of a diesel loco which weighed four and a half tons met with a fatal accident when his locomotive crashed through the rear wall of the engine shed and dropped nine feet into the yard below killing the driver. Rail traffic ceased c/1964

Gauge : 2ft 0in

-	4wDM	RH 168437	1933	New	(1)
LIZ	4wDM	RH 177606	1936	New	(3)
-	4wDM	OK		(a)	(2)
-	4wPM			(a)	s/s

(a) origin and identity unknown.
(1) to Rugby Portland Cement Co Ltd, Halling, Kent, after 4/4/1940.
(2) to West Thurrock Works, for store.
(3) buried on site and abandoned, c1/1969.

UNITED GLASS (ENGLAND) LTD
KEY WORKS, Harlow
E179
TL 462119

Standard gauge sidings served this works on the south of the ex-GER line about ½ mile west of Harlow Station. Rail traffic ceased and system dismantled.
"Machinery Market" 31/8/1967, for sale – 11 ton Planet 0-4-0D (Dorman 3DL) std gauge – Key Glassworks Ltd, Edinburgh Way, Temple Fields, Harlow.

Gauge : 4ft 8½in

-	4wDM	FH	3596	1953	(a)	(1)

(a) ex Charlton Works, London, /1967.
(1) to Harlow Development Corporation, Kings Head Field, Harlow, for preservation, 10/1971.

E.W.J. UPTON
PONDS FARM SAND & BALLAST PITS, Wennington
L180
TQ 550800

Company listed in HMSO list of quarries for 1931 until at least 1957/58, listed in 1938 as haulage contractors of Brooklyn, Tavistock Road, E18.
Track removed.

Gauge : 2ft 0in

-	4wPM (exDM)	Austro-Daimler	(a)	(1)

(a) fitted with O.D.M.A.G. engine 24100, here derelict 16/8/1950.
(1) scrapped c/1953, after 9/1952.

VAN DEN BERGHS & JURGENS LTD
PURFLEET WORKS N181
Jurgens Ltd, 23/11/1914 until 31/12/1936　　　　　　　　　　　　　　TQ 569772

A margarine factory set up by Jurgens Ltd during WW1, construction started 7/1917 production commenced 4/1918 – prior to this, margarine was imported from Holland. The company merged in Holland with Van Den Berghs Ltd in 1928. The works had a jetty on the Thames and materials including timber for margarine boxes were imported via the river. A locomotive was occasionally loaned to the adjacent BP, Purfleet, site, returning here at the end of each day for re-charging with steam. Rail traffic ceased c/1968

Gauge : 4ft 8½in

No.1		0-4-0F	OC	AB	1492	1916	(a)	(1)
No.2		0-4-0F	OC	AB	1493	1916	(a)	(1)
-		0-4-0F	OC	OK	8141	1916	(b)	Scr 4/1959

(a)　ex MoM, Gretna, Dumfriesshire, /1919.
(b)　ex Oss Factory, Holland, c/1924.
(1)　to Laporte Chemicals Ltd, Luton, Bedfordshire, 9/1968 for storage.

VANGE BRICK FIELDS
VANGE BRICKWORKS K182
Robert L. Curtis, by 1894　　　　　　　　　　　　　　　　　　　　　　　TQ 720866

Brickworks established between 1886 and 1894 by Robert Leabon Curtis of Vange Hall. A narrow gauge tramway ran from a point close to Fobbing Crossing (TQ 711862) on the former LT&SR railway and connected with the works and a wharf on Vange Creek (TQ 729863). Curtis owned six 90ton barges which could carry 35 to 44,000 bricks each, these were also used to carry ashes, coal, chalk and sand for the brick works. The works were taken over in 1914 by the Military and was dismantled at the end of the First World War. Alfred Perkins Curtis, J.P. who lived in Empire House, Vange succeeded his father as Lord of the Manor and principal landowner when Robert L. Curtis died on 19th October 1918. The Vange Brick Fields owned by A.P. Curtis had closed by 1922 but it was not until 1935 that the machinery and plant were advertised for sale.

Gauge : narrow

　　　　　　　　　　　　　　4wPM　　　　　　　　　　　　(a)　　　s/s c/1935

(a)　origin and identity unknown.

VOPAK TERMINAL PURFLEET LTD
OLIVER ROAD, West Thurrock N183
Van Ommeren Tank Terminal Purfleet Ltd until 17/3/2000　　　　　　　TQ 576766
Thames Matex Ltd incorporated 17/11/1964 until 8/10/1996

Works on the north bank of the River Thames served by standard gauge sidings connecting with the ex-LT&SR line between Purfleet and Grays Stations. Rail traffic ceased from c/1993.

Gauge : 4ft 8½in

-	0-4-0DM	AB	395	1956	(a)	(2)
No.20	0-4-0DM	JF 4210003		1949	(b)	(1)
-	0-6-0DH	RSHD/WB 8343		1962	(c)	(3)

Note: Class 03 0-6-0DM locos were hired from BR in the period 1975-8 when both the company's own locos were unserviceable, these included, 03081 in 3/1976, 03047 in 2/1977 & 10/1978.

(a) ex BR (ER), D2953, 6/1966;
     to BP Refinery (Kent) Ltd, Grain Refinery, Kent, hire, /1967; returned 24/9/1968;
     to Shell-Mex & BP Ltd, Purfleet, c/1969; returned c/1970, by 2/1971.
(b) ex Shell-Mex & BP Ltd, Purfleet, c/1969, here again 9/1971.
(c) ex West Midlands Gas Board, Coleshill, Warwickshire, c4/1973.

(1) to Shell-Mex & BP Ltd, Purfleet, c3/1971.
(2) to South Yorkshire Railway, Chapeltown, Sheffield, 15/12/1985.
(3) to Eden Valley Railway Trust, Warcop Station, Cumbria c2/2003.

CHARLES WALL LTD
GLOBE WHARF RUBBISH SHOOT   M184
TQ 622764

A rubbish shoot in operation by 7/1926 and probably much earlier near the site of the old Grays Sewage Works. Refuse was received by barge, the tip had a siding from the light railway which ran from Globe Wharf on the Thames at (TQ 622763) to Wall's Globe Works . An engine shed was constructed at the wharf and is shown on the 25in OS map for 1920, use of locomotives unknown but the tip was probably shunted by a locomotive from Charles Wall's fleet, the line from the wharf to Globe Works closed in 7/1941.

WAR OFFICE

NORTH WEALD BASSETT AERODROME, near Harlow   E185
TL 488044

This aerodrome, 3½ miles south-east of Harlow, was opened in 1916. The locomotive was presumably used on construction work.
Gauge : 2ft 0in

-	0-4-0WT	OC	HC	1170	1918	New	(1)

(1) later (by /1926) with John Dickinson Ltd, contractors, Bolton, Lancashire.

COALHOUSE FORT, East Tilbury   M186
TQ 690768

A traction engine type locomotive believed to be an Aveling & Porter was employed to move guns and was used in 1914 to remove two remaining 6in guns from Coalhouse Fort to Cliffe Fort across the river, the standard gauge line ran to a jetty on the Thames. There was around 600 yards of track here, and some remains in situ. Coalhouse Fort is now a museum and an original flat wagon has been rescued from the moat and preserved inside the buildings.

Reference: Industrial Locomotive No.55
Gauge : 4ft 8½in

2-2-0WT G   AP?   (a)   s/s

(a) origin and identity unknown.

WAR DEPARTMENT
ROYAL ENGINEERS, ABBEY MILLS   C187
TQ 387821 approx

Royal Engineers stores Depot located in Cody Road, Canning Town, became CSD127 on 1/6/1920, depot closed 1/1/1923.

"Machinery Market" 2/9/1921, Bradshaw Brown to auction 26/9/1921 at RE Stores CSD 127, Abbey Mills, London – plant including 22in gauge trench loco.

Gauge : 4ft 8½in

4211	NAPOLEON	0-4-0ST	OC	HL	3247	1917	New	(2)
	AJAX	[0-4-0ST	OC	MW	983	1886]	(a)	(1)
98	4109	0-4-0ST	OC	MW	780	1881	(b)	(3)

(a) if identified correctly, earlier Thomas Docwra & Son.
(b) ex MoM National Projectile Factory, Hackney Marshes, London, /1921.

(1) possibly to MoM, Bramley Depot, Hampshire, c/1918.
(2) to S. Pearson, Sennar Dam Construction, Egypt, by /1922,
    later Royal Engineers, Suez, used on the Delta Barrage,
    later works shunter, Railway Works, ME. RE, Suez, 10/1951.
(3) to William Jones Ltd, dealers, London, /1923.

ROYAL ENGINEERS, PITSEA   K188
TQ 736865

Depot located on or near the former British Explosives Syndicate site. This site was occupied in 1921 by the Sea Transport Stores Depot part of the Marine Division of the Board of Trade. The Engineers Depot is believed to have been operated by No.114 REME Anti-Aircraft workshops of Anti Aircraft Command. Spares for the locomotive listed below were consigned in 9/1943 and 11/1943 to a location given as Thames Haven A.A site, information taken from makers records.

Gauge : 60cm

-		4wDM		HE	1975	1939	(a)	(1)

(a) ex Royal Engineers, Shoeburyness, Depot, by 18/8/1943.

(1) to George Cohen Sons & Co Ltd; later with Desborough Clay & Pigment Ltd, Desborough, Northamptonshire, by 7/1953

ROYAL ENGINEERS, PURFLEET   N189
TQ 565776

Depot operated by the 42nd Transportation Company on premises of the Purfleet Deep Wharf & Storage Co Ltd requisitioned during WW1 and additional locos obtained. Large numbers of WDLR locomotives and surplus materials were stored and sold by auction from this depot after WW1. In 1920 Mr T. D. Abbott works manager of the Motor Rail & Tram Car Co Ltd visited Purfleet and purchased a large number of surplus petrol Trench Tractors, engines, gearboxes and other parts on behalf of the company. Thos W. Ward Ltd also purchased large amounts of scrap metal including wagon spare parts, rail, points and loco spares. The depot became CSD 125 on 23/1/1920 and closed 1/11/1922.

"Surplus" 2/6/1919 60cm gauge locos for sale:- at RE Depot Purfleet, forty two HE 4-6-0T 9½x12; twenty Blw 4-6-0T 9x12; one Alco 2-6-2T 9x14 and five Hudson 0-6-0WT 6½x12, also ninety three 20hp and thirty three 40hp MR 4wp locos, also 2ft gauge; five KS 0-4-0ST 6x9 at RE Purfleet.

"Surplus" 1/7/1919, for sale std gauge 180hp tractors two laying at Purfleet (packed in cases for shipment).

"Locomotive News & Railway Notes" 10/9/1919, gives Blw 4-6-0T here in 1918 as ROD No's 517, 538, 560, 590, 633, 722, 749, 752, 791, 800, 878.

"Locomotive News & Railway Notes" 10/10/1919 lists two Alco 2-6-2T as 1263, 1287.

"Contract Journal" 28/1/1920, G N Dixon to sell 24/2/1920 for Min of Munitions at Purfleet – 15 40hp Simplex 60cm gauge; 12 HE 4-6-0T 60cm gauge; 12 Blw 4-6-0T 60cm gauge & one Baldwin-Westinghouse electric mining loco 60cm gauge.

"Contract Journal" 9/2/1921, For sale by private treaty by MoM – loco lying Purfleet – one Blw mining type electric loco, weight 14000lbs, 2ft gauge, with trolley pole & gathering reel.

"Contract Journal" 24/8/1921, Fuller Horsey to sell 6/9/1921 re MoM at Purfleet Depot – plant inc 20 4-6-0 locos 2ft gauge by Blw & Alco, and a 2ft gauge mining type loco, (has two 20hp motors).

"Engineering" 11/2/1921, for sale by private treaty by Disposal Board, one Baldwin/Westinghouse mining type electric loco 2ft 0in gauge – weighs 14000lbs, fitted with 20hp 500v motor, trolley pole and gathering reel, lying at Purfleet, (possibly Blw 44298 built 1916).

"Surplus" 1/3/1922, for sale by tender, at Purfleet Depot. One 45hp Caledon petrol railway lorry type R, No.379 (5ft gauge), three ditto No.376, No.377, No.380 (8ft gauge), one 60cm petrol rail inspection car, large quantity of loco spares all types.

Gauge : 4ft 8½in

	DENMARK	0-4-0ST	OC	MW	1442	1899		
		reb		MW		1913	(a)	(1)
	MORNINGTON	0-4-0ST	OC	MW	1619	1904	(a)	(1)
	PURFLEET	0-4-0ST	OC	TW	420	1876		
		reb		HE *		1898	(b)	(1)
	-	0-4-0ST	OC	TW	373	1873		
		reb		*		1897	(c)	(4)
No.33	MILDRED	0-6-0ST	OC	AE	1763	1917	New	(3)
	ADJUTANT	0-6-0ST	OC	MW	1913	1917	New	(2)

*   rebuilt from 0-4-0WT.

(a)   ex Purfleet Deep Wharf & Storage Co Ltd, /1914.
(b)   ex Bass Ratcliff & Gretton Ltd, Burton, Staffordshire, 8, 4/1917.
(c)   ex Bass Ratcliff & Gretton Ltd, Burton, Staffordshire, 5, 4/1917.

(1) to Purfleet Deep Wharf & Storage Co Ltd, /1919.
(2) to Cannock & Rugeley Colliery Co Ltd, Staffordshire, c/1919.
(3) to Mendip Mountain Quarries Ltd, Vobster, Somerset, by 6/1920.
(4) to Mapperley Colliery Co Ltd, Stanley Colliery, Derbyshire.

Gauge : 60cm

2350	4-6-0T	OC	HE	1322	1918	(a)	(1)
2351	4-6-0T	OC	HE	1323	1918	(a)	(2)
3201	0-6-0WT	OC	HC	1318	1918	New	(3)
3202	0-6-0WT	OC	HC	1319	1918	New	(4)
3203	0-6-0WT	OC	HC	1373	1919	New	(5)
3204	0-6-0WT	OC	HC	1374	1919	New	(6)
3205	0-6-0WT	OC	HC	1375	1919	New	(7)
3206	0-6-0WT	OC	HC	1376	1919	New	(8)

(a) ex WDLR

(1) to KS, 7/1920; new plates affixed (KS 4212/1920);
thence to Palestine Electricity Corporation Ltd, Haifa, Palestine, 6/1928.
(2) to KS, 7/1920; new plates affixed (KS 4213/1920);
thence to Penmaenmawr & Welsh Granite Co Ltd, Trevor Quarry, Caernarvonshire, 11/1921.
(3) later Nott Brodie & Co, Bristol, by /1929.
(4) later Sir Lindsay Parkinson & Co, Haydock, Lancashire, /1929.
(5) to Solitude Estate, Mauritius, by 7/1930.
(6) to Sudan.
(7) to Empresa Carbonifera do Douro, Minas de Pejao, Pedorido, Portugal.
(8) to Oporto, Portugal.

ROYAL ENGINEERS, WARLEY   E190

Spares for all three locos were ordered from RH for delivery to Royal Engineers, Long Ridings, Roundwood Ave, Hutton Mount; near Shenfield. Spares for 218007 were delivered in 11/1947 to 869 Plant Squadron RE, Stapleford Tawney, Airfield. Locos possibly used for construction works at RAF Station Stapleford Tawney, but further details unknown.

Gauge : 60cm

4wDM	RH	211662	1942	(a)	s/s
4wDM	RH	217968	1942	(a)	(2)
4wDM	RH	218007	1943	(a)	(1)

(a) original allocation, not known; here by 30/11/1944.

(1) to Cox & Danks Ltd, Park Royal, Middlesex, after 11/12/1947, by 25/5/1948.
(2) to Cox & Danks Ltd, Park Royal, Middlesex, by 10/12/1947.

UNITED STATES ARMY, HAINAULT DEPOT   D191
                                                              TQ 450920

During WW2, the War Office took over the new Central Line Depot at Hainault and installed sidings adjacent. The Depot was established by 756 Railway Shop Battalion, United States Army Transportation Corps. 155 Railway Workshop Company were employed in 1944 to

assemble wagons built in the USA and shipped to England in kit form; later replaced in 9/1944 by 955 Company. The Depot also used a number of rail mounted steam cranes.
Reference : Railways to Victory, Vic Mitchell, Middleton Press, 1998.

Gauge: 4ft 8½in

	4wDM		RH	218047	1943	(a)	(1)

(a)   ex MoS, Highbridge, Somerset, by 29/6/1943.

(1)   to A.M.W.D., RAF Honington, Bury St. Edmunds, Suffolk, by 23/2/1948.

WARD FERROUS METALS LTD
SILVERTOWN MACHINERY WORKS, Thames Road, Silvertown   B192
Thos. W. Ward Ltd until 28/5/1982   TQ 415799

Business established in Sheffield in 1878 by Thomas W. Ward, the company being incorporated 19/5/1904. The works here (later a scrapyard) was opened in 1906. This entry attempts to list the locomotives which were not at this site for re-sale – see also the entry for Thos. W. Ward Ltd in Dealers section. The company was taken over by Rio Tinto Zinc in 1982 and announced the rundown and closure of its scrap metal business in 2/1983. Rail traffic ceased by 5/9/1991.

"Contract Journal" 6/1/1909 for sale, direct driven cement works single cylinder loco, std gauge 10x12 – T.W. Ward, Silvertown (possibly AP 1121 ex Tunnel Portland Cement Co).

 "Machinery Market" 9/7/1920 for sale – plant inc MW 4w coup loco 12x18 – T.W. Ward, Silvertown.

Gauge : 4ft 8½in

	HARBORO	0-4-0ST	OC	MW	901	1885	(a)	(1)
	(MAX)	0-4-0ST	OC	HC	1337	1918	(b)	(3)
	KING GEORGE	0-4-0ST		HL	2839	1910	(c)	(2)
	(THAMES)	0-4-0DM		JF	4210076	1952	New	Scr c3/1984
No.3434		0-4-0DM		JF	22934	1941	(d)	(4)
TWA 4304		4wDM		RH	237929	1946	(e)	(5)
	-	4wDM		FH	3641	1953	(f)	(6)
No.20		0-4-0DM		JF	4210003	1949	(g)	Scr 7/1980
S75		4wDM		RH	305314	1951	(h)	Scr 7/1983
No.10		4wDM		FH	3900	1959	(j)	Scr 7/1983
	-	0-4-0DM		HC	D1009	1956	(k)	Scr c5/1987
	-	4wDM		RH	398611	1957	(l)	(7)
	-	0-4-0DH		HC	D1291	1964	(m)	Scr c1988
10332		0-4-0DH		RR	10189	1964	(n)	Scr 9/1989
	SUSAN	4wDH		TH	176v	1966	(o)	(8)

(a)   ex J. T. Firbank Ltd, Park Royal Plant Depot, Middlesex, /1907.
(b)   ex British Celanese Ltd, Spondon, Derbyshire, 8/1921.
(c)   ex Thos. W. Ward Ltd, Charlton Works, Sheffield, 39564,
       on hire 14/9/1938 – 12/1939.
(d)   ex Titan Works, Grays Works, loan, /1959.
(e)   ex South-Eastern Gas Board, Dover Gas Works, Kent, c/1961.
(f)   ex M & S Commercials Ltd, North Woolwich, by 9/1971,
       earlier South Eastern Gas Board, Clynne Gap Works, Hastings, Sussex.
(g)   ex Shell Mex & BP Ltd, Purfleet, c5/1973.

(h)   ex Tinsley Works, Sheffield, c6/1975;
      earlier CEGB, Drax Power Station, West Yorkshire.
(j)   ex MoD, Navy Dept, Portsmouth Dockyard, Hampshire, 10/1975.
(k)   ex Brindle Heath Scrapyard, Salford, Lancashire, 20/8/1980.
(l)   ex Birds Commercial Motors Ltd, Long Marston, Worcestershire, after 5/1984 by 12/1985.
(m)   ex John Mowlem, Welham Green, Hertfordshire, /1985.
(n)   ex Thos. W. Ward Ltd, Tinsley, South Yorkshire, 11/1986.
(o)   ex Booth Roe Metals, Rotherham, South Yorkshire, 5/5/1989.

(1)   to North Thames Gas Board, Beckton Gas Works, /1909.
(2)   to Midland Iron Co Ltd on hire, 18/12/1939 – 3/1940.
(3)   derelict /1953 scrapped by 6/1953.
(4)   returned to Thos. W. Ward Ltd, Grays, /1960.
(5)   to Peter Lind & Co Ltd, Kingsnorth contract, Kent, c/1964.
(6)   to Columbia Wharf, Grays, c8/1974.
(7)   to R. and A. Jenkins, Fransham Station, Great Fransham, near Swaffham, Norfolk, 14/2/1987.
(8)   to Shropshire Loco Collection after 31/8/1992 by 26/4/1993.

The following locos are known to have been broken up here for scrap :

JJ197	WEAR	0-4-0ST	OC	MW			(a)	Scr /1930
		reb		HL				
	-	2-4-2ST	OC	NBH	17223	1906	(b)	Scr
	ELSIE	0-4-0ST	OC	MW	1671	1905	(c)	Scr
	No.1	0-4-0ST	OC	MW	635	1877	(d)	Scr
	No.3	0-4-0ST	OC	MW	1039	1887	(d)	Scr
	BLACK	0-4-0WT	OC	KS	3048	1917	(e)	Scr
	-	0-4-0ST	OC				(f)	Scr
111	-	0-6-0ST	IC	RSHN	7111	1943	(g)	Scr
		reb		WB	7077	1955		

(a)   ex Titan Works, Grays, for sale 12/1926.
(b)   ex WD Shoeburyness, /1933.
(c)   ex Tate & Lyle Ltd, Silvertown, /1934.
(d)   ex Samuel Wiilliams (Dagenham Dock) Ltd, Dagenham Dock, /1937.
(e)   ex Imperial Chemical Industries Ltd, Silvertown Works, 11/1957.
(f)   ex Imperial Chemical Industries Ltd, Silvertown Works.
(g)   ex WD Bicester, Oxfordshire, 11/1960.

Gauge : 2ft 0in

4-6-0	OC	HE		(a)	s/s
4-6-0	OC	HE		(a)	s/s
4-6-0	OC	HE		(a)	s/s
4-6-0	OC	HE		(a)	s/s
4-6-0	OC	HE		(a)	s/s

(a)   ex War Department, for sale 12/1926.

THOS. W. WARD LTD
COLUMBIA WHARF, Grays   N193
TQ 612776

"The Times" of 28/3/1904 carried an advert; Messrs G.A. Wilkinson & Son to auction on Friday April 15th 1904 Columbia Wharf situated at Grays, with siding and frontage to the River Thames of about 260ft a side dock of about 250ft with jetty 375ft long into the river and two houses the whole containing about 4½ acres. A further auction took place on Monday May 26th 1919; Fuller Horsey & Co, offered for sale Columbia Wharf with jetty and railway siding having a frontage to the Thames of 250ft and occupying a ground area of about 5 acres. A ship breaking yard opened here in 1919, (private siding agreement dated 7/12/1920); the first vessel to arrive was the torpedo boat destroyer H.M.S. LIFFEY sold to Ward on 26/6/1919, in 5/1921 Thos. W. Ward Ltd made a bulk purchase of 113 Admiralty vessels of which twenty torpedo destroyers went to Columbia Wharf and Rainham for scrap.

Gauge : 4ft 8½in

AMW No 246		0-4-0DM	JF	23003	1943	(a)	s/s by 2/1971
No.1		4wDM	FH	3641	1953	(b)	Scr
		0-4-0DM	JF	4210072	1952	(c)	Scr
09362		4wDM	RH			(d)	Scr
7		0-4-0DM	AB	419	1957	(e)	Scr c3/1984
	HENGIST	0-4-0DH	RSHD	8367			
			WB	3212	1962	(f)	(1)
2		0-4-0DH	EEV	D1122	1966	(f)	(1)

(a) ex Esso Petroleum Co Ltd, Purfleet, /1965, by 29/5/1965.
(b) ex Silvertown Machinery Works, c8/1974, here on 18/12/1977.
(c) ex British Gas Corporation, Reading Gas Works, Berkshire, 8/5/1976.
(d) ex Titan Works, Grays, by 18/12/1977.
(e) ex Bridge Metals, Basildon, 6/1979.
(f) ex CEGB, Croydon, Surrey, c12/1981, after 22/3/1981 by 23/2/1982.

(1) to Resco Railways Ltd, Woolwich Works, London, c5/1983 by 10/6/1983.

The following locos are known to have been scrapped here.

A. M. W. & B. 103		0-6-0ST	IC	MW	212	1866	(a)	Scr c/1949	
		Reb	HC			1907			
		0-4-2ST	OC	KS	4065	1920	(b)	(1)	
011	9	0-6-0ST	OC	HC	1510	1923	(c)	(2)	
110		0-6-0ST	IC	RSHN	7108	1943	(d)	(3)	
122		0-6-0ST	IC	HE	2896	1943	(d)	(3)	
142		0-6-0ST	IC	WB	2739	1944	(d)	(3)	
126		0-6-0ST	IC	HC	1762	1944	(e)	Scr	
162		0-6-0ST	IC	HE	3213	1945	(f)	Scr 3/1961	
173		0-6-0ST	IC	WB	2775	1945	(f)	Scr 3/1961	
120		0-6-0ST	IC	AB	2182	1944	(g)	Scr	
	STAR	0-4-0ST	OC	AB	1940	1927	(h)	Scr	
	P.H.B.	0-4-0ST	OC	HL	3760	1932	(h)	Scr	
	GEORGE	0-4-0ST	OC	AB	1281	1912	(h)	Scr	
	PLANET	0-4-0ST	OC	AB	747	1894	(h)	Scr	
	THURWHIT	0-4-0ST	OC	P	1734	1927	(h)	Scr	
	-	4wDM		MR	5755	1948	(j)	(4)	

(a) ex WD Shoeburyness, by 23/6/1949, earlier Air Ministry, Halton Camp, Bucks.
(b) ex James W Perkins & Co Ltd, Busch Corner, Isleworth, Middlesex, c20/5/1955 (after 9/5/1955, by 28/5/1955): (sold by Ronald Clarke & Co, Brixton to Thurrock Chalk & Whiting Co Ltd but sent here for scrap), still here 15/9/1956.
(c) ex WD, Shoeburyness, by 11/3/1959.
(d) ex WD, Bicester, Oxfordshire, 6/1959.
(e) ex WD, Bicester, Oxfordshire, 1/1961.
(f) ex WD, Shoeburyness, 12/1960.
(g) ex WD, Shoeburyness, 8/1963.
(h) ex Thurrock Chalk & Whiting Co Ltd, West Thurrock, 3/1966.
(j) ex Esso Petroleum Co Ltd, Purfleet c/1967 for scrap.

(1) Scr after 9/1956 by 4/1957.
(2) Scr after 2/1960 by 5/1960.
(3) Scr after 9/1960.
(4) Scr after 9/1971, by 4/1972.

WENNINGTON SAND & BALLAST CO LTD
SAND PITS, Wennington
L194
TQ 548804

Company registered on 30/8/1921 and listed in quarry directories for 1929 until at least 1948, with pits on the Arterial Road. Company went into voluntary liquidation 14/11/1969.
"Contract Journal", 14/10/1953, Hy Butcher to auction 26/10/1953 re Wennington Sand & Ballast Co Ltd, Rainham – plant including track and 2ft gauge diesel locos by RH, 10hp, 16/20hp & 18/21hp.

Gauge : 2ft 0in

-		4wPM	MR	6011	1930	New	s/s
-		4wDM	RH	164347	1933	(a)	(1)
-		4wDM	RH	175408	1935	(a)	(1)
-		4wDM	RH	211608	1941	(a)	(1)

(a) ex Freshwater Sand & Ballast Co Ltd, Fingringhoe.

(1) to Barham & Tait Ltd, Dovers Corner, Rainham, c8/1950.

SAMUEL WEST LTD
BEACH PIT, Lee-over-Sands, near St Osyth
H195
TM 098125

Company registered 13/4/1904 at 40 Trinity Square, EC, to acquire the business carried on by C. Tuff, F. Miskin and S. West as "Tuff and Miskin London Branch" and at Gravesend as "Tuff Miskin and West". Pits listed in quarry directories by 1929. Track removed and site derelict by 1958, company went into voluntary liquidation 21/4/1960.

Gauge : 2ft 0in

-		4wPM	FH	1737	1931	New	(1)
-		4wPM	FH	1771	1931	New	(1)
-		4wPM	MR	9102	1942	(a)	(2)

(a) ex Forestry Commission, Pickering, North Yorkshire.

(1) derelict /1952; s/s by /1958.
(2) derelict /1952; s/s after 8/1958.

WHITEHALL SECURITIES CORPORATION LTD
GIBBS WHARF, Grays   **N196**
Thames Land Co Ltd until /1945   Loco Shed TQ 596773

The Thames Land Co Ltd was registered 1/11/1916 as a subsidiary of S. Pearson & Sons, to acquire and develop Thames-side properties, which included the Gibbs Works of APCM (1900) Ltd . A standard gauge railway connected jetties to quarries although by 10/1948 only the lower section from jetties to a timber yard north of the A126 road was still in use. Locomotives were stabled in a two-road shed at the wharf, the company also shunted the works of Thos Hedley Ltd until the latter acquired their own locomotive. Rail operations replaced by road transport. Whitehall Securities Corporation Ltd was registered on 30/12/1907 with Sir Weetman D. Pearson, Bart, MP (Viscount Cowdray) as president.

Gauge : 4ft 8½in

	MARFLEET	0-6-0ST	IC	HE	550	1892	(a)	(2)
No.142	WOOLWICH	0-6-0ST	IC	MW	841	1882	(a)	Scr 7/1955
	WEST THURROCK	0-4-0ST	OC	P	1707	1926	(b)	(3)
	CUNARDER	0-6-0ST	OC	HE	1690	1931	(c)	(1)

(a)   ex S. Pearson & Son Ltd, Queen Mary Reservoir (MWB) (1921-1924) contract, Middlesex, /1926.
(b)   ex P, 7/1939;
earlier Butterley Co Ltd, Codnor Park, Nottinghamshire.
(c)   ex Edmund Nuttall, Sons & Co Ltd, Colnbrook Plant Depot, Buckinghamshire, hire, 4/1944.

(1)   returned to Edmund Nuttall, Sons & Co Ltd, Colnbrook Plant Depot, Buckinghamshire, 8/1944.
(2)   to Tunnel Portland Cement Co Ltd, West Thurrock, 2/1955.
(3)   to BPCM, Wouldham Works, West Thurrock, 3/1955.

SAMUEL WILLIAMS (DAGENHAM DOCK) LTD
DAGENHAM DOCK   **P197**
Samuel Williams & Sons Ltd until /1977   TQ 490820
Samuel Williams & Sons (1906) Ltd until 31/1/1907
Samuel Williams & Sons Ltd until 26/11/1906
Samuel Williams & Sons from 1886 until 31/12/1897

This lighterage firm, established in 1855, had a depot in Battersea from 1863 and purchased land on Dagenham Marsh in 1887 from the liquidator of Dagenham Dock Co. The marsh was then filled in with London rubbish and the spoil from underground railway construction, which was transported onto the site using a branch railway from the LT&SR and a pier left by the dock company. Over the years a complex wharfingers business was built up, including barge building and repairing, coal storage, sawmill operations, lighterage and tugs, dredging, cranes, piles, fuel oil, shipping and roadstone. Some parts of the site were let to other firms but remained served by the Dock's own railway system. Between 1909 and 1914 four factories were built and let to tenants. This led to the expansion of the rail system, the purchase of three more locomotives and the building of a new locomotive shed. On 4/2/1944 the Ministry of War Transport (Coal Division) requested from the Railway Executive Committee the loan of a 13" loco for their agents Samuel Williams & Sons. The LNER agreed they could loan a Y3 class Sentinel, however it's not known if the offer was accepted. Radio telephones were introduced on all locomotives in 1954 and 22 miles of track were in daily use in 1955. Following the acquisition of the whole share capital of John Hudson Fuel & Shipping Ltd in 1962, the company changed its name on 10/12/1962 to Williams Hudson Ltd which became a holding company. Samuel Williams & Sons Ltd (a new company registered 3/12/1962) became a subsidiary of Williams Hudson Ltd. Samuel Williams (Dagenham Dock) Ltd went into voluntary liquidation on 20/11/1980.

Rail traffic ceased and all locos were out of use. All staff were made redundant from 5/12/1980, an auction took place on 24–26/6/1981 which included one RH and the four remaining FH locos.

Reference : "Railway Bylines" M. Millichip, October 1997 & February 1998.

Gauge : 4ft 8½in

No.	Name	Type		Builder	Works No.	Year	Notes	Disposal
–		0-6-0ST	IC	SS			(a)	s/s
			reb	Merryweather				
–		0-6-0ST	IC	SS			(a)	s/s
			reb	Merryweather				
No.1		0-4-0ST	OC	MW	635	1877	(b)	(5)
No.2		0-4-0ST	OC	MW	883	1883	(c)	(8)
No.3		0-4-0ST	OC	MW	1039	1887	New	(4)
* –		0-4-0ST	OC	MW	585	1876	(d)	s/s
* –		0-4-0ST	OC	MW	856	1883	(e)	s/s
No.4		0-6-0ST	IC	MW	641	1877	(f)	(17)
No.6	(SWANSEA)	0-6-0ST	IC	HE	1	1865	(g)	(6)
No.7	DEVONPORT	0-6-0T	IC	HE	401	1886	(h)	(2)
5		0-4-0ST	OC				(j)	(1)
[No.8?]	[ACCRINGTON]	0-6-0ST	IC	[MW	951	1885]	(k)	(3)
No.5		4wVBT	VCG	S	5735	1926	New	(9)
No.7		0-6-0ST	IC	MW	1488	1900	(l)	(12)
No.8	BOMBAY	0-6-0ST	IC	MW	1674	1906	(m)	(9)
No.1		0-6-0ST	IC	MW	1590	1903	(n)	(16)
No.3	EDGWARE	0-6-0ST	IC	MW	2045	1926	(o)	(12)
No.9		0-6-0ST	IC	MW	1617	1903		
			reb	YE		1922	(p)	(9)
14	MARYHILL	0-4-0ST	OC	P	1606	1923		
			reb	Sheppard		1939	(q)	(13)
10		0-6-0ST	IC	HC	1526	1924	(r)	(10)
No.11	DUNRAVEN	0-6-0ST	IC	MW	1190	1890		
			reb	Sheppard		1940	(s)	(7)
12		0-4-0ST	OC	AB	1129	1907	(t)	(15)
15		0-6-0ST	IC	HC	1676	1937	(u)	(11)
20		0-6-0DM		HC	D680	1949	New	s/s c/1971
21		0-6-0DM		HC	D701	1949	New	(14)
22		0-6-0DM		HC	D702	1949	New	(14)
23		4wDM		FH	3722	1955	New	(21)
24		4wDM		FH	3768	1955	New	(18)
25		4wDM		FH	3799	1956	New	(18)
26		4wDM		FH	3813	1956	New	(19)
27		4wDM		FH	3945	1960	New	(20)
(28)		4wDM		FH	3997	1963	New	(20)
28		4wDH		FH	3949	1960	(v)	(20)
	CHRISTOPHER	0-4-0DH		RH	437364	1961	(w)	(20)

* one of these locos was presumably 5.

The following cranes were noted here in 9/1945
1 Grafton 480, 7 Grafton 1861, 14 Grafton 2416.

(a)   ex Lucas & Aird, Tilbury Docks (1884-1886) contract.
(b)   ex J. Aird & Co, contractors, /1887.
(c)   ex Lucas & Aird, Tilbury Docks (1884-1886) contract, /1887.
(d)   ex Kirk & Randall Ltd, contractors, Woolwich, London.
(e)   ex Kirk & Parry, contractors, Tilbury, 6/1888.
(f)   ex C.D. Phillips, dealer, Newport, Gwent, /1889;
     earlier Wm Rigby, Market Weighton-Driffield (1887-1890) contract, North Yorkshire, JESSIE.
(g)   ex Sir John Jackson, Grays Plant Depot, /1909.
(h)   ex Sir John Jackson, Grays Plant Depot, after 9/1914, by 4/1918, earlier WD Bulford & Larkhill Camps (1914-1916) contract, Wiltshire.
(j)   origin and identity unknown, here by /1916.
(k)   there is much confusion regarding this loco, if ever here then ex Sir John Jackson, Grays Plant Depot, after 8/1921. **Note,** MW 951 was listed in the original Sir John Jackson auction catalogue for 8/1921 as ACCRINGTON and is likely to have still carried that name if/when acquired by Samuel Williams & Sons Ltd at or after the 1921 auction, if the loco was ever here it most likely carried the No.8. The name DEVONPORT has been associated with this loco, but this is believed to be confusion with HE 401 DEVONPORT. The alternative is that this loco was never here and was sold at the Jackson auction of 8/1921 to Prentce Bros, Burwell, Cambridgeshire, there by 5/1928.
(l)   earlier Davies Middleton & Davies, Caerphilly (RR) (1912-1915) contract, Mid Glamorgan; to here, 7/1929.
(m)   ex C.J. Wills & Sons Ltd, Becontree Housing Estate (LCC) (1921-1934) contract, Essex, BOMBAY, /1932.
(n)   ex H. Arnold & Son Ltd, Dinting Tunnel (LNER) (1931) contract, Cheshire, 3, /1934.
(o)   ex C.J. Wills & Sons, St. Helier Housing Estate (LCC) (1929-1936) contract, Surrey, /1936.
(p)   ex John W. Wilson, contractor, Birmingham, via YE, /1937.
(q)   ex Sheppard & Sons Ltd, Bridgend, /1939; earlier Harecastle Collierys Ltd, Staffordshire.
(r)   ex Ableson (Engineers) Ltd, Sheldon Plant Depot, Birmingham, by 3/1940, originally McAlpine, No.41.
(s)   ex Sheppard & Sons Ltd, Bridgend, /1940; earlier Cribbwr Fawr Collierys Ltd, Mid Glamorgan.
(t)   ex Pumpherston Oil Co Ltd, Midlothian, /1940.
(u)   ex Ableson & Co (Engineers) Ltd, Sheldon Plant Depot, Birmingham, by 6/1942, originally McAlpine, No.86.
(v)   ex TH,26/4/1978; rebuilt from 4wDM;
     earlier Tees & Hartlepool Port Authority, Grangetown, Cleveland.
(w)   ex Patent Shaft Steel Works Ltd, West Midlands, 10/1980.

(1)   s/s, /1924.
(2)   to Thos. W. Ward Ltd, after /1925, s/s.
(3)   If ever here thence to Thos. W. Ward Ltd, Grays, later to Prentice Bros, Burwell, Cambridgeshire, by 5/1928.
(4)   to Thos. W. Ward Ltd, Silvertown, /1937.
(5)   to Thos. W. Ward Ltd, Silvertown, /1937.

(6) to Thos. W. Ward Ltd, thence to NCB Shawcross Colliery, West Yorkshire, 8/1949.
(7) dismantled by 26/3/1946, scrapped /1949.
(8) to Thos. W. Ward Ltd, Grays, for scrap, 5/1947.
(9) to George Cohen, Sons & Co Ltd, Canning Town, for scrap, 1/1957.
(10) to George Cohen, Sons & Co Ltd, Canning Town, for scrap, /1957.
(11) to George Cohen, Sons & Co Ltd, Canning Town, for scrap, c12/1958.
(12) to George Cohen, Sons & Co Ltd, Canning Town, for scrap, /1959, by 22/7/1959.
(13) to George Cohen, Sons & Co Ltd, Canning Town, for scarp, /1960.
(14) to Vancouver Wharf, Canada, 5/1960.
(15) to H. & J.R. Saunders & Co Ltd, Leyton Station, for scrap, 11/1960.
(16) to H. & J.R. Saunders & Co Ltd, Leyton, 7/1961.
(17) preserved on site; then to Alan Bloom, Bressingham Hall, Norfolk, 31/10/1970.
(18) to Resco (Railways) Ltd, Woolwich, London, 11/1981.
(19) to Thames Metal Co Ltd, Greenwich, London, c11/1981.
(20) auctioned 24/6/1981 to 26/6/1981, scrapped c12/1986.

Locos used on land reclamation.
Gauge : 2ft 0in

1	4wPM	MR	5005	1929	New		(2)
2	4wPM	MR	4711	1936	New		(3)
3	4wPM	MR	4721	1937	New		(3)
4	4wPM	MR	4722	1937	New		(4)
5	4wPM	MR	1234	1918		(a)	(1)
6	4wPM	MR				(b)	(3)
7	4wPM	MR				(b)	(1)
8	4wPM	MR				(b)	(3)
9	4wPM	MR				(b)	(3)
10	4wPM	H				(b)	(1)

(a) ex WDLR 2955.
(b) origin and identity unknown.

(1) s/s after 3/1947.
(2) to East Acton Brick Co Ltd, Middlesex, via George W. Bungey Ltd, Hayes, Middlesex, /1951.
(3) to George W. Bungey Ltd, Hayes, Middlesex, /1951.
(4) s/s after 3/1953.

The following locos were also used here on hire as detailed :
Gauge : 4ft 8½in

	-	0-6-0ST	IC	MW	2005	1921	(a)		(1)
	WOODCROFT	0-6-0ST	IC	HC	1583	1926	(b)		(5)
	-	0-6-0ST	IC	HC	1061	1914	(c)		(2)
	ALICE	0-6-0ST	OC	AE	1460	1903	(d)		(4)
	GREENHITHE	0-4-0ST	OC	P	1742	1927	(e)		(3)
P16	KILINDINI	0-6-0ST	OC	WB	2168	1921	(f)		(11)
SL 2	CUNARDER	0-6-0ST	OC	HE	1690	1931	(g)		(12)
	SOUTHAMPTON	0-6-0ST	OC	HE	1647	1931	(h)		(7)
	FORTH	0-6-0ST	OC	AB	1844	1924	(j)		(9)
	BOBBY	0-6-0ST	OC	HC	1593	1927	(k)		(8)

		TRAFFORD PARK	0-6-0ST	OC	HE	1689	1931	(l)	(6)
		BRAMLEY No.6	0-6-0ST	OC	HE	1644	1929		
				reb	HE		1938	(m)	(10)
	EN 3	NUTTALL	0-6-0ST	OC	HE	1686	1931	(n)	(13)
		TW 3434	0-4-0DM		JF	22934	1941	(o)	(14)

(a)   ex Thos. W. Ward Ltd, Grays, TW317.
(b)   ex John Mowlem & Co Ltd.
(c)   ex Sir Lindsay Parkinson & Co Ltd, contractor, by 7/1943.
(d)   ex Ford Motor Co Ltd, Dagenham, for repairs by 11/1931.
(e)   ex BPCM, Greenhithe, Kent, by 30/9/1945.
(f)   ex Pauling & Co Ltd, by 30/9/1945; returned ; and again /1954, by 1/1955.
(g)   ex Edmund Nuttall, Sons & Co Ltd, Colnbrook Plant Depot, Buckinghamshire, 1/1945; returned 9/1945; and again 11/1954.
(h)   ex John Mowlem & Co Ltd, by 26/3/1946.
(j)   ex John Mowlem & Co Ltd, by 26/3/1946.
(k)   ex John Mowlem & Co Ltd, Chingford Reservoir contract, /1947.
(l)   ex John Mowlem & Co Ltd, /1948.
(m)   ex Thos. W. Ward Ltd, Grays, 6/1952.
(n)   ex Edmund Nuttall, Sons & Co Ltd, Colnbrook Plant Depot, Buckinghamshire, 7/1952; returned 4/1953; and again 10/1954.
(o)   ex Thos. W. Ward on hire, here 28/5/60.
(1)   returned to Thos. W. Ward Ltd, Grays.
(2)   returned to Sir Lindsay Parkinson & Co Ltd.
(3)   returned to BPCM, Greenhithe, Kent.
(4)   returned to Ford Motor Co Ltd, Dagenham, by 5/1932.
(5)   returned to John Mowlem & Co Ltd, 8/1945.
(6)   returned to John Mowlem & Co Ltd, by /1949?
(7)   returned to John Mowlem & Co Ltd, 2/1949.
(8)   returned to John Mowlem & Co Ltd, by 7/1949.
(9)   returned to John Mowlem & Co Ltd, by 29/7/1949.
(10) returned to Thos. W. Ward Ltd, Grays, by 1/1953.
(11) returned to Pauling & Co Ltd, /1955 after 18/2/1955.
(12) returned to Edmund Nuttall, Sons & Co Ltd, Colnbrook Plant Depot, Buckinghamshire, 3/1955.
(13) returned to Edmund Nuttall, Sons & Co Ltd, Colnbrook Plant Depot, Buckinghamshire, 4/1956.
(14) returned to Thos. W. Ward Ltd, by 9/1960.

SIDNEY-WILMOT LTD
Great Horkesley, near Colchester   H198

Company registered 27/9/1928 as carriers & forwarding agents at Nayland, Suffolk, directories for 1933 & 1937 list Sidney-Wilmot Ltd as haulage contractors at The Grove, Great Horkesley; company went into voluntary liquidation 10/11/1960. The company operated gravel pits at Thorington Street (TM 010350) which is just over the county boundary in Suffolk.

Gauge : 2ft 0in

| | 4wDM | RH | 187099 | 1938 | New | (1) |

(1)   to Alresford Sand & Ballast Co Ltd, Alresford Creek, near Colchester, after 22/12/1942 by 25/6/1943.

WIVENHOE SAND, STONE & GRAVEL CO LTD
WIVENHOE WORKS

H199
TM 047224

The company was formed 21/7/1925 to take over the business of Frank Pertwee. A small quarry, which was opened early in 1920, with a short internal narrow gauge system running between the pits and screening plant. The line was originally horse worked and returned to horse working after disposal of the locomotive. The tramway was replaced by road transport; the company was sold to the Tilbury Group in 1960 and later to Redland Aggregates in 1985, and was dissolved 27/10/1998.

Reference: The Narrow Gauge No.154.

Gauge : 2ft 0in

-	4wPM	L	3854	1931	New	s/s by /1937

SECTION 2
CONTRACTORS LOCOMOTIVES
LONDON & CONTINENTAL RAILWAYS
CHANNEL TUNNEL RAIL LINK, Dagenham   **PC1**

The CTRL project was designed and managed by Rail Link Engineering (RLE) a consortium of Bechtel Ltd, Arup Group Ltd, Systra & Sir William Halcrow & Partners Ltd. Work began on section 2 of the CTRL in 2001, this comprised of the construction of 39.4km of the high speed line from Ebbsfleet (Kent) to St. Pancras (London), mainly in tunnel. Contract 220 – (Nishimatsu-Cementation Skanska JV) – for 7.5km of twin tunnel from Kings Cross to Stratford; Contract 230 – (Skanska) – 1km long Stratford Box for the new Stratford International Station; Contract 240 – (Costain/Skanska/Bachy Soletanche JV) – 4.7km of twin tunnel between Stratford and Barrington Road, Newham; Contract 250 – (Nuttall-Wayass/Freytag-KierJV) – 5.3km twin tunnel from Barrington Road to Dagenham; Contract 320 – (Hochtief/J. Murphy & Sons JV) – 2.5km twin tunnel from Thurrock Marshes to Swanscombe, Kent. A new servicing & maintenance Depot was established at Temple Mills and the line designated HS1 (High Speed 1) opened 14/11/2007.

Gauge : 4ft 8½in

14029	0-6-0DH		Sdn	1965	(a)	(2)
	4wDM	R/R	UCA	2004	(b)	(9)
2062 CAT 3406	4wDH		Jenbach		(b)	(4)
(D9504)	0-6-0DH		Sdn	1964	(c)	(1)
ETI 41 (68811)	4wDHR		Perm 001	1987	(d)	(5)
20189	Bo-BoDE		EE 3670	1967	(e)	(3)
			VF D1065	1967		
08588	0-6-0DE		Crewe	1959	(f)	(6)
73114	Bo-BoDE/RE		EE 3582	1966	(g)	(7)
			EEV E352	1966		
DX53 CXA	2w-4DMR	R/R	SRS/Volvo		(h)	(8)
N442 OGJ	4wDM	R/R	SRS/Volvo		(h)	(8)
N443 OGJ	4wDM	R/R	SRS/Volvo		(h)	(8)

(a)   ex Stratrail Ltd on hire 19/7/2004;
      to Nene Valley Railway, for repairs, 19/9/2004; returned, 23/2/2005.
(b)   ex ACT Joint Venture, CTRL Swanscombe site, Kent, after 5/10/2004, by 16/10/2004.
(c)   ex Stratrail Ltd, on hire, 21/11/2004, transferred from Swanscombe site, Kent.
(d)   ex Elec-Track Installations Ltd, Middlewich, Cheshire, on hire, by 26/2/2005.
(e)   ex Wabtec Rail, South Yorkshire, 22/1/2006, on hire from R.T. Rail, Cheshire.
(f)   ex R.T. Rail, Cheshire, on hire, 28/9/2006.
(g)   ex The Battlefield line, Shackerstone, Leicestershire, 24/4/2007.
(h)   ex SRS Rail System Ltd, Bolsover, Derbyshire, on hire.

(1)   to Nene Valley Railway, for repairs, 18/3/2005, returned 7/11/2005;
      to NVR 13/1/2006, returned 2/3/2006;
      to Nene Valley Railway for repairs 25/1/2007.
(2)   to CTRL Swanscombe site, Kent, 9/2005, returned;
      to Nene Valley Railway for repairs 14/4/2006, returned 7/9/2006;
      to Nene Valley Railway, 29/9/2006, returned 24/1/2007;
      to Nene Valley Railway 24/4/2007 off hire.

(3) returned to Wabtec Rail, South Yorkshire, 27/2/2006, off hire.
(4) exported c10/2006 by 11/2006.
(5) returned to Elec-Track Installations Ltd, by 22/10/2006 off hire.
(6) to Wabtec Rail Ltd, South Yorkshire, by 6/11/2006.
(7) to The Battlefield Line, Shackerstone, Leicestershire, c24/7/2007.
(8) returned to SRS Rail System Ltd, Bolsover, Derbyshire, off hire.
(9) assumed returned to UCA, Merskem, Belgium.

PLANT YARD, Choats Manor Way, Dagenham   PC2
TQ 481828

Locomotives stored at this plant yard pending disposal.
Gauge : 4ft 8½in

4wDH		Dtz 57871	1965		
	reb	Newag 132	2004	(a)	(1)
4wDH		(German)			
	reb	Newag 130	2004	(b)	(1)
4wDH		(German)			
	reb	Newag	2004	(b)	(1)
4wDH		Dtz			
	reb	Newag	2004	(b)	(1)

(a) originally Tanklarger & Transportmittel GmbH VTG, Hamburg, Germany, rebuilt by Newag, Oberhausen, Germany, /2004.
(b) origin unknown, rebuilt by Newag, Oberhausen, Germany, /2004.
(1) s/s after 13/5/2007, by 1/12/2007, (possibly returned to Newag, Oberhausen, Germany).

WILLIAM JOHN ADCOCK
EAST TILBURY CONTRACT   MC3
Construction works at the New Battery in East Tilbury village (Coalhouse Fort), details uncertain. The Grays and Thurrock Gazette for 23/11/1889 carried a report of a fatal accident at the New Fort at East Tilbury. James Joiner of Eythorne near Dover was riding on a truck full of ballast being hauled or propelled by a new engine being used for the first time that day when he was pitched off with fatal results. The contractor for the works was **William John Adcock** of Dover, who died on 1/4/1907.
Reference : Industrial locomotive No.55.
Gauge : 4ft 8½in

tank loco		(a)	s/s

(a) origin and identity unknown, engine said to have been "cased in".

JOHN AIRD & SONS
BECKTON CONTRACT   AC4

Construction of Beckton Gasworks for the North Thames Gas Board, 1868 – 1870.
Gauge : 4ft 8½in

      KINGSTON   0-4-0ST   OC   MW   221 1866   (a)   (1)

(a)   ex Tooting-Wimbledon (LBSCR/LSWR) (1865-1868) contract, Surrey.

(1)   later Lucas & Aird, Tilbury Dock (1884-1886) contract, by 10/1883.

AMEC CONSTRUCTION LTD
CLACTON CONTRACT   HC5

Company registered 9/7/1992. Contract for the construction of a sewer from Clacton to Holland on Sea, comprised 4.6km of 2.1m diameter tunnel, contract sum £10.5m, work started c3/1997, due for completion in 1998.

Six trains were used during construction hauled by CE battery locos, details unknown.

THURROCK CONTRACT   NC6

Contract for the construction of a cable tunnel for the National Grid from West Thurrock to Littlebrook. Four trains using battery locos were used during construction for the removal of earth, details unknown. Tunnel completed 1/2003.

HAROLD ARNOLD & SON
COLCHESTER CONTRACT   HC7

Construction of Colchester Asylum, - 1912.
Gauge : 2ft 0in

      MASHAM   0-6-0ST   OC   TG   366 1904   (a)   (1)

(a)   ex Harrogate Corporation Water Works, Masham, North Yorkshire, 5/1912.

(1)   advertised for sale by R.H. Longbotham, Wakefield, ("Machinery Market" 4/5/1917); Later at A. Batchelor Ltd, Halling, Kent.

HAROLD ARNOLD & SONS LTD
BARKING CONTRACT   LC8
Harold Arnold & Sons until 28/7/1916

Construction of the Metropolitan District Railway, Barking-Upminster Extension, 1931-1932. This involved the construction of a new double-track line (7¾ miles) to the north of the existing LT&SR railway. The Grays & Tilbury Gazette of 14/2/1931 reported that fifteen tractors and 140 tip wagons were in use on five miles of light railway and 300 men and five petrol navvies were employed on the works. The new line opened to traffic on 12/9/1932.

Gauge : 4ft 8½in

1		4wVBT	VCG AtW	117 1930	[New?]	(1)
(11)		0-6-0tank			(a)	s/s
	(UPNEY)	0-6-0tank			(a)	s/s

(a) origins and identities not known.
(1) to Dinting Tunnel (LNER) (1931) contract, Cheshire.

Gauge : 2ft 0in

-	4wPM	MR	5075 1930	New	s/s
-	4wPM	MR	5076 1930	New	(1)
-	4wPM	MR	5077 1930	New	s/s
-	4wPM	MR	5078 1930	New	s/s
-	4wPM	MR	5223 1930	New	s/s
-	4wPM	MR	5224 1930	New	(2)

(1) later with City of Leeds Waterworks Dept, Holbeck, Leeds, Yorkshire, by 25/8/1948.
(2) later with Sussex & Dorking United Brick Co Ltd, Old Park Brickworks, Farnham, Surrey, by 26/10/1948.

SIR WILLIAM ARROL & CO LTD
ROYAL ALBERT DRY DOCK CONTRACT   AC9

Contract for works at the Royal Albert Dry Dock for ship repairers R & H Green & Silley Weir Ltd, in 1921, exact details unknown.

Gauge: 4ft 8½in

TRUDY	0-6-0ST	OC	HE	574 1893	(a)	(1)

(a) ex Sir William Arrol & Co Ltd, Bowling Petrol depot contract, Dunbartonshire, after 5/1920, by 3/1921.
(1) s/s after 10/1921.

BALFOUR BEATTY & CO LTD
PITSEA CONTRACT   KC10

Construction of sewage treatment works on Pitsea Marsh (TQ 736870) and a trunk sewer for Basildon Development Corporation c1952-1953. The contract was for the construction of the power house and pumping station, settling and aeration tanks and sludge drying beds; also 2,800 yards of 45in diameter concrete pipes laid in a heading to a maximum depth of 53ft and 1,580 yards of concrete pipes of 15in to 45in diameter laid in open cut trenches. The contract also included the construction of roads, pipelines and ancillary works. On a visit to the Balfour Beatty Plant Depot at Brimsdown Middlesex on 7/11/1953 it was stated that six locomotives were out on a sewage works contract at Pitsea, the identity of the locos has not been established.

BALFOUR BEATTY
ROMFORD OVERHEAD LINE MAINTENANCE DEPOT, Romford   EC11
TQ 499880

Locomotives stored at this Plant Depot between contracts.

Gauge : 4ft 8½in

WALTER	4wBE	CE	B4427A	2006	(a)	(1)
ANNE	4wBE	CE	B4427B	2006	(a)	(1)
LOU	4wBE	CE	B4427C	2006	(a)	(1)
KITTY	4wBE	CE	B4427D	2006	(a)	(1)

(a) ex Waterloo & City Line refurbishment contract, 8/2006 by 23/8/2006.

(1) to East London Railway Extension contract after 7/1/2008 by 12/1/2008.

JOHN BAND
GRAYS CONTRACT   NC12

John Band a builder of Chancellor Road, Southend and previously of Grays, had a contract for the construction of Grays Sewage Works, 7/1892 - 2/1894. John Band died 1/9/1902.

Reference: Industrial Locomotive No.55 p144 - ILS

Gauge : 4ft 8½in

	VBT	(a)	(1)

(a) origin and identity unknown.

(1) for sale, 2/1894 and 6/1894; s/s.

HENRY BOOT & SONS (LONDON) LTD
Henry Boot & Sons Ltd until 26/11/1919
Henry Boot & Son until 18/4/191

CANNING TOWN CONTRACT   CC13

Contract for the construction of Canning Town Glass Works for British Glass Industries Ltd.

"Contract Journal" 9/11/1921 for sale – plant inc 2ft gauge petrol loco – H. Boot & Sons (London) Ltd, Stephenson Street, Canning Town.

CHARLES BRAND & SON LTD

CENTRAL LINE CONTRACT   DC14

Contract for construction of the Central Line between Wanstead and Gants Hill – 1938; the contract was signed 31/5/1938 at a price of £500,000 and was for 1½ miles of tunnel to be built in concrete segments, use of locomotives unknown.

DARTFORD CONTRACT   NC15
Contract for the construction of a pilot tunnel for the Dartford Road Tunnel commenced 6/1936. In 4/1937 an arrangement was made with Thurrock Chalk & Whiting Co Ltd to use their line and build a siding off it to receive materials to the shaft on the Essex side of the River Thames, construction of the main tunnel was delayed by war. Use of locos unknown.

BRASSEY, OGILVIE & HARRISON
BISHOPS STORTFORD CONTRACT   FC16
Construction of the Bishops Stortford & Dunmow Railway, 2/1864 - 10/1866 (although the line did not open for traffic until 22/2/1869).

"The Engineer" 23/4/1869 Augustus Portway to auction 4/5/1869 re Brassey Ogilvie & Harrison, Bishops Stortford & Dunmow Rly, at Gt Dunmow station – plant inc nearly new tank loco by Gilks Wilson & Co.

Gauge : 4ft 8½in

    - .   0-4-0T   GW   160 1863   New   (1)

(1)   a tank loco by Gilkes, Wilson, built c/1864, was for sale by auction on 4/5/1869.

BRASSEY & MACKENZIE
COLCHESTER CONTRACT   HC17
Construction of the Eastern Union Railway in 1844-1846; an unidentified loco was used during the construction of an embankment between Colchester and the River Stour in 1845.

CAFFIN & CO LTD
WALTON-ON-NAZE CONTRACT   HC18
Company registered 29/3/1921 as Railway contractors and engineers at 25 Craven Street, Strand, WC. This contract was for the construction of a new line from Frinton-on-Sea to Walton-on-Naze, 1 mile (on a new alignment to replace a route threatened by coastal erosion), for the LNER, 1929.

Gauge : 4ft 8½in

    -                        0-6-0ST   IC   MW   1576 1903   (a)   s/s

(a)   ex Jersey Marine (GWR) (1928) contract, Swansea, Glamorgan, by 5/1929.

SHOEBURYNESS CONTRACT   JC19
Shoeburyness Sidings Extension construction for LMSR, 1932.

Gauge : 4ft 8½in

    FRANK   0-6-0ST   IC   MW   1642 1904
                                        reb   MW   1913   (a)   (1)

(a)   ex Hornsey Marshalling Yard (LNER) (1930) contract, Middlesex, after 8/11/1930.

(1)   to Leigh-on-Sea New Station (LMSR) (1932-1934) contract.

LEIGH-ON-SEA CONTRACT   JC20

Construction of a new station (on a new site) at Leigh-on-Sea and widening the line to Chalkwell for the LMSR, 1932-1933. The new station at Leigh-on-Sea opened on 1/1/1934.
Gauge : 4ft 8½in

-	0-6-0ST	OC	BH	889	1890	(a)	(1)
FRANK	0-6-0ST	IC	MW	1642	1904		
		reb	MW		1913	(b)	(2)

(a) earlier H. Arnold & Sons Ltd, Bedlam Royal Hospital (LCC) (1928-1930) contract, Shirley, Surrey.
(b) ex Shoeburyness Sidings (LMSR) 1932) contract.
(1) to Park Royal Trading Estate (1933) contract, Middlesex, after 10/1932.
(2) [possibly later High Wycombe (GWR) (1938) contract, Bucks;]
    later Ruislip-Northolt (GWR) (1938-1940) widening contract, Middlesex.

CHRISTIANI & NIELSEN LTD
GRAYS PLANT DEPOT, Rectory Rd, Grays   NC21
Christiani & Nielsen until 10/6/1932   TQ 624783

This well established Danish firm of contractors have employed locos on a number of contracts. Some of these locos may have passed through this depot in addition to those listed below.
Gauge : 2ft 6in/750mm

-	4wDM	RH 7002/0566/3	1965	(a)	(1)	
-	4wDM	RH 7002/0566/4	1965	(a)	(1)	
-	[4w?]DM	PMA#	13403	(b)	s/s	

\# PMA = Pendershaab Mashinfabric A/S.

(a) ex unknown location (New to Christiani & Nielsen, Spain).
(b) ex unknown location (New to Christiani & Nielsen, Spain).
(1) auctioned 18/5/1971; s/s.

Gauge : 2ft 0in

-	4wPM	MR	915	1918	(a)	(1)

(a) origin unknown, originally WDLR.
(1) to John J. Shardlow & Co contractors, Leicester.

BARKING CONTRACT   PC22

Contract for the construction of a jetty and pumphouse at Barking Power Station c/1935, several 2ft gauge petrol locos are said to have been used on the contract.

M.J. CLANCY & SONS LTD
BRENTWOOD CONTRACT EC23

Company registered 24/3/1958. Work at Brentwood, details unknown, 1983-1984.
Gauge : 2ft 0in

| | 4wBE | WR M7556 1972 | (a) | (1) |

(a) ex Track Supplies & Services Ltd, Old Wolverton, Buckinghamshire, hire, 7/1983.

(1) returned to Track Supplies & Services Ltd, Old Wolverton, Buckinghamshire, off hire, c/1984.

S.A. CLARK LTD
BARKINGSIDE DC24

Company (registered 20/12/1924) and listed as manufacturers and dealers in bricks & tiles etc with premises on part of the former site occupied by Muirhead, Macdonald, Wilson & Co Ltd. In 1937 the company amalgamated with B. Finch & Co Ltd, manufacturing, sanitary & heating engineers of Belvedere Road, Lambeth, who in 1938 moved their registered office to Belvedere Works, Barkingside. The company was reported as having contracts with the War Office these may have originated with S.A.Clark Ltd, it being stated that they were obtaining orders in connection with the rearmament programme; Mr S.A.Clark became managing director of the new company. What the loco was doing at Rochester has not been established; the name MAYBURY is believed to be associated with Sir Henry Maybury, Director General of the Roads Department of the Ministry of Transport which was formed in 1919 and took over the powers and duties of the Road Board; Maybury was in France during WW1 in charge of road building for the Army.

Gauge : 4ft 8½in

| No.3 MAYBURY | 0-4-0ST OC | DK | 9976 1901 | (a) | (1) |

(a) ex Cadbury Bros Ltd, Bournville, West Midlands, No.3.

(1) to Rochester Wharf, Kent by 8/1930, later Settle, Speakman & Co, Queenborough, Kent.

CLEVELAND BRIDGE AND ENGINEERING CO LTD
TILBURY DOCK CONTRACT MC25

Construction of a 1,140ft passenger landing stage for ocean going liners on the riverside at Tilbury at a cost of £350,000, 1927-1929.
Reference: Locomotive Magazine 15/7/1930.
Gauge : 4ft 8½in

| | 0-4-0ST OC | BLW 45285 1917 | (a) | (1) |

(a) ex Thos. W. Ward Ltd, Grays, TW177, /1928.

(1) to Thos. W. Ward Ltd, Sheffield, 37691, /1930.

COCHRANE & SONS
PARKESTON QUAY CONTRACT   HC26

Construction of New Iron Quay and other works at Parkeston Quay, 1882 -1883.
Gauge : 4ft 8½in

-		0-4-0VBT	OC	Balmforth	(a)	(1)

(a) origin unknown.

(1) sold by auction with other plant, 14/9/1883; s/s.

COLCHESTER COUNCIL
COLCHESTER BY PASS CONTRACT   HC27

Colchester Council purchased £4000 worth of plant from J.C. Oliver in 1930 for the construction of a by-pass road. Council minutes show the council purchased a further loco from Oliver in 1931. The contract was completed in 1933, after which J.C. Oliver purchased surplus plant.

Gauge : 2ft 0in

-	4wPM	FH	1677	1930	New	s/s
-	4wPM	FH	1678	1930	New	s/s
-	4wPM	FH	1779	1931	New	s/s
-	4wPM				(a)	s/s
-	4wPM	[FH/MR?]			(b)	s/s

(a) origin and identity unknown, photographic evidence shows this loco to be similar to an "Excelsior" petrol tractor of the type fitted with a 'V' skip body.

(b) origin and identity unknown, photographic evidence shows this to be a 20hp "Simplex" loco.

CONCRETE PILING LTD
BARKING CONTRACT   PC28

Company registered 23/12/1911 as contractors for concrete piling and concrete foundations and other works with an address at 43 Broadway Court, Victoria Street, Westminster. Construction of foundations of Barking Power Station, Creekmouth. (TQ 465819) (1926-1930) for the County of London Electric Supply Co Ltd.

Gauge : 4ft 8½in

WOLF	0-4-0ST	OC	MW		(a)	s/s
LION	0-4-0ST	OC	MW		(a)	s/s

(a) origin and identity unknown.

Gauge : 2ft 0in

-	4wPM	MR	6014	1931	New	s/s
-	4wPM	MR	6017	1931	New	s/s

RICHARD COSTAIN LTD
Richard Costain & Sons until 10/5/1923
BECKTON CONTRACT   AC29
Contract for upgrading the Northern Outfall Works, Beckton, with new sewage tanks for London County Council.
Gauge : 2ft 0in

		4wDM	MR	5877	1935	New	(1)
-		4wDM	MR	5878	1935	New	(2)

(1)   to Dan & Stone Ltd, Marsh Works, Ham Road, Faversham, Kent, by 4/5/1938.
(2)   later ICI Ltd, Hindlow Limeworks, Derbyshire.

DAGENHAM (ELM PARK) CONTRACT   LC30
Contract for the construction of the Elm Park housing estate c/1934; the two locos listed below were delivered to Wye Bridge Farm (TQ 519857).
Gauge : 2ft 0in

-		4wDM	MR	5638	1934	(a)	s/s
-		4wDM	MR	5640	1933	(a)	s/s

(a)   originally Petrol Loco Hirers, Bedford, sold to Greenham Plant (Hiring) Co Ltd, 9/3/1934, despatched here 10/3/1934.

HORNCHURCH CONTRACT   LC31
Contract for Hornchurch Sewage Works (TQ 513850?) in Rainham Rd, Hornchurch.
Gauge : 2ft 0in

		4wDM	MR	5876	1935	New	s/s

COSTAIN CIVIL ENGINEERING LTD
ROYAL VICTORIA DOCK DRAINAGE SCHEME   AC32
Main site and compound at gate 1, North Woolwich Road, TQ 413803. Work was in progress in 1989.
Gauge : 1ft 6in

S137	4wBE	CE	5911A	1972	(a)	(1)
CONTEX 2	4wBE	CE	5940B	1972	(b)	(1)
	4wBE	Greensburg			(c)	(1)
	4wBE	Greensburg			(c)	(1)

(a)   originally C.V. Buchan.
(b)   originally A. Streeter.
(c)   ex Jay-Dee, Detroit, USA, /1986.

(1)   later Wetherby Plant Yard, North Yorkshire.

COSTAIN, SKANSKA, BACHY SOLETANCHE JOINT VENTURE
CHANNEL TUNNEL RAIL LINK PHASE 2, Contract 240　　　　　　　　　PC33

Construction of 4.7km of twin bore tunnels of 7.15m diameter from Stratford to Barrington Road, Newham, (2001 – 2004).
Gauge : 900mm

		4wDH	Schöma	5554	1998	(a)	(2)
	-	4wDH	Schöma	5555	1998	(a)	(1)
	-	4wDH	Schöma	5556	1998	(a)	(2)
	-	4wDH	Schöma	5557	1998	(a)	(1)
	-	4wDH	Schöma	5558	1998	(a)	(1)
	-	4wDH	Schöma	5559	1998	(a)	(1)

(a)　ex Schöma, Germany, /2002.
(1)　returned to Schöma, Germany.
(2)　returned to Schöma, Germany, thence to UTE Canal de Navarra, Spain, /2004.

CROWLEY, RUSSELL & CO LTD
BECKTON CONTRACT　　　　　　　　　　　　　　　　　　　　　　AC34

Contract for new settling tanks and foundations for new electric pumps at Northern Outfall Works, Beckton, for London County Council, contract completed c4/1952.
Gauge : 2ft 0in

	-	4wDM	MR	7305	1938	New	(1)

(1)　delivered new to the Northern Outfall Works in 1938, noted on this contract, 4/1950; may have been used elsewhere during WW2 and then returned here.

GRAYS PLANT DEPOT, Rectory Road, Grays　　　　　　　　　　　　NC35

Gauge : 2ft 0in

	-	4wDM	MR	7305	1938	(a)	s/s
	-	4wDM	MR	7454	1939	(b)	s/s
	-	4wDM	RH	178994	1936	(c)	s/s
	-	4wDM	RH	187046	1937	(d)	s/s

(a)　New to Northern Outfall works contract Beckton for London County Council; here on 9/7/1958 and 22/7/1959.
(b)　New to unknown contact Arbroath, Scotland; here on 9/7/1958 and 22/7/1959.
(c)　New on hire to unknown contract at Bothwell, Lanarkshire, Scotland; here on 22/7/1959.
(d)　New to Barclay, Ross & Hutchinson Ltd, Aberdeen, for Aberdeen County Council, Rocks Quarry, Balmedie, near Aberdeen, Scotland; here on 9/7/1958.

DEMOLITION & CONSTRUCTION CO LTD
Company registered 19/7/1923, as a subsidiary of **Cementation Co Ltd.**
BECKTON CONTRACT　　　　　　　　　　　　　　　　　　　　　　　　　　　AC36
Contract at Northern Outfall, for London County Council, 6/1956 – 4/1957, for the construction of a new pier.
Gauge : 2ft 0in

L62	4wDM	HE	4345	1952	(a)	(2)
L61	4wDM	HE	4300	1950	(a)	(1)
TW2861	4wDM	RH	217963	1942	(b)	(3)

(a)　originally Higgs & Hill Ltd (possibly Singapore contract).
(b)　ex Thos. W. Ward Ltd, Grays, hire.

(1)　to Folkestone Sea Wall repair contract, Kent, by 9/1958.
(2)　to Folkestone Sea Wall repair contract, Kent, by /1959.
(3)　returned to Thos. W. Ward Ltd, off hire.

EASTERN COUNTIES RAILWAY
LONDON – BRENTWOOD CONTRACT　　　　　　　　　　　　　　　　　　　EC37
Construction of the Eastern Counties Railway 1838 – 1843.

Several contractors were employed on the works, these were – James Munday (contract No.1), B.& N. Sherwood (No's. 2,3,4, 7,8,9, & 11), George Munday (No.5), John Woodcock (No.6 & No.20), J.Curtis (No.6d & No.18), John Burge (No.10), John Jay (No.19), James & George Munday (No.21).

John Braithwaite was appointed chief engineer to the ECR in 1836. In 2/1838 an order was placed for the construction of two locomotive engines with a further order for two more placed in 3/1838 – these locos were to be built by Braithwaite Milner & Co under the superintendence of the engineer John Braithwaite. In 4/1838 the first locomotive boiler was delivered from Fairbairn & Co with a second boiler and locomotive frame delivered from the same firm in 5/1838. In 6/1838 a contract was agreed with Braithwaite Milner & Co for a further three engines & tenders to be delivered within four months, a further contract for three more engines was agreed later the same month – these are thought to be 2-2-0 passenger engines. On 26/6/1838 the engineer reported to the Board that the first engine to be ordered was now ready for use and in 7/1838 this loco was in use on construction of the Stratford Embankment. Also in 7/1838 locomotive boilers and frame work were delivered from the Horsley Iron Co at a cost of £435-6s-0d. The second locomotive ordered was delivered on 7/8/1838 with the third engine delivered in 9/1838.

On 11/12/1838 the engineer reported to the board that during the period from 13/11 – 11/12 he had a total number of 2224 men employed on the works and four locomotive engines at work; by 2/1839 six engines were employed day and night.

The line opened from Devonshire Street to a temporary terminus at Romford on 18/6/1839, and fully opened to Romford 20/6/1840 and to Brentwood 1/7/1840.

Reference : National Archives, Rail 186.

Gauge: 5ft 0in

ESSEX	0-4-0	IC	#	1838	New	s/s
MIDDLESEX	0-4-0	IC	#	1838	New	s/s
NORFOLK	0-4-0	IC	#	1838	New	s/s
SUFFOLK	0-4-0	IC	#	1838	New	s/s

| | | 0-4-0 | IC | # | 1839 | New (a) | (1) |
| | | 0-4-0 | IC | # | 1839 | New (b) | (2) |

Built by Braithwaite, Milner & Co.

(a)   converted to standard gauge, 141.
(b)   converted to standard gauge, 142.

(1)   to Mr Brassey, contractor, 1/1849.
(2)   to H.P. Burt, Blackfriars Road, London, 1/1849.

ESSEX COUNTY COUNCIL
BENFLEET CONTRACT   JC38
Construction of the Benfleet By Pass Road during 1937-1938. In 1930 Essex County Council approved a scheme to construct a by pass road from the Anchor Inn at the top of Vicarage Hill to a point near the railway station at the junction of Station Road and the High Street. In 1931 the council authorised the purchase of land for the scheme but construction was postponed until about 1937. Photographic evidence exists of a narrow gauge Simplex loco and V skips in use tipping concrete during construction.
Gauge : narrow

| | | [4wPM/DM?] | [MR/FH?] | | (a) | s/s |

(a)   origin and identity unknown.

FINETURRET LTD
Incorporated 2/7/1975
ROYAL VICTORIA DOCK DRAINAGE SCHEME   AC39
Contract for construction work on the Royal Docks drainage scheme, 1989, exact details unknown.
Gauge : 1ft 6in

| L13 | | 4wBE | CE | 5965B 1973 | (a) | (1) |

(a)   ex Miller Construction, Rugby, Warwickshire, 19/4/1989.

(1)   returned to Miller Construction, Rugby, by 13/9/1990.

W. & C. FRENCH LTD
The firm of W. & C. French was founded in 1870 by William and Charles French and was incorporated as a private company on 23/3/1931 to acquire the business of Public Works and Haulage Contractors and Brick Makers. The company changed its name on 9/11/1973 when it merged with J.L. Kier & Co Ltd to form French Kier Holdings Ltd.

LANGHAM VALLEY RESERVOIR CONTRACT   HC40
Construction of Langham Valley Reservoir for South Essex Waterworks Co, 1930.
Gauge : 2ft 0in

		0-4-0ST	OC	KS	4020 1919	(a)	s/s
		0-4-0ST	OC	KS	4022 1919	(b)	(1)
		0-4-0ST	OC	KS	4273 1922	(c)	(1)
		0-4-0ST	OC	KS	4274 1922	(c)	(1)

(a) ex William Jones Ltd, dealer, Greenwich, London;
    earlier Stewart & McDonnell, Kingston By- Pass contract, Surrey, /1927.
(b) earlier Hackney Marsh Sewage Works contract, London.
(c) ex Foundation Co Ltd, Brimsdown Power Station/Plant Depot, Middlesex, 18/6/1930.
(1) to Buckhurst Hill Plant Depot on completion of contract.

SHELLHAVEN CONTRACT   KC41

Contract at Shellhaven in 1958, details unknown.
Gauge: 2ft 0in

| 2975 | | 4wDM | RH | 187120 | 1938 | (a) | (1) |

(a) ex Buckhurst Hill Plant Depot, by 10/1958.
(1) returned to Buckhurst Hill Plant Depot.

STRATFORD CONTRACT   CC42

Contract in Bisson Road Stratford in 1934, details unknown.
Gauge : 2ft 0in

L259		4wPM	MR	5414	1934	New	s/s
L149		4wPM	MR	5415	1934	New	s/s
L260		4wPM	MR	5416	1934	New	s/s

BUCKHURST HILL PLANT DEPOT   DC43

Plant Depot (at TQ 404940) located in Epping New Road where locos were stored and repaired between contracts. Depot closed c/1965.
Gauge: 2ft 0in

		0-4-0ST	OC	KS	4022	1919	(a)	Scr /1932
		0-4-0ST	OC	KS	4273	1922	(a)	(1)
		0-4-0ST	OC	KS	4274	1922	(a)	Scr /1932
		0-4-0ST	OC	KS	4162	1921	(b)	Scr /1932
L9		4wDM		OK	4241	1930	New	s/s
L107		4wPM		MR	5322	1931	(c)	s/s
L158		4wPM		MR	5340	1931	(c)	s/s
L202		4wPM		MR	5344	1931	(c)	(3)
LE 14	3208	4wBE		WR	1199	1938	New	(7)
3209		4wBE		WR	1200	1938	New	s/s
.		4wBE		WR	1211	1938	New	s/s
.		4wBE		WR	1212	1938	New	(11)
LD 1087		4wDM		MR	7486	1940	New	(2)
-		4wDM		MR	7487	1940	New	(4)
L1140	-	4wDM		MR	7465	1940	New	(6)
LD 1194		4wDM		MR	7472	1940	New	(8)
L191	LOCO 191	4wDM		OK	5129	1933	New(d)	(10)
-		4wPM		L	34521	1949	New	(12)

LD 2973 L5011	4wDM	RH	179012	1936	(e)	s/s
LD 2974	4wDM	RH	181815	1936	(e)	(7)
LD 2977	4wDM	RH	189993	1938	(f)	(5)
LD 2978	4wDM	RH	189997	1938	(f)	s/s
2975	4wDM	RH	187120	1938	(g)	(9)
2976	4wDM	RH	187123	1938	(g)	(7)
3010	4wBE	WR	5659	1957	(h)	s/s
LD 1198	4wDM	MR	7473	1940	(j)	(7)

(a) ex Langham Valley Reservoir (1930) contract.
(b) ex Stonebridge Park, North Circular Road (1928) contract, Middlesex, 1/1929.
(c) ex Petrol Loco Hirers, Bedford, 20/10/1934.
(d) ex Buckhurst Hill Brickworks, c/1947, here 10/1958.
(e) originally WD, Corsham, Wiltshire, here by 11/4/1957.
(f) ex WD, Liphook Depot, Hampshire, 11/4/1957.
(g) originally WD, Corsham, Wiltshire, here by 11/4/1957.
(h) ex Kilburn contract Middlesex, by 8/6/1957.
(j) ex Luxborough Lane Brickworks, Chigwell, by 10/1958.

(1) to A.M. Carmichael, Fort William-Inverness road (1932-1934) contract, Inverness-shire.
(2) to Elstree contract, Hertfordshire by 11/12/1954 and returned, to Woodford Brickworks.
(3) to Sir Lindsay Parkinson & Co Ltd, Thames Haven contract by 27/2/1955.
(4) to Mill Lane Aggregates Ltd, Cheshunt, Herts, by 8/1958.
(5) to LCC Crossness contract, Kent, 3/1959.
(6) to Tractor Hire (Contractors) Ltd, Romney Marsh (1962-1964) contract, Kent.
(7) to Harlow New Town Depot, by 4/1965.
(8) to Elstree contract Hertfordshire by 1/1954 and returned, to Harlow New Town Depot by 4/1965.
(9) to Shellhaven contract by 10/1958 and returned, to Harlow New Town Depot by 4/1965.
(10) to Chigwell UDC, Buckhurst Hill, /1971.
(11) to Harlow Plant Depot by 9/1971.
(12) to Chigwell Urban District Council, Luxborough Lane Sewage Works.

HARLOW NEW TOWN PLANT DEPOT EC44

Yard (at TL 460120) adjacent to the ex-GER Broxbourne - Bishops Stortford line, ½ mile west of Harlow Station, where locos were maintained and stored between uses at other sites.
Gauge : 2ft 0in

LD 2974	4wDM	RH	181815	1936	(a)	s/s
LD 2976	4wDM	RH	187123	1938	(a)	s/s
LD 7	4wDM	RH	187120	1938	(a)	s/s
LD 1194	4wDM	MR	7472	1940	(a)	s/s
LD 1198	4wDM	MR	7473	1940	(a)	s/s
LE 13	4wBE	WR			(a)	s/s
LE 14	4wBE	WR	1199	1938	(a)	(2)
118679	4wDM	MR	21620	1957	(b)	s/s
119520	4wBE	CE	5339	1967	New(c)	s/s
102878	4wBE	WR	1212	1938	(d)	(1)

(a) ex Buckhurst Hill Depot, by 4/1965.
(b) ex GLC, Perry Oaks Sewage Works, Middlesex, /1968.
(c) here 9/1971.
(d) earlier Buckhurst Hill Depot, here 9/1971.
(1) to T. & A.M. Kilroe Ltd ? 11/1971, otherwise s/s.
(2) to Tickhill Plant Ltd, Tickhill, Doncaster, South Yorkshire.

Gauge : 1ft 6in

-	0-4-0BE	WR	6600	1962	New(a)	(2)
108534	0-4-0BE	WR	6702	1962	New	(1)
109496	0-4-0BE	WR	6711	1963	(b)	(2)

(a) here by 4/1965, to Willment Bros, Leytonstone contract on hire by 1/1971, returned by 8/1971.
(b) ex Ipswich contract, Suffolk.
(1) to Willment Bros, Leytonstone contract on hire 1/1971, returned by 10/1971, s/s.
(2) to T. & A.M. Kilroe Ltd, contractors, 11/1971.

GEORGE FURNESS
PLAISTOW & BARKING CONTRACT   AC45

Contract let 10/1860 for construction of the Northern Outfall Sewer and outfall works at Barking. By 4/1861 a wharf had been constructed at the Barking outfall site, by 6/1861 a tramway had been laid from Plaistow to Barking. "The Times" of 26/5/1862 reported a visit by the Institute of Civil Engineers who inspected the whole of the works, being conveyed for part of the route in ballast trucks; it was reported that five locomotive engines were in use with 500 trucks and 2,300 men employed.

Reference: Industrial Locomotive No.76.

"The Engineer", 27/1/1865, Fuller Horsey to auction 28/2/1865 re plant of Northern Outfall Sewer, at Plaistow – including 11in, 6w coupled tank loco by MW, 1862, and two tank locos by GE, 10in & 12in.

Gauge : 4ft 8½in

NORTHERN OUTFALL	0-6-0ST	MW	44	1862	New	(1)
-	tank loco	GE			(a)	s/s
-	tank loco	GE			(a)	s/s

(a) origin and identity unknown.
(1) later Logan & Hemingway, Penarth-Cadoxton (TVR) (1887-1888) contract, Cardiff, PHOENIX.

EDMUND GABBUTT
WOODFORD CONTRACT   DC46

Construction of the Claybury Mental Asylum at Woodford for the Middlesex County Council. The contractor built a temporary railway line, which crossed two roads and the River Roding, about 1¼ miles in length from Woodford Station (GER) to the site to convey construction materials. This line had been completed by 2/1890 when main construction of the hospital commenced. Plant was offered for sale in 8/1893 on completion of the contract, and the line was subsequently dismantled.

Reference: Industrial Locomotive No.39, No.40, No.85.
Gauge : 4ft 8½in

FLEETWOOD   0-6-0ST   IC   HC   313   1888   (a)   (1)

(a)   originally Thomas Riley, Preston & Wyre Railway contract, Fleetwood, Lancs; here by 9/1890.
(1)   for sale by auction, 19-20/9/1893; s/s.

JOHN GILL (CONTRACTORS) LTD
BARKING CONTRACT   PC47

Contract for extensions to Barking Power Station 1930 – 1931. Four 12in steam locos are said to have been used on this contract.
Gauge : 4ft 8½in

	0-4-0ST	OC	AE	1610	1912	(a)	(1)
	0-4-0ST	OC	AE	1874	1921	(b)	(2)
	0-4-0ST	OC	AE	1875	1921	(b)	(2)

(a)   ex Thos. W. Ward Ltd, Titan Works, Grays, by 2/1931.
(b)   ex Thos. W. Ward Ltd, Titan Works, Grays, by 11/1930.
(1)   to Thos. W. Ward Ltd, Titan Works, Grays.
(2)   to Thos. W. Ward Ltd, Titan Works, Grays, after 1/1931.

W. & J. GLOSSOP LTD
WALTHAMSTOW CONTRACT   CC48

Contract for Waltham Forest Borough Council for the construction of the Southern Branch Sewer of the Highams Area Flood Relief Scheme; the contract was awarded in 1972 to W. & J. Glossop Ltd at a price of £627,979. The loco listed below was delivered to a site in Hookers Road, off Blackhorse Lane, Walthamstow, TQ 358896.

A visit to M. & H. Tunnel & Civil Engineering at Thurcroft in 3/1973, recorded that WR 4818 was on hire to "Kelly Goldhorne" at Walthamstow; however it is thought that this information is in error as no organisation or company of that name has been traced at the time of writing.

Reference : Contract Journal 6/7/1972.
            Narrow Gauge News No.83, 6/1973.

Gauge : 2ft 0in

	-	4wBE	WR	4818	1951	(a)	(1)

(a)   ex M. & H. Tunnel & Civil Engineering Co Ltd, Thurcroft, West Yorkshire, by 3/1973.
(1)   returned to M. & H. Tunnel & Civil Engineering Co Ltd, later Delta Construction Ltd, Godalming, Surrey.

GRANTRAIL LTD
TEMPLE MILLS CONTRACT　　　　　　　　　　　　　　　　　　　CC49
Contract for ballasting sidings for new diesel depot at Temple Mills, 2001.
Gauge: 4ft 8½in
　　　　GR 5087　　　　　　0-4-0DE　　　　RH　425478　1959　　(a)　　　　(1)

(a)　ex Scunthorpe Plant Depot, Lincolnshire, or another contract.

(1)　to Scunthorpe Plant Depot, Lincolnshire, or another contract.

J. G. GRAY
BRAINTREE CONTRACT　　　　　　　　　　　　　　　　　　　　FC50
Contract for the construction of a new weaving shed for Courtaulds Ltd, whose mill in Chapel Hill, Braintree was acquired by Samuel Courtauld & Co in 1843.
Gauge : 2ft 0in
　　　　　　-　　　　　　　　4wPM　　　　　MR　　4028　1926　　New　　　s/s

WILLIAM HANSON
SAFFRON WALDEN CONTRACT　　　　　　　　　　　　　　　　　FC51
Saffron Walden Railway construction. 1865-1866, opened 23/11/1865.
Gauge : 4ft 8½in
　　　　　LITTLE EASTERN　　0-4-0ST　OC　MW　　143　1865　　New　　(1)

(1)　to Barrow-in-Furness Docks (1870-1872) contracts, Lancashire.

T.W. HEATH (PUBLIC WORKS) LTD
FRINTON-ON-SEA CONTRACT　　　　　　　　　　　　　　　　　HC52
Construction of sea wall and sewer outfall for Frinton UDC, 1924 -1925.
Gauge : 2ft 0in
　　　　　　-　　　　　　　　4wPM　　　　Austro-Daimler　　　　(a)　　　(1)

(a)　origin and identity unknown.

(1)　for sale on completion of contract, 5/1925.

HIGGS & HILL LTD
LOUGHTON CONTRACT　　　　　　　　　　　　　　　　　　　　CC53
Company registered 7/2/1898 to acquire and carry on the business of builders and contractors carried on under the style of "Higgs and Hill" at Crown Works, South Lambeth. Contract for construction of 1600 houses including roads, sewers & services on the Debden Housing Estate in Loughton for London County Council 1946 – 1950.

"Contract Journal" 6/12/1950 - for sale – 1935 Planet diesel loco 2ft gauge 20/24hp reconditioned 1946, type 2DL – Higgs & Hill Ltd, LCC Housing Estate, Loughton.

Gauge : 2ft 0in

		4wDM	FH	1935	(a)	s/s

(a) origin and identity unknown.

HOLLOWAY BROTHERS (LONDON) LTD
TILBURY CONTRACT   MC54

Construction of foundations for Tilbury Power Station for Central Electricity Authority, c3/1953 - c6/1954.

Gauge : 2ft 0in

	4wDM	FH	2198	1939	(a)	(2)
LOC 5	4wDM	FH	2199	1939	(a)	(2)
LOC 6	4wDM	FH	2280	1940	(b)	(2)
-	4wDM	FH	2282	1940	(b)	(2)
-	4wDM	FH	2283	1940	(b)	(2)
-	4wDM	FH	2286	1940	(b)	(2)
-	4wDM	FH	2287	1940	(b)	s/s by 6/1954
LOC 9	4wDM	FH			(b)	(2)
LOC 10	4wDM	FH			(b)	(2)
LOC 16	4wDM	MR	9409	1948	(c)	(3)
LOC 17	4wDM	MR	9410	1948	(c)	(3)
LOC 18	4wDM	MR	9411	1948	(c)	(3)
LOC 19	4wDM	MR	9412	1948	(c)	(1)
LOC 20	4wDM	MR	9413	1949	(c)	(1)
LOC 21	4wDM	MR	9414	1949	(c)	(1)
LOC 22	4wDM	MR	9415	1949	(c)	(1)
LOC 23	4wDM	MR	9416	1949	(c)	(1)
LOC 24	4wDM	MR	9417	1949	(c)	(1)
LOC 25	4wDM	MR	9418	1949	(c)	(1)

Note that LOC 9 and LOC 10 in the list above may in fact be included in FH 2282 – 2287.

(a) originally Littlebrook Power Station contract, Dartford, Kent; here by 20/6/1953.
(b) originally Kirkby (ROF No.7) contract, Liverpool; here by 20/6/1953.
(c) originally Stewart's Lane, Battersea, (1949) contract, London.

(1) gone by 6/1954, later West Thurrock Power Station (1959) contract.
(2) to M.E. Engineering Ltd, Cricklewood, Middlesex, by 6/1954.
(3) later Beckton (1957) contract.

BECKTON CONTRACT   AC55

Construction of a river wall for the North Thames Gas Board 1955-1957, contract value £519,000.

Gauge : 4ft 8½in

LOC 28	4wDM	FH	3700	1955	New	(1)

(1) to St. Albans plant depot, Hill End, St. Albans, Hertfordshire, c/1957 by 6/4/1957.

Gauge : 2ft 0in

-	4wDM	MR	9409	1948	(a)	(1)
-	4wDM	MR	9410	1948	(a)	(1)
-	4wDM	MR	9411	1948	(a)	(1)

(a) ex Tilbury Power Station (1953-1954) contract.

(1) later West Thurrock Power Station (1959) contract.

WEST THURROCK CONTRACT   NC56

Construction work for the CEGB at West Thurrock Power Station in 1959.
Gauge : 2ft 0in

LOC 19	4wDM	MR	9412	1948	(a)	(4)
	4wDM	MR	9414	1949	(a)	(4)
	4wDM	MR	9415	1949	(a)	(1)
LOC 23	4wDM	MR	9416	1949	(a)	(6)
LOC 24	4wDM	MR	9417	1949	(a)	(2)
	4wDM	MR	9418	1949	(a)	(1)
	4wDM	MR	9409	1948	(b)	(1)
	4wDM	MR	9411	1948	(b)	(3)
	4wDM	MR	21286	1959	New(c)	(7)
	4wDM	MR	21287	1959	New(c)	(7)
	4wDM	MR	9410	1948	(d)	(5)
	4wDM	MR	9413	1949	(e)	(4)

(a) ex another contract, here 8/6/1959.
(b) ex another contract, here 8/6/1959, earlier Beckton (1957) Contract.
(c) ex Diesel Loco Hirers Ltd, Bedford, 11/6/1959 on hire.
(d) ex another contract, here 14/7/1959, earlier Beckton (1957) contract.
(e) ex another contract, here 14/7/1959.

(1) later to Joseph Arnold & Sons Ltd, Leighton Buzzard, Bedfordshire, c/1964, after 6/1962.
(2) later to MR, by 7/1964;
thence to Bell Rock Gypsum Ltd, Stanton in the Vale, Nottinghamshire, 13/7/1965.
(3) to Thos. E. Gray & Co Ltd, Burton Latimer, Northamptonshire, 4/1965.
(4) later to MR 5/1965, dismantled for spares.
(5) later to MR 6/1965, dismantled, engine sold.
(6) to MR ?, thence to Bell Rock Gypsum Ltd, Stanton in the Vale, Nottinghamshire, c/1965.
(7) returned to Diesel Loco Hirers Ltd, from hire.

HOLME & KING
SHENFIELD CONTRACT   GC57

Construction of the Shenfield - Wickford railway (10 miles) for GER. The line was authorised on 16/7/1883 and opened on 19/11/1888. The continuation from Wickford to Southend was built by Walter Scott & Middleton.

"Contract Journal", 9/11/1887 - Contractors locos for sale by Holme & King, Billericay, one 14in 0-6-0, two 13in 0-6-0, one 12in 0-6-0, one 12in 0-4-0, four 10in 0-4-0, one 9in 0-4-0.

"Contract Journal", 21/11/1888 – A.T. Crow to sell 29/11/1888 at yard Mount Nessing, Essex for Holme & King re GER Shenfield & Wickford Rly contract, including five locos 0-4-0 and 0-6-0 tanks, cyl 9in, 10in, 12in, 14in by MW, FW, HE and P.

"The Engineer", 23/11/1888 – A.T. Crow to auction 29/11/1888 re Holme & King, GER Shenfield & Wickford Rly, at yards Mount Nessing, near Shenfield – plant including five 6 & 4w coupled locos, 9in, 10in, 12in, & 14in by MW, HE, FW, & P.

Gauge : 4ft 8½in

PHYLLIS	0-4-0ST	OC	HE	365	1885	New	(2)
BEATRICE	0-4-0ST	OC	HE	366	1885	New	(2)
ALICIE	0-4-0ST	OC	P	442	1885	New	(4)
-	0-4-0ST	OC	HE	229	1879	(a)	(3)
-	0-6-0ST	IC	MW	898	1883	(b)	s/s
KING ARTHUR	0-6-0ST	IC	MW	636	1876	(c)	(6)
ALICIA	0-6-0ST	IC	MW	318	1870	(d)	(1)
-	0-6-0ST	OC	FW	343	1877	(e)	(5)
-	[0-4-0ST	OC			?]	(f)	s/s
-	[0-4-0ST	OC			?]	(f)	s/s

(a) ex John Garlick, contractor, Birmingham;
  earlier John Knowles, Woodville, Derbyshire, JOHN KNOWLES, until 5/1881.
(b) ex Thos. Nelson & Co, Roath Dock (1883-1887) contract, Cardiff, No.17 HAMILTON.
(c) ex J.C. Lang, Bodmin Road-Bodmin (GWR) (1886-1888) contract, Cornwall, by /1886.
(d) originally Benton & Woodiwiss, contractors, Warrington, Lancashire.
(e) identity subject to confirmation; if correct then:
  ex Dransfield & Co, Leeds or Knottingley (1875-1879) contracts, West Yorkshire.
(f) origins and identities not known.

(1) to James Brier, Son & Wilson, Beeston-Batley (GNR) (1888-1890) contract, West Yorkshire; thence to MW;
  thence to Isle of Axholme Light Railway, Lincolnshire, No.7 BLETCHER, after 7/1896.
(2) to Relf & Pethick, Lydford-Devonport (PD&SWJR) (1887-1890) contract, Devon.
(3) possibly to Mackay & Son and Morrison & Mason, Scotland, ELLA;
  later to Wm Kennedy Ltd, contractors, Lanarkshire, by 7/1899.
(4) later Brighouse widening (LYR) (1899-1905) contract, West Yorkshire.
(5) later Abram Kellett & Sons, Frankley Reservoir (1898-1904) contract, Birmingham.
(6) later at Low Laithes Colliery Co Ltd, Wakefield, West Yorkshire, by 6/1928.

HUSSEY, EGAN & PICKMERE LTD
PRITTLEWELL CONTRACT   JC58

Company of civil and public works engineers, (registered 10/7/1930) at 174 Corporation Street, Birmingham. Contract for the construction of Prittlewell Sewage Works (TQ 881876), Southend, c/1931.

Gauge : 2ft 0in

		4wDM	MR	5612	1931	New	(1)
		4wDM	MR	5613	1931	New	(2)

(1) to St Albans Sand & Gravel Co Ltd, Nazeing Pits, 5/1937.
(2) to St Albans Sand & Gravel Co Ltd, Smallford Pits, Hertfordshire, 5/1937.

Both locos probably returned to the Birmingham Plant Depot before sale.

INDUSTRIAL CONSTRUCTIONS LTD
BENFLEET CONTRACT   JC59

Company registered 8/7/1920 as contractors for the construction of reinforced concrete buildings and structures. Construction of sewage disposal works (TQ 766864) and laying of sewers for Benfleet Urban District Council, work commenced in 2/1930 for completion by 11/1931.

Gauge : 2ft 0in

		4wPM	MR	5063	1930	New	s/s
		4wPM	FH	1697	1930	New	s/s

A. JACKAMAN & SON
PARKESTON QUAY EXTENSION CONTRACT   HC60

Extension of Parkeston Quay for the GER, work included new berths, the extension of the quay being about 1000ft and included the erection of a new shed 440ft long by 60ft wide to include two tracks running through it. Work commenced in the summer of 1906 and was due for completion by the autumn of 1908. When the locomotive "AJAX" arrived its first job was the removal of spoil from Parkeston Hill to the area of the new quay construction.

Gauge : 4ft 8½in

AJAX	0-6-0ST	IC	MW	212	1866		
		reb	HC		1907	(a)	(1)

(a) ex HC, 8/4/1907; earlier Whitaker Bros Ltd, contractors, No.20.
(1) later Air Ministry Stores Depot, Watford, Hertfordshire.

JOINT COMMITTEE FOR RIVER LEE FLOOD RELIEF SCHEME (LEE CONSERVANCY BOARD and WEST HAM UDC)
RIVER LEE FLOOD RELIEF SCHEME   PC61

Work began in 1931-1932 using direct labour to improve the river, including filling in redundant channels and constructing new ones in an area near to the junction with the Thames.

Gauge : 2ft 0in

-	4wPM	MR		(a)	(1)
-	4wPM	MR		(b)	(1)
-	4wPM	MR		(c)	(1)

(a) ex George Cohen Sons & Co Ltd, on hire 6/1931.
(b) ex George Cohen Sons & Co Ltd, on hire 11/1931.
(c) ex George Cohen Sons & Co Ltd, on hire 3/1932.

(1) returned to George Cohen Sons & Co Ltd, off hire c12/1932.

SIR JOHN JACKSON LTD
GRAYS PLANT DEPOT   NC62

Company listed as Engineers and Contractors for Public Works, founded by Sir John Jackson in 1876 and incorporated on 17/8/1898. Yard acquired by Thos. W. Ward Ltd, /1924.

"The Engineer" 8/4/1921 Geo N Dixon to sell 24/5/1921 at Grays Plant Depot for Sir John Jackson Ltd, inc two 0-6-0ST std gauge by MW & HE, five 0-4-0ST std gauge by HL, HE, Barclay etc, ten 0-4-0ST 3ft gauge by WB, A Baird & Sons and Barclay.

"Contract Journal" 10/8/1921, GN Dixon to sell 30/8/1921 re Sir John Jackson at Grays Depot – fifteen locos 0-4-0 & 0-6-0 std gauge & 3ft gauge.

"Machinery Market" 15/5/1925 for sale – plant inc 10in & 12in locos – Sir J Jackson Ltd, London SW1.

Gauge : 4ft 8½in

	NOTTINGHAM	0-6-0ST	IC	HE	1	1865	(a)	(1)
	DEVONPORT	0-6-0T	IC	HE	401	1886	(c)	(2)
	MANCHESTER	0-6-0ST		HE	437	1888	(c)	s/s
	ACCRINGTON	0-6-0ST	IC	MW	951	1885	(a)	(4)
148	WESTMINSTER	0-6-0T	IC	SS	3474	1888		
		reb		Keyham		1903	(a)	(6)
711	VICTORIA	0-4-0ST	OC	AB	887	1901	(a)	(9)
	TAMAR	0-4-0ST	OC	HE	360	1884	(b)	(5)
	GLASGOW	0-4-0ST	OC	AB	185	1877	(d)	(3)
	ST MONANS	0-4-0ST		AB	279	1885	(e)	(10)
98	DOVER	0-4-0ST	OC	AB	891	1901	(f)	(8)
197	WEAR	0-4-0ST	OC	MW				
		reb		HL			(g)	(7)

(a) earlier Devonport Dockyard (Admiralty) (1896-1907) contract, Devon.
(b) earlier Ferrol Dockyard (1906-1910) contract, Spain.
(c) earlier Devonport Dockyard (Admiralty) (1896-1907) contract, Devon, by 9/1914.
(d) earlier WD Heytesbury Military Railway, Wiltshire in 2/1916;
    earlier Devonport Dockyard (Admiralty) (1896-1907) contract, Devon.
(e) earlier at WD, Heytesbury Military Railway, Wiltshire.
(f) [possibly ex Vancouver Harbour contract, Canada;
    and earlier Simonstown Naval Base contract, South Africa];
    earlier Devonport Dockyard (Admiralty) (1896-1907) contract, Devon.
(g) origin and identity unknown;
    hired to Chas Wall Ltd, Grays; returned.

(1)   to Samuel Williams & Sons Ltd, Dagenham Dock, No6. /1909.
(2)   to WD Bulford & Larkhill Camps (1914-1916) contract, Wiltshire, returned, to Samuel Williams & Sons Ltd, Dagenham Dock, by 4/1918.
(3)   for sale 5/1921; s/s.
(4)   to Samuel Williams & Sons Ltd, Dagenham Dock, after 8/1921.
(5)   to H. Covington & Sons Ltd, Tilbury Marshes Rubbish Shoot, /1923.
(6)   to Thos. W. Ward Ltd, Grays, TW 46, /1924.
(7)   to Thos. W. Ward Ltd, Grays, /1924.
(8)   to Thos. W. Ward Ltd, Grays, TW 44, /1924;
       thence to John Spencer & Sons (1928), Newburn-on-Tyne, Northumberland, /1932..
(9)   to Thos. W. Ward Ltd, Grays, /1924, thence to Inverkiething Yard, Fife, after /1926 by /1935.
(10) to Thos. W. Ward Ltd; thence to Cafferata & Co Ltd, Beacon Hill Works, Newark, Nottinghamshire, by 2/1926.

"Machinery Market" 5/6/1908 for sale – plant inc 3ft gauge locos 9in by Baird, 9in by WB, 9in by BH – R. A. King dealer, London.

Gauge : 3ft 0in

PARKER	0-4-0ST	WB	1657	1902	(a)	s/s
WALSHAW DEAN	0-4-0ST	WB	1567	1899	(a)	s/s
LOCH LEVEN	0-4-0IST	WB	1504	1897	(a)	s/s
BRANCKER	0-4-0IST	WB	1116	1889	(a)	s/s
GYP	0-4-0T	AB	761	1895	(a)	(1)
RANNOCH	0-4-0ST	Baird	335		(a)	s/s
GLENCOE	0-4-0ST	Baird	334		(a)	s/s
ALSTON	0-4-0IST	WB	1434	1894	(a)	(1)
BUNTY	0-4-0IST	WB	1480	1897	(a)	(1)
THE NIDD	0-4-0IST	WB	1423	1893	(a)	s/s

(a)   ex Kinlochleven (British Aluminium) (1904-1910) contract, Argyll.

(1)   to P.& W. Anderson Ltd, Torrington-Halwill Jct (ND&CLR) (1922-1924) contract, Devon.

KIRK & PARRY
BARKING – PITSEA CONTRACT   LC63

Construction of Barking - Pitsea railway for LT&SR. The contract for Barking-Upminster (7¾ miles) was let on 6/6/1883 and it opened to traffic on 1/5/1885. That for Upminster-Pitsea (10½ miles) was let on 23/10/1884 and the section opened throughout on 1/6/1888. Kirk & Parry, contractors for both sections, were reported to have used eight locomotives on the work.

"Contract Journal" 6/6/1888, A T Crow to sell 20-21/6/1888 for Kirk & Parry at yards E Horndon and Pitsea on completion LT&SR Barking & Pitsea contract, includes five MW 0-6-0 and 0-4-0, and two six-wheel main line locos and tenders by LNWR.

"The Engineer" 15/6/1888, A T Crow to auction 20-21/6/1888, re Kirk & Parry, LT&SR Barking & Pitsea line – plant including five 6 & 4w coupled locos by MW and two main line locos & tenders by LNWR.

"The Engineer" 11/10/1889, Fuller Horsey to auction 29/10/1889, re Kirk & Parry at yard East Horndon Station – plant including one 15in 6w coupled loco & tender, two 6w coupled & two 4w coupled tank locos 8in to 13in.

All locos for sale by auction on completion of contract, 20-21/6/1888.

Gauge : 4ft 8½in

-		0-4-0	IC	BCK		1845	(a)	s/s
-		0-6-0	IC	NG	64	1847	(b)	s/s
5		0-4-0ST	OC	MW	886	1883	New	s/s
6		0-4-0ST	OC	MW	856	1883	New	(3)
7		0-4-0ST	OC	MW	888	1883	New	(2)
8		0-4-0ST	OC	MW	900	1884	New	(1)
9		0-6-0ST	OC	MW	927	1885	New	(4)

(a) earlier Lincoln-Navenby-Honington Jnc, GNR contract, originally LNWR, 1175, by 25/10/1865.
(b) originally LNWR, 1150, by 26/1/1866.
(1) to Cheadle Railway (1888-1890) contract, Staffordshire, 5/1888.
(2) to Boythorpe Colliery Co Ltd, Boythorpe Colliery, Derbyshire, [6/1888?]
(3) to Samuel Williams & Sons, Dagenham Dock, 6/1888.
(4) to J.T. Firbank, Winchester contract, Hampshire, ELY, 6/1888.

KIRK & RANDALL
Established 1850
TILBURY DOCKS CONTRACT   MC64

Construction of Tilbury Dock for East and West India Docks Co, 1882-1884. Work commenced with the first sod cut on 8/7/1882. Difficulties were encountered with the excavations which led to a dispute between the firm and the East and West India Docks Co, who terminated the contract about 7/1884 when Kirk and Randall were evicted from the site. A new contract was let 10/1884 to Lucas & Aird who completed the work. The resulting court case and arbitration ended on 6/12/1888 when Sir Frederick Bramwell found in favour of Kirk & Randall and ordered the Dock Company to pay their claim and all costs. Shortly afterwards the East and West India Docks Co went into liquidation. A report in the "Times" for 26/7/1883 stated that Kirk & Randall had 35 locomotives, 40 steam cranes, two winding engines, 7 steam navvies and 1000 side tip wagons in use on the contract on 35 miles of railway, the whole area of operations covering 320 acres of land. Records of Grafton Cranes Ltd show that between 4/1883 & 4/1884, Kirk & Randall ordered forty five 2ton capacity steam cranes for delivery to Tilbury.

Gauge : 4ft 8½in

	RIBBLESDALE	0-6-0ST	IC	MW	663	1878	(a)	(1)
	CLITHEROE	0-4-0ST	OC	MW	672	1878	(a)	(3)
	HAMPTON	0-6-0ST	IC	MW	694	1878	(b)	(1)
1	BRUCE	0-6-0ST	IC	HE	234	1880	(c)	(1)
2	WALLACE	0-6-0ST	IC	HE	235	1880	(c)	(1)
4	ANT	0-4-0ST	OC	MW	78	1863	(d)	(1)
	CALDEW	0-4-0ST	OC	MW	225	1867	(e)	(1)
5	TILBURY	0-4-0ST	OC	MW	554	1876	(f)	(1)
6		0-4-0ST	OC	MW	585	1876	(g)	(2)
20		0-4-0ST	OC	MW	861	1883	New	(1)
21		0-4-0ST	OC	MW	862	1883	New	(1)

22		0-4-0ST	OC	MW	867	1883	New	(1)
23		0-4-0ST	OC	MW	882	1883	New	(1)
24		0-4-0ST	OC	MW	883	1883	New	(1)
25		0-4-0ST	OC	MW	618	1876	(f)	(1)
27		0-6-0ST	IC	MW	872	1883	New	(1)
28		0-6-0ST	IC	MW	873	1883	New	(1)
	BAID	0-6-0ST					(h)	s/s
	GROSVENOR	0-6-0T					(h)	(1)

(a) ex Grand Junction Canal Co, Slough Branch construction (1879-1882), Berkshire.
(b) ex Wm. Webster, Hampton Pumping Station (1878-1882) contract, Middlesex.
(c) ex Easton Gibb & Co, Upper Barden Reservoir (1876-1882) contract, North Yorkshire.
(d) earlier Forcett Limestone Co Ltd, Forcett Quarry, North Yorkshire, ANT.
(e) use on this contract subject to confirmation; if correct then-
ex E.W. Goodenough, dealer, London;
earlier Thos. Nelson, Standedge Second Tunnel (LNWR) (1868-1871) contract,
West Yorkshire, CALDEW.
(f) ex Henry Ward, Tilbury, earlier Kirk & Parry, Liverpool.
(g) earlier Kirk & Parry, Liverpool.
(h) origin and identity not known.

(1) to Lucas & Aird, Tilbury Docks (1884-1886) contract, /1884.
(2) to Samuel Williams & Sons, Dagenham Dock.
(3) to J Wilson & Sons, Edge Hill, Liverpool.

JOHN LAING & SON LTD
THAMES HAVEN CONTRACT KC65
Contract at Shell Thames Haven for construction of foundations for a new ammonia plant in 1957.
Gauge : 2ft 0in

	4wDM	RH	387819	1955	(a)	s/s
	4wDM	RH			(b)	s/s

(a) ex Dolgarrog contract North Wales, by 6/12/1957.
(b) origin and identity unknown.

WEST THURROCK CONTRACT NC66
Contract for construction work at West Thurrock Power Station in 1959.
Gauge : 2ft 0in

2	4wDM	RH	375354	1954	(a)	s/s
5	4wDM	RH	375361	1955	(b)	(1)

(a) earlier Trawsfynydd Reservoir (1954-1956) contract, North Wales, here by 8/6/1959.
(b) earlier Coedty Reservoir (1955-1957) contract, North Wales, here by 8/6/1959.

(1) to Harrisons & Crosfield exported to Sungei Boaia Estate, Klang, Malaysia.

LEE CONSERVANCY BOARD
CUSTOM HOUSE CONTRACT   AC67

Contract near Custom House station, further details unknown.
Gauge : 2ft 0in

	4wDM	OK	6194	1935	(a)	(1)

(a)   ex William Jones Ltd, dealer, Greenwich, London, on hire by 13/12/1957.

(1)   returned to William Jones Ltd.

LESLIE & CO LTD
ILFORD CONTRACT   DC68

Company registered 13/11/1894 to acquire as a going concern the business of builders and contractors under the respective titles of Leslie & Co, Aldin Bros & Davies. Construction of Goodmayes Mental Hospital, Ilford for West Ham Borough Council, contract value £209,531, work commenced in 1898, hospital opened 1/8/1901. A line ran from GER Permanent Way sidings at Goodmayes Station to the site in Barley Lane.
Gauge : 4ft 8½in
No details available.

LILLEY - WADDINGTON LTD
HARVEY ROAD PLANT DEPOT, Basildon   KC69

Plant Depot located at TQ 736904, closed c/1980, company dissolved 22/3/1994..
Gauge : 2ft 0in

	4wDM	MR	22031	1959	(a)	(4)
	4wDM	MR	22032	1959	(a)	s/s
DIGGER	4wDM	MR	8882	1944	(a)	(3)
	4wDM	RH	277573	1949	(a)	(3)
	4wDM	MR	9263	1947	(b)	(2)
	4wDM	MR	8696	1941	(c)	(1)
(PN 66/4/5)	0-4-0BE	WR	3219	1945	(d)	s/s
(PN 66/6/5)	4wBE	WR	4352	1950	(d)	s/s
66/4/1	0-4-0BE	WR	4475	1950	(d)	s/s
66/4/2	0-4-0BE	WR	4476	1950	(d)	s/s
R3 (PN 66/6/6)	4wBE	WR	4897	1952	(d)	s/s
66/4/7	4wBE	WR	D6878	1964	(d)	s/s
66/4/8	4wBE	WR	D6879	1964	(d)	s/s
66/6/3 6	4wBE	WR	6097	1959	(e)	s/s
	4wBE	WR	1071	1930	(f)	s/s
S2	4wBE	WR			(g)	s/s
66/6/7 W3	4wBE	WR			(g)	s/s
W4	4wBE	WR			(g)	s/s
6	4wBE	WR			(g)	s/s
7	4wBE	WR			(g)	s/s
W8	4wBE	WR			(g)	s/s
W9	4wBE	WR			(g)	s/s
	4wDM	RH			(g)	s/s

(a) ex Darenth Valley contract, Farningham, Kent, c9/1972.
(b) ex Darenth Valley contract, Farningham, Kent, c11/1972.
(c) ex Darenth Valley contract, Farningham Kent.
(d) ex Harlesden or Ealing Plant Depot, Middlesex, by /1973.
(e) earlier Taylor Woodrow (Plant) Ltd, Southall, Middlesex.
(f) origin unknown.
(g) origin and identity unknown.

(1) to Alan Keef Ltd, Cote Farm, Bampton, Oxfordshire, c4/1973.
(2) to Alan Keef Ltd, Cote Farm, Bampton, Oxfordshire, exported to Singapore c5/1973.
(3) to Alan Keef Ltd, Cote Farm, Bampton, Oxfordshire, c/1972.
(4) s/s after 29/5/1980.

Gauge : 1ft 6in

66/4/6		0-4-0BE	WR	3788	1948	(a)	s/s
66/4/3		0-4-0BE	WR	4579	1950	(a)	s/s
-		0-4-0BE	WR	4580	1950	(a)	s/s
		4wBE	CE	5827A	1970	(a)	s/s
		4wBE	CE	5827B	1970	(a)	s/s
		4wBE	CE	5920A	1972	(b)	s/s
		4wBE	CE	5920B	1972	(b)	s/s
		4wBE	WR			(c)	s/s
		4wBE	WR			(c)	s/s
		4wBE	WR			(c)	s/s
		4wBE	WR			(c)	s/s

(a) ex Harlesden or Ealing Plant Depot, Middlesex, by /1973.
(b) originally Parnell Plant, Rugby, Warwickshire.
(c) origin and identity unknown.

PETER LIND & CO LTD
Peter Lind & Co until 31/12/1927

BECKTON CONTRACT   AC70

Contract at Northern Outfall Sewage Works, Beckton, for London County Council, 1931.

Gauge : 2ft 0in

214	4wPM	FH	1731	1931	New	(1)

(1) to another Lind contract; later to M.E. Engineering Ltd, Cricklewood, Middlesex, 3/1955; thence to J. Balfour, Northern Ireland.

AVELEY PLANT DEPOT   NC71
                                                                     TQ 557787

Gauge : 4ft 8½in

PL1806		4wDM	RH 195856 1939	(a)	s/s after 9/1971
PL1805	TWA 4304	4wDM	RH 237929 1946	(b)	s/s after 9/1971

(a)   ex Kingsnorth, Power Station contract, Kent, after 7/1965; formerly A.E.I. (Lamp & Lighting) Ltd, Chesterfield, Derbyshire, via Thos W. Ward Ltd, Templeborough, Sheffield, Yorkshire (WR).
(b)   ex Kingsnorth, Power Station contract, Kent, after 7/1965, earlier Thos W. Ward Ltd, Silvertown.

Gauge : 2ft 0in

PL 1230   4wDM   MR   7309 1938   (a)   (1)

(a)   ex Folkington Reservoir contract, Eastbourne, Sussex, after 6/1970, here by 2/1971.

(1)   sold to Mr Norrington of Grays, c2/1971.

LONDON COUNTY COUNCIL
BARKING CONTRACT   AC72

Northern Outfall Sewer Widening, Barking to Old Ford, in 1901-1906.
"Contract Journal" 30/10/1907 Fuller Horsey to sell 26/11/1907 for London County Council at site near Beckton Gasworks and LCC pier at west side of creek – plant inc three 0-4-0ST 3ft 6in gauge by MW & Barclay.
"Contract Journal" 6/5/1908 clearance sale on completion Northern Outfall and Lower Level Sewer Works, by order London County Council at various depots, Stratford, Old Ford, Beckton and Fulham, plant inc one 0-4-0 std gauge and three 0-4-0 3ft 6in gauge locos, sale 26-27/5/1908.

Gauge : 4ft 8½in

35	BELVEDERE	0-4-0ST	OC	MW	1318	1896	(a)	(2)
36	TALBOT	0-4-0ST	OC	MW	1279	1895	(b)	(1)
	STRATFORD	0-4-0ST	OC	AB	992	1904	New	(3)

(a)   ex Crossness Works, London, c/10/1902.
(b)   ex J. Wardell & Co, dealers, 7/1903;
      earlier S. Pearson & Son Ltd, Wootton Bassett-Chipping Sodbury (GWR) (1897-1903) contract, Gloucestershire/Wiltshire.

(1)   to Northern Middle Level Sewer Construction, London, 1906-1908.
(2)   later at Crossness construction, London, c10/1907.
(3)   to Charles Wall Ltd, Chingford Reservoir (MWB) (1908-1913) contract, c/1908.

Gauge : 3ft 6in

47	LEA	0-4-0ST	OC	B	272	1880	(a)	(2)
48	THAMES	0-4-0ST	OC	MW	1552	1902	New	(1)
91	BARKING	0-4-2WT	OC	CF	1151	1897	New?	(3)

(a)   ex Thos. W. Ward Ltd, Sheffield, 5/1901.

(1)   to J. Stiff & Sons, dealers, London, /1908.
      thence B. Fayle & Co Ltd, Norden Clay Mines, Corfe Castle, Dorset, 5/1909.
(2)   possibly purchased by J.F.Wake, thence to B. Whitaker & Sons Ltd, Pool Quarry, Pool-in-Wharfedale, West Yorkshire, c/1909.
(3)   to Wm. Murphy, contractor Akwapim branch contract, Gold Coast Railway, Africa, c/1909.

LONDON AND ST. KATHARINE DOCK COMPANY
ROYAL ALBERT DOCK EXTENSION CONTRACT AC73

Extension to the Royal Albert Dock 1884 – 1886. New works undertaken by the Dock Company itself under the direction of Colonel B.H. Martindale, RE, general manager with Mr R. Carr as constructing engineer. Mr Thomas (assistant engineer) acted as contractor for labour on behalf of the company. Works included construction of a new entrance lock and extension of Gallions Basin from 12 to 15¼ acres and a new river wharf of 1,250ft. Work commenced on 12/5/1884 and was completed in 5/1886 at a cost of £250,000. It seems likely that the locos mentioned below may not have been sold at auction and remained in the dock company's fleet.

"The Times" 17/9/1887 Fuller, Horsey, Sons, & Cassell are instructed by the London and St. Katharine Dock Company to auction on 12/10/1887 a portion of plant and machinery used by them in the completion of the Docks, including six four wheel coupled saddle tank locomotive engines by Shanks, Ruston & Proctor, the Falcon Company and other makers, with cylinders from 7 to 10in diameter and 15 to 20in stroke.

Gauge : 4ft 8½in

0-4-0ST	OC	MW	893	1884	New	(1)
0-4-0ST	OC	MW	905	1884	New	(1)
0-4-0ST	OC	HE	343	1884	New	(1)
0-4-0ST	OC	FE		1884	New	(1)
0-4-0ST	OC	RP	c1870		(a)	(1)
0-4-0ST	OC	Shanks		1872	(b)	(1)

(a) origin and identity unknown.
(b) possibly ex Alfred Giles, Southampton Dock Co, Hampshire.
(1) to Royal Albert Dock locomotive fleet on completion of contract, after 10/1887.

LUCAS & AIRD
VICTORIA DOCK EXTENSION CONTRACT AC74

Construction of an extension to the Victoria Dock for the London and St. Katharine Dock Co. Work started in 1875 and was completed on 6/5/1880 with the dock opening on 24/6/1880 when it was named the Royal Albert Dock. A visit here in 7/1877 recorded sixteen locos in use whilst "The Engineer" for 2/7/1880 stated that seventeen locos had been used on this contract.

"The Times" 9/2/1881 reported that a disastrous fire broke out in No.5 shed in the Victoria Docks on 8/2/1881; the fire spread to other warehouses including No.1, 2, 3 & 4 sheds destroying tea, rice, jute & sugar. A large number of railway trucks were destroyed and 12 barges laden with grain were burnt out, the intense heat caused railway tracks to buckle and damaged jetties. The stern of the barque ASCALON which had arrived from Shanghai laden with straw braid, cotton & silk was completely burnt out and the cabin gutted.

"The Times" 7/5/1881, Salvage – Three Locomotive Steam Engines damaged by fire – Fuller Horsey Sons & Co are instructed to sell by auction on Friday May 13, at the Albert Docks, near the basin North Woolwich side, two Locomotive Steam Engines for 4ft 8½in gauge and one Contractors Locomotive Steam Engine with vertical boiler.

An article in the "Locomotive Magazine" for January 1902 stated that two locomotives belonging to Lucas & Aird which were stored in a shed on the dock were badly damaged by fire; these are thought to be the two locomotives which were auctioned on May 13th 1881.

The two locos were CHELSEA and the Longridge loco which were repaired and hired to the Dock Company, the latter becoming No.8 in the dock fleet.
Gauge : 4ft 8½in

	RUSHTON	0-6-0ST		MW	588	1876	New	(1)
	VICTORIA	0-6-0ST		MW	589	1876	New	(3)
	DUVAL	0-4-0ST		MW	606	1876	New	(4)
		0-6-0ST		MW	607	1876	New	s/s
		0-6-0ST		MW	619	1876	New	s/s
	-	0-6-0ST	IC	MW	621	1876	New	s/s
10		0-4-0ST	OC	MW	624	1876	New	s/s
	-	0-4-0ST	OC	MW	625	1876	New	s/s
		0-4-0ST		MW	631	1876	(c)	s/s
		0-4-0ST		MW	690	1878	New	(5)
	CHELSEA	0-6-0ST	IC	Bton?	c1865		(a)	(2)
	-	0-6-0ST	IC	Longridge	1847			
				reb Bton	1860, 1865		(b)	(2)

(a) identity unknown, earlier at Portsmouth contract, Hampshire;
(b) ex Portsmouth contract, /1876; originally LBSCR 104, earlier 75.
(c) new to Admiralty Portsmouth Dockyard, Hampshire.

(1) later Beadle Bros/Wm Cory & Sons Ltd, Erith Wharf, Kent, SURREY, c/1880.
(2) to London & St. Katharine Dock Co, on hire after 5/1881.
(3) to Holwell Ironworks Ltd, Holwell Ironworks, Leicestershire, HOLWELL No.11.
(4) later C.D. Philips, dealers, Newport, Gwent, DUVAL; thence s/s.
(5) later Walter Scott Ltd, Leeds Steel Works, Hunslet, Leeds, Yorkshire.

TILBURY DOCKS CONTRACT   MC75

Construction of Tilbury Dock for East and West India Dock Co, 1884-1886.
Completion of contract taken over from Kirk & Randall (who ceased work in 7/1884); construction resumed on 27/10/1884, dock opened 17/4/1886.
"The Times" for 19/4/1886 carried a report on the construction of the docks by Lucas & Aird; in which it was stated that 54 locomotives, 35 portable engines, 46 pumping engines, 207 steam cranes & pile engines, six steam excavators, five dredgers, 1,650 wagons and 4,500 men were employed on this contract, total engines in steam 250 and 38 miles of temporary track laid.
"The Times", 3/10/1885 – Fuller, Horsey, Sons & Cassell to auction 11/11/1885 re East & West India Dock Co, Tilbury Docks works – first portion of plant and machinery including 15 locos 4w & 6w coupled locomotive engines, 8in to 14in by SS, MW & HE.
"The Times", 6/2/1886 – Fuller, Horsey, Sons & Cassell to auction 10/2/1886 re East & West India Dock Co, Tilbury Docks works – second portion of plant and machinery including 6w coupled saddletank locomotive engine by HE 14x18, one ditto 11x16, 4w coupled ditto 9x14, three ditto 8x14 all by MW.
"The Times", 29/5/1886 – Fuller, Horsey, Sons & Cassell to auction 17/6/1886 re East & West India Dock Co, Tilbury Docks works – third portion of plant and machinery including three 4w coupled oc saddletank locomotive engines 8x14 by MW.
"The Times", 21/8/1886 – Fuller, Horsey, Sons & Cassell to auction 14/9/1886 re East & West India Dock Co, Tilbury Docks works – fourth portion of plant and machinery including nine 4w coupled oc saddletank locomotive engines 10x16 by MW, HE & other makers.
"The Times", 8/1/1887 – Fuller, Horsey, Sons & Cassell to auction 7/2/1887 re East & West India Dock Co, Tilbury Docks works – fifth portion of plant and machinery including three 6w

coupled saddletank locomotive engines by SS with 11x20, seven 4w coupled by MW & other makers with 8 & 10in cyls 14 & 16in strokes. (Three locos from this sale sold to the Aire & Calder Navigation – See IL.137 p.225)

"The Times", 18/6/1887 – Fuller, Horsey Sons & Cassell to auction 12/7/1887 re East & West India Dock Co, Tilbury Docks works – sixth portion of plant and machinery including seven 4w coupled saddletank locomotive engines by MW & other makers, 8in to 10in dia cyls by 14in to 16in stroke & locomotive traction engine by AP.

"The Times", 25/2/1888 – Fuller, Horsey, Sons & Cassell to auction 2/3/1888 re East & West India Dock Co, remaining portion of plant and machinery used in construction of Tilbury Docks – including five 6w coupled ic saddletank locomotive engines 11in cyls and eight 4w coupled oc ditto with cyls 7in to 11in.

Gauge : 4ft 8½in

	Name	Type	IC/OC	Maker	Works#	Year	Note	Ref
	RIBBLESDALE	0-6-0ST	IC	MW	663	1878	(a)	(1)
	HAMPTON	0-6-0ST	IC	MW	694	1878	(a)	(1)
	BRUCE	0-6-0ST	IC	HE	234	1880	(a)	(1)
	WALLACE	0-6-0ST	IC	HE	235	1880	(a)	(8)
4	ANT	0-4-0ST	OC	MW	78	1863	(a)	s/s
5	TILBURY	0-4-0ST	OC	MW	554	1876	(a)	(2)
18		0-6-0ST					(e)	s/s
19		0-6-0ST					(e)	s/s
20		0-4-0ST	OC	MW	861	1883	(a)	(14)
21		0-4-0ST	OC	MW	862	1883	(a)	s/s
22		0-4-0ST	OC	MW	867	1883	(a)	(3)
23		0-4-0ST	OC	MW	882	1883	(a)	(4)
24		0-4-0ST	OC	MW	883	1883	(a)	(7)
25		0-4-0ST		MW	618	1876	(a)	(15)
26		0-4-0ST					(e)	s/s
27		0-6-0ST	IC	MW	872	1883	(a)	(1)
28		0-6-0ST	IC	MW	873	1883	(a)	(1)
	CALDEW	0-4-0ST	OC	MW	225	1867	(b)	(5)
42		0-6-0ST	IC	MW	858	1882	(c)	(6)
52		0-6-0ST	IC	MW	841	1882	(c)	(6)
100		0-4-0ST	OC	MW	221	1866	(d)	(6)
124		0-4-0ST					(e)	(9)
132				MW			(e)	s/s
	-	0-4-0ST	OC	MW			(e)	(10)
	JUMBO	0-4-0ST	OC	HH			(e)	s/s
	GROSVENOR	0-6-0T					(e)	s/s
	BLACK DIAMOND	0-4-0ST					(f)	s/s
	ANT	0-4-0ST	G				(g)	s/s
	-	0-6-0ST	IC	SS reb Merryweather			(h)	(11)
	-	0-6-0ST	IC	SS reb Merryweather			(h)	(12)
	-	0-6-0ST	IC	SS reb Merryweather			(h)	(13)
	-	0-6-0ST	IC	SS reb Merryweather			(h)	(13)
	-	0-6-0ST	IC	SS reb Merryweather			(h)	s/s

(a) ex Kirk & Randall, with site, 10/1884.
(b) use on this contract subject to confirmation; if correct then -
    ex Kirk & Randall, with site, 10/1884.
(c) ex Esher-Guildford (etc.) (LSWR) 91882-1885) contract, Surrey.
(d) earlier John Aird & Sons, Beckton Gas Works (1868-1870) contract.
(e) origin and identity unknown.
(f) origin and identity unknown, in use as a winding engine.
(g) origin and identity unknown, chain driven.
(h) ex Merryweather; rebuilds of SS locos of the Larmanjat type 1-2-1 built new for Lisbon Steam Tramways Co Ltd.(five of SS 2254/5, 2270-75, 2286-93).

(1) to East & West India Dock Co, with site, /1886.
(2) later with Chas Brand & Son, [possibly Mawcarse-Bridge of Earn (1887-1890) contract, Perthshire];
    thence to Shipley Coal Co, Coppice & Woodside Collieries, Ilkeston, Derbyshire.
(3) later S. Pearson & Son Ltd, Blackwall Road Tunnel (1890-1897) contract, London, JIM.
(4) to S. Pearson & Son, Empress Dock (1886-1890) contract, Southampton, Hampshire, SOLENT.
(5) to J.T. Firbank, Oxted-Groombridge (LBSCR) (1885-1888) contract, Surrey, CALDEW.
(6) later John Aird & Sons, Staines Reservoirs (1898-1902) contract, Middlesex.
(7) to Samuel Williams & Sons, Dagenham Dock, /1887.
(8) to East & West India Dock Co, East & West India Dock, Poplar, by 30/6/1893.
(9) to Suakin - Berber Railway contract, Sudan [*could be MW 862 duplicated* ??].
(10) to Beadle Bros, Erith, Kent, ERITH.
(11) possibly to Beadle Bros, Erith, Kent, DEVON; otherwise s/s.
(12) possibly to Beadle Bros, Erith, Kent; otherwise s/s.
(13) to Samuel Williams & Sons, Dagenham Dock.
(14) to Craven Lime Co Ltd, Craven Quarry, Settle, North Yorkshire.
(15) to Trafford Park Estates, Manchester.

TOTTENHAM – FOREST GATE CONTRACT CC76

Contract for the construction of the Tottenham – Forest Gate railway (4 miles) for the Midland and LT&SR. The contract was let to Lucas & Aird on 4/5/1891 at a price of £264,422, 5s-9d; the line opened to traffic on 1/7/1894.

A visit to the works by the Society of Engineers in 10/1893 found four locomotives in use. These locomotives are as yet un-identified.

MARPLES RIDGEWAY & PARTNERS LTD
DAGENHAM CONTRACT PC77

Contract for the extension of the Briggs Motor Body Works, 1956 – 1957.
Gauge : 2ft 0in

R L 8	4wDM	FH	3709	1954	(a)	(1)
	4wDM	OK	10253	1939	(b)	(2)

(a) New to Marples Ridgeway & Partners Ltd, possibly ex another contract.
(b) ex William Jones Ltd, Greenwich, London, on hire by 15/9/1956.
(1) s/s after 16/3/1957.
(2) returned to William Jones Ltd off hire, later General Plant Reconstruction Ltd, Aberthaw Power Station contract, Glamorgan, 3/1957.

HENRY MARTIN
CHELMSFORD CONTRACT GC78

Contract for the extension of the Hoffmann Manufacturing Co Ltd factory at Chelmsford. The contractor Henry Martin of Northampton used a small locomotive to haul building materials during construction of the works in the latter part of WW1, further details unknown.
Gauge : ?

| | | | tank | | | (a) | s/s |

(a) origin and identity unknown.

MAY GURNEY & CO LTD
COLCHESTER PLANT DEPOT, Great Horkesley HC79
TL 979301

Company registered 7/10/1926 to carry on the business of public works contractors and engineers formerly carried on by R. J. May at Trowse Newton, Norfolk.
Gauge : 3ft 0in

| MG/S 100 | GORDON | 4wDM | MR | 10160 | 1950 | (a) | (1) |

(a) ex Southend Pier contract c2/1986.
(1) to Sackers Ltd, Claydon, Ipswich, Suffolk, for scrap c4/1989.

SOUTHEND PIER CONTRACT JC80

Contract in 1985 for the re construction of the 3ft 6in gauge railway to 3ft at a cost of £1.3m, work commenced 11/1984.
Gauge : 3ft 6in

| MG/S 100 | GORDON | 4wDM | MR | 10160 | 1950 | (a) | (1) |

(a) ex Track Supplies & Services Ltd, Wolverton, Buckinghamshire, by 27/12/1984, earlier Fisons Ltd, Swinefleet, Yorkshire (ER).
(1) returned to Track Supplies & Services Ltd, re gauged to 3ft, returned here /1985, to Great Horkesley plant Depot c2/1986.

SIR ROBERT McALPINE & SONS LTD
[Reference: The Story of McAlpine Steam Locomotives, J.B.Latham, 1993.]

TILBURY DOCKS CONTRACT   MC81

Improvement works at Tilbury Docks for Port of London Authority, 1926 - 1929.

Works included the construction of a new entrance lock 1,000ft long and 110ft wide and provision of a new dry dock 750ft long, 110ft wide; other works included an increase of the main area of the dock by 15 acres and diversion of the railway to the south side of the main dock and construction of a new passenger landing stage on the river, contract awarded 15/9/1926 at a price of £2,400,000, the official opening took place on 26/9/1929.

A visit here by civil engineers in 1928 reported thirty steam locos and twenty one petrol locos in use on twenty miles of standard gauge and five miles of 2ft gauge track.

Reference : "The Times" 27/7/1923 & 7/9/1923.

Gauge : 4ft 8½in

11	0-4-0ST	OC	HC	532	1899	(a)	(19)
23	0-6-0ST	IC	HC	653	1903	(b)	(21)
18	0-4-0ST	OC	HC	534	1900	(c)	(22)
25	0-6-0ST	IC	HC	664	1905	(c)	(23)
26	0-6-0ST	IC	HC	888	1909	(c)	(2)
30	0-6-0ST	IC	HC	1011	1912	(c)	(20)
46	0-6-0ST	IC	HC	1539	1924	(d)	(5)
43	0-6-0ST	IC	HC	1528	1924	(e)	(17)
40	0-6-0ST	IC	HC	1525	1924	(f)	(9)
41	0-6-0ST	IC	HC	1526	1924	(g)	(4)
45	0-6-0ST	IC	HC	1538	1924	(h)	(6)
42	0-6-0ST	IC	HC	1513	1924	(j)	(11)
47	0-6-0ST	IC	HC	1494	1923	(k)	(14)
49	0-6-0ST	IC	HC	1511	1923	(l)	(17)
44	0-6-0ST	IC	HC	1524	1924	(m)	(3)
48	0-6-0ST	IC	HC	1510	1923	(m)	(15)
33	0-6-0ST	IC	HC	1029	1913	(n)	(12)
32	0-6-0ST	IC	HC	1028	1913	(o)	(1)
31	0-6-0ST	IC	HC	1026	1913	(p)	(16)
53	0-6-0ST	IC	MW	1560	1902	(q)	(18)
54	0-6-0ST	IC	HC	1601	1927	New	(10)
55	0-6-0ST	IC	HC	1585	1927	New	(13)
56	0-6-0ST	IC	HC	1602	1927	New	(8)
57	0-6-0ST	IC	HC	1586	1927	New	(7)

(a) earlier at Great Stanney Plant Depot, Cheshire, /1921.
(b) ex A2 (Watling Street) road reconstruction, Dartford - Strood, Kent, after /1923.
(c) ex St Lawrence - Ramsgate railway (SR) contract, Kent, /1926.
(d) ex A41 (Watford By-pass) road contract, Hertfordshire, 20/7/1926.
(e) ex St Lawrence - Ramsgate railway (SR) contract, Kent, 12/8/1926.
(f) ex Ellesmere Port (Manchester Ship Canal) contract, Cheshire, 6/9/1926.
(g) ex A4123 (Birmingham - Wolverhampton) road contract, Birmingham, 2/10/1926.
(h) ex Great Stanney Plant Depot, Cheshire, 4/10/1926.
(j) ex A4123 (Birmingham - Wolverhampton) road contract, Birmingham, 5/10/1926.
(k) ex Ellesmere Port (Manchester Ship Canal) contract, Cheshire, 21/10/1926.

(l)   ex Sir W.G. Armstrong Whitworth & Co Ltd, Wakefield contract, West Yorkshire, 21/10/1926.
(m) ex Great Stanney Plant Depot, Cheshire, 29/10/1926.
(n)   earlier Wilbraham Estate contract, Manchester, /1921; to here 9/11/1926.
(o)   ex Ellesmere Port (Manchester Ship Canal) contract, Cheshire, /1927.
(p)   earlier A2 (Watling Street) road reconstruction, Dartford - Strood, Kent; then exhibited at Wembley Exhibition, /1923-4; to here 31/1/1927.
(q)   ex Great Stanney Plant Depot, Cheshire, 12/7/1927.

(1)   to West India Dock (PLA) contract, London, /1927.
(2)   to Southampton Docks contract, Hampshire, /1928.
(3)   to Southampton Docks contract, Hampshire, 12/6/1928.
(4)   to Great Stanney Plant Depot, Cheshire, 17/8/1928.
(5)   to Southampton Docks contract, Hampshire, 11/9/1928.
(6)   to Great Stanney Plant Depot, Cheshire, 1/11/1928.
(7)   to Southampton Docks contract, Hampshire, 19/12/1928.
(8)   to Wimbledon- Sutton railway (SR) contract; 28/1/1929.
(9)   to Great Stanney Plant Depot, Cheshire, 30/1/1929.
(10) to Great Stanney Plant Depot, Cheshire, 11/2/1929.
(11) to Southampton Docks contract, Hampshire, 11/2/1929.
(12) to Southampton Docks contract, Hampshire, 12/4/1929.
(13) to Southampton Docks contract, Hampshire, 16/4/1929.
(14) to Southampton Docks contract, Hampshire, 25/4/1929.
(15) to Southampton Docks contract, Hampshire, 11/5/1929.
(16) to Wimbledon- Sutton railway (SR) contract; 10/8/1929.
(17) to Great Stanney Plant Depot, Cheshire, 31/8/1929.
(18) to Southampton Docks contract, Hampshire, 31/8/1929.
(19) to Dunston Power Station contract, Newcastle-upon-Tyne, by c/1931.
(20) to Southampton Docks contract, Hampshire, by /1932.
(21) to Great Stanney Plant Depot, Cheshire, by 14/11/1933;
       thence to Hayes Plant Depot, Middlesex where scrapped, 5/1/1934.
(22) to Great Stanney Plant Depot, Cheshire, by 9/1934; s/s.
(23) to Great Stanney Plant Depot, Cheshire, by /1936;
       thence to Hayes Plant Depot, Middlesex, /1936; s/s.

Gauge : 3ft 0in
```
51              0-4-0ST   OC  HC   1536  1924   (a)   (1)
52              0-4-0ST   OC  HC   1537  1924   (b)   (2)
39              0-4-0ST   OC  HC   1534  1924   (c)   (3)
```

(a)   ex A1 (Barnet By-pass) road contract, Hertfordshire, 2/9/1926.
(b)   ex A58 (Crompton Way) road contract (Bolton Corporation), Lancashire, 27/9/1926.
(c)   ex Great Stanney Plant Depot, Cheshire, 13/10/1927.

(1)   to A4123 (Birmingham - Wolverhampton) road contract, Birmingham, 2/3/1927.
(2)   to A4123 (Birmingham - Wolverhampton) road contract, Birmingham, 9/3/1927.
(3)   to Great Stanney Plant Depot, Cheshire, 11/10/1928.

ROMFORD CONTRACT   EC82

Widening of the Romford - Gidea Park railway for LNER, 1/1930 -9/1931.
Gauge : 4ft 8½in

31	0-6-0ST	IC	HC	1026	1913	(a)	(2)
56	0-6-0ST	IC	HC	1602	1927	(b)	(1)

(a)   ex Wimbledon - Sutton railway (SR) contract, Surrey, /1930.
(b)   ex Wimbledon - Sutton railway (SR) contract, Surrey, 1/4/1930.

(1)   to Irlam contract (Manchester Ship Canal), Irlam, Lancs, 12/2/1931.
(2)   to Irlam Steelworks Extension contract (Lancashire Steel Corporation), Irlam, Lancs, 25/2/1931.

SHENFIELD CONTRACT   EC83

Widening of the Gidea Park to Shenfield railway for the LNER, 2/1932 - 1/1/1934.
Reference : British Railways Illustrated Annual No.5, 1996.
  "50 petrol tractors were used" of which as yet only the following have been identified.
Gauge : 2ft 0in

-	4wDM	MR	5614	1932	New	(7)
-	4wDM	MR	5620	1932	New	(6)
-	4wDM	MR	5621	1932	New	(7)
-	4wPM	MR	4592	1930	New	(7)
-	4wPM	MR	4593	1930	New	(7)
-	4wPM	MR	4594	1930	New	(7)
-	4wPM	MR	4596	1932	New	(7)
-	4wPM	MR	4597	1932	New	(7)
-	4wPM	MR	4598	1932	New	(4)
	4wPM	MR	4599	1932	New (c)	(7)
	4wPM	MR	4600	1932	New (c)	(7)
	4wPM	MR	4601	1932	New (c)	(7)
	4wPM	MR	4602	1932	New (c)	(7)
	4wPM	MR	4603	1932	New (c)	(7)
	4wPM	MR	4604	1932	New (c)	(7)
	4wPM	MR	4605	1932	New (c)	(7)
	4wPM	MR	4606	1932	New (c)	(7)
	4wPM	MR	4607	1932	New (c)	(7)
	4wPM	MR	4608	1932	New (c)	(7)
	4wPM	MR	4609	1932	New (c)	(7)
	4wPM	MR	4610	1932	New (c)	(4)
	4wPM	MR	4611	1932	New (c)	(4)
	4wPM	MR	4612	1932	New (c)	(7)
	4wPM	MR	4613	1932	New (c)	(5)
	4wPM	MR	4614	1932	New (c)	(5)
	4wPM	MR	4615	1932	New (c)	(5)
	4wPM	MR	4616	1932	New (c)	(7)
	4wPM	MR	4617	1932	New (c)	(4)
	4wPM	MR	4618	1932	New (c)	(7)
	4wPM	MR	4619	1932	New (c)	(5)
	4wPM	MR	4620	1932	New (c)	(7)

4wPM	MR	4621	1932	New (c)	(7)
4wPM	MR	4622	1932	New (c)	(7)
4wDM	RH	164338	1932	(a)	(1)
4wDM	RH	165365	1932	(b)	(2)
4wDM	RH	166009	1932	New	(3)

(a) ex RH, Lincoln, on trial 15/2/1932 to work with a Simplex loco.
(b) ex RH, Lincoln, on trial by 14/3/1932.
(c) New to Sir Robert McAlpine & Sons Ltd, delivered to an unspecified location in the London area in /1932, believed this contract, use here unconfirmed.

(1) returned to RH 3/3/1932, to E. Allman & Co, General Engineers, Birdham Road, Chichester, Sussex, by 21/3/1932, to John Heaver Ltd, Whyke Gravel Pits, Chichester, Sussex by 19/12/1933.
(2) to RH, Wood Works, Lincoln, after 10/10/1932, thence to W. Dennis & Sons Ltd, Nocton Estate, Lincolnshire, by 1/11/1932.
(3) to War Office, R.E. Aldershot, Hampshire, by 4/1934.
(4) to Ebbw Vale contract, Monmouthshire, /1938.
(5) possibly to Ebbw Vale contract, Monmouthshire, /1938.
(6) to unknown location ; MR had spares order from "Wigglesworths", 6/2/1950.
(7) to another contract, otherwise s/s.

LEE VALLEY CONTRACT   DC84

Contract for construction of conduit between Banbury, Lockwood & Maynard reservoirs for the Metropolitan Water Board in 1934, petrol locos used details unknown.

ROYAL VICTORIA DOCK CONTRACT   AC85

Extension of the Royal Victoria Dock for the Port of London Authority, 1936 -1937.
Gauge : 3ft 0in

No.50	0-4-0ST	OC	HC	1535	1924	(a)	(1)
No.51	0-4-0ST	OC	HC	1536	1924	(b)	(2)

(a) ex Hayes Plant Depot, Middlesex, /1936 (after 21/3/1936).
(b) ex Hayes Plant Depot, Middlesex, 11/8/1936.

(1) to Hayes Plant Depot, Middlesex, /1937.
(2) to Hayes Plant Depot, Middlesex, 25/9/1937.

CHALKWELL CONTRACT   JC86

Prittle Brook Flood Diversion Scheme, Chalkwell Section, 1972-1973.
Gauge : 2ft 0in

1	4wBE	WR	J7274	1969	(a)	(1)
2	4wBE	WR	J7272	1969	(a)	(1)
3	4wBE	WR	J7208	1969	(a)	(1)
4	4wBE	WR	J7282	1970	(a)	(1)
5	4wBE	WR	J7206	1969	(a)	(1)
6	4wBE	WR	J7273	1969	(a)	(1)

(a) ex another McAlpine contract or plant depot.

(1) to Kettering Plant Depot, Northamptonshire, by 8/4/1973.

McCORMICK & SONS
HUTTON CONTRACT   GC87

Contract for the construction of a childrens home for Poplar Council, (TQ 617952), the home later passed to London County Council.

"The Engineer" 22/2/1907, F Warman to auction 7/3/1907 re McCormick & Sons on completion of Poplar Childrens Home, Hutton, Essex – plant including two full gauge four coupled contractors locomotives.

Reference: Industrial Locomotive No.88.

Gauge : 4ft 8½in

[FANNY   0-4-0ST   OC   HG   246   1866]   (a)   s/s

(a) if identified correctly, ex HC after rebuild 6/9/1904, earlier Carnforth Hematite Iron Co, Lancashire.

METROPOLITAN WATER BOARD
CHINGFORD CONTRACT   DC88

Work to raise the bank height at King George V Reservoir, Chingford, 1939 -1943, authorised on 14/10/1938.

Gauge : 2ft 0in

-                  4wDM   OK   6504   1936   (a)   (1)

(a) ex William Jones Ltd, dealers, Greenwich, London, on hire, 6/1939; purchased 12/1940.

(1) to another two MWB sites on direct labour contracts; then to Shepperton Depot, Middlesex, c/1955.

MILLER CONSTRUCTION LTD
Incorporated 23/12/1963
ROYAL VICTORIA DOCK DRAINAGE SCHEME   AC89

Contract for construction of the Royal Docks drainage & sewerage scheme, 1988, exact details unknown.

Gauge : 1ft 6in

    L15   4wBE   CE   B0109A 1973   (a)   (1)
    L16   4wBE   CE   B0109B 1973   (a)   (1)
    L17   4wBE   CE   5431 1968   (a)   (1)

(a) ex Rugby Plant Depot, Warwickshire, by 8/4/1988.

(1) returned to Rugby Plant Depot by 20/6/1989.

BRENTWOOD CONTRACT EC90

Contract for Brentwood Urban District Council, /1986, details unknown.
Gauge : 1ft 6in

L13	4wBE	CE	5965B	1973	(a)	(1)
	4wBE	WR	L801	1983	(b)	(2)
	4wBE	WR	544901	1983	(c)	(2)

(a)   ex Rugby plant Depot, Warwickshire, 28/7/1986.
(b)   ex Welham Plant Ltd, Cambridgeshire, 1/8/1986.
(c)   ex Welham Plant Ltd, Cambridgeshire, 8/9/1986.

(1)   returned to Rugby plant Depot, 12/11/1986.
(2)   returned to Welham Plant Ltd.

FRED MITCHELL & SON LTD
CLACTON CONTRACT HC91

Contract at Clacton on Sea, details unknown.
"Contract Journal" 19/8/1931 for sale or hire at Clacton on Sea – plant including one 20hp Simplex 2ft gauge – Fred Mitchell & Son Ltd, Manchester.
Gauge : 2ft 0in

	4wPM	MR		(a)	s/s

(a)   possibly one of either MR3855 or 3858 built 1928.

MITCHELL BROS, SONS & CO LTD
TILBURY CONTRACT MC92

Work at Tilbury Power Station for CEGB, c/1966.
Gauge : 2ft 0in

MBS 236	4wBE	LMM	1066	1950	(a)	(1)
MBS 237	4wBE	LMM	1072	1950	(a)	(1)
MBS 238	4wBE	LMM			(a)	(1)
MBS 247	4wBE	LMM			(a)	(1)
MBS 386	4wBE	LMM	1049	1950	(b)	(1)
MBS 387	4wBE	LMM	1053	1950	(b)	(1)

(a)   ex Tickhill Plant Depot, South Yorkshire, c/1966;
        earlier Edmund Nuttall Ltd, contractors, until /1965.
(b)   ex Tickhill Plant Depot, South Yorkshire, c/1966;
        earlier M.E. Engineering Ltd, dealers, Cricklewood, Middlesex.

(1)   to Victoria Line Underground (LT) contract, (via Tickhill Plant Depot ?), /1968.

MORGAN SINDALL, VINCI CONSTRUCTION, BACHY SOLETANCHE J.V.
LEE TUNNEL CONTRACT   AC93
Contract for the construction of 6.4km of tunnel for Thames Water; this deep level tunnel 7m in diameter and 75m deep will run from Abbey Mills pumping station to Beckton treatment plant and is designed to carry 16 million tonnes of sewage and rainwater annually. Work started 30/9//2010 with completion in 2014 at an estimated total cost of £635m. Use of locomotives unknown at time of writing.

JOHN MOWLEM & CO
BECKTON CONTRACT   AC94
Construction of the Northern Outfall sewage treatment works 1887 – 1889. The contract was let in 1/1887 and included construction of a pier, tramway and twelve cottages.
Reference: Industrial Locomotive No.76.

Gauge :4ft 8½in

	-	0-4-0ST	OC	MW	1014	1887	New	(1)
	BARKING	0-4-0ST	OC	HE	388	1887	New	(2)

(1) later Mountsorrel Granite Co Ltd, Mountsorrel Quarries, Leicestershire.
(2) later Williams & Son Ltd, contractors, Langley, Buckinghamshire, by 6/1901;
    to Ebbw Vale Steel, Iron & Coal Co Ltd, Ebbw Vale Works, Monmouthshire, by /1917.

UPMINSTER CONTRACT   LC95
Construction of Grays - Upminster railway for LT&SR. The contract for Grays-Upminster (5 miles) was let to Mowlem on 4/7/1889 and for Upminster-Romford (3 miles) in 6/1890. The Grays section opened on 1/7/1892 and the Romford line on 7/6/1893.

Gauge : 4ft 8½in

	GREENWICH	0-4-0ST	OC	MW	887	1883	(a)	(4)
	ALBERT	0-6-0ST	IC	MW	902	1884	(a)	(1)
	GROSVENOR	0-6-0ST	IC	HE	288	1884	(a)	(2)
18	GUERNSEY	0-4-0ST	OC	MW	1241	1892	New	(3)

(a) ex Bournemouth (LSWR) (1884-1886) contract, Hampshire, c/1887.
(1) to Albert Batchelor Ltd, Halling, Kent.
(2) to Walter Scott & Co Ltd, Charwelton - Quainton Road (MS&LR) (1894-1898) contract, Northamptonshire, by /1895.
(3) to Gibbs & Co, West Thurrock.
(4) to Bott & Lewis Jones, Tattenham Corner branch (SECR) (1895-1901) contract, Surrey, by /1900.

JOHN MOWLEM & CO LTD
John Mowlem & Co until 8/6/1903

DAGENHAM CONTRACT   PC96

Contract for a large jetty and water circulating tunnels for the power house at the Ford Motor Works, 1931-1932, on which 2ft gauge petrol locos were used.

LEYTON CONTRACT   CC97

Construction of the tunnels for the extension of the LT Central Line "Tube" from Mile End to Leyton. The contract was placed 10/1936 and work started the same month; the tunnels were completed by about 1940 (but not opened to traffic to Stratford until 3/12/1946 and on to Leytonstone on 5/5/1947).

Gauge : 2ft 0in

-	4wBE	WR	1060	1937	New	s/s
-	4wBE	WR	1061	1937	New	s/s
-	4wBE	WR	1062	1937	New	s/s
-	4wBE	WR	1063	1937	New	s/s
-	4wBE	WR	1064	1937	New	s/s
-	4wBE	WR	1065	1937	New	s/s
-	4wBE	WR	1066	1937	New	s/s
-	4wBE	WR	1067	1937	New	s/s

CHINGFORD CONTRACT   DC98

Construction of William Girling Reservoir, Chingford (TQ 369945) for Metropolitan Water Board authorised in 7/1935 at a cost of £682,156. In 7/1937 an embankment slip occurred during construction and a further slip took place in December that year. As a result the reservoir had to be redesigned increasing the cost of construction to an estimated £2,334,000. Delays caused by WW2 meant work on the reservoir resumed in 1946 and was not completed until 1951, it being officially opened on 4/9/1951.

Narrow gauge internal combustion locos are said to have been used on clay haulage.

Gauge : 4ft 8½in

BOBBY	0-6-0ST	OC	HC	1593	1927	(a)	(4)
SOUTHAMPTON	0-6-0ST	OC	HE	1647	1931	(b)	(1)
MILLBROOK	0-6-0ST	OC	HE	1648	1931	(c)	(6)
TRAFFORD PARK	0-6-0ST	OC	HE	1689	1931	(d)	(4)
PENN	0-6-0ST	IC	MW	1539	1902	(e)	
	reb		YE		1931		s/s after 12/1947
SHIRLEY	0-4-0ST	OC	AB	1993	1932	(f)	(3)
ITCHEN	0-4-0ST	OC	AE	2037	1931	(g)	(4)
WOODCROFT	0-6-0ST	IC	HC	1583	1926	(h)	(2)
FORTH	0-6-0ST	OC	AB	1844	1924	(j)	(4)
NEWPORT	0-6-0ST	IC	MW	1665	1905	(k)	(5)

(a) ex Millbrook Graving Dock (Mowlem/Nuttall Joint) (SR) (1931-1935) contract, Southampton, Hampshire (per George Cohen, Sons & Co Ltd, NM 109, NM8, /1936);
to Ministry of Fuel & Power, West Hallam Disposal Point, Derbyshire, /1945; returned;
to Samuel Williams & Sons Ltd, Dagenham Dock, /1947; returned.

(b) ex Millbrook Graving Dock (Mowlem/Nuttall Joint) (SR) (1931-1935) contract, Southampton, Hampshire (per George Cohen, Sons & Co Ltd, NM 102, NM2, /1936);
to a ROF contract; returned, by 10/10/1943.

(c) ex Millbrook Graving Dock (Mowlem/Nuttall Joint) (SR) (1931-1935) contract, Southampton, Hampshire (per George Cohen, Sons & Co Ltd, NM 111, /1936);
to another contract, /1942; returned by 4/1947.

(d) ex Millbrook Graving Dock (Mowlem/Nuttall Joint) (SR) (1931-1935) contract, Southampton, Hampshire (per George Cohen, Sons & Co Ltd, NM 106, NM6, /1936);
to Ministry of Fuel & Power, Bennerley Disposal Point, Nottinghamshire, by 1/1945; returned, after 4/1946;
to Samuel Williams & Sons Ltd, Dagenham Dock, /1947; returned.

(e) ex Millbrook Graving Dock (Mowlem/Nuttall Joint) (SR) (1931-1935) contract, Southampton, Hampshire (per George Cohen, Sons & Co Ltd, NM 111, /1936), here on 10/10/1943.

(f) ex Dover Train Ferry Dock (Mowlem/Nuttall Joint) (SR) (1934-1937) contract, Kent;
to Swynnerton (ROF) (1939-1945) contract, Staffordshire, /1939; returned by 1/1943;
to County of London Electricity Supply Co, Creekmouth, /1943; returned, /1946.

(g) ex Dover Train Ferry Dock (Mowlem/Nuttall Joint) (SR) (1934-1937) contract, Kent;
to Swynnerton (ROF) (1939-1945) contract, Staffordshire, /1939; returned /1941;
to Metropolitan Water Board, Lea Bridge, /1941; returned /1948.

(h) ex Joseph Pugsley & Sons Ltd, dealers, Stoke Gifford, Bristol, /1937;
(earlier C.J. Wills & Sons Ltd, Becontree Housing (LCC) (1920-1934) contract);
to Swynnerton (ROF) (1939-1945) contract, Staffordshire, /1940; returned by 10/10/1943.

(j) ex Joseph Pugsley & Sons Ltd, dealers, Stoke Gifford, Bristol, /1937;
(earlier C.J. Wills & Sons Ltd, Becontree Housing (LCC) (1920-1934) contract);
to Swynnerton (ROF) (1939-1945) contract, Staffordshire, /1939; returned /1943;
to Samuel Williams & Sons Ltd, Dagenham Dock, /1943; returned /1946.

(k) ex Welham Green Plant Depot. Hertfordshire, after c/1942;
earlier Swynnerton (ROF) (1939-1945) contract, Staffordshire.

(1) to Welham Green Plant Depot, Hertfordshire, after 10/10/1943.
(2) to Welham Green Plant Depot, Hertfordshire, /1944.
(3) to North Metropolitan Electric Power Supply Co, Brimsdown Power Station, /1947.
(4) to Welham Green Plant Depot, Hertfordshire, /1948.
(5) frames returned to Welham Green Plant Depot, Hertfordshire, /1948 (and scrapped).
(6) to Welham Green Plant Depot, Hertfordshire, by 4/1949.

SHELL HAVEN CONTRACT   KC99

Construction of new installation at Shell Haven for Shell-Mex & B.P. Ltd, 1949.
Gauge : 4ft 8½in
    MILLBROOK   0-6-0ST   OC   HE   1648 1931 (a)   (1)

(a) ex Welham Green Plant Depot, Hertfordshire, by 24/6/1949.

(1) returned to Welham Green Plant Depot, Hertfordshire, by 7/1950.

WEST THURROCK POWER STATION CONTRACT   NC100

Thames-side Power Station (at TQ 590770) with standard gauge rail connections used during construction for the delivery of materials. Commissioned in 10/1962.
Gauge : 4ft 8½in

| ED 8 | | 0-4-0ST | OC | AB | 2353 | 1954 | (a) | (1) |

(a)   ex CEGB Tilbury Power Station, 9/3/1959.
(1)   to CEGB Bow Power Station, c8/1965.

WILLIAM MUIRHEAD MACDONALD WILSON & CO LTD
Wm Muirhead & Co Ltd until 2/11/1918
Wm Muirhead & Co until 21/1/1910

BARKINGSIDE PLANT DEPOT   DC101

Company changed its name from Wm Muirhead & Co Ltd 2/11/1918 whose directors in 1919 were listed as Thomas Muirhead, Percy Charles Organ & Charles Fison. Depot established on 16 acres of land at Barkingside, had sidings and a rail connection with the Wanstead to Ilford loop at Barkingside Station. The company had previously operated a depot at Fulham, West London, until 1921; went into voluntary liquidation and liquidator appointed 6/9/1928.

"Machinery Market" 6/11/1925 for sale – three 10hp McP petrol locos – Muirhead, McDonald Wilson & Co.

"Machinery Market" 18/6/1926 for sale – plant inc three 2ft gauge locos – Muirhead, McDonald, Wilson & Co Ltd, Barkingside.

"Contract Journal" 13/7/1927 The Plant Depot, Barkingside, for sale std gauge 0-6-0 P 14in; 0-6-0 MW 12in; 0-6-0 12in Vulcan; 0-4-0 MW 9in; also a McP petrol loco 2ft gauge.

"Contract Journal" 25/7/1928 for sale – std gauge 17in 0-6-0 AE; 14in 0-6-0 P; 12in 0-6-0 VF; 12in 0-6-0 MW; 9in 0-4-0 MW; also 2ft gauge locos 6in 0-4-0 KS & 6in 0-4-0 Koppel – The Plant Depot, Barkingside.

Gauge : 4ft 8½in

AVON	0-6-0ST	IC	MW	738	1881	(a)	s/s
-	0-6-0ST	IC	AE	1016	1874	(b)	(5)
-	0-6-0(ST)		P			(c)	(1)
-	0-6-0(ST)		MW			(c)	(2)
-	0-4-0(ST)		MW			(c)	(3)
-	0-6-0(ST)		VF			(c)	(4)
-	0-4-0(ST)		MW			(c)	(4)

(a)   earlier Charles Brand & Son, Hendon contract, Middlesex.
(b)   earlier MoM Stratton, Swindon, Wiltshire.
(c)   origin & identity unknown.
(1)   for sale 11/1926, 7/1927 and 7/1928, s/s.
(2)   for sale 11/1926, 7/1927 and 7/1928 (may be AVON above), s/s.
(3)   for sale 11/1926, s/s.
(4)   for sale 7/1927 and 7/1928, s/s.
(5)   for sale 7/1928, s/s.

Gauge : 2ft 0in
Amongst other 2ft 0in gauge locos for sale here were :

		0-4-0ST	OC	KS	[4247 1922?]	(a)	(3)
	-	0-4-0WT	OC	OK		(b)	(2)
	-	0-4-0PM		McP		(b)	(1)

(a) earlier Southend Road contract.
(b) origin and identity unknown.
(1) for sale 7/1927, s/s.
(2) for sale 7/1928, s/s.
(3) for sale 7/1928, to Organ & Co Ltd, Barkingside, by 5/1929.

SOUTHEND ROAD CONTRACT   DC102

Construction of the Southend Arterial Road, Barkingside - Rayleigh, for the Ministry of Transport, 1921-1925. Construction work began in 12/1921; the road was thirty miles long and was opened by Prince Henry on 25/3/1925.

"Contract Journal", 9/5/1923, for sale – re London – Southend Road – 14 steam locos, 2ft gauge – Ministry of Transport, Hall Lane, Upminster.

"Contract Journal", 26/11/1924, Fuller Horsey to sell 9-10/11/1924 for Ministry of Transport on completion of contracts new London Southend Road - including 12in MW 0-6-0 std gauge, and 18 2ft gauge locos by WB, KS, and Hudson, also for sale in this auction were three Deutz 20hp, 24in gauge petrol tractors and two "Excelsior" tractors at various depots between Woodford & Rayleigh.

Reference : Industrial Locomotive No.60, No.61, No.62, No.64.

Gauge : 2ft 0in

-	0-4-0ST	OC	KS	2459	1915	(a)	(5)
-	0-4-0ST	OC	KS	2460	1915	(a)	(6)
-	0-4-0ST	OC	KS	2469	1916	(b)	(23)
-	0-4-0ST	OC	KS	4002	1918	(c)	(7)
-	0-4-0ST	OC	KS	4005	1918	(c)	(8)
-	0-4-0ST	OC	KS	4162	1921	(d)	(14)
81	0-4-0ST	OC	KS	4163	1921	New	(9)
145	0-4-0ST	OC	KS	4246	1922	New	(1)
146	0-4-0ST	OC	KS	4247	1922	New	(24)
147	0-4-0ST	OC	KS	4248	1922	New	(1)
148	0-4-0ST	OC	KS	4249	1922	New	(12)
165	0-4-0ST	OC	KS	4250	1922	New	(16)
166	0-4-0ST	OC	KS	4251	1922	New	(16)
171	0-4-0ST	OC	KS	4252	1922	New	(18)
172	0-4-0ST	OC	KS	4253	1922	New	(22)
151	0-4-0ST	OC	KS	4254	1922	New	(13)
152	0-4-0ST	OC	KS	4255	1922	New	(12)
153	0-4-0ST	OC	KS	4256	1922	New	(10)
154	0-4-0ST	OC	KS	4257	1922	New	(12)
155	0-4-0ST	OC	KS	4258	1922	New	(16)
156	0-4-0ST	OC	KS	4259	1922	New	(12)
157	0-4-0ST	OC	KS	4260	1922	New	(11)

158	0-4-0ST	OC	KS	4261	1922	New	(15)
159	0-4-0ST	OC	KS	4262	1922	New	(17)
160	0-4-0ST	OC	KS	4263	1922	New	(20)
161	0-4-0ST	OC	KS	4264	1922	New	(12)
162	0-4-0ST	OC	KS	4265	1922	New	(16)
173	0-4-0ST	OC	KS	4267	1922	New	(19)
174	0-4-0ST	OC	KS	4268	1922	New	(21)
175	0-4-0ST	OC	KS	4269	1922	New	(22)
176	0-4-0ST	OC	KS	4270	1922	New	(13)
	0-4-0ST	OC	WB	2047	1918	(e)	(4)
	0-4-0ST	OC	WB	2073	1918	(f)	(2)
	0-4-0ST	OC	WB	1663	1902	(g)	s/s
	0-4-0ST	OC	WB	2048	1918	(h)	(3)

(a) ex McDonald Gibbs & Co, Didcot, Oxfordshire.
(b) ex Home Grown Timber Committee, Totton, Hampshire.
(c) ex Air Ministry, RAF Henlow, Bedfordshire.
(d) ex Ilford Housing Estate contract.
(e) ex H Boot & Sons, National Shipyard, Chepstow, Monmouthshire, here by 17/1/1922.
(f) ex Ilford Housing Estate contract by 12/4/1922.
(g) ex Old Delabole Slate Co Ltd, Delabole, Cornwall, here by 27/5/1922.
(h) ex Ilford Housing Estate contract by 17/8/1922.

(1) to Thos. W. Ward Ltd, c1925.
(2) to Holborough Cement Co Ltd, Snodland, Kent, by 16/7/1925.
(3) to George Cohen, Sons & Co Ltd, by 19/2/1926.
(4) to Holborough Cement Co Ltd, Snodland, Kent, by 26/11/1926.
(5) to Thos. W. Ward Ltd, Grays, TW 151, /1927.
(6) to Thos. W. Ward Ltd, Grays, TW 152, /1927.
(7) to Thos. W. Ward Ltd, Grays, TW 146, /1927.
(8) to Thos. W. Ward Ltd, Grays, TW 148, /1927.
(9) to Cement Industries, Asheham, /1927.
(10) to Thos. W. Ward Ltd, Grays, TW 149, /1927.
(11) to Thos. W. Ward Ltd, Grays, TW 147, /1927.
(12) to Stewart & McDonnell Ltd, Kingston by Pass contract, Surrey, by 6/1927.
(13) to H Coxhead & Co Ltd, Crow Park Bridge contract, Sutton-on-Trent, Nottinghamshire, by 8/1928.
(14) to W.& C. French Ltd, North Circular Road contract, Middlesex, by 10/1928.
(15) to Robert Jenkins & Co Ltd, Engineers, Rotherham, South Yorkshire, by 12/1928.
(16) to Devon County Council, by 1929.
(17) to Aubrey Watson & Co Ltd, contractors, by 1/1929.
(18) to David Thompson & Sons Ltd, Balbriggan Waterworks contract, Ireland, by 1/1929.
(19) to Aubrey Watson & Co Ltd, by 3/1929.
(20) to Ham River Grit Co, Ham Pits, Surrey, by 1929?
(21) to William Jones Ltd, dealers, Greenwich, London, by 4/1931.
(22) to Sir W.G. Armstrong Whitworth & Co Ltd, contract for construction of East Coast Main Trunk Railway, New Zealand, via J.J. Niven & Co (agents).
(23) to R. Stroud & Son, Goodmayes.
(24) possibly to Barkingside Plant Depot.

ILFORD CONTRACT
DC103

Construction of Ilford Housing Estate for the City of London Corporation, 4/1920 - 21/3/1922. Part of the site was taken over by Muirhead as a plant depot.

Reference :   Industrial Locomotive No.30, No.32, No.38.

Gauge : 4ft 8½in

SHARPNESS	0-6-0T	IC	SS	3472	1889	(a)	(1)

(a)   earlier Sir John Jackson Ltd, Bulford Camp (WD) (1914-1916) contract, Wiltshire.

(1)   to Nott, Brodie & Co Ltd, Portway Road contract, Bristol, after 7/1924.

Gauge : 2ft 0in

DIANA	0-4-0WT	OC	HC	1132	1916	(a)	(3)
SYLVIA	0-4-0ST	OC	KS	4162	1921	New	(4)
-	0-4-0ST	OC	WB	2048	1918	(b)	(2)
	0-4-0ST	OC	WB	2073	1918	(c)	(1)

(a)   ex Rees Roturbo Ltd, Ponders End, Middlesex, 1/1921.
(b)   ex MoM, Neasden Depot, Middlesex, c5/1920, to R.H. Neal & Co Ltd, Ealing, Middlesex; thence to WB for overhaul 16/7/1920, despatched to here per R.H. Neal & Co Ltd, 19/7/1921.
(c)   ex MoM Royal Arsenal, Woolwich, for sale 5/1920; to R.H. Neal & Co Ltd, Ealing, Middlesex; thence to WB for overhaul 10/7/1920, despatched to here per R.H. Neal & Co Ltd, 26/7/1921.

(1)   to Barkingside contract by 12/4/1922.
(2)   to Barkingside contract by 17/8/1922.
(3)   to George Cohen, Sons & Co Ltd, Canning Town Depot, 7/1924.
(4)   to Barkingside contract.

WM. MUNRO
HALSTEAD CONTRACT
FC104

Construction of Colne Valley & Halstead Railway, Chappel - Halstead, 1856-1860, opened 16/4/1860.

Reference : From Construction to Destruction, An Authentic History of the Colne Valley and Halstead Railway, Edward P. Willingham, Halstead and District Local History Society, 1989.

Gauge : 4ft 8½in

CAM	2-4-0T	OC	GE		New	(1)
COLNE	2-4-0T	OC	GE		New	(1)

(1)   to Haverhill contract, /1860.

HAVERHILL CONTRACT
FC105

Construction of Colne Valley & Halstead Railway, Halstead - Haverhill, 1860-1863, Halstead – Hedingham opened 1/7/1861, Hedingham – Great Yeldham opened 26/5/1862, Yeldham – Haverhill opened 10/5/1863. The Essex Telegraph of 15/2/1862 reported that the works on the line were far advanced between Castle Hedingham & Great Yeldham and that the line would be ready for government inspection in the course of three weeks. The land up to Haverhill was in the possession of the contractor who had **three locomotives** and several hundred men working night and day to accomplish the opening throughout by the first of May.

"Halstead Gazette" 9/1863, At Haverhill Suffolk, to Railway Contractors, Blacksmiths, Brickmakers and others, second and last clearance sale, to be sold at auction by Mr F.C. Smith and Samuel J. Surridge at the yard adjoining railway station on Friday 4th September next at 12 o'clock, by order of the proprietor, William Munro Esq, who has completed the Colne Valley Railway, the remaining part of his valuable railway plant including two good Locomotive Engines, the COLNE and the CAM by England & Co.

Gauge : 4ft 8½in

CAM	2-4-0T	OC	GE	(a)	(1)
COLNE	2-4-0T	OC	GE	(a)	(1)

(a) ex Chappel contract, /1860, hired to CV&HR, /1863.

(1) to Brightlingsea Railway contract.

BRIGHTLINGSEA CONTRACT   HC106

Construction of Brightlingsea Railway, -1866, opened 18/4/1866.

The locomotives listed below may have been used on this contract.

Gauge : 4ft 8½in

CAM	2-4-0T	OC	GE	(a)	(1)
COLNE	2-4-0T	OC	GE	(a)	(1)

(a) ex Halstead - Haverhill Railway contract.

(1) [possibly to Athenry & Ennis Railway construction, Co. Clare, Ireland ?,] (1863-1865) contract, otherwise s/s.

WALTON CONTRACT   HC107

Construction of the Tendering Hundreds Railway, opened to Weeley 28/7/1866 and to Walton, opened 17/5/1867.

"The Engineer" 31/5/1867, Fenn Cook & Fenn to auction 5/6/1867 re Tendering Hundreds Railway, Weeley to Walton on Naze, at Weeley – plant including excellent tank loco and a tender loco METEOR, the construction of which is of most expensive description.

Gauge : 4ft 8½in

	tank	(a)	s/s
METEOR	tender engine	(b)	s/s

(a) possibly TRENT ex Newcastle & Carlisle Railway, at Stratford Works for disposal by 2/1869.

(b) possibly METEOR ex Newcastle & Carlisle Railway, built by Edward Bury & Co, /1835, sold to Fossick & Hackworth, /1861, thought to be the 12in, 4ft diameter, 4wheel coupled METEOR advertised for sale 21/10/1864 by Isaac Bigland, Stockton on Tees, metal broker.

J. MURPHY & SONS LTD
Company registered 28/2/1951

WOODFORD CONTRACT DC108

Construction of sewerage tunnels, c1973.
Gauge : 2ft 0in

-	0-4-0BE	WR	M7550	1972	(a)	(1)

(a)   ex Cannock Plant Depot, Staffordshire, 2/1973.
(1)   to another contract or Kentish Town plant Depot, London.

STANSTEAD CONTRACT FC109

Contract for the construction of the Stanstead Airport Rail Link for British Rail. The contract was for 1.6km of tunnel under the main runway, contract sum £13.5m, work commenced c5/1988, breakthrough of the tunnel was achieved 23/10/1989.
Gauge : 2ft 6in

-	4wBE	CE	B1534B	1977	(a)	(1)

(a)   ex Welham Plant Ltd, St Neots, Cambridgeshire, 7/1988.
(1)   to CE for regauging to 2ft 0in by 3/1990.

WEST HAM CONTRACT CC110

Construction of a cable tunnel for London Electricity form West Ham to Greenwich c/1997-1998 used three trains hauled by CE battery locos, details unknown.

LEA VALLEY CONTRACT DC111

Contract for the construction of 12km of cable tunnels to replace overhead power lines in the Lower Lea Valley. The two tunnels one of 2.5m and the other 4m in diameter will be 20metres underground, the 4m tunnel will have a monorail installed for inspection and maintenance, tunnelling commenced in 6/2006 and was expected to be completed in 2007.
Gauge : 600mm

-	4wDH	Schöma	5713	2001	(a)	(1)
-	4wDH	Schöma	5714	2001	(a)	(1)
-	4wDH	Schöma	5715	2001	(a)	(1)
-	4wDH	HE	9346	1994	(b)	(2)
-	4wDH	HE	9348	1994	(b)	(2)
-	4wDH	HE	9351	1994	(b)	(2)

(a)   ex Schöma, Germany, /2006.
(b)   ex L H Group Services Ltd, Barton-under-Needwood, Staffordshire, 8/8/2006.
(1)   returned to Schöma, Germany.
(2)   to LH Group Services Ltd, Barton-under-Needwood, Staffordshire, 7/2007.

NISHIMATSU-CEMENTATION, SKANSKA JOINT VENTURE
CHANNEL TUNNEL RAIL LINK PHASE 2, Contract 220   **CC112**

Contract for the construction of twin bore 7.5km tunnels of 7.14m diameter from Stratford to Kings Cross, (2001 – 2004).
Gauge : 900mm

-	4wDH	Schöma	4415	1980	(a)	(1)
-	4wDH	Schöma	5424	1995	(a)	(2)
-	4wDH	Schöma	5724	2002	New	(3)
-	4wDH	Schöma	5725	2002	New	(3)
-	4wDH	Schöma	5726	2002	New	(4)
-	4wDH	Schöma	5727	2002	New	(4)
-	4wDH	Schöma	5728	2002	New	(4)
-	4wDH	Schöma	5729	2002	New	(4)

(a)   ex Schöma, Germany, /2003.
(1)   returned to Schöma, Germany, thence to NECSO/Metro, Madrid, Spain, /2006.
(2)   returned to Schöma, Germany, thence to Sacyr/Abdalajis West, Spain, /2006.
(3)   returned to Schöma, Germany.
(4)   returned to Schöma, Germany, thence to S.E.L.I, Ethiopia, /2004.

EDMUND NUTTALL, SONS & CO (LONDON) LTD,
Company registered 20/9/1935
ILFORD CONTRACT   DC113

Construction of tunnels for the extension of the LT Central Line "Tube" from Gants Hill to Newbury Park. Contract let in 8/1938; the tunnels were completed by 1940 and were then used for a munitions factory by the **Plessey Co Ltd** (see the Industrial section) during WW2, and completed for railway use post-war.
Gauge : 2ft 0in

-	4wBE	GB	1602	1938	New	s/s

Use of other battery locomotives is probable; identities not yet known.

BECKTON CONTRACT   AC114

Contract for new sedimentation tanks at Beckton Sewage Works for London County Council 1950-1954.
Construction of the Beckton Outfall Sewer, Jenkins Lane, Barking for the Metropolitan Water Board, 1950-1952.
Gauge : 2ft 0in

EN 62	4wDM	RH	187113	1937	(a)	(3)
EN 39	4wDM	RH	189948	1938	(b)	(3)
EN 68	4wDM	RH	187111	1937	(c)	(1)
EN 45	4wDM	RH	200780	1941	(d)	(4)
EN 46	4wDM	RH	200781	1941	(e)	(2)

(a) ex Colnbrook Plant Depot, Bucks, after 10/1949, by 22/4/1951;
(b) ex Colnbrook Plant Depot, Bucks, after 22/9/1949, by 22/4/1951.
earlier Tarbet (North of Scotland Hydroelectric Board) contract, Dumbarton, 9/1949;
(c) ex Colnbrook Plant Depot, Bucks, after 10/1949, by 5/9/1951.
earlier Warminster (WD) contract, Wiltshire, until after 6/7/1943.
(d) ex Colnbrook Plant Depot, Bucks, after 5/1949, by 16/3/1952;
earlier at Cliff Quay Power Station contract, Ipswich, Suffolk, /1945-1949.
(e) ex Colnbrook Plant Depot, Bucks, after 10/10/1952; originally Hirwaun (ROF) (1941-1943) contract, Mid Glamorgan.

(1) later to Pauling & Co Ltd, Park Royal Plant Depot, Middlesex, after 16/3/1952, by 18/10/1952.
(2) later to Colnbrook Plant Depot, Bucks, by 1/1954; s/s.
(3) later to Standard Brick & Sand Co Ltd, Redhill, Surrey, after 11/9/1951, by 6/1958.
(4) later to Sykes & Son (Poole) Ltd, Creekmoor Potteries, Poole, Dorset, by c/1962.

PURFLEET CONTRACT   NC115

Purfleet - Dartford Road Tunnel contract. c1961.

Gauge : 4ft 8½in

| - | 0-6-0ST | OC | HE | 1686 | 1931 | (a) | (1) |
| SOUTHERN | 0-6-0ST | OC | HE | 1688 | 1931 | (a) | (1) |

(a) ex Colnbrook Plant Depot, Buckinghamshire.

(1) to Tunnel Portland Cement Co Ltd, West Thurrock, (/1961).

GREAT SAMPFORD (RAF) CONTRACT   FC116

Work for the RAF at Great Sampford Aerodrome which was operational from 14/4/1942 .

Gauge : 2ft 0in

| - | 4wDM | RH | 189941 | 1937 | (a) | (1) |

(a) ex Colnbrook Plant Depot, Buckinghamshire, by 15/3/1941;
earlier Glascoed (ROF) contract, Monmouthshire, until after 5/11/1940.

(1) to ROF Hirwaun contract Mid Glamorgan, after 18/4/1941, by 20/5/1941.

ROYAL VICTORIA DOCK CONTRACT, London   AC117

Work carried out at the Co-operative Wholesale Society site in /1944, details uncertain.

Gauge : 2ft 0in

| - | 4wDM | RH | 189952 | 1938 | (a) | (1) |
| - | 4wDM | RH | 200782 | 1941 | (b) | (2) |

(a) ex Glascoed ROF contract, Monmouthshire, by 12/7/1939.
(b) ex Glascoed ROF contract, Monmouthshire, after 23/12/1941, by 27/3/1944.

(1) later Richard Wild, Fletcher Bank Quarries, Ramsbottom, Lancashire, by 17/12/1945.
(2) to Colnbrook Plant Depot, Bucks, EN 47, after 27/3/1944, by 3/4/1944; s/s after 10/10/1952.

NUTTALL, WAYSS & FREYTAG-KIER JOINT VENTURE
CHANNEL TUNNEL RAIL LINK PHASE 2, Contract 250   PC118

Contract for the construction of 5.3km of twin bore tunnels of 7.15m diameter between Ripple Lane, Dagenham and Barrington Road, Newham, (2001 – 2004).
Gauge : 900mm

-		4wDH	Schöma	5560	1998	(a)	(1)
-		4wDH	Schöma	5561	1998	(a)	(1)
-		4wDH	Schöma	5562	1998	(a)	(1)
-		4wDH	Schöma	5610	1999	(a)	(3)
5		4wDH	Schöma	5611	1999	(a)	(3)
6		4wDH	Schöma	5615	1999	(a)	(2)

(a) ex Schöma /2002.

(1) returned to Schöma, Germany.
(2) returned to Schöma, Germany after 16/5/2004 by 28/8/2004.
(3) returned to Schöma, Germany, thence to AMEC plc, Docklands Light Railway extension, /2006.

ORGAN & CO LTD
FERRO WORKS, Horns Road, Barkingside   DC119

Company registered 18/1/1929 as public and municipal works contractors & engineers with Mr P.C.Organ as managing director and listed in directories for 1933 with an address at River Road, Barking. The "Contract Journal" 15/7/1931 advertised an auction at the LNER goods yard Barkingside which stated the company was moving to a new yard at Fleetmouth, Barking. (Percy, Charles Organ of Mile House, Blackheath, London, had been appointed in 1919 as a director of Wm Muirhead Macdonald Wilson & Co Ltd).
Gauge : 2ft 0in

	0-4-0ST	OC	KS	4247	1922	(a)	(1)

(a) earlier William Muirhead, Macdonald Wilson & Co Ltd, Southend Road contract, by 5/1929.

(1) to Hall & Co Ltd, Mitcham, Surrey.

"Contract Journal" 22/5/1929, for sale – std gauge 0-4-0ST 12x20 – Organ & Co, dealers, Barkingside.
Gauge : 4ft 8½in

-	0-4-0ST	OC			(a)	s/s

(a) origin & identity unknown.

J. PARKINSON & SONS (BLACKPOOL) LTD
Jacob Parkinson & Sons until 10/3/1902
GRAYS CONTRACT   NC120
Construction of Purfleet By-Pass road (Purfleet – Tilbury) 1923 – 1926.
Gauge : 2ft 0in

-	4wPM	MR	2210	1922	New	(a)	s/s
-	4wPM	MR	2214	1922	New	(b)	s/s

(a) reconstructed from MR 940(?), ex WDLR, per William Jones Ltd, dealers, Greenwich, London.
(b) reconstructed from MR 293/1916 ex WDLR, per William Jones Ltd, dealers, Greenwich, London.

SIR LINDSAY PARKINSON & CO LTD
Company registered 13/7/1937
THAMES HAVEN CONTRACT   KC121
Contract in 1955 for building a sea wall for London & Thames Haven Oil Wharves Ltd.
Gauge : 2ft 0in

LP202	4wDM	MR	5344	1931	(a)	s/s

(a) here by 27/2/1955, earlier W.& C. French Ltd.

BASILDON CONTRACT   KC122
Construction of a new factory for the production of agricultural tractors for the Ford Motor Company. The works, off Cranes Farm Road, Basildon, was built on a 100 acre site; work started 2/4/1962 and was completed 29/2/1964. The factory had the capacity to build 500 engines per week and assemble 300 tractors per day, the first prototype was completed 15/5/1964 and the last tractor from the Dagenham Works was completed in 7/1964, full production at Basildon commenced in 8/1964.
Gauge : 2ft 0in

	4wDM	RH	226288	1944	(a)	s/s

(a) originally MoS.

S. PEARSON & SON LTD
S. Pearson & Son until 12/8/1897
Samuel Pearson c/1850 until 1856
WALTHAMSTOW RESERVOIRS CONTRACT   DC123
Construction of the Banbury (TQ 362914) and Lockwood (TQ 353905) Reservoirs, Walthamstow, for the East London Waterworks Co, 1899 – 1904.
Gauge : 4ft 8½in

No.95	ALWILDA	0-4-0ST	OC	P	798	1899	New	(3)
No.97	ARAB	0-4-0ST	OC	P	800	1899	New	(3)
No.100	LOCKWOOD	0-6-0ST	IC	HE	716	1900	New	(2)

No.101	BANBURY		0-6-0ST	IC	HE	717	1900	New	(4)
No.102	CHINGFORD		0-6-0ST	IC	HE	718	1900	New	(1)
	WALTHAMSTOW		0-6-0ST	IC	MW	1208	1890	(a)	(2)
	ANNIE		0-4-0ST	OC	MW	1135	1891	(b)	(5)

(a) possibly ex Pethick Bros, contractors (location unknown);
earlier Logan & Hemingway, Beighton-Chesterfield (MS&LR) (1890-1892) contract, Derbyshire, No.29.

(b) ex Wootton Bassett-Patchway (GWR) (1897-1903) contract, Wiltshire/Gloucestershire.

(1) to Valetta Harbour Breakwater contract, Malta, by 10/1903.
(2) to Valetta Harbour Breakwater contract, Malta, by 11/1903.
thence to King George V Dock (1906-1914) contract, Hull, East Yorkshire, /1909.
(3) to Wouldham Cement Co, Grays, by /1901.
(4) to Valetta Harbour Breakwater contract, Malta, by 3/1906;
thence to Dover Harbour (Admiralty) (1897-1914) contract, Kent, after 8/1910.
(5) to West Thurrock Plant Depot.

ROYAL ALBERT DOCK (EXTENSION) CONTRACT   AC124

Extension of the Royal Albert Dock for the Port of London Authority, 1912 – 1918. This contract (dated 18/7/1912) was suspended with work unfinished, due to WW1. The contract being surrendered on 2/8/1918, was completed after the war by PLA direct labour. The PLA took over what plant and locos they needed from Pearson, mostly on hire. These included, SOMERFORD, LONDONDERRY, WOOLWICH, LADYSMITH, MARFLEET, BANBURY, and No.8.

The new dock was opened by His Majesty on 8/7/1921 and named King George V Dock; an auction of plant and machinery took place 28/2-1/3/1922.

Gauge : 4ft 8½in

	CORSTON		0-6-0T	IC	MW	1196	1890	(a)	(1)
	ABBEY		0-6-0ST	OC	P	939	1902	(a)	(7)
71	SOMERFORD		0-6-0ST	OC	P	720	1898	(a)	(3)
No.45	JOE		0-4-0ST	OC	HE	620	1895	(a)	(2)
	QUEEN		0-6-0ST	IC	P	958	1902	(a)	s/s
	SIR WILLIAM CRUNDALL		0-6-0ST	IC	P	1241	1910	(a)	(8)
	LADYSMITH		0-6-0ST	OC	HC	560	1900	(a)	s/s
81	THE AUDITOR		0-6-0ST	OC	BH	1105	1895	(b)	(4)
	BARKING		0-6-0ST	IC	MW	858	1882	(c)	(3)
142	WOOLWICH		0-6-0ST	IC	MW	841	1882	(d)	(3)
8	BECKTON		0-6-0ST	IC	HC	833	1910	(e)	(9)
99	LONDONDERRY		0-6-0ST	OC	P	806	1900	(f)	(3)
	MARFLEET		0-6-0ST	IC	HE	550	1892	(g)	(6)
No.53	CANADA		0-6-0T	IC	HE	630	1895	(h)	s/s
No.101	BANBURY		0-6-0ST	IC	HE	717	1900	(j)	(3)
	WYE		0-4-0ST	OC	HE	420	1887	(k)	s/s
	WILLIE		0-4-0ST	OC	HC	477	1897	(l)	(5)

(a) ex Dover Harbour (Admiralty) (1897-1914) contracts, Kent.
(b) ex Dover Harbour (Admiralty) (1897-1914) contracts, Kent, by 5/1913.
(c) ex West Thurrock Plant Depot.
(d) earlier Lucas & Aird, Fulham Plant Depot, London;
earlier Lucas & Aird, Staines Old Reservoir (1890-1899) contract, Middlesex.
(e) ex Stockport Corporation, Kinder Reservoir (1903-1913) construction, Hayfield, Derbyshire, after 9/1911.
(f) ex King George V Dock (1906-1914) contract, Hull, Yorkshire (ER).
(g) ex King George V Dock (1906-1914) contract, Hull, Yorkshire (ER), by 7/1913.
(h) earlier Salina Cruz contract, Mexico, after 2/1907.
(j) ex Dover Harbour (Admiralty) (1897-1914) contracts, Kent, by 8/1917.
(k) ex William Rigby & Co, Plumstead contract c/1915 contract, after 16/5/1916, here by 31/10/1919.
(l) use on this contract probable but not confirmed;
earlier C. Baker & Sons, Newdigate Colliery (1897) contract, Warwickshire.

(1) hired to Sir Robert McAlpine & Sons Ltd, unknown location, /1918.
to John F. Wake & Co Ltd, dealers, Darlington, Co. Durham, for repairs;
thence to Sunbury & Littleton Reservoirs (Metropolitan Water Board) (1919-1924) contract, Middlesex.
(2) to Griffiths Wharf Co, Silvertown, hire, /1918.
(3) to Sunbury & Littleton Reservoirs, (Metropolitan Water Board) (1919-1924) contract, Middlesex.
(4) to North Lincolnshire Iron Co Ltd, Frodingham Ironworks, Scunthorpe, Lincolnshire, c/1919.
(5) if here, then to Gretna Factory (Ministry of Munitions) (1915-1917) contract, Dumfriesshire.
(6) to Sunbury & Littleton Reservoirs, (Metropolitan Water Board) (1919-1924) contract, Middlesex, after 11/1921, by 1/1922.
(7) to Topham, Jones & Railton.
(8) to Ministry of Munitions, National Filling Factory No.3, Park Royal, Middlesex.
(9) to Sir E.W. Moir, /1931[where ?].

WEST THURROCK DEPOT   NC125

Plant Yard, St Clements Wharf, TQ 595771?, West Thurrock.
"Machinery Market" 7/2/1913 for sale – plant inc 12x18 HC and 10x16 MW – S Pearson & Son, London SW.

Gauge : 4ft 8½in

	ANNIE	0-4-0ST	OC	MW	1135	1891	(a)	(2)
	BARKING	0-6-0ST	IC	MW	858	1882	(b)	s/s
	MARFLEET	0-6-0ST	IC	HE	550	1892	(c)	(1)
142	WOOLWICH	0-6-0ST	IC	MW	841	1882	(c)	(1)
39	MARGAM	0-4-0ST	OC	MW	1306	1895	(d)	(2)

(a) earlier at Walthamstow Reservoirs (1899-1904) contract.
(b) ex Lucas & Aird, Fulham Plant Depot, London, 42;
earlier Lucas & Aird, Staines Old Reservoir (1890-1899) contract, Middlesex,
to Royal Albert Dock Extension (1912-1918) contract;
ex Sunbury & Littleton Reservoirs, (Metropolitan Water Board) (1919-1924) contract, Middlesex, to BPCM Wouldham Works on loan, returned.

(c) ex Sunbury & Littleton Reservoirs, (Metropolitan Water Board) (1919-1924) contract, Middlesex.
(d) ex Wootton Bassett-Patchway (GWR) (1897-1903) contract, Wiltshire/Gloucestershire.

(1) to Thames Land Co, West Thurrock, c/1926.
(2) to Wouldham Cement Co, West Thurrock, /1909.

PERRY & CO (BOW) LTD
Perry & Co until 9/12/1908
TREDEGAR WORKS, Bow CC126

This firm, established in the 1830's by John Perry, did a large amount of building work and civil engineering projects including railway works; the firm owned its own brickfield.

Herbert Henry Bartlett had carried on a business of contractors and engineers for a number of years at Tredegar Works, Bow, under the style of Perry & Co, (Bartlett became sole partner upon the death of Perry's sons). The firm was incorporated on 9/12/1908 as Perry & Co (Bow) Ltd with a registered office at 56 Victoria Street, London, SW1, with Herbert Henry Bartlett and Hardington Arthur Bartlett as the first directors. A receiver was appointed on 1/7/1929, this remained in force until 3/1936; the company was dissolved by the Board of Trade in 1941 and notice to this effect was published in the London Gazette dated 27/1/1942.

"Contract Journal" 24/3/1926 Fuller Horsey to sell 20/4/1926 on expiration of lease for Perry & Co (Bow) Ltd, at Tredegar Works, Bow; plant including four 18in gauge electric locos and two 6hp petrol locos.

TILBURY DOCK CONTRACT MC127

Construction of the deep water riverside Cargo Jetty for the Port of London Authority 1912-1918 was awarded to Perry & Co (Bow) Ltd at a price of £105,466-17s. Work on this contract was interrupted by WW1, the contract was surrendered on 2/8/1918 upon payment of £100,000 which included the sale to the PLA of all plant and materials. The PLA completed the work by direct labour; use of locos unknown but Perry & Co are known to have owned Koppel locos, the loco listed below may have worked on this contract.

"Machinery Market" 3/2/1922, Port of London Authority, plant & machinery used in the construction of the King George V dock – Fuller Horsey Sons & Cassell to sell by auction Tuesday February 28th and Wednesday March 1st, contractors plant laying partly within Royal Albert Dock, North Woolwich and partly at Tilbury Dock, including **Koppel 24in gauge** loco and 40 steel Jubilee side tip wagons 24in gauge. (This plant was that used on the Pearson contract at King George V dock and by Perry & Co at Tilbury, both contracts were surrendered to the PLA on 2/8/1918).

PETO, BETTS & BRASSEY
THAMES HAVEN CONTRACT
KC128

Constuction of the Thames Haven Railway (later LT&SR) from Stanford-le-Hope to Thames Haven, (3miles 67chains) for the Thames Haven Dock & Railway Company. An agreement was signed 31/8/1853 for construction of a single track railway and wharves at a price of £49,500, work began c3/1854, the line opened 7/6/1855, some materials were delivered by rail from the LT&SR, use of locomotives unknown.

Reference : "The Thames Haven Railway", Peter Kay, 1999.

LONDON TILBURY & SOUTHEND CONTRACT
XC129

Construction of the London Tilbury & Southend Railway (LT&SR) for the LT&SR Joint Committee (formed from the Eastern Counties Railway and the London & Blackwall Railway), the contract was awarded to Peto, Betts & Brassey at a price of £400,000 (later increased to £425,000) the agreement signed 18/8/1852. Work began 9/1852 on the section from Forest Gate to Tilbury which opened 13/3/1854. The remainder of the line was opened in stages, Tilbury to Stanford-le-Hope 14/8/1854, Stanford-le-Hope to Leigh on Sea 1/7/1855, Leigh on Sea to Southend 1/3/1856. The partnership also had a contract to build the line from Stratford to Hackney Wick for the Eastern Counties Railway. An agreement was made with the ECR to run spoil trains from the LT&SR line for use in constructing the Hackney line using ECR locos on hire, use of contractors locos unknown.

Reference : "The London Tilbury & Southend Railway", Peter Kay, 1996 & 1997.

A.D. PIERCE (CONTRACTS) LTD
Henham
EC130

"Contract Journal", 12/4/1956, for sale – 2ft gauge Simplex diesel loco (National engine) – A.D. Pierce (Contracts) Ltd, Henham, Herts.

Gauge : 2ft 0in

		4wDM	FH	2281	1940	(a)	s/s

(a) ex M.E. Engineering Ltd, Cricklewood, Middlesex.

THOMAS D. RIDLEY
PARKESTON QUAY CONTRACT
HC131

Construction of a sea wall and sluice at Ramsay Bay for the GER. In 1878 the contract was let to **Samuel Coote Ridley**, work started 5/1878; by 9/1878 two locos, 130 men and a Ruston Proctor steam navvy were at work. In 1879 the contract was transferred to Thomas D. Ridley.

Gauge : 4ft 8½in

		0-4-0ST	OC	HC	195	1879	New	s/s
	PARKESTON	0-4-0ST	OC	HC	220	1882	New	(1)
		0-4-0ST	OC	MW	704	1878	New	s/s
		0-4-0ST	OC	JF	2850	1876	(a)	s/s
		0-4-0ST	OC	D&S	58c1875		(b)	s/s

(a) new to S.C. Ridley, Tottenham, Middlesex.
(b) ex HC 4/11/1879,
    earlier Butlin Bevan & Co Ltd, Irthlingborough Ironworks, Northamptonshire.
(1) to Baldry & Yerburgh, MS&LR Annesley – Staveley (1890-1892) contract.

Five 0-4-0ST by MW and others were auctioned 5-6/10/1881. One 10in cyl HC (presumed HC 220) and one 8in cyl JF (presumed JF 2850) were auctioned on completion of the contract, 1/9/1886.

WALTER SCOTT & MIDDLETON
SOUTHEND AND SOUTHMINSTER BRANCHES CONTRACTS   GC132

Construction for the GER of the Wickford-Southend line (11 miles) and the Wickford-Southminster branch (12 miles) together with the Woodham Ferrers to Maldon line (4 miles). Work started about 1885; the Southminster branch opened to goods on 1/6/1889 (passenger 1/7/1889) and the Southend and Maldon lines opened on 1/10/1889. Walter Scott & Co obtained the contract to construct forty miles of railway in Essex for the GER, Thomas Middleton was placed in charge of the Burnham section, upon completion he took charge of maintenance; whilst in execution of his duty in this position he met with a fatal accident at Wickford. Having stepped out of the way of an approaching goods train he was knocked down by an empty carriage train on an adjoining line and died from his injuries on 20/2/1891.

Reference: The Carlisle Journal, 27/2/1891.
   Branch Line to Southminster, Dennis L Swindale, Halstead Press Ltd, 1981.

Gauge : 4ft 8½in

C H STANTON	0-6-0ST	OC	BH	758	1885	New	(1)
NEWCASTLE	0-6-0ST	OC	HC	237	1883	(a)	(1)

(a) originally Thomas Nelson & Co, contractors, Carlisle.

(1) to Walter Scott & Co, Woodford Halse-Quainton Rd (MS&LR) (1894-1898) contract, Northamptonshire.

G. SHELLABEAR & SON LTD
Company registered 4/9/1916
LANGFORD RESERVOIR CONTRACT   GC133

Construction of Langford Reservoir and pumping station (TL 832089) for Southend Waterworks Co, 1926-1928. The works included the construction of two reservoirs each holding thirty million gallons of water and covering an area of 10.5 acres, the reservoirs were gravity fed from intakes on the Rivers Chelmer and Blackwater and supplied water to Southend via a pipeline 14 miles long, the pumping station opened in 1929 and closed in 1963 and is now a museum. Photographic evidence exists of a loco being delivered to the site driven under its own steam along the streets of Maldon.

Gauge : 4ft 8½in

CAMEL	0-4-0CT	OC	HL	2402	1898	(a)	(1)
		reb	Wake		1917		
-	0-4-0ST	OC	AB	1018	1905	(b)	(1)
-	0-4-0ST	OC	MW	1106	1888	(c)	(1)
-	0-6-0ST	OC	AE	1672	1914	(d)	(1)
-	0-6-0ST	IC	MW	1561	1902	(e)	(2)

(a) ex Stanlee Ship Breaking & Salvage Co Ltd, Dover, 3/1926.
(b) earlier Admiralty, Chepstow, Monmouthshire.
(c) ex P.L.A. 35, /1926.
(d) earlier William Jones Ltd, dealers, Greenwich, London, c/1926.
(e) earlier MoM Quedgeley, Gloucestershire, after 5/1925, by 7/1928.

(1) to Park Royal Plant Depot, Middlesex.
(2) to Northfleet Deep Water Wharf Ltd, Kent, by 5/1930.

ROYAL ALBERT DOCK CONTRACT   AC134

Contract for the enlargement of the Western Dry Dock at Royal Albert Dock, /1912, was awarded to G. Shellabear & Son of Plymouth at a price of £55, 525-3s-4d. The use of locomotives unknown.

SILVER END DEVELOPMENT CO LTD
SILVER END CONSTRUCTION   FC135

Construction of Silver End Garden Village, near Witham, 3/1926 - 6/1930.

The village was constructed for the workforce of the **Crittall Manufacturing Company Ltd** by it's founder Francis Henry Crittall J.P. This company of metal window manufacturers was established in 1889 with a works in Braintree. A subsidiary company the Silver End Development Co was set up to build the Garden Village and a new factory at Silver End on 220 acres of land purchased in October 1925. Plans were drawn up to build 100 houses per year by direct labour; 300 men were employed and a temporary railway was used during construction with 600 houses built within five years. Francis Henry Crittal made his home in the centre of the village at the 'Manors', he died at sea aged 74 whilst returning from a cruise to the West Indies on 9th March 1935.

At least one ex-WDLR 2ft gauge 20hp MR 4wPM was used. MR supplied spares on 22/10/1927 and 29/1/1931 and orders imply that at least three such locos were in use here.

SOUTH ESSEX WATERWORKS COMPANY
ABBERTON RESERVOIR CONSTRUCTION   G/HC136

Construction of Abberton Reservoir, near Colchester. 1935-1939, construction was undertaken by direct labour.

"Contract Journal", 19/10/1938, for sale – four Fowler diesel locos 40hp, three KS steam 0-4-0 6", two KS steam 0-4-2, all 2ft gauge – South Essex Waterworks, Abberton Reservoir, near Colchester.

"Contract Journal", 21/6/1939 for sale – three 40hp diesel locos & one 6in steam loco 2ft gauge – South Essex Waterworks, near Colchester.

Gauge : 2ft 0in

No.3	WASP	0-4-0ST	OC	KS	4001	1918	(a)	(1)
	HARTHOPE	0-4-2ST	OC	KS	1291	1915	(b)	(2)
	BURNHOPE	0-4-2ST	OC	KS	1144	1911	(c)	(3)
	IRESHOPE	0-4-2ST	OC	KS	1142	1911	(d)	(3)
	GNAT	0-4-0ST	OC	KS	4291	1923	(e)	(1)

MIDGE	0-4-0ST	OC	KS	4290	1923	(f)	(4)
ABBERTON	4wDM		JF	21293	1936	New	(5)
LAYER	4wDM		JF	21294	1936	New	(5)
PELDON	4wDM		JF	21295	1936	New	(5)
ROMFORD	4wDM		JF	21408	1936	New	(6)

(a) ex Durham City Water Board, Wearhead, Co. Durham, 80, by /1935.
(b) ex Durham City Water Board, Wearhead, Co. Durham, 78, by /1935.
(c) ex Durham City Water Board, Wearhead, Co. Durham, 74, by /1935.
(d) ex Durham City Water Board, Wearhead, Co. Durham, 76, by /1935.
(e) ex Durham City Water Board, Wearhead, Co. Durham, 73, by /1935.
(f) ex Durham City Water Board, Wearhead, Co. Durham, 72, by /1935.

(1) for sale 7/1938;
    later Pauling & Co, Crymlyn Burrows Plant Depot, Swansea, by 7/1942.
(2) later Eastwoods Cement Ltd, Barrington Works, Cambridgeshire.
(3) to County Borough of Hastings, Darwell Hole Reservoir, Sussex, /1938.
(4) to APCM, Holborough Works, Kent, 1. /1939.
(5) to Alpha Cement Ltd, Cliffe-at-Hoo Works, Kent, /1939.
(6) to Corby (Northants) & District Water Co, Eyebrook Reservoir, Northamptonshire.

STOUR SCHEME   HC137

Locos possibly used on the Stour Valley trunk pipeline and reservoirs construction, 1929-1932. The work undertaken by direct labour comprised a water intake from the River Stour at Langham Mill (6 miles north of Colchester) with a treatment plant there. A pipeline ran to Tiptree covered reservoir thence to Danbury covered reservoir, and continued to Herongate reservoir and onward to supply the Romford and Ilford areas, a distance of 45 miles. These reservoirs were not major storage reservoirs.

Gauge : 2ft 0in

-	4wPM	HU	36459	1929	New (a)	s/s
-	4wPM	L	3145	1930	New ?	s/s
-	4wPM	L	3289	1930	New ?	s/s
-	4wPM	L	3834	1930	New ?	(1)
-	4wPM	FH				
		BgE	2046	1930	New (a)	s/s
	4wDM	OK	4031	1930	(b)	(2)
	4wDM	OK			(c)	s/s

(a) new to this company; assumed to this site.
(b) origin unknown, here by 16/11/1930.
(c) origin and identity unknown.

(1) to Crendon Gravel & Concrete Co Ltd, Long Crendon, Buckinghamshire.
(2) later Thanet Brick Co Ltd, Ramsgate, Kent.

SOUTHEND-ON-SEA BOROUGH COUNCIL
SOUTHEND CONTRACT   JC138

Construction of new sea wall by direct labour, from 1929.
Gauge : 2ft 0in

		4wPM	OK		1929	New	(1)
		4wPM	FH				
			Bg	1805	1932	New	(1) #

\# Built to 1ft 8in gauge.

(1)   advertised for sale 10/1956; s/s.

The OK loco was purchased new in 1/1929 for the sea wall construction. Bg 1805 was built under sub-contract from FH but no FH works number is known; this loco may have been used on the sea wall contract. One, or both of the locos, was subsequently used in the construction of a new park.

L.J. SPEIGHT LTD
DAGENHAM CONTRACT   PC139

Construction of Dagenham Dock Works (TQ 495815) for Ford Motor Co Ltd, 1929-1932, work included concrete floors and drainage.
Gauge : 4ft 8½in

No.21		0-4-0ST	OC	HE	515	1890		
			reb	HE		1915	(a)	(1)
22		0-4-0ST	OC	HC	309	1889	(b)	(2)

(a)   ex J. & B. Martin, Crayford, Kent, by 2/1931, earlier NFF Crossgates, Leeds, Yorkshire, until c11/1918.
(b)   ex H. Arnold & Son, Doncaster Plant Depot, South Yorkshire, by 5/1930.
(1)   to Ford Motor Co, Dagenham Dock. /1932.
(2)   to Bedford Gas Co, Bedford Gas Works, Bedfordshire, /1932, MARY.

Gauge : 2ft 0in

		4wPM	FH	1645	1929	New	s/s
		4wPM	FH	1656	1930	New	s/s

S.S. STOTT & SONS LTD
ILFORD CONTRACT   DC140

Company listed as mechanical handling engineers of Laneside Foundry, Haslingden, Lancashire. Makers records show the locomotive listed as delivered new to Ilford Sewage Works.
Gauge : 2ft 0in

	- -	4wDM	RH	354028	1953	New	(1)

(1)   later Greater London Council, Gascoigne Road sewage pumping station, Barking.

A. STREETER & CO LTD
A. Streeter & Co until 30/1/1912
BARKING CONTRACT CC141
Construction of sewers from Barking to Manor Park, c1972 -1973.
Gauge : 2ft 0in

-	4wBE	CE	5940A	1972	New	(1)
-	4wBE	CE	5940B	1972	New	(1)
-	4wBE	CE	5940C	1972	New	(1)
-	4wBE	CE	5940D	1972	New	(1)
-	4wBE	CE	5961A	1972	New	(1)

(1)   to A. Streeter & Co Ltd, Godalming, Surrey, or to another contract.

ROYAL VICTORIA DOCK DRAINAGE SCHEME, PHASE 1 AC142
Construction of 5.2km of foul and surface water sewers for the London Docklands Development Corporation, 1986.
Gauge :

-	4wBE	CE	(a)	(1)
-	4wBE	CE	(a)	(1)

(a)   identity unknown, ex Welham Plant Ltd, Cambridgeshire, c7/1988.

(1)   returned to Welham Plant Ltd?

TARMAC/MONTCOCOL, JOINT VENTURE
JUBILEE LINE EXTENSION, Stratford Market Depot CC143
Contract for track work at Stratford Market Depot for the Jubilee Line Extension from Green Park to Stratford. The line is 16km long in twin bore tunnels with twelve stations; contracts were signed in 11/1993 and the line opened 22/12/1999 at a cost of £3.5 billion.
Gauge : 4ft 8½in

81	538010	4wDM	R/R	Unimog			
				166200	1991	(a)	s/s
83	538009	4wDM	R/R	Unimog			
				166228	1991	(b)	s/s
MYFANWY		0-4-0DH		RSHD 8366	1962	(c)	(1)
				WB 3211	1962		
Q166 BCH		4wDM	R/R	Unimog		(d)	(2)

(a)   origin unknown
(b)   ex Yorkshire Engine Company Ltd, Long Marston, Warwickshire, 6/11/1995.
(c)   ex Alan Keef Ltd, Lea Line, Herefordshire, 1/1996.
(d)   ex Balfour Beatty Ltd.

(1)   to Tarmac Trackwork Ltd, Heathrow Express Rail Contract, 24/7/1996 returned 15/11/1996, to Centrac, Rugby, Warwickshire, after 21/3/1998 by 8/4/1998.
(2)   later Balfour Beatty Railway Engineering Ltd, Tinsley Park Depot, Sheffield, South Yorkshire.

WM. TAWSE LTD
CLACTON CONTRACT
HC144

Contract for sea defence works in 1967 used RH locos for moving concrete.

TAYLOR WOODROW CONSTRUCTION LTD
Company registered 14/12/1936
RAINHAM CONTRACT
PC145

Pipelaying contract west of Rainham Station and near to Ford's Works, Dagenham.
Gauge : 2ft 0in

| | - | 4wDM | RH 432664 1959 | (a) | (1) |

(a)   earlier on Rheidol Hydro-Electric Scheme contract, Cardigan, /1959-c/1961, here 4/1963.

(1)   later G.W. Lewis (Tileries) Ltd, Rosemary Tileries, Cannock, Staffordshire, by 3/1967.

TOPHAM, JONES & RAILTON LTD
Topham, Jones & Railton established 1893, incorporated 28/12/1905
TILBURY DOCK CONTRACT
MC146

Contract for the extension of Tilbury Docks 1912 – 1917 was awarded to Topham Jones & Railton Ltd at an estimated price of £292,000, but amounted to £318,036-17s. Works, which included an additional 1,480ft of quay, continued at a slow pace during WW1 due to labour shortage.

Gauge : 4ft 8½in

| 8 | GOWY | 0-6-0ST | IC | MW | 1119 | 1889 | (a) | (1) |

(a)   ex Kings Dock, Swansea Embankment contract, South Wales, /1912.

(1)   to Crymlyn Burrows Plant Depot, Swansea, South Wales;
thence to Oxfordshire Ironstone railway contract, Banbury, Oxfordshire, 26/8/1917.

WEST THURROCK BALLAST PITS
NC147

Workings to the north of Tunnel Portland Cement Co's chalk quarry, to supply sand and ballast for the Tilbury Dock contract.
Reference: Industrial Locomotive No.67
Gauge : 4ft 8 ½in

| | SANKEY | 0-4-0ST | OC | MW | 1088 | 1888 | (a) | (1) |

(a)   earlier Ocean Dock (LSWR) (1908-1911) contract, Southampton, Hampshire.

(1)   later Edge Hill Light Railway, Warwickshire, 6/1922.

G. PERCY TRENTHAM LTD

Company registered 31/3/1913 to take over from 1/1/1913 the business of a public works contractor carried on by G.P. Trentham at Winchester House, Victoria Square, Birmingham.

DAGENHAM CONTRACT   PC148

Contracts for various construction works at the Ford Motor Works, 1929-1932; these included removal of large quantities of rubbish from the site, (previously part of Flower & Everett's tip) and levelling the site. A 2ft gauge Simplex loco and thirty tip wagons were used from 1929. Other work included construction of a standard gauge service railway to the site for construction materials; other contracts included piling, building and miscellaneous construction work.

SOUTH HORNCHURCH PLANT DEPOT   LC149

Gauge : 2ft 0in

	-	4wDM	OK	4501	1931	(a)	(1)

(a)   origin unknown, here by 2/6/1956.

(1)   later on a contract at Crawley New Town, Sussex, to F. J. Dartnell, Upminster by 9/9/1967.

TROLLOPE & COLLS LTD
TILBURY CONTRACT   MC150

Construction of housing estate for Tilbury Urban District Council, Tilbury, -1924.

Gauge : 2ft 0in

Four 2ft gauge 4wPM by MR were auctioned on completion of contract, 30/4/1924. Three were likely to be :

	-	4wPM	MR	856	1918	(a)	s/s
	-	4wPM	MR	867	1918	(a)	(1)
	-	4wPM	MR	1985	1920	(a)	(2)

(a)   ex Welwyn Garden City construction contract, Hertfordshire.

(1)   to MR, 1/5/1924; rebuilt as MR 3689/1924, later West Riding County Council, Calverley & Rodley Station, Yorkshire.

(2)   to MR, 1/5/1924; rebuilt as MR 3690/1924, later Melville Dundas & Whitson, Wolverhampton, Staffordshire.

CHARLES WALL LTD
GLOBE WORKS and PLANT DEPOT, Grays   MC151

Charles Wall, along with his brother George, was a partner in the firm of **Wall Brothers**, who commenced business in 1866 as builders and contractors with premises at Carlton Works, Carlton Road, Kentish Town, London. Charles left the firm in 8/1873; George joined by another brother John, continued to trade in partnership as Wall Brothers until John died in 1883 thereafter George was joined by his son. Amongst other building projects, Wall

Brothers were noted for building forty schools for the London School Board; George Wall went bankrupt in 9/1890 with debts of £25,360.

Company registered 17/2/1904 at 53 Upcerne Road, Chelsea, to acquire and amalgamate the businesses of C. Wall at Globe Works and H. Wall as Henry Wall and Co at Carlton Works, Kentish Town. Globe works was acquired in 1896 and located on the site of the Globe Cement, Brick, Whiting & Chalk Co's Works. An auction at Grays on 11/5/1915 included four-wheel and six-wheel locos and two 2ft 4in gauge locos.

The "Times" carried an advert for Fuller, Horsey, Sons & Cassell for an auction to take place 18/10/1933 for the freehold waterside estate with a frontage to the Thames of 938ft, with deep water jetty and an area of about 102 acres with standard gauge sidings from the L.M.& S. railway. The estate comprises of about 50 acres of building land the remainder occupied by Whiting Works, Carbonate of Lime Works, Chalk Quarries, Foundry & Sand pits etc. The estate of the old established business of Charles Wall Ltd, the well known contractors. Company went into voluntary liquidation on 22/3/1966.

Gauge : 4ft 8½in

JUMBO	0-4-0ST	OC	HH		1883	(a)		(6)
ALICE	0-4-0WT	ICG	#			(b)		Scr
HENRY APPLEBY	0-6-0ST	IC	HE	45	1870	(c)		Scr /1934
ANNIE	0-6-0ST	OC	HE	593	1893	(d)		Scr /1923
PARANA	0-4-0ST	OC	MW	1141	1889	(e)		(2)
PORTSMOUTH	0-6-0ST	IC	HE	4	1865	(f)	(1)	s/s
FELSPAR	0-4-0ST	OC	MW	1846	1914	(g)		(10)
SWANSEA	0-6-0ST	IC	MW	595	1876	(h)		Scr /1934
		reb	Price, Wills & Reeve, 1902					
DAGENHAM	0-4-0ST	OC	HC	1564	1925	(h)		(8)
LORD MAYOR	0-4-0ST	OC	HC	402	1893	(j)		(9)
WEAR	0-4-0ST	OC	MW					
		reb	HL			(k)		(11)
KAPPA	0-4-0ST	OC	CF	1164	1898	(l)		(7)
AMBERGATE	0-6-0ST	IC	MW	1069	1888	(m)		Scr /1934
JACOB WILSON	0-6-0ST	OC	BH	496	1879	(m)		(4)
DEVON	0-6-0ST	IC	HC	442	1895	(m)		(5)
STRATFORD	0-4-0ST	OC	AB	992	1904	(m)		(3)

\# thought to have been built at Pendleton Ironworks, Manchester, and in the ownership of BH during 1886-1896, possibly named RUSSIA. Not confirmed.

(a) ex Globe Cement, Brick, Whiting & Chalk Co, with site;
to Chingford contract; returned.
(b) ex unknown location in the North of England;
to Napsbury contract, Hertfordshire, /1903; returned /1905.
(c) ex Charles Williams, dealer, Morriston, West Glamorgan, /1908;
earlier J.T.Firbank, Fishguard Harbour (GWR) (1894-1906) contract, Pembrokeshire;
to Chingford contract; returned.
(d) ex Chingford contract after 8/1909.
(e) earlier Pauling & Co Ltd, High Wycombe (GW/GC Jt) (1901-1907) contract, Buckinghamshire, No.17 PARANA.
(f) earlier Charles Williams, dealer, Morriston, West Glamorgan; here by 4/1910
earlier J.T.Firbank, Fishguard Harbour (GWR) (1894-1906) contract, Pembrokeshire.
to Chingford contract; returned after 7/1910, by 9/1912.
(g) ex George Cohen Sons & Co Ltd, Canning Town. CP 119. /1933-34.

(h) ex C.J. Wills & Sons, Becontree contract, (per Cohen), by 6/1935.
(j) hired from George Cohen, Sons & Co Ltd, Canning Town, CP 473. /1934.
(k) hired from Sir John Jackson, Grays.
(l) hired from Associated Portland Cement Manufacturers Ltd, Gibbs Works, Grays.
(m) ex Chingford contract.

(1) to Chingford contract by 7/1910, returned by 9/1912.
(2) later at E.J. & W. Goldsmith Ltd, Grays, rubbish shoot, after /1910.
(3) to H. Covington & Sons Ltd, Tilbury Marshes rubbish shoot, /1913.
(4) to Fenton Colleries Ltd, Fenton, Staffordshire, 1, c/1914.
(5) later to Dick, Kerr & Co Ltd, Littleton Reservoir contract, Sunbury, Middlesex, by /1915.
(6) scrapped after 11/5/1915 when for sale.
(7) returned to APCM Brooks Works, Grays, by /1920.
(8) to Grays Co-operative Society Ltd. /1940.
(9) to Purfleet Wharf & Sawmills Ltd loan, returned, thence returned to George Cohen Sons & Co Ltd, Canning Town.
(10) to George Cohen Sons & Co Ltd, Canning Town.
(11) returned to hirer.

Gauge : 2ft 0in

4wDM	RH 174526	1935	New	(1)
4wDM	RH 223716	1944	New	(2)

(1) s/s after 21/4/1944.
(2) to Fence Houses Brickworks Ltd, County Durham, after 20/9/1950 by 14/1/1955..

CHINGFORD CONTRACT DC152

Construction of King George's Reservoir, Chingford, for Metropolitan Water Board. 1908-1913. The contract was let in 1/1908, at a price of £340,859; work commenced in 3/1908. The reservoir covered an area of 416 acres 1¾ miles long with a capacity of 3,000 million gallons. During construction, which involved the diversion of the River Lea, twelve miles of standard gauge and two miles of narrow gauge railway were laid. The completed reservoir had an estimated total cost of £550,000 and was officially opened by King George V on 15/3/1913.

"Contract Journal", 21/2/1912 – "for sale at new reservoirs, Ponders End – Eight locos four-coupled and six-coupled, 10in to 12in, Chas Wall Ltd, Chingford".

It is said ten locomotives were brought to contract from Wall's Works Depot at Grays and returned there afterwards.

Gauge : 4ft 8½in

Name	Type			No.	Year		
CALDEW	0-4-0ST	OC	MW	225	1867	(a)	(5)
ANNIE	0-6-0ST	OC	HE	593	1893	(b)	(1)
STRATFORD	0-4-0ST	OC	AB	992	1904	(c)	(3)
DEVON	0-6-0ST	IC	HC	442	1895	(d)	(6)
AMBERGATE	0-6-0ST	IC	MW	1069	1888	(e)	(6)
LIMERICK	0-6-0ST	IC	HE	457	1888	(f)	(4)
PORTSMOUTH	0-6-0ST	IC	HE	4	1865	(g)	(2)
JACOB WILSON	0-6-0ST	OC	BH	496	1879	(h)	(6)
HENRY APPLEBY	0-6-0ST	IC	HE	45	1870	(j)	(6)
JUMBO	0-4-0ST	OC	HH		1883	(j)	(6)

(a) ex Charles Williams, dealer, Morriston, West Glamorgan;
     earlier J.T.Firbank, Fishguard Harbour (GWR) (1894-1906) contract, Pembrokeshire.
(b) ex John F. Wake & Co Ltd, dealer, Darlington, Co. Durham, after 6/1903;
     earlier S. Pearson & Sons Ltd, Wootton Bassett-Patchway (GWR) (1897-1903) contract, Wiltshire/Gloucestershire.
(c) ex London County Council, Barking contract c/1908.
(d) here by /1909; earlier Bott & Stennett, Tattenham Corner (SER) (1895-1901) contract, Surrey.
(e) ex Orson Wright & Co, Ambergate Reservoir construction, Derbyshire, after 7/1909.
(f) ex Robert McAlpine & Sons, Peterhead Harbour (1905-1909) contract, Aberdeenshire, by 9/1909.
(g) ex Globe Works after repairs, by 6/1910.
(h) earlier Walter Scott, contractor.
(j) ex Globe Works, Grays.

(1) to Globe Works, Grays, after 8/1909.
(2) to Globe Works, Grays, by 9/1912.
(3) to Globe Works, Grays.
(4) later John Cochrane & Sons Ltd, Stonebridge Park Contract, Middlesex, by 10/1913
(5) to Exors of Chas Murrell, London, Rainham & Barking rubbish shoots.
(6) to Globe Works, Grays.

Gauge : 2ft 4in

	DENIS	(0-4-2)T	OC	Ch. Wall	(a)	s/s
		(2-4-0)		Ch. Wall	(a)	s/s
33	("LUMPY TOM")	0-4-0VBT	HCG	Ch. Wall #	(a)	s/s
	ERICA	2-4-0WT	OC	Ch. Wall	(a)	s/s
	CHANNEL DRIVER	4wTG		Ch. Wall	(a)	s/s
30	("COFFEE STALL")	4wTG		Ch. Wall $	(a)	s/s

# carried boiler with plate, John Fraser & Son.
$ carried boiler with plate, Plenty 1288.

(a) built by Charles Wall Ltd.

WALSH BROTHERS (TUNNELLING) LTD
FLOOD RELIEF SCHEME, Ingatestone   GC153

Contract for 500m of 1.5m diameter tunnel for flood relief in Ingatestone High Street, work commenced 9/1988 and was completed 5/1989. This company was incorporated 8/2/1977 and went into liquidation on 4/4/1995.

Gauge : 1ft 6in

                    4wBE   CE B2200B 1979   (a)   (1)

(a) earlier Sheridan (Plant Hire) Ltd, Nechells Green, Birmingham, West Midlands, after /1982.
(1) to CE for rebuild under job B3990, 8/1993.

HENRY WARD
Tilbury
MC154

Henry Ward M.I.C.E. born Lambeth 1849 was a civil engineer; directories for 1884 list him at 110 Cannon Street EC. As an expert witness he gave evidence for Kirk & Randall during the arbitration hearing in connection with the Tilbury Docks contract, 1884-1888. In his capacity as a civil engineer, Henry Ward may have been involved with the excavations or trial borings of the soil on Tilbury Marshes and may have given evidence in this regard to the arbitration hearing in support of Kirk & Randall. It would appear that there was some connection between Henry Ward, Kirk & Randall and the Tilbury Docks contract but further details are unknown. Henry Ward of Whitehall Court, SW, died aged 87 on March 6th 1936. Between 5/1884 & 3/1885 eight steam cranes were ordered from Grafton & Co of Bedford for delivery to Henry Ward at the Royal Albert Docks, further details unknown.

Reference : "The Times" 9/3/1936

"Engineer" 15/9/1882 – wanted – good second-hand tank loco 8in to 10in – Hy.Ward, 61 Old Broad Street, EC.

"Engineer" 21/3/1884 – wanted – 10in & 13in tank locos – Hy.Ward, 110 Cannon Street, London

Gauge : 4ft 8½in

		0-4-0ST	OC	MW	554	1876	(a)	(1)
		0-4-0ST	OC	MW	618	1876	(a)	(1)
	WESTERHAM	0-4-0ST	OC	MW	718	1880	(b)	(2)

(a) ex Kirk & Parry, Liverpool.
(b) earlier Charles Chambers, Westerham Railway (1879-1881) contract, Dunton Green, Kent, WESTERHAM.

(1) to Kirk & Randall, Tilbury Docks, contract
(2) to Milford Docks Co, Milford Haven, Pembrokeshire.

WHITAKER ELLIS LTD
CHELMSFORD CONTRACT
GC155

Company registered 16/8/1920 as public works contractors at 121 Victoria Street, SW1. Contract in Butts Green Road, Howe Green, near Chelmsford.

Gauge : 2ft 0in

		4wPM		MR	7061	1938	New	s/s

C.J. WILLS
Established 1892
WOODFORD CONTRACT
DC156

Construction of Woodford - Ilford railway (the Fairlop Loop, 6¼ miles) for GER. Work started 1900 and the line opened on 1/5/1903.

Gauge : 4ft 8½in

	SOUTHSEA	0-4-0ST	OC	HE	215	1879	(a)	(1)

Several other locomotives by MW, HC & P were used on this contract, details not known.
(a)   ex Price & Wills, Barry Dock (1894-1898) contract, South Glamorgan, by 6/1901.
(1)   to Grimsby Plant Depot, Lincolnshire; later Becontree Housing (1920-1934) contract.

C.J. WILLS & SONS LTD
C J Wills & Sons until 1912
C J Wills established 1892 until 1905

BARKING CONTRACT   PC157

Construction of a new section of the London - Tilbury Road, Rippleway, Barking, 1921-1924.
"Machinery Market" 11/4/1924 Fuller Horsey to auction 2/5/1924 re C J Wills & Sons Ltd, Barkingside – plant inc two MW 12in 6w coup locos.
Gauge : 4ft 8½in

DUNSTON	0-6-0ST	IC	MW	1446	1899	(a)	(1)
STANLOW	0-4-0ST	OC	MW	1017	1887	(b)	(2)
-	0-6-0ST		MW			(c)	s/s

(a)   earlier H.M. Nowell, Enfield-Cuffley (GNR) (1906-1910) contract, Hertfordshire, DUNSTON.
(b)   earlier Price & Wills, Tuxford-Lincoln (LD&ECR) (1894-1896) contract, Nottinghamshire.
(c)   origin and identity unknown.
(1)   to Becontree Housing (LCC) (1920-1934) contract, after 2/1924.
(2)   to Willam Rigley & Sons Ltd, Bulwell Wagon Works, Nottinghamshire, after 2/1924, by /1927.

BECONTREE CONTRACT   PC158

Construction of Becontree Housing Estate for London County Council, 1920-1934. Work commenced 9/1920. To serve the site a long railway ran from Chadwell Heath to a jetty on the Thames at TQ 821479 serving gravel pits and having two bridges over the LMS railway and a level crossing over Ripple Road.
"The Times" 9/9/1933, Fuller, Horsey, Sons & Cassell to auction 19/9/1933, contractors plant, machinery and material lying at the Depot Chadwell Heath, Essex, including ten steam locos, 6 and 4 wheel, 10 to 14 inch, three loco steam cranes 2 to 5 tons and other plant including 200, 10 ton ballast wagons; 120 side tip wagons and 1,000 tons flat bottom rails.
Reference : "Railway Bylines" Ian P. Peaty, June 2005.
Gauge : 4ft 8½in

SOUTHSEA	0-4-0ST	OC	HE	215	1879	(a)	(7)
BARRY	0-6-0ST	IC	HC	440	1896	(b)	(9)
SWANSEA	0-6-0ST	IC	MW	595	1876		(6)
	reb				1902	(c)	
PARTINGTON	0-6-0ST	IC	VF	1236	1888	(c)	(1)
LORD MAYOR	0-4-0ST	OC	HC	402	1893	(d)	(12)
FRANCES	0-4-0ST	OC	HC	435	1895	(d)	(5)
SOMERTON	0-4-0ST	OC	HC	656	1903	(d)	(8)

BOMBAY	0-6-0ST	IC	MW	1674	1906	(d)	(2)
DUNSTON	0-6-0ST	IC	MW	1446	1899	(e)	(13)
FORTH	0-6-0ST	OC	AB	1844	1924	(f)	(10)
BECONTREE	0-4-0ST	OC	HC	1563	1925	New	(3)
DAGENHAM	0-4-0ST	OC	HC	1564	1925	New	(4)
CECIL LEVITA	0-6-0ST	IC	HE	1499	1926	New	(11)

(a) ex Grimsby Plant Depot, Lincolnshire;
earlier Woodford-Ilford (GER) (1900-1903) contract.
(b) earlier Price, Wills & Reeves, Bombay Dry Dock (1906-1911) contract, India.
(c) earlier Price, Wills & Reeves, Immingham Dock (1906-1913) contracts, Lincolnshire.
(d) earlier Price, Wills & Reeves, Spurn Head (WD) (1915-1918) contract, East Yorkshire.
(e) ex Barking (Rippleway Road) (1921-1924) contract.
(f) ex Sir William Arrol & Co Ltd, Avonmouth Dock (1924-1926) contract, Bristol.

(1) to St. Helier Housing (LCC) (1929-1936) contract, Surrey.
(2) to Samuel Williams & Sons Ltd, Dagenham Dock, 8, /1932.
(3) to Hudsons, Crayford, Kent, rubbish tip, by 11/1933;
later C.W. Walker Ltd, Midland Ironworks, Donnington, Shropshire, by 11/1939.
(4) to Charles Wall Ltd, Grays, by 6/1935.
(5) to H. Covington & Sons, Wennington, /1934.
(6) to George Cohen, Sons & Co Ltd, dealers, Canning Town, CW 204, /1934.
(7) to George Cohen, Sons & Co Ltd, dealers, Canning Town, CW 207, /1934.
(8) to George Cohen, Sons & Co, dealers, Canning Town CP 467 (CW 206). /1934.
(9) to George Cohen Sons & Co Ltd, dealers, Canning Town CP 466 (CW 202), /1934.
(10) to George Cohen, Sons & Co, dealers, Canning Town CW 203, /1934.
(11) to George Cohen, Sons & Co, dealers, Canning Town CW 201, /1934.
(12) to George Cohen Sons & Co Ltd, dealers, Canning Town, CP 473.
(13) to Joseph Pugsley & Sons Ltd, dealers, Bristol (per Cohen CW 205);
thence to George Wimpey & Co Ltd, Perivale-Park Royal (LPTB) contract, Middlesex.

WILLMENT BROS
LEYTONSTONE CONTRACT   DC159

Tunnel construction contract, c/1971 with a Depot in Crownfield Road, a shaft was sunk at the junction of Crownfield Road and Leytonstone High Road to a tunnel which had a 2ft gauge line, a further shaft was sunk at the end of Selby Road to a tunnel with a 1ft 6in gauge line.

Gauge : 2ft 0in

-	0-4-0BE	WR	6600	1962	(a)	(1)
-	0-4-0BE	WR	6702	1962	(a)	(2)

(a) ex W. & C. French Ltd, contractors, hire, by 1/1971.

(1) returned to W. & C. French Ltd, off hire, by 8/1971.
(2) returned to W. & C. French Ltd, off hire, by 10/1971.

Gauge : 1ft 6in

432/28	4wBE	CE	5431	1968	(a)	(2)
-	4wBE	CE	5858	1971	(b)	(1)

(a) ex Anglo-Scottish Plant Ltd, Cambridgeshire.
(b) ex CE on hire 12/1/1971.
(1) to Willment Bros, Erncroft Works, Isleworth, Middlesex.
(2) returned to Anglo-Scottish Plant Ltd.

GEORGE WIMPEY & CO LTD
Company registered 2/7/1919
BRAINTREE SEWERAGE CONTRACT, 1925-1926 FC160

Gauge : 2ft 0in

       -                 0-4-0ST   OC   (a)   s/s

(a) origin and identity unknown.

GEORGE WYTHES
HARWICH CONTRACT HC161

Construction of the Harwich Branch for the Eastern Union Railway, the contract, let to George Wythes, was sealed 29/12/1852, the estimated cost of the works was £80,000 with a further £2000 for maintenance; the line opened 10/8/1854. Further details unknown.
Reference : National Archives, Rail 186/10, 187/9.

Unknown Contractor
COLCHESTER CONTRACT HC162

Engines possibly used by William Munro on the construction of the Colne Valley Railway (1860-1863) and/or the Tendering Hundreds Railway, Hythe – Wivenhoe – Walton (1863-1867), and/or by Munro & Frederick Furniss on the Wivenhoe – Brightlingsea Railway (1865-1866).
Reference : National Archives, Rail 227/133.
              Locomotive Magazine, 1902 & 1906.

Gauge : 4ft 8½in

8	TYNE	0-4-2ST	IC	RWH	217 1836	(a)	(1)
9	EDEN	0-4-2	IC	RS	162 1836	(b)	(1)

(a) earlier Newcastle & Carlisle Railway, sold to Christopher Tarn for £550, 6/8/1860.
(b) earlier Newcastle & Carlisle Railway, sold to Mr Shrimpton for £750, /1860, another source quotes this loco as a 2-4-0.

(1) believed scrapped at Stratford Works; one of these two engines believed to be TYNE was noted at Stratford Works in the 1860s. On 4/2/1869 it was recorded by the GER board that an old saddle back tank engine had been standing in Stratford Yard for some years; the Locomotive Superintendent had contacted the managers of the Waveney Valley & Colne Valley companies but was unable to establish who owned the engine. Subsequently Mr Adams of the NLR had valued the engine as being worth £85 for breaking up. On 31/3/1869 the GER board issued instructions for the engine to be broken up.

SECTION 3
BUILDERS, REPAIRERS AND DEALERS

JOHN S. ALLEN & SON LTD
MARDYKE WORKS, St Marys Lane, Cranham, near Upminster   **LD1**
TQ 598871

Company registered 8/8/1957 with an address at 15 Church Road, Wimbledon SW19 as importers & exporters and dealers in merchandise of all kinds.

"Machinery Market" 3/1/1963 for sale – two RH diesel locos 2ft gauge (one 2VSOL, one Lister engine) ex Depot London, John S Allen & Son Ltd, London SW19.

"Machinery Market" 5/5/1966 for sale – RH 20DL 2ft gauge diesel loco – John S Allen & Son Ltd, London.

Gauge : 2ft 0in

	-	4wDM	RH	256194	1948	(a)	(1)
	-	4wDM	RH	226264	1943	(b)	(2)
	-	4wDM	RH	226298	1943	(b)	(2)
	-	4wDM	RH	235712	1945	(b)	(2)
	-	4wDM	RH	381703	1955	(c)	(1)
83		4wDM	RH	287665	1951	(d)	(1)
98		4wDM	RH	433493	1958	(d)	(1)
	-	4wDM	MR	7994	1947	(e)	(1)
	-	4wDM	MR	10031	1948	(e)	(1)
	-	4wDM	MR	10362	1953	(e)	(1)

(a)   ex Milton Hall (Southend) Brick Co Ltd, Star Lane Brickworks, Great Wakering, c/1966.
(b)   ex Pen-Yr-Orsedd Slate Quarry, Nantlle, Caernarvonshire, c3/1967.
(c)   ex Geo Wimpey & Co Ltd, Hayes Plant Depot, Middlesex, by 5/1967.
(d)   ex Geo Wimpey & Co Ltd, Hayes Plant Depot, Middlesex, by 9/1967.
(e)   ex Inns & Co Ltd, Waterford Workshops, Hertfordshire.

(1)   exported to Singapore.
(2)   exported to Singapore, 6/1973.

Gauge : 2ft 2¾in

-	4wDM	RH	202005	1940	(a)	(1)

(a)   ex BR Boston PW Depot, Lincolnshire, via Rundle & John Philips & Co Ltd.
(1)   exported to Singapore.

Gauge : 2ft 6in

-	4wDM	MR	26009	1965	(a)	(1)

(a)   ex Stockport Corporation, Cheadle Heath Sewage Works, Cheshire, 8/1973.
(1)   exported to Singapore.

BARHAM & TAIT LTD
Dovers Corner, Arterial Road, Rainham   LD2

Company registered 15/9/1951 as engine and plant hirers at 97b South Street, Romford and whose directors were J.F.Barham & W.L.Tait. This company would appear to have some connection with Peters & Barham gravel pit owners and the Freshwater Sand & Ballast Co Ltd , an associated company at this same location was **Essex Welding Co** whose registered office in 1958 was given as 33 Eastern Road, Romford; this company went into voluntary liquidation on 21/7/1958.

"Contract Journal", 28/1/1954 for sale, plant including 10hp RH diesel loco – Barham & Tait Ltd.

Gauge : 4ft 8½in

| | - | 0-4-0ST | HC | 1442 | 1921 | (a) | | Scr |

(a)   ex Wm Cory & Son Ltd, Rainham rubbish shoot, here by 14/4/1951.

Gauge : 2ft 0in

	-	4wDM	RH	211608	1941	(a)	(1)
	-	4wDM	RH	164347	1933	(a)	s/s
	-	4wDM	RH	175408	1935	(a)	(1)

(a)   ex Wennington Sand & Ballast Co Ltd, Wennington, c8/1950.

(1)   to Currall Lewis Ltd 12/1953, seen on contract at Kings Norton 18/12/1954.

BRIDGE METALS (BASILDON) LTD
Archers Field, Basildon   KD3
                                                                    TQ 735904

Company of scrap metal merchants, went into voluntary liquidation 11/2/1992.

Gauge : 4ft 8½in

| | 0-4-0DM | AB | 419 | 1957 | (a) | (1) |

(a)   ex WD Shoeburyness, 12/1978.

(1)   to Thos. W. Ward Ltd, Silvertown 6/1979.

B.R.T. SECURITIES
Earles Colne Airfield   FD4

Dealers in government surplus stock, premises taken over by C.A. Blackwell (Contractors) Ltd.

Gauge : 4ft 8½in

| MED YARD No.2 | 0-4-0ST | OC | HL | 3683 | 1927 | (a) | s/s |

(a)   ex MOD Chatham Dockyard, Kent, c/1958.

Gauge : 2ft 6in

| | 4wBE | GB | 3586 | 1948 | (a) | s/s c/1964 |

(a)   ex Admiralty, Chattenden and Upnor Railway, Kent, by 12/1962.

CHAMBERLAIN INDUSTRIES LTD
Company registered 20/6/1935
ARGALL WORKS, Argall Avenue, Leyton CD5

Company were users and hirers of plant, earlier traded as Commercial Structures Ltd under the trade name "Staffa Services".

"Contract Journal" 24/1/1945 for hire – RH diesel locos 2ft gauge – Commercial Structures Ltd, Staffa Works, Leyton.

"Contract Journal" 11/2/1948 for sale – plant inc two Simplex 20hp petrol locos 2ft gauge – Commercial Structures Ltd, Staffa Works, Leyton.

"Machinery Market" 10/8/1951 for sale – RH diesel loco 2ft gauge 33hp – C.I. Ltd, Leytonstone.

"Machinery Market" 13/4/1956 for sale – plant inc RH 20DL loco engine No 224096 2ft gauge – Chamberlin Industries Ltd, Leyton, still for sale 29/8/1957.

Gauge : 2ft 0in

-	4wDM	RH			(a)	s/s
-	4wDM	RH			(b)	(1)
-	4wDM	RH			(b)	(1)
	4wDM	RH	203014	1941	(c)	s/s

(a) possibly ex Geo. W. Bungey Ltd, here on 14/7/1956.
(b) origin & identity unknown.
(c) earlier Gee, Walker & Slater Ltd, ROF Pontrilas, Herefordshire, by 24/2/1945
(1) sold to Thos. W. Ward Ltd, Sheffield, South Yorkshire, 24/4/1945, resold where lying to McDermott & Moore Ltd, Manchester.

Note: The loco advertised for sale in 4/1956 & 8/1957 was fitted with engine number 224096 - this engine was originally fitted to RH 217992 which was noted at Geo W Bungey's works in 7/1955 and again in a dismantled state in 8/1957.

F.J. DARTNELL LTD
Upminster LD6

"Machinery Market" 3/12/1964 for sale – 18in gauge 0-4-4-0 D HE 4524/1954 (88hp McLaren engine) & 2ft gauge 16hp Montania diesel loco – F.& J. Dartnell, Mardyke Works, Upminster.

Gauge : 2ft 0in

	4wDM	OK	4501	1931	(a)	s/s

(a) ex G. Percy Trentham, South Hornchurch Plant Depot, by 3/1963.

Gauge : 1ft 6in

CARNEGIE	0-4-4-0DM	HE	4524	1954	(a)	(1)

(a) ex Royal Arsenal, Woolwich, London, by 12/1964.
(1) to M.E. Engineering Ltd, Cricklewood, Middlesex, 2/1966.

ENGINEERING SERVICES
BROOKLYN FARM, Horndon-on-the Hill   KD7
                                                TQ 668844

Company engaged in the transport of railway rolling stock from the 1980s, the two locomotives listed below were stored on behalf of private owners whilst effectively being, "in transit"

Gauge : 4ft 8½in

No.57   689/167   0-6-0DH   RR   10214   1964   (a)   (1)

(a)   ex Stirling Metro-Rail, c12/1995.

(1)   to Gloucestershire and Warwickshire Railway, Toddington, Gloucestershire, 28/1/2000.

Gauge : 2ft 0in

9   0-6-0DM   Bg/DC 2395 1953   (a)   (1)

(a)   ex South Johnstone Co-op Sugar Milling Association, Queensland, Australia, 4/2/2005.

(1)   to Lynton & Barnstaple Railway Association, Woody Bay, Devon, c9/2005.

ESSEX IRON & STEEL CO LTD
THOBY PRIORY INDUSTRIAL ESTATE, Thoby Lane, Mountnessing   GD8

Company with registered office at 143 Battersea Rise, London, SW11. Closed c12/1992, The company purchased RH437362, 457303, 512463 and possibly 512464 from Purfleet Thames Terminal Ltd c6/1992, it is not known if any of these locos were moved to this yard, see entry for Mr Lunnon.

Gauge : 4ft 8½in

    -                     0-4-0DE   RH   512842 1965   (a)   (1)

(a)   ex Reed Paper & Board (UK) Ltd, Kent, c/1990 after 24/5/1990.

(1)   to Mr Lunnon, Stondon Hall Farm, Stondon Massey, by c12/1992.

FOGDENS LTD
Southend   JD9

Fogdens were Motor Engineers (registered 29/1/1921) with an address in Chalkwell Park and a works and stores in St Helens Road, Southend. The loco listed below was delivered to Stanford-le-Hope station. The company may have acted as agents in purchasing the loco for an unknown customer possibly one of the gravel pits in the Stanford-le-Hope area, information taken from makers records.

Gauge : 600mm

    -                     4wPM   MR   5615   1932   New   (1)

(1)   to Thos W. Ward Ltd, Grays, by 14/7/1939, later with Eastwoods Ltd, Orton Works, Yaxley, near Peterborough, by 12/1941.

JACK GEVERTZ LTD
THE WINDMILL, Aythorpe Roding

ED10
TL 593143

Company of General and Machinery Merchants, went into voluntary liquidation on 31/7/1982.

Gauge : 2ft 0in

AD 904	2w-2PMR	Wkm 3403 1943	(a)	(1)
AMW 193	4wDM	RH 200513 1940	(b)	s/s
AMW 202	4wDM	RH 200800 1940	(b)	(2)

(a) ex MoD Eastriggs, Dumfriesshire, Scotland, c/1972.
(b) ex MoD (Air Ministry), Dinton, Wiltshire, /1975.

(1) to Brockham Museum, Surrey, after 27/7/1975 by 31/10/1975.
(2) to Raynesway Plant/Balfour Beatty & Co Ltd, Dartford Tunnel contract, Kent, c4/1976.

WALTER HANCOCK
High Road, Stratford

CD11

Hancock was a builder of road steam vehicles during the years 1824-1836; in 1840 he designed and built a locomotive for the Eastern Counties Railway. An advert appeared in the "Times" for 29/7/1842; Mr W.H. Dean will sell by auction on the premises at Stratford on Thursday August 11th at 11, the stock in trade and plant of Mr W. Hancock, engineer, comprising six steam omnibuses and two railroad locomotive steam engines.

KEYMAN, PEARSON, PARKER LTD
Walthamstow
earlier Keyman & Pearson (1963) Ltd

CD12
TQ 374884

Scrap merchants in Queens Road goods yard, Walthamstow, also with an address in 1983 at Bowen Wharf, 40 River Road, Barking. Company went into voluntary liquidation 4/3/1983.

Gauge : 4ft 8½in

-	4wDM	RH 421416 1958	(a)	Scr

(a) ex T & M Beaton Bros, Chiswick, after /1970, earlier NTGB, Lea Bridge Gas Works.

P.W. LEEMAN LTD
East Tilbury

MD13
TQ 673777

Company of scrap metal merchants (registered 20/1/1976) with a yard in Station Road, Low Street, East Tilbury.

Gauge : 4ft 8½in

-	0-4-0DM	RH 252687 1947	(a)	Scr
-	0-4-0DE	RH 433676 1960	(a)	Scr
-	0-4-0DE	RH 412716 1957	(b)	Scr

(a) ex Tunnel Portland Cement Ltd, West Thurrock Works, by 12/11/1978.
(b) ex Purfleet Deep Wharf & Storage Co Ltd, Purfleet, by 12/11/1978.

Mr LUNNON
STONDON HALL FARM, Stondon Massey

ED14
TQ 575016

Locomotives stored at this private site for a while pending sale.

Gauge : 4ft 8½in

1	0-4-0DH	RH 437362 1960	(a)	Scr /2004
2	0-4-0DH	RH 457303 1963	(a)	Scr /2004
3	0-4-0DH	RH 512463 1965	(a)	(1)
	0-4-0DE	RH 512842 1965	(b)	(2)

(a)   ex Essex Iron & Steel Co Ltd, Mountnessing, after 6/1992 by 24/6/1993, originally Purfleet Deep Wharf & Storage Co Ltd.
(b)   ex Essex Iron & Steel Co Ltd, Mountnessing, c12/1992, by 3/1993. originally Reed Paper & Board (UK) Ltd, Kent.

(1)   to Paradise Wildlife Park, Broxbourne, Hertfordshire, after 9/1/1995, by 14/4/1996.
(2)   to Great Eastern Traction, Hardingham Station, Norfolk, 26/2/1999.

MAY & BUTCHER LTD
Heybridge
GD15

Company registered 19/4/1911 to carry on a business of timber and general merchants, ship and boat builders and government stores contractors and to adopt an agreement with A.E. May and A. Butcher.

"The Times" 2/10/1935, for sale by auction by direction of May and Butcher Ltd, the extensive freehold mercantile property with a total area of nearly five acres and over 600ft of river frontage; comprising the fully equipped sawmill and extensive timber yard, industrial premises and complete wood-working plant and machinery, electric power plant, loading wharf and patent slipway with hauling gear adjacent to the Chelmer Blackwater canal; also three vessels, the steel built hull, "Sir Robert Hay", twin screw tug "Basins Pride" and wood built hulk "Gloria", will be sold at auction by C.M. Stanford & Son at The Swan Hotel, Maldon, on Wednesday 16th October 1935.

"Contract Journal" 20/9/1922, for sale – steam loco AERONAUT well tank type, 2ft gauge, nearly new, Hudson 1183 – May & Butcher Ltd, Heybridge, Maldon, Essex.

Gauge : 2ft 0in

AERONAUT	0-4-0WT OC	HC	1183 1918	(a)	(1)

(a)   originally Air Ministry Kidbrooke, London.

(1)   possibly to E.R. Cole (dealers) Wood Lane, London, later Buckland Sand & Silica Co, Surrey.

MAYER-NEWMAN & CO LTD
Plaistow

CD16
TQ 396830

Scrap Yard located on the former site of Plaistow Locomotive Shed.
Gauge : 4ft 8½in

No.1	0-4-0T	OC	N	4397	1891	(a)	Scr
No.3	0-4-0ST	OC	RSHN	7309	1946	(a)	Scr
No.4	0-4-0ST	OC	AB	1666	1920	(a)	Scr

(a)   ex North Thames Gas Board, Bromley-By-Bow Gas Works, /1963.

McEWAN, PRATT & CO LTD
MURRAY WORKS, Wickford
McEwan, Pratt & Co until 16/12/1909

GD17
TQ 754929

Company formed about 1905 by Robert McEwan, Arthur Pratt and Robert Davison with an address at 44 Devonshire Chambers, London, EC – opened a Motor Works in Wick Lane, Wickford. The firm specialised in the construction of petrol locomotives and railcars. McEwan Pratt & Co Ltd was registered on 16/12/1909 and went into receivership on 5/3/1912. During 1912 the works was run down with some orders being transferred to Baguley Cars Ltd (established in 1911) who in 1913 bought assets of the company including the working drawings. McEwan Pratt & Co Ltd was wound up on 5/2/1914. The works was sold and in 1914 was occupied by D.J. Smith & Co Ltd, General Automobile and Marine Engineers and re named Compton Works.

Reference : Baguley Locomotives 1914 – 1931, Rodney Weaver, IRS, 1975.
The Railway Products of Baguley - Drewry Ltd and its Predecessors, Allen Civil & Roy Etherington, Industrial Railway Society, 2008.

DAVEY PAXMAN & CO LTD
Colchester
Davey Paxman & Co until 26/4/1898

HD18

Company founded in 1865 as general engineers supplying boilers, portable engines, winding engines etc. They are best known in recent years for their diesel engines, which were first built in 1926, versions of which powered ships and locomotives. Their records indicate that they supplied three locomotives in 1880-1882; two of which were of eighteen inch gauge for the diamond mines at Kimberley in South Africa, but no precise details are available.
Reference : Industrial Locomotive No.59.

E.L. PITT & CO (COVENTRY) LTD
Canning Town

CD19
TQ 393816

Scrap Yard located in the goods yard at Canning Town Station, Liverpool Road E16, in existence by 1955 closed by 1968. An associated company R. Fenwick & Co Ltd operated a scrap yard located at Blackwall Goods Depot TQ 390812 from c/1969, closed by 1971.

Gauge : 4ft 8½in

No.1		0-4-0ST	OC	AE	2002	1930	(a)	Scr 11/1964
	RODNEY	0-4-0ST	OC	AE	2003	1933	(b)	Scr after 3/4/1965
	ANGLO-DANE	0-4-0ST	OC	P	1318	1913	(c)	Scr
	FOLA	0-4-0ST	OC	P	1806	1930	(c)	Scr
	POLAND	0-4-0ST	OC	P	1994	1940	(c)	Scr

(a)   ex CEGB Battersea Power Station, London, /1964 by 31/10/1964.
(b)   ex A.E. Reed & Co Ltd, Aylesford Paper Mills, New Hythe, Kent, c12/1964.
(c)   ex Tunnel Portland Cement Co Ltd, West Thurrock, c11/1965.

SAIL & STEAM ENGINEERING LTD
BRIGHTLINGSEA SHIPYARD   HD20
Shipyard Services Ltd until 1987

A company of Rail and Marine Engineers who provided services for the restoration and rebuilding of railway locomotives for a number of preservation groups, company went into voluntary liquidation 8/8/1990.

Gauge : 4ft 8½in

18		0-6-0ST	IC	HE	3809	1954	(a) (1)

(a)   ex D. Frazer, Tillicoultry, Clackmannanshire, 11/1/1986.
(1)   to North Norfolk Railway, Sheringham, Norfolk, 10/1988.

H. & J.R. SAUNDERS & CO LTD
Leyton   CD21

Company registered 26/3/1955 as steel & metal merchants had a scrap yard located adjacent to Leyton (Midland Road) Station. Company went into voluntary liquidation on 21/5/1981.

"Machinery Market" 13/7/1956 for sale 12ton std gauge loco P /41 – H & J R Saunders & Co Ltd, Railway goods yard, Leytonstone.

"Machinery Market" 15/1/1959 for sale – P 7x12 0-4-0 – H & J R Saunders & Co Ltd, dealer, Leytonstone.

Gauge : 4ft 8½in

-	0-4-0ST	OC	P	2015	1941	(a)	Scr
-	0-4-0ST	OC	P	936	1902	(b)	Scr
12	0-4-0ST	OC	AB	1129	1907	(c)	Scr
No.1	0-6-0ST	IC	MW	1590	1903	(d)	Scr

(a)   ex RoF Hayes, Middlesex, 5/56, by 5/10/1956.
(b)   ex CEGB Bow, /1957 by 14/4/1957.
(c)   ex Samuel Williams Ltd, Dagenham Dock, 11/1960.
(d)   ex Samuel Williams Ltd, Dagrenham Dock, 7/1961.

W. SIMMONS & SON
Great Bentley   HD22

"Machinery Market" 2/9/1965 for sale – 44hp std gauge RH diesel loco – T.G. Simmons, Great Bentley, Essex.

Gauge : 4ft 8½in

    YARD No43   4wDM   RH 221639 1943   (a)   (1)

(a)   ex Admiralty, RNAD Wrabness, c/1964, by 3/1965.

(1)   to Brown & Tawse Ltd, West Horndon, c/1966.

STEEL & ALLOY SCRAP CO LTD
Pier Road, North Woolwich   AD23
                                                                                                 TQ 431798

A scrap yard visited in 4/1971 when identified as M & S Commercial, however no firm or company of this name has been traced at time of writing. The locos listed are believed to have been located in North Woolwich Station goods yard on the site of the Steel & Alloy Scrap Co Ltd. In 3/1985 a visit to the same yard found two more locos, the proprietor of the yard given as Mr Albert Wakefield, possibly trading as Pier Metals. The London Gazette of 30/4/1973 carried a notice of a court hearing regarding Mr Albert Alexander Wakefield of Hornchurch, trading as Steel & Alloy Scrap, the company was stuck off by notice in the London Gazette 3/7/1973.

Gauge : 4ft 8½in

    No.1   4wDM   FH 3641 1953   (a)   (1)
    BOUNTY   4wDM   RH 476142 1963   (b)   Scr 18/3/1985
    HORNBLOWER 0-4-0DE   RH 416211 1957   (c)   Scr 3/1985

(a)   ex South Eastern Gas Board, Glyne Gap Gas Works, Bexhill, Sussex, c/1970, (after 20/10/1969, by 4/1971).

(b)   ex Reed Paper & Board (U.K.) Ltd, Aylesford Paper Mills, New Hythe, Kent, 15/3/1985.

(c)   ex Reed Paper & Board (U.K.) Ltd, Aylesford Paper Mills, New Hythe, Kent, 19/3/1985.

(1)   to Thos. W. Ward Ltd, Silvertown, by 9/1971.

THAMES IRON WORKS & SHIPBUILDING CO LTD
Blackwall   AD24
                                                                                                 TQ 394810

These works were established in 1835 by Joseph Ditchburn and Charles Mare, who were primarily shipbuilders but also did structural steelwork and constructed railway lines in South Africa and elsewhere. They were notable for building Britain's first ironclad battleship, the WARRIOR, in 1860; the firm also built steam cranes and are known to have built an electric loco for the City & South London Railway in 1898. The firm was incorporated as a limited company on 5/1/1857 with an office address at Orchard Yard, Blackwall. The works closed on 21/12/1912.

Reference : Shipbuilders of the Thames and Medway, Philip Banbury, David & Charles, 1971.

TILBURY MACHINERY & IRONWARE CO

MD25

"Grays and Tilbury Gazette", 15/5/1886 carried a report that the works of the Tilbury Machinery and Ironware Company were undertaking repairs to a locomotive from the Globe brickfields which included the manufacture of new castings and turning of the tire's on the iron wheels.

W.T. TOWLER & SON LTD
PLAISTOW IRON WORKS, Plaistow

CD26

W.T. Towler & Son until 9/5/1905
TQ 400832 approx

Company established by William Thomas Towler & Henry Francis Towler as boiler makers and engineers with a works in London Road, Plaistow, and a registered office at Carpenters Lane, Stratford. The firm offered for sale a locomotive in 1903; company went into voluntary liquidation 27/1/1912.

"Machinery Market", 1/1/1903 for sale – 11in, 4 wheel coupled loco, standard gauge – Towler Ironworks, Plaistow.

Gauge : 4ft 8½in

                [0-4-0ST?]   (a)   s/s

(a)   origin and identity unknown.

UNIVERSAL STEAM TRAM-CAR CONSTRUCTION COMPANY LTD
ABBEY LANE WORKS, West Ham

CD27

Company formed 5/1879, with offices at 35 Finsbury Circus, London EC, and works in West Ham, Stratford. The company was formed for the purpose of purchasing, working and dealing in the invention of Mr Joseph Apsey, for improvements in Steam Tram-Cars. The Apsey steam cars were designed to carry 44 passengers, viz,:- 22 inside and 22 outside, when loaded its weight was about 10 tons. The steam cars could work an incline of 1 In 20 and curves of 30ft radius and could be driven from either end; they were stated to emit neither steam nor smoke to frighten horses. A single line of rails was laid around the boundary of the works in a circular form where a car was exhibited to the public. The company went into voluntary liquidation 16/6/1882.

"The Times" 21/7/1883, Fuller Horsey, Sons & Cassell are instructed to sell by auction on Tuesday July 24th, on the premises, Abbey Lane Works, West Ham; the machinery, tools and stores, including four tram engines, three boilers, nine tons of tram rails and two steam tram cars, also a 10 horse-power portable steam engine by Ransome and Sims.

T. WAKEFIELD (SCRAP METALS)
Romford   ED28
Romford Scrap & Salvage Co Ltd, until c /1970   TQ 493878

Scrap metal merchants of 288, Crow Lane, Romford, locomotives stored here for a while until sold.
Gauge : 4ft 8½in

NC1	0-4-0DM	HC	D963 1956	(a)	(1)
NC2	0-4-0DM	HC	D964 1956	(b)	(2)
D1001	0-4-0DM	HC	D894 1954	(c)	s/s
D2002	0-6-0DM	HC	D760 1951	(d)	(5)
D2006	0-6-0DM	HC	D916 1956	(e)	(6)
D2005	0-6-0DM	HC	D915 1956	(e)	(4)
D2007	0-6-0DM	HC	D917 1956	(d)	(3)
D2008	0-6-0DM	HC	D918 1956	(d)	(3)

(a) ex Newton Chambers & Co Ltd, Thorncliffe Ironworks, Chapeltown, near Sheffield, per Pounds (Shipbreakers & Shipowners) Ltd, Portsmouth, 7/1971.
(b) ex Newton Chambers & Co Ltd, Thorncliffe Ironworks, Chapeltown, near Sheffield, per Pounds (Shipbreakers & Shipowners) Ltd, Portsmouth, 8/1971.
(c) ex Port of Bristol Authority, Portishead, Somerset, c4/1973.
(d) ex Port of Bristol Authority, Portishead, Somerset, c12/1973.
(e) ex Port of Bristol Authority, Portishead, Somerset, c12/1973, after 28/11/1973.

(1) exported to Sobemai NV, Maldegem, Belgium, after 10/1971,
     later with Töleries Delloye-Matthieu SA, Marchin, Belgium.
(2) exported to Sobemai NV, Maldegem, Belgium, after 10/1971.
(3) to W. R. Cunis Ltd, Great Coldharbour Rubbish Shoot, Rainham, c12/1973.
(4) exported to Sobemai NV, Maldegem, Belgium,
     later with Georges et Cie SA, Marchienne, Charleroi, Belgium, by 9/1999.
(5) exported to Sobemai NV, Maldegem, Belgium; later G. Goossens, Halle.
(6) exported to Sobemai NV, Maldegem, Belgium.

THOS. W. WARD LTD
TITAN WORKS, Orsett Road, Grays   ND29
                                                                            TQ 613786

Plant Depot and workshops where locos were repaired and stored for hire and resale, locos were also overhauled for other customers. Site acquired in 1924.

NOTE: The TW prefix in this list denotes locos at or for sale from TITAN WORKS. Some locos may not physically have been at this location and may have been sold as and where lying. In some cases the TW number may not have been carried on the loco.

"Contract Journal" 15/6/1932 advert included for sale at Titan Works, 6 wheel 13x18 MW built 1921; 4 wheel 12x18 WB built 1921; and Atw built 1927.

"Contract Journal" 23/3/1938 advert included for sale at Grays, 2ft gauge 10hp diesel Montania and 4/6hp Lister petrol loco 2ft gauge.

"Contract Journal" 29/6/1938 advert included for sale at Grays, 6 wheel 13x18 MW; and Jung 10hp 2ft gauge diesel.

"Contract Journal" 24/7/1940, for sale 2ft gauge 20hp Howard loco (Blackstone engine) - T W Ward, Grays.

"Contract Journal" 31/7/1940, for sale 2ft gauge 10hp Planet petrol loco - T W Ward, Grays.

"Contract Journal" 27/11/1940 for sale by dealer T W Ward, Grays – plant inc 25hp HC petrol loco 2ft gauge.

"Contract Journal" 25/2/1942, for sale by dealer T W Ward, Grays – plant inc 40hp std gauge MH.

"Machinery Market" 30/7/1954 for sale – 0-4-0D RSH 6991/40 150hp, 0-4-0D FCH 2914 30/40hp, 0-4-0ST RSH 7040/39 14x22, 0-6-0ST HE oc 14x20, 0-6-0ST ic 13x18 – T W Ward, London WC2.

Gauge : 4ft 8½in

TW 44 (JJ98)
     DOVER 0-4-0ST OC AB 891 1901
ex Sir John Jackson Ltd, Grays, with site /1924, for sale at Grays, 12/1926.
to John Spencer & Sons (1928) Ltd, Newburn-on-Tyne, Northumberland, /1932.

TW 46 (JJ148)
     WESTMINSTER 0-6-0T IC SS 3474 1888
ex Sir John Jackson Ltd, Grays, with site /1924, for sale at Grays, 12/1926.
s/s

TW 103 MAX 0-4-0ST OC HC 1337 1918
ex British Celanese Ltd, Spondon, Derbyshire.
to Silvertown Machinery Works.

TW 105 (JJ197)
     WEAR 0-4-0ST OC MW
                                reb HL
ex Sir John Jackson Ltd, Grays, with site /1924.
for sale at Silvertown Machinery Works, 12/1926.

 JJ711 VICTORIA 0-4-0ST OC AB 887 1900
ex Sir John Jackson Ltd, Grays, with site /1924, for sale at Grays 12/1926.
to Inverkeithing Shipbreaking Yard, Fife, Scotland.

TW 153 NAPIER 0-6-0T OC HE 882 1905
ex WD Lydd, Kent, /1927, WD27461.
scrapped /1933.

TW 159 0-6-0T OC HE
origin and identity unknown
s/s

TW 177 0-4-0ST OC BLW 45285 1917
ex Willys Overland Crossley Ltd, Heaton Chapel, Cheshire, 15/9/1927.
to Cleveland Bridge and Engineering Co Ltd, Darlington, Co. Durham, /1928.

TW 178 CWM TÂF 0-6-0ST IC HC 1466 1921
ex Cardiff Corporation Water Dept, 30/9/1927.
possibly on hire to Sheffield Corporation Waterworks, Ewden Valley Reservoir construction c/1927, to Perry & Co (Bow) Ltd, Bromborough Dock, contract, Cheshire, by 3/1928, to Sir Lindsay Parkinson & Co Ltd, Grimsby Dock contract, Lincolnshire, /1932, possibly to Sir Lindsay Parkinson & Co Ltd, ROF Chorley (1936) contract, Lancashire; returned; to Bryn Hall Colliery Co Ltd (loan) by 12/1939, returned; to John Delaney Ltd, Horton in Ribblesdale Limeworks, Yorkshire, (WR), by 1/1946.

TW 179 LLWYN-ON 0-6-0ST IC HC 1429 1920
ex Cardiff Corporation Water Dept, 30/9/1927.
to Perry & Co (Bow) Ltd, Bromborough Dock, contract, Cheshire, by 3/1928; returned; to Sir Lindsay Parkinson & Co Ltd, Grimsby Dock, contract, Lincolnshire, /1932, to Sir Lindsay Parkinson & Co Ltd, ROF Chorley (1936) contract Lancashire; returned; to Bryn Hall Colliery Co Ltd (loan) by 1/1941; returned; to BSC Ipswich, Suffolk, by 11/1946.

TW 316 ANIK 0-6-0ST IC MW 2004 1921
ex Bombay Port Trust, Bombay, India, 7/3/1929.
to BSC Wissington, Norfolk, on hire on several occasions 1935/36 then purchased c5/1938.

TW 317 P55 LAURIE
(WADALA) 0-6-0ST IC MW 2005 1921
ex Bombay Port Trust, Bombay, India, 7/3/1929.
to C. Fison & Co Ltd, Acton contract, Middlesex, hire /1930, returned, to BSC Ipswich, Suffolk, hire /1932, returned /1934, to Harrisons (London) Ltd, Purfleet, hire 1/1946-1/1947, Tunnel Portland Cement, West Thurrock, /1950, returned by 10/1950, to Samuel Williams (Dagenham Dock) Ltd (no date), NTGB, Mill Hill Gas Works, Middlesex, hire, /1950s, here 16/3/1952, still here 29/8/1955, still here 24/9/1960, s/s.

TW 318 0-6-0ST IC MW 2006 1921
ex Bombay Port Trust, Bombay, India, 7/3/1929.
to BSC Wissington, Suffolk, on hire on several occasions 1932-1937 until purchased in 5/1938.

TW 319 (TW325) 0-4-0ST OC HL 3424 1919
ex Bombay Improvement Trust, Matunga, Bombay, India, 26/1/1929.
to Highley Mining Co Ltd, Kinlet Colliery, Shropshire, /1929.

TW 320 (TW326) 0-4-0ST OC HL 3425 1920
ex Bombay Improvement Trust, Matunga, Bombay, India, 26/1/1929.
to Farnley Fireclay Co Ltd, Leeds, Yorkshire (WR), /1929.

TW 321 (TW319) 0-4-0ST OC HL 3467 1920
ex Bombay Improvement Trust, Matunga, Bombay, India, 26/1/1929.
to Hamsterley Colliery Ltd, Co. Durham, /1933.

TW 322 (TW320) 0-4-0ST OC HL 3468 1920
ex Bombay Improvement Trust, Matunga, Bombay, India, 26/1/1929.
to Thos. W. Ward Ltd, Tinsley Works, Sheffield, South Yorkshire, /1929.

TW 323 TITAN (TW321)
       (KOLVADA)   0-4-0ST   OC   HL   3469 1920
originally Bombay Improvement Trust, 26/1/1929.
to BSC Bury St Edmunds, Suffolk, on hire /1932 returned /1933, to Scotts Shipbuilding &
Engineering Co Ltd, Greenock, Renfrewshire, /1933.

TW 324 (TW322)   0-4-0ST   OC   HL   3477 1921
ex Bombay Improvement Trust, Matunga, Bombay, India, 26/1/1929.
to Dorman Long (Steel) Ltd, Newport Works, Middlesborough, North Yorkshire, /1933.

TW 325 (TW323)   0-4-0ST   OC   HL   3478 1921
ex Bombay Improvement Trust, Matunga, Bombay, India, 26/1/1929.
to Ketton Portland Cement Co Ltd, Rutland, /1929.

TW 326 (TW324)   0-4-0ST   OC   HL   3479 1921
ex Bombay Improvement Trust, Matunga, Bombay, India, 26/1/1929.
to Ketton Portland Cement Co Ltd, Rutland, /1929.

TW 327 COLUMBIA (TW334)
       (WORLI)   0-4-0ST   OC   AE   1874 1921
ex Bombay Improvement Trust, Matunga, Bombay, India, 26/1/1929.
to John Gill (Contractors) Ltd, Barking, by 11/1930, returned to Titan Works, by 6/1931.
to APCM Crown & Quarry Works, Frindsbury, Kent.

TW 328 (TW333) MATUNGA   0-4-0ST   OC   AE   1875 1921
ex Bombay Improvement Trust, Matunga, Bombay, India, 26/1/1929.
to John Gill (Contractors) Ltd, Barking, by 11/1930, returned to Titan Works, by 7/1931.
to RPC Totternhoe, Bedfordshire, 12/1/1938.

TW 329 (TW327)   0-4-0ST   OC   AE   1876 1921
ex Bombay Improvement Trust, Matunga, Bombay, India, 26/1/1929.
to Crane-Bennett Ltd, Nacton Heath Works, Ipswich, Suffolk, 7/10/1930.

TW 330 (TW329)   0-4-0ST   OC   AE   1877 1921
ex Bombay Improvement Trust, Matunga, Bombay, India, 26/1/1929.
to Settle Speakman & Co Ltd, Kent, /1930, returned.
to Tunnel Portland Cement Co Ltd, West Thurrock, 11/11/1930.

TW 331 (TW328)   0-4-0ST   OC   AE   1878 1921
ex Bombay Improvement Trust, Matunga, Bombay, India, 26/1/1929.
to Constable, Hart & Co Ltd, Matlock, Derbyshire; on hire, returned to Thos W. Ward Ltd,
Charlton Works, Sheffield, South Yorkshire, plant No. 40373, 5/1930.

TW 332   0-4-0ST   OC   AE   1879 1921
ex Bombay Improvement Trust, Matunga, Bombay, India, 26/1/1929.
to Borough of Morecambe & Heysham, Morecambe Gas Works, Lancashire, by 6/1933.

TW 333 (TW331)   0-4-0ST   OC   AE   1880 1921
ex Bombay Improvement Trust, Matunga, Bombay, India, 26/1/1929.
To Mirvale Chemical Co Ltd, Mirfield, Yorkshire (WR), 7/3/1930.

TW 334 (TW330)   0-4-0ST   OC   AE   1881 1921
ex Bombay Improvement Trust, Matunga, Bombay, India, 26/1/1929.
Ashmore Benson Pease & Co Ltd, Parkfield Works, Stockton on Tees, by 3/1932.

TW 388 ASHBURY   0-4-0ST   OC   AE   1610 1912
originally Ashbury Railway Carriage & Iron Co Ltd, Openshaw, by /1930.
to John Gill (Contractors) Ltd, Barking, by 2/1931, returned, to Titan Works by 6/1931.
to Bennerley Slag & Tarmacadam Co, Bennerley Ironworks site, Awsworth, Nottinghamshire, by 5/1935.

TW 405 PHILIP   0-4-0ST   OC   AE   1631 1912
originally Baldwins Ltd, Netherton Furnaces, Netherton, Worcestershire, until 3/1929, here by 4/1931.
to Eldon Colliery, Bishop Auckland, by 2/1934.

TW 491 ALBION   0-4-0ST   OC   WB   2178 1921
originally Salsette – Trombay Railway, Bombay, India.
to Cement Industries Ltd, Rodmell Works, Lewes, Sussex, /1932.

TW 493 PARK   0-6-0T   OC   KS   3078 1917
ex R.H. Neal & Co Ltd, Ealing, Middlesex, by 1/1932.
to Abdon Clee Stone Quarrying Co Ltd, Shropshire, by 11/1934.

TW 494 ROYAL   0-6-0ST   OC   AB   1577 1918
ex R.H. Neal & Co Ltd, Ealing, Middlesex.
to Purfleet Deep Wharf & Storage Co Ltd, Purfleet, /1936.

TW 507 YORK   4wVBT   VCG Atw   102   1927
ex Henry Leatham & Sons Ltd, Hungate Flour Mills, York.
to WD Eskmeals, Cumberland.

TW 554   0-6-0ST   OC   AE
origin and identity unknown, for sale 5/1935.
s/s

TW 559 FORWARD   0-6-0ST   OC   AE   1655 1913
ex Checkland & Co Ltd, Coleorton Colliery, Leicestershire, after 3/1933, by 4/1934.
to National Smelting Co Ltd, Avonmouth, c/1936.

TW 570   0-4-0ST   OC   DK   1889
origin & identity unknown.
scrapped /1933.

TW 965   0-4-0ST   OC   AB   1281 1912
ex New Westbury Iron Co Ltd, Wiltshire.
to Thurrock Chalk & Whiting Co Ltd, West Thurrock, /1939.

TW 1917 L163 L194   0-4-2ST   OC   KS   3129 1918
ex Harrisons (London) Ltd, Purfleet Wharf, 4/1945.
Scrapped /1953.

TW 2568   0-4-0ST   OC   P   [1739] 1928
if identified correctly ex Co-operative Wholesale Society Ltd, Silvertown Flour Mills, here 11/1949, thence to Thos. W. Ward Ltd, Inverkeithing c/1949, loco advertised for sale in May 1950 with cylinders 14x22 built 1928, P 1739 was class W6 with 14x22 cylinders built 1928.

TW 2694 L54 BRAMLEY No6   0-6-0ST   OC   HE   1644   1929
earlier WD Bramley, Hampshire, 71660, advertised for sale 5/1950, here 16/3/1952.
to Samuel Williams (Dagenham Dock) Ltd, hire 6/1952 returned by 1/1953, to Tunnel Portland Cement Ltd, West Thurrock, hire by 25/1/1953, returned /1953, to Tunnel Portland Cement Ltd, Pitstone, Buckinghamshire, hire /1953 by 10/1953, returned /1954, still here 29/8/1955 & 24/8/1957, to NCB Harworth, Bircotes, Nottinghamshire, by 9/1958, returned, still here 24/9/1960, s/s.

TW 2695   0-4-0BE   Electromobile   W247   1928
ex WD Shoeburyness 6/1948.
for sale in "Contract Journal" 3/1949, 9/1949 & 12/1949, s/s.

TW 2919 L179   0-4-0DM   RSHN   6991   1940
ex Air Ministry Henlow, Bedfordshire, here 16/3/1952.
to Darwin Paper Mill Co Ltd, Lancashire, c/1950 ? still here 19/9/1954.

TW 2971   0-4-0ST   OC   P   2047   1943
ex Metal Recovery Produce Co, Eaglescliffe, Co. Durham, by 1/1951.
to S & L, Newcastle, Australia, /1952.

TW 2995 L5 P61   4wDM   FH   2914   1944
ex Phoenix Timber Co, Rainham /1951, still here 29/8/1955,
to Pittrail Ltd, Blenheim & Woodstock branch track lifting contract, Oxfordshire, thence to Aldridge Depot, Staffordshire.

TW 3098   4wPM   MR   5752   1939
ex APCM Dunstable, Beds, after 8/1951,here 16/3/1952.
to Effra Sales & Service Ltd, Apedale, Newcastle-under-Lyne, North Staffordshire, possibly by 16/10/1953.

TW 3415 P52 TYNE   0-4-0ST   OC   RSHD   7040   1941
ex Wm Cory, Erith, Kent, 7/1954, here 19/9/1954 & 29/8/1955.
to John Riley & Sons Ltd, Hapton, Lancashire, /1956.

TW 3434 (400 S)   0-4-0DM   JF   22934   1941
ex BR Southampton Docks, Southern Region, here 9/1954, still here 24/8/1957 working, to Silvertown Machinery Works /1959 returned /1960, to Samuel Williams Dagenham Dock on hire by 28/5/1960, returned by 24/9/1960.
to Eagre Construction Ltd, Scunthorpe, Lincolnshire, after 24/9/1960, by /1961.

TW 3570 AMW No.188 0-6-0DM   JF   22885   1940
ex Air Ministry, Halton Camp, Buckinghamshire, after 5/1955 by 5/1960.
to Eagre Construction Co Ltd, Scunthorpe, Lincolnshire, after 24/9/1960 by 7/2/1961.

0-6-0ST   OC   AE   1819 1918
ex WD Shoeburyness, 7/1947.
to NCB Cortonwood, Brampton, Yorkshire, /1948.

0-4-0ST   OC   P   2049 1944
ex Metal Recovery Produce Co, Eaglescliffe, Co. Durham, by 1/1951.
to Wm Doxford Ltd, Pallion, Sunderland, Co. Durham, 4/1951.

L167   CROSTON No.2   0-4-0DM   JF   22987 1942
ex ROF Ulnes Walton, Lancashire, here 16/3/1952 & 6/2/1954, returned after repairs.

P95   EAST RIGGS No.1   0-4-0DM   JF   22947 1942
ex ROF East Riggs, Dumfriesshire, here 16/3/1952 & 6/2/1954, returned after repairs, 4/1954.

P140   DUNHAM No.1   0-4-0DM   JF   22978 1942
ex ROF Dunham Hill, Cheshire, /1953, by 20/6/1953.
to BIS Middleton Towers, Norfolk, /1963.

ALWILDA   0-4-0ST   OC   P   798 1899
ex BPCM Wouldham Works here for repair on 19/9/1954 still here 12/3/1955
returned to BPCM Wouldham Works by 14/7/1955

4wDM   RH   349041 1953
ex George Carter & Son Ltd, Yokesford Hill, Romsey, Hampshire, by 2/1971,
earlier Pauling and Company Ltd, Park Royal Depot, Middlesex.
to E. Thomas & Co Ltd, Ponsanooth Plant Depot, Cornwall, 1/1973.

0-6-0ST   OC   HC   823 1908
ex East Kent Railway, 2.
to Purfleet Deep Wharf & Storage Co Ltd , Purfleet, 12/1945.

SHEPPERTON   0-6-0ST   OC   P   1616 1923
ex S Pearson & Son Ltd, Staines Reservoir contract Middlesex, by 15/7/1927.
to Willys Overland Crossley Ltd, Stockport, Cheshire, 15/9/1927.

SUNBURY   0-6-0ST   OC   P   1617 1923
ex S Pearson & Son Ltd, Staines Reservoir contract, Middlesex, by 15/7/1927.
to R.H. Neal & Co Ltd, Ealing, Middlesex, 10/9/1927.

0-4-0ST   OC   WB   2172 1921
originally Salsette – Trombay Railway, Bombay, India .
to NTGB Kensal Green Gas Works, London, /1929 by 29/6/1929.

0-6-0ST   IC   MW   951 1885
possibly ex Samuel Williams & Sons Ltd, Dagenham Dock, /1924.
to Prentice Bros, Burwell, Cambridgeshire, by 5/1928.

0-4-0ST   OC   MW   883 1883
ex Samuel Williams & Sons Ltd, Dagenham Dock, 5/1947, s/s.

09362   4wDM   RH
origin and identity unknown, here 12/9/1971.
to Columbia Wharf, by 18/12/1977, s/s

Gauge : 2ft 0in
TW 107   4-6-0T   OC   HE   1252 1916
ex WD Purfleet.
purchased by Lawler Ayres & Co (London agents for Hunslet) by 9/1927 for Palestine Electric Corporation., rebuilt by HE, delivery promised 8/11/1927.

TW 146   0-4-0ST OC   KS   4002 1918
ex William Muirhead Macdonald Wilson & Co Ltd, /1927.
to New Gorsllan Colliery Co Ltd, Swansea, by 2/1929.

TW 147   0-4-0ST OC   KS   4260 1922
ex William Muirhead Macdonald Wilson & Co Ltd, /1927.
to Devon County Council, Wilminstone Quarry, Tavistock, Devon, /1929.

TW 148   0-4-0ST OC   KS   4005 1918
ex William Muirhead Macdonald Wilson & Co Ltd, /1927.
for sale 4/1933 by Frank Young & Son Ltd, as lying at Cartford Bridge, Little Eccleston, Lancashire, later Castle Firebrick Co Ltd, Ewloe Green, Flintshire, by 5/1935.

TW 149   0-4-0ST OC   KS   4256 1922
ex William Muirhead Macdonald Wilson & Co Ltd, /1927.
to Devon County Council, Wilminstone Quarry, Tavistock, Devon, /1929.

TW 150   0-4-0ST OC   KS
Identity unknown, ex William Muirhead Macdonald Wilson & Co Ltd, /1927, s/s.

TW 151   0-4-0ST OC   KS   2459 1915
ex William Muirhead Macdonald Wilson & Co Ltd, after /1927, by 7/1931, possibly to Sheffield Corporation, Ewden Beck Works, South Yorkshire, by 4/1932.

TW 152   0-4-0ST OC   KS   2460 1915
ex William Muirhead Macdonald Wilson & Co Ltd, after /1927, by 6/1931, to Thos. W. Ward Ltd, Albion Works, Sheffield, South Yorkshire, by 5/1932.

TW 313   4wPM   "Hercule"
origin and identity unknown.
s/s

TW 370   4wPM   [MR/FH?]
origin and identity unknown.
s/s

TW 371   4wPM   [MR/FH?]
origin and identity unknown
s/s

TW 733   4wDM   OK
origin and identity unknown, for sale 3/1938, s/s.

TW752   4wDM   Jung
origin and identity unknown, for sale 3/1938, s/s.

TW760   4wDM   OK
origin and identity unknown, for sale 3/1938, s/s.

TW769   4wPM   L
origin and identity unknown, for sale 3/1938, s/s.

TW 2044   4wDM   MR   4587 1928
originally Buckland Sand & Silica Co Ltd, Betchworth Surrey, here 12/4/1950.
to APCM Bobbing Clay Pit, Kent, by c/1952.

TW 2374 L27   4wDM   RH 203012 1941
ex Moorfoot (Aggregates) Ltd, Reymerston gravel pits, Dereham, Norfolk, by 28/8/1947
s/s after 5/12/1951.

TW 2405 P9   4wDM   RH 211626 1941
ex MoS, HGTPD Horncott Saw Mill, Boyton, Launceston, Cornwall, here by 3/3/1948 , to
Cementation Co Ltd , Doncaster, South Yorkshire, by 6/2/1950, returned by 18/5/1951,
here 19/9/1954 with plant numbers P137 TW 2408, s/s after 29/8/1955.

TW 2406 P8 P138   4wDM   RH 211649 1942
originally MoS, here by 12/2/1948, here 19/9/1954 with plant numbers P138 TW 2409, s/s.

TW 2407 L117 P94   4wDM   RH 179872 1936
ex River Great Ouse Catchment Board, Cambridgeshire, No.12, by 9/2/1949.
to Proserpine Co-operative Sugar Milling Association Ltd, Queensland, Australia, after
24/9/1960.

TW 1615 L28   4wDM   FH
Identity unknown, ex McAlpine No.3, here 12/4/1950, s/s.

TW 1341   4wPM   Austro-Daimler
Identity unknown, ex T.C. Jones & Co Ltd, London, PN712 here 12/4/1950, s/s.

TW 1701   4wPM   MR
origin and identity unknown, s/s.

TW 1703   4wPM   MR   5454 1935
earlier with Leightons Engineers Ltd, Potters Bar, Hertfordshire, until c6/1939, here by
12/4/1950.
s/s.

TW 1695 L31   4wPM   MR   5294 1931
earlier with Geo Wimpey & Co Ltd, 10/1937, here 12/4/1950.
s/s.

TW 1697 L24   4wPM   MR   7225 1938
ex George Wimpey & Co Ltd, No.32 here 12/4/1950, originally Greenham Plant Hire Ltd, Middlesex.
s/s.

TW 2860   4wDM   RH   221604 1943
originally MoS, here 12/4/1950 & 29/8/1955.
to Demolition & Construction Co Ltd, on hire, s/s.

TW 2897 P12 P10   4wDM   RH   177641 1936
ex Ipswich Sand & Gravel Co Ltd, Suffolk, after 1/1949 by 6/1950.
here 20/6/1953 & 29/8/1955, here 24/9/1960, s/s

TW 2861 P11 L119   4wDM   RH   217963 1942
originally MoS by 12/4/1950; to Demolition & Construction Co Ltd, Beckton contract, on hire after 29/8/1955, by 4/1957; returned by 24/8/1957; s/s after 8/6/1959.

TW 2961   4wDM   RH   213824 1942
ex Ministry of Works Park Royal, Middlesex, by 6/6/1951.
s/s

TW 2893 L112 P11   4wDM   HE   2924   1944
ex County Borough of Hastings, Sussex, by 16/3/1952.
s/s after 24/9/1960.

TW 2894 L113 P10   4wDM   HE   2946   1944
ex County Borough of Hastings, Sussex, by 6/2/1954.
s/s after 24/9/1960.

TW 3128   4wDM   RH   186327 1937
ex River Great Ouse Catchment Board, Cambridgeshire, by 14/2/1952.
to Feltham Sand & Gravel Co Ltd, Southall, Middlesex, after 28/5/1952, by 5/1953.

                            4wDM   RH   166009 1932
ex War Dept, Gravel Pits, Upper Hale, Farnham, Surrey, after 13/6/1938 by 29/10/1945
exported to Sungei Besi Mines Ltd, Tin Mines, Sungei Way, Selangor, Malaya, after 9/11/1945 by 5/12/1946.

                            4wDM   RH   187083 1937
ex WD Corsham, Wiltshire, after 6/5/1940, by 26/8/1947.
to Marley Tile (Aveley) Co Ltd, Aveley, after 2/9/1947 by 17/3/1948.

                            4wDM   RH   202980 1941
ex Morfoot (Aggregates) Ltd, Reymerston gravel pits, Dereham, Norfolk, by 28/8/1947.
exported to Calcutta, India, by 12/1948.

                            4wDM   RH   198265 1939   #
originally WD Corsham, Wiltshire, by 16/3/1952, to WD, seen at Liphook auction, Hampshire, 28/3/1957,
later with John Heaver Ltd, Whyke gravel pits, Chichester, Sussex, by /1962, auctioned 10/1962, s/s.

| | 4wDM | RH | 211632 | 1941 | # |

originally MoS. Here by 16/3/1952, to WD, at Liphook Auction, Hampshire, 28/3/1957 s/s.

| | 4wDM | RH | 211635 | 1941 | # |

originally MoS. Here by 16/3/1952, to WD, at Liphook Auction, Hampshire, 28/3/1957 s/s.

| | 4wDM | RH | 211646 | 1942 | # |

originally MoS. Here by 24/5/1951, to WD, at Liphook Auction, Hampshire, 28/3/1957 s/s.

| | 4wDM | RH | 218014 | 1943 |

originally MoS, ex Arnolds (Branbridge) Ltd, Branbridges, Paddock Wood, Kent after 4/4/1946, by 9/12/1946, s/s

| | 4wDM | RH | 235622 | 1945 | # |

originally MoS. Here by 8/5/1951, to WD, at Liphook Auction, Hampshire, 28/3/1957 s/s.

| | 4wDM | RH | 226267 | 1944 | # |

originally MoS. Here by 8/5/1951, to WD, at Liphook 28/3/1957, identified by engine 234282.
later with A. Streeter & Co, Godalming, Surrey.

| | 4wPM | MR | 5615 | 1932 |

ex Fogdens Ltd, Southend, by 14/7/1939.
later with Eastwoods Ltd, Orton Works, Yaxley, near Peterborough, by 4/12/1941.

| | 4wPM | HC | P251 | 1925 |

ex Surrey County Council by 2/1939.
s/s.

Noted here on 16/3/1952. However, locos carrying these works numbers were seen amongst many others for sale at an auction at WD Liphook on 28/3/1957. (It is thought these locos were sent here by the War Department for repairs.)

Gauge : 60cm

| | 4wDM | HE | 1764 | 1934 |

Originally John Cochrane & Sons Ltd, Assiut Barrage contract, Egypt, here by 22/6/1942. s/s after 23/10/1943.

Gauge : 2ft 6in

| TW 297 | 0-4-0ST | OC | WB | 2103 | 1919 |

earlier E. M. Jellett, Easton Neston Ironstone Quarries, Northamptonshire, by /1928.
later Hendy Merthyr Colliery Co Ltd, Hendy Merthyr Colliery, West Glamorgan, by 5/1931.

| TW 335 | 0-4-0T | OC | HC | 1336 | 1918 |

originally Leeds Forge Co Ltd, Leeds, Yorkshire, here 19/7/1935.
s/s.

Gauge : 3ft 0in

TW 124   ADVANCE   0-4-2ST   OC   KS   876   1904
ex John Moffat, Brentford Contract, Middlesex, /1927.
to Tilmanstone (Kent) Collieries Ltd, Tilmanstone colliery, Kent, /1929.

S.P. WHEELER & SONS
HYTHE QUAY, Colchester   HD30
TM 016242

Scrap metal merchants.
Gauge : 2ft 0in

-	4wDM	RH 172889 1934	(a)	Scr
-	4wDM	RH 175399 1935	(a)	Scr
-	4wDM	RH 178256 1936	(a)	Scr
-	4wDM	RH 187101 1937	(b)	(1)
-	4wDM	RH 393327 1956	(c)	(2)

(a)   ex Rowhedge Sand & Ballast Co Ltd, c/1966, here 9/9/1967.
(b)   ex Rowhedge Sand & Ballast Co Ltd, c/1966.
(c)   ex Hall & Ham River Ltd, Brightlingsea Pits, c4/1967.

(1)   to Talbot Garage, Knowle Quarry, Much Wenlock, Shropshire, c4/1967.
(2)   to P.D. Nicholson and sent to Alan Keef Ltd, Cote Farm, Bampton, Oxfordshire, 16/8/1974.

WICKHAM RAIL ENGINEERING LTD
RAIL, MARINE, INDUSTRIAL ENGINEERS, CROSS ENGINEERING WORKS, Thorrington   HD31
TQ 087203

Gauge : 4ft 8½in

226		0-4-0DM	VF 5261 1945			(5)
			DC 2180 1945	(a)		
No.4	11792	0-4-0DH	JF 4220008 1959	(b)	(1)	
2268		0-4-0DE	YE 2686 1958	(b)	(1)	
D99	11790	0-4-0DM	VF D77 1948			
			DC 2251 1948	(c)	(2)	
	DX 98201A	4wDMR	Plasser 52465A 1982	(d)	(4)	
97703		4wDE	Matisa 2654 1975			
		reb	Kilmarnock 1988	(e)	(3)	
97701		4wDE	Matisa 2655 1975			
		reb	Kilmarnock 1986	(f)	(1)	

(a)   ex Mangapps Farm Railway Museum, Burnham-on-Crouch, by 6/1999.
(b)   ex Tunbridge Wells & Eridge Railway, Kent, by 25/7/1999.
(c)   ex East Kent Railway, Shepherdswell, Kent, by 27/1/2000.
(d)   ex R.D. Geeson, Ripley, Derbyshire c/2000 by 5/2000.
(e)   ex Railtrack, Dalmeny Forth Bridge CE Workshops, Scotland, 30/9/2001.
(f)   ex Railtrack, Dundee, Tay Bridge, Dundee, Scotland, c/2002.

(1) scrapped after 4/4/2004 by 5/6/2004.
(2) to Rother Valley Railway, Robertsbridge, East Sussex, 13/10/2005.
(3) to Rother Valley Railway, Robertsbridge, East Sussex, 26/9/2005.
(4) scrapped on site c/2005 by 26/3/2006.
(5) to Mangapps Farm Railway Museum, Burnham-On-Crouch, 3/9/2006.

UNCORRELATED DATA

"Machinery Market" 26/6/1925 for sale – two BEV 18/24in gauge locos; DuCroo 6hp petrol tractor; petrol loco 2ft gauge by Honeywill Bros – Ordell Plant Co, Bow.

"Machinery Market" 17/3/1944 for sale – three 2ft gauge electric locos (have been converted from battery to live rail pickup) by BEV & Booth – S Cohen & Sons, London Road, Barking.

"Machinery Market" 7/11/1947 for sale – plant inc 10hp 2ft gauge diesel locos – R.G. Gray, Romford.

"Machinery Market" 6/5/1949 H Butcher to auction 19/5/1949 at Burwell Road, Leyton, London E10 – plant inc Simplex locos.

"Contract Journal" 7/6/1956 for sale – plant inc two 2ft gauge Simplex petrol locos – Hippersons Supplies Ltd, Dagenham.

"Machinery Market" 1/10/1970 for sale – two RH 3ft gauge LBT type locos 1966 31hp 2YDAL engines – M.C. Plant & Equipment Ltd, 543 Rayleigh Road, Benfleet.

SECTION 4
PRESERVED LOCOMOTIVES

BRITISH POSTAL MUSEUM, STORE & ARCHIVE
Debden DP1
TQ 447964

Museum and Store located at Unit 7, Imprimo Park, Lenthall Road, Debden.
Gauge : 2ft 0in

		4w Atmospheric Car		1861	(a)	
807		2w-2-2-2wRER	EE/DK 807	1932	(b)	
809		2w-2-2-2wRER	EE/DK 809	1931	(c)	(1)
21		2w-2-2-2wRER	GB 420461/21	1981		
			HE 9120	1981	(d)	

(a) ex National Railway Museum, Leeman Road, York, 13/9/2006.
(b) ex Science Museum, Annexe, Wroughton, Wiltshire.
(c) ex Royal Mail Letters Ltd, Post Office Railway, London.
(d) ex Royal Mail Letters Ltd, Post Office Railway, London, 18/4/2008.

(1) to National Railway Museum, York, by 3/1/2009.

BUTLINS LTD
CLACTON HOLIDAY CAMP HP2
Later Atlas Park TM 168139

Pleasure line of 680yds in the shape of a balloon loop with a station at TM168139 opened 27/5/1939 and temporarily closed during WW2, final closure took place at the end of the 1954 summer season.

Gauge : 1ft 9in

| 6201 | PRINCESS ELIZABETH | 4-6-2DH | S/O | HC | D611 1938 | (a) | (1) |
| 6203 | PRINCESS MARGARET ROSE | 4-6-2DH | S/O | HC | D612 1938 | (a) | (2) |

(a) ex Glasgow Exhibition, Glasgow, 5/1939.

(1) to Butlins Ltd, Pwllheli Camp, Gwynedd, North Wales, 4/1953.
(2) to Butlins Ltd, Ayr Camp, Ayrshire, c4/1956.

Pleasure line within this holiday camp, opened c/1960, the line was originally 600yds long and constructed as a balloon loop, this was later re laid c/1963-1964 as a 450yd circuit with a station at TM164136. Camp closed 18/10/1983 but was re opened 19/4/1984 by Amusement Enterprises Ltd as Atlas Park but closed 9/1984. Line closed and lifted.

Gauge : 2ft 0in

			0-4-0DM	Bg	3232	1946	(a)	(2)
			4wDM	MR	22070	1960	(b)	(1)
148	C.P.HUNTINGTON		4w-2-4wPH	Chance 76.50148-24		1976	New	(3)

(a) ex pleasure railway at Markeaton Park, Derbyshire, c/1971.
(b) ex Alan Keef Ltd, Cote Farm, Bampton, Oxfordshire, hire, 7/1974;
earlier Redland Flettons Ltd, Kempston Hardwick Brickworks, Bedfordshire.

(1) returned to Alan Keef Ltd, off hire.
(2) to Alan Keef Ltd, Cote Farm, Bampton, Oxfordshire, /1977.
(3) sold and exported to Denmark c/1985.

CANVEY CASINO LTD
Canvey Island   JP3
Company registered 23/2/1934   TQ 802825

Alexander Harold Beaumont and Ernie Madle formed a partnership in 1932 with an idea to build an amusement park to entertain visitors to Canvey; the partners purchased 200ft of frontage on Canvey seafront. The amusement park called the "Casino" opened on 10/6/1933. Two locos were purchased for use on a narrow gauge railway which was intended to run east along the seafront from the casino to Seaview Road, a distance of about ¾ of a mile; however the project was abandoned and the locos scrapped during the Second World War. The casino later closed and was demolished in 1993.

Gauge : 2ft 11½in

1	0-4-0WT	OC	SS	4114	1895	(a)	Scr
2	0-4-0WT	OC	SS	4115	1895	(a)	Sc r

(a) ex Commercial Gas Co, Poplar Gas Works, London, c/1937.

J. CARTER
North Fambridge   GP4

Locomotive stored for preservation at a private location.

Gauge : 2ft 0in

-	4wDM	RH	217999	1942	(a)	(1)

(a) earlier Featherby's Brickworks, near Rochford.

(1) to Leighton Buzzard Narrow Gauge Railway, Bedfordshire, c10/1975.

CHIGWELL URBAN DISTRICT COUNCIL
THAXTED ROAD PLAYGROUND, Buckhurst Hill   DP5
                                                                                    TQ 421948

Locomotive displayed for some years in a childrens playground.

Gauge : 2ft 0in

-	4wDM	OK	5129	1933	(a)	(1)

(a) ex W. & C. French Ltd, contractors, Harlow, /1971.

(1) to Brian Gent ?, for scrap?, 21/12/1979.

COLNE VALLEY RAILWAY
CASTLE HEDINGHAM STATION   FP6
TL 774362

Preservation operation which from 1973 has re-established a working railway on the site of the closed station and the trackbed of the Colne Valley Light Railway.

Gauge : 4ft 8½in

ARMY 190 WD 190 (CASTLE HEDINGHAM)		0-6-0ST	IC	HE	3790	1952	(a)
72		0-6-0ST	IC	VF	5309	1945	(b)	(1)
40		0-6-0T	OC	RSHN	7765	1954	(c)	(2)
1875	BARRINGTON	0-4-0ST	OC	AE	1875	1921	(d)
No.1		0-4-0ST	OC	HL	3715	1928	(e)
		4wDM		Lake & Elliot	c1924		(f)
		reb		Ford		1997	
	VICTORY	0-4-0ST	OC	AB	2199	1945	(g)
	YD No.43	4wDM		RH	221639	1943	(h)
(No.1)	D2041	0-6-0DM		Sdn		1959	(j)
		4wDM		FH	3147	1947	(k)
	-	0-4-0DM		AB	349	1941	(l)
		2w-2PMR		Wkm	1946	1936	(m)
60	JUPITER	0-6-0ST	IC	RSHN	7671	1950	(n)
D2184		0-6-0DM		Sdn		1962	(o)
03030		4wDM	R/R	Unilok	2109	1980	(p)
		0-4-0DM		RH	281266	1950	(q)
D3476		0-6-0DE		Dar		1957	(r)	(3)
68706		4wDHR		Perm	011	1986	(s)
68030	(DB965369)	2w-2DMR		Matisa	D8.005	1971	(t)
		reb		Kilmarnock		1979	
D3255		0-6-0DE		Derby		1956	(u)	(4)
No.1		0-4-0ST	OC	RSHN	7817	1954	(v)
	JENNIFER	0-6-0T	OC	HC	1731	1942	(w)
	(PLUTO)	4wDM		FH	3777	1956	(x)

(a) ex Stour Valley Railway Preservation Society, Chappel & Wakes Colne Station, 7/9/1973.
(b) ex NCB, South Durham Area, South Hetton Colliery, Co. Durham, 12/1973.
(c) ex North Norfolk Railway, c5/1974.
(d) ex Main Line Steam Trust, Loughborough, Leicestershire, 11/1974.
(e) ex South Eastern Steam Centre, Ashford, Kent, c6/1976.
(f) ex Lake & Elliot Ltd, Braintree, 1/2/1977.
(g) ex South Eastern Steam Centre, Ashford, Kent, 24/4/1978.
(h) ex Brown & Tawse (Tubes) Ltd, West Horndon, c9/1979.
(j) ex CEGB, Rye House Power Station, Hertfordshire. 15/1/1981.
(k) ex D.G. Owen, Llandudno, Gwynedd, 22/2/1982.
(l) ex D G Owen, Llandudno, Gwynedd, c20/3/1984.
(m) ex Stour Valley Railway Preservation Society, Chappel & Wakes Colne Station, c4/1984.
(n) ex Stour Valley Railway Preservation Society, Chappel & Wakes Colne Station, c12/1985.

(o) ex Southern Depot Co, Southend-on-Sea, 17/10/1986.
(p) ex Ciba-Geigy Plastics Ltd, Cambridgeshire, 23/4/1996.
(q) ex Great Eastern Railway Trust (1989) Ltd, Norfolk, c17/4/1998.
(r) ex South Yorkshire Railway, Meadowhall, Sheffield, South Yorkshire, 1/12/2000.
(s) ex Elec Trac Installations, Rutherglen, Scotland, 11/12/2004.
(t) ex Railtrack, Stranraer, Scotland, 20/12/2004.
(u) ex Brighton Railway Museum, Preston Park, Brighton, East Sussex, 9/9/2008.
(v) ex Midland Railway – Butterley, Derbyshire, by 9/11/2008.
(w) ex Llangollen Railway plc, North Wales, after 10/8/2008, by 7/12/2008.
(x) ex World Naval Base, The Historic Dockyard, Chatham, Kent, 11/12/2009.

(1) to Llangollen Railway plc, North Wales, 21/4/2006.
(2) to Weardale Railway Trust, Wolsingham, Co. Durham, 2/11/2006.
(3) to T.J. Thomson & Son Ltd, Stockton, Co. Durham, 3/3/2009.
(4) to Tim Ackerley, private location, North Yorkshire, 27/8/2009.

ROGER CRAVEN
private location   XP7

Gauge : 3ft 0in

-	4wDMF	RH 418803 1957	(a)	

(a) ex Higham Ferrers Locomotives, Wellingborough, Northants, 1/11/2006.

Gauge : 600mm

4	4wDM	Moës	(a)	
17	4wDM	Moës	(a)	
21	4wDM	Moës	(a)	(1)
	4wDM	Moës	(b)	
	4wDM	Moës	(b)	

(a) ex Rumpst Brickworks, Belgium, c/2003.
(b) ex? Belgium, c3/2008.

(1) to Leadhills & Wanlockhead Railway, Leadhills, South Lanarkshire, 20/8/2005.

Gauge : 750mm

5	4wDH	HE 8829 1979	(a)	
16	4wDH	HE 9079 1984	(a)	
15	4wDH	HE 9080 1984	(a)	
10	4wDM	Moës	(b)	(2)
11	4wDM	Moës	(b)	
12	4wDM	Moës	(b)	(2)
13	4wDM	Moës	(b)	
27	4wDM	Moës	(b)	
P13351	4wDH	HE 6660 1965	(c)	(1)
-	4wDMF	RH 338438 1953	(d)	(3)
-	4wDMF	RH 375693 1954	(d)	

(a) ex BAE Systems, Bishopton, Renfrewshire, c11/2002.
(b) ex Rumst Brickworks, Belgium, c/2003.
(c) ex MoD, Dean Hill, Wiltshire, /2004 by 1/4/2004.
(d) ex Higham Ferrers Locomotives, Wellingborough, Northamptonshire, 6/9/2006.

(1) to Lynton & Barnstaple Railway Association, Woody Bay, Devon, 2/8/2005.
(2) to S. Thomason, Cambridgeshire, 6/9/2006.
(3) scrapped by 4/2010.

EAST ANGLIAN RAILWAY MUSEUM
CHAPPEL STATION, near Colchester FP8
Stour Valley Railway Preservation Society TL 898289

A preservation group which commenced in 1971 and which occupies the former goods yard and goods shed on the east side of Chappel Station.

Gauge : 4ft 8½in

		4wPM		MR	2029	1920	(a)	
68067	GUNBY	0-6-0ST	IC	HE	2413	1941	(b)	(5)
60	JUPITER	0-6-0ST	IC	RSHN	7671	1950	(c)	
AMW No.144	JOHN PEEL (PAXMAN)	0-4-0DM		AB	333	1938	(d)	
190		0-6-0ST	IC	HE	3790	1952	(e)	(2)
8410/45	BELVOIR	0-6-0ST	OC	AB	2350	1954	(f)	(15)
		0-6-0T	OC	RSHN	7597	1949	(g)	
8310/54	PENN GREEN	0-6-0ST	IC	RSHN	7031	1941	(h)	
		0-4-0ST	OC	AE	1875	1921	(j)	(1)
		4wWE		KS	1269	1912	(k)	(10)
No.11	STOREFIELD	0-4-0ST	OC	AB	1047	1905	(l)	
No.13		0-6-0T	OC	RSHN	7846	1955	(m)	(3)
	JUBILEE	0-4-0ST	OC	WB	2542	1936	(n)	
		4wDM		FH	3294	1948	(o)	
23		0-4-0DH		JF	4220039	1965	(p)	
No.2	(D2279)	0-6-0DM		DC	2656	1960		
				RSHD	8097	1960	(q)	
	JEFFREY	0-4-0ST	OC	P	2039	1943	(r)	
		0-4-0ST	OC	P	1438	1961	(s)	
1144	OLWEN	0-4-0ST	OC	RSHN	7058	1942	(t)	
		0-4-0ST	OC	AB	1865	1926	(v)	(9)
	WILLIAM MURDOCK	0-4-0ST	OC	P	2100	1949	(w)	(11)
FORD No.01		0-6-0DM		RSHD	7897	1958		
				DC	2611	1958	(x)	(7)
FORD No.02		0-6-0DM		RSHD	8098	1960	(x)	(7)
				DC	2657	1960		
		4wDMR		BD	3706	1975	(y)	(12)
	BIRKENHEAD	0-4-0ST	OC	RSHN	7386	1948	(z)	(11)
	KING GEORGE	0-6-0ST	IC	HE	2409	1942	(aa)	
	ROBERT	0-6-0ST	OC	AE	2068	1933	(ab)	

(RT 960232)	2w-2PMR	Wkm			(ac)	(13)
(960239)	2w-2PMR	Wkm	1583	1934	(ad)	(14)
960721	2w-2PMR	Wkm	1946	1936	(ad)	(4)
TR 37 PWM 2787 A25M	2w-2PMR	Wkm	6896	1954	(ae)	
(PWM 2830)	2w-2PMR	Wkm	5008	1949	(af)	(13)
960208 LNER 338	2w-2PMR	Wkm	626	1934	(ag)	(13)
A37W	2w-2PMR	Wkm	8502	1960	(ah)	(13)
A34W	2w-2PMR	Wkm	8501	1960	(aj)	(6)
DB 965080	2w-2PMR	Wkm	7595	1957	(ak)	(8)
	2w-2BER	TS&S NO/1023		1985	(al)	(13)
	2w-2PMR	Bg/DC	1895	1950	(am)	(16)

(a) ex APCM, Norman Works, Cambridge, 2/1971.
(b) ex British Steel Corporation, Tubes Division Minerals, Harlaxton Ironstone Quarries, Lincolnshire, 10/2/1971.
(c) ex British Steel Corporation, Tubes Division Minerals, Buckminster Ironstone Quarries, Lincolnshire, 2/1971.
(d) ex Hill & Sons Ltd, Botley, Hampshire, 2/1971; earlier MoDAD, Ruislip, Middlesex.
(e) ex MoDAD, Shoeburyness, 22/6/1971.
(f) ex British Steel Corporation, Tubes Division Minerals, Woolsthorpe Ironstone Quarries, Leicestershire, 1/1972.
(g) ex CEGB, Rye House Power Station, Hertfordshire. 3/1972.
(h) ex South Cambs. Rural & Industrial Steam Museum, Heydon, Cambridgeshire, 8/1972.
(j) ex Rugby Portland Cement Co Ltd, Barrington, Cambridgeshire, c9/1972.
(k) ex CEGB, York Power Station, York, /1972.
(l) ex South Cambs. Rural & Industrial Steam Museum, Heydon, Cambridgeshire, 2/1974.
(m) ex CEGB, Hams Hall Power Station, Warwickshire, 16/8/1974; stored at CEGB, Rye House Power Station, Hertfordshire.
(n) ex South Cambs. Rural & Industrial Steam Museum, Heydon, Cambridgeshire, c7/1976.
(o) ex CEGB, Barking Power Station, c12/1978.
(p) ex Shell UK Ltd, Shell Haven, c4/1980.
(q) ex CEGB, Rye House Power Station, Hertfordshire. c3/1981.
(r) ex Triad Arts Centre, Bishops Stortford, Hertfordshire. 12/6/1981.
(s) ex J.M. Walker, High Cogges, Oxfordshire, c6/1981.
(t) ex Gwili Railway, Bronwydd Arms, Carmarthen, South Wales, 20/7/1995, returned 18/11/1995.
(v) ex GWR Preservation Group, Southall, Greater London, c8/1997, after 29/6/1997 by 25/8/1997.
(w) ex GWR Preservation Group, Southall, Greater London, after 29/6/1997 by 1/1998.
(x) ex Ford Motor Co, Dagenham, 24/9/1998.
(y) ex GWR Preservation Group, Southall, Greater London, after 17/5/1998 by 14/8/1999.
(z) ex Swindon & Cricklade Railway, Swindon, Wiltshire, 12/2000.
(aa) ex Gloucestershire Warwickshire Steam Railway plc, Gloucestershire, 22/3/2005 on loan.
(ab) ex Newham Borough Council, Meridian Square, Stratford, 9/3/2008.
(ac) ex BR (ER?) by 7/4/1973.
(ad) ex BR (ER) by 8/4/1973.
(ae) ex BR Newtown, Montgomeryshire 5/1982.
(af) ex Severn Valley Railway, Shropshire, after 9/11/1986 by 25/9/1988.

(ag) ex Bygone Heritage Village Flegborough, Norfolk, after 12/4/1998 by 3/4/1999.
(ah) ex West Somerset Railway, Minehead, Somerset, after 12/4/1998 by 3/4/1999.
(aj) ex Foxfield Railway, Staffordshire, by 29/8/1998.
(ak) ex private location, Darlington, Co. Durham, via Bowes Railway, Tyne & Wear, 6/8/1999 earlier Timothy Hackworth Museum, Shildon, Co Durham, until c11/1998.
(al) ex Bovis Civil Engineering Ltd, Holborn contract, London, 7/8/1999.
(am) ex MoD Shoeburyness, c/2003.

(1) to Main Line Steam Trust, Loughborough, Leicestershire, 8/1973.
(2) to Colne Valley Railway, 7/9/1973.
(3) to North Downs Steam Railway, Higham, Kent, 19/6/1982.
(4) to Colne Valley Railway, Castle Hedingham. c4/1984.
(5) to Gwili Railway, Bronwydd Arms, Carmarthen, South Wales, 21/7/1995.
(6) to Mangapps Farm Railway Museum, Burnham-on-Crouch, after 14/8/1999 by 15/9/1999.
(7) to North Norfolk Railway, Norfolk, 16/2/2000.
(8) to Mangapps Farm Railway Museum, Burnham-on-Crouch, 5/3/2000.
(9) to Ribble Steam Railway, Lancashire, c5/7/2002.
(10) to Roger Harvey, Walton on Naze, after 9/1999 by c7/2001.
(11) to GWR Preservation Group Ltd, Southall, Greater London, 17/11/2005.
(12) to GWR Preservation Group Ltd, Southall, Greater London, 16/4/2006.
(13) to Royal Deeside Railway, Banchory, Milton of Crathes, Aberdeen, Scotland, 26/11/2006.
(14) scrapped by 11/2006.
(15) to Rutland Railway Museum, Cottesmore, Rutland, 2/10/2008.
(16) to G.W.R. Preservation Group Ltd, Southall, Greater London, c/2006.

Gauge : 60cm

RTT/767162		2w-2PMR	Wkm	3235	1943	(a)	(1)

(a) ex Yorkshire Engine Co Ltd, Long Marston, Warwickshire, 8/1999.

(1) to Royal Deeside Railway, Banchory, Milton of Crathes, Aberdeen, Scotland, 11/2006.

P. ELMS
73, Crow Lane, Romford   EP9

Locomotives stored for preservation at this private location.

Gauge : 3ft 6in

WOTO	0-4-0ST	OC	WB	2133	1924	(a)	(1)
SIR TOM	0-4-0ST	OC	WB	2135	1925	(a)	(2)

(a) ex British Industrial Callenders Cables Ltd, Belvedere. Kent, c4/1969.

(1) to Alan Keef Ltd, Lea Line, Ross-on-Wye, Herefordshire, 6/5/1988.
(2) to John Quentin, Herefordshire, after 5/1988, by 9/1991.

EPPING & ONGAR RAILWAY
Ongar
EP10
TL 551039

A preservation operation based at Ongar Station.
Gauge : 4ft 8½in

D1995	0-4-0DM	VF D293		
		DC 2566	1955	(a)
	4wDM	RH 398616	1956	(b)
	4wDM	RH 512572	1965	(b)
95	4wDM	RH 466625	1962	(c)
03170	0-6-0DM	Sdn	1960	(d)

(a) ex Imperial Bus Co, Rainham, 1/4/2005,
    earlier South Midland Railway, Dunstable, Bedfordshire.
(b) ex Trackwork Ltd, Doncaster, South Yorkshire.
(c) ex Docklands Light Railway Ltd, London, 7/10/2009.
(d) ex The Battlefield Line, Shackerstone, Leicestershire, c17/9/2010.

GLENDALE FORGE
Monk Street, Thaxted
FP11
TL 612287

Manufacturers of blacksmiths tools and wrought ironwork, works with a pleasure line.
Gauge : 2ft 0in

	ROCKET	0-2-2+4wPH	S/O	Group 4, B'ham	1970	(a)
145	C.P. HUNTINGTON	4w-2-4wPH	S/O	Chance 76-50145-24	1976	(b)
		4wDM		RH		(c)

(a) ex Kiln Park, Tenby, Dyfed, c/1982, after 6/8/1980.
(b) ex Alan Keef Ltd, Lea Line, Herefordshire, 5/1990.
(c) either RH 217973 1942, or 213853 1942, ex Alan Keef Ltd, 14/12/1990.

HARLOW DEVELOPMENT CORPORATION
LOWER MEADOW PLAY CENTRE, Harlow
EP12
TL 452078

Locomotive displayed in a playground off Paringdon Rd, Southern Way.
Gauge : 4ft 8½in

-	4wDM	FH	3596	1953	(a)	(1)

(a) ex United Glass Containers Ltd, Harlow, 10/1971.
(1) sold or scrapped after 11/7/1991 by 10/1992.

ROGER HARVEY
Walton-on-Naze HP13

Gauge : 4ft 8½in

		4wWE	KS	1269	1912	(a)

(a) ex East Anglian Railway Museum, Chappel & Wakes Colne Station, after 9/1999 by c7/2001.

Gauge : 900mm

(RR009)		4wDH Rack	HE	9282	1988	(a)	(1)
RS106	MARY	4wDH	RFSK	L106	1989	(a)	(1)

(a) ex TML Sevington, Kent, c/1998 (after 9/10/1997).
(1) to Alan Keef Ltd, Lea Line, Ross-on-Wye, Herefordshire, (per R.D. Darvill), 15/6/2005.

JAYWICK MINIATURE RAILWAY
JAYWICK SANDS, near Clacton HP14
C F Parsons

Pleasure Line opened 31/7/1936 and closed 9/1939, re opened after WW2 for the 1949 season only.

Gauge : 1ft 6in

No.1	CENTURY	4-2-2	OC	#		1911	(a)	(1)
No.2		4wVBT		Jaywick		1939	New	(2)
-		4wDM		OK	6707	1936	(b)	(3)

\# Built by Regent Street Polytechnic/ E Notter.

(a) ex C.F. Parsons.
(b) ex M.E. Engineering, Cricklewood, Middlesex, c/1949 re gauged from 60cm.
(1) to A L Bird, Cambridge, /1946.
(2) to Fairy Glen Miniature Railway, New Brighton, Cheshire, /1947.
(3) Derelict by 3/7/1951, possibly to W.H. Collier & Co Ltd, Marks Tey Brickworks, otherwise s/s.

LYNN TAIT GALLERY, THE OLD FOUNDRY
Leigh-on-Sea JP15
TQ 889891

Gauge : 3ft 6in

21		4wRER	A.C.Cars	1949	(a)

(a) ex Talycafn Tramway Museum, Talycafn, North Wales, originally Southend Pier Railway.

MANGAPPS FARM RAILWAY MUSEUM
Southminster Road, Burnham-on-Crouch   **GP16**
                                                                      TQ 944980

A preservation operation which has been developed since 1987 by Mr J. Jolly on his farm one mile north of Burnham-on-Crouch. A station building from the Mid-Suffolk Light Railway at Horham has been rebuilt on this site, which had not earlier had any rail connections, together with stock storage shed, sidings and a demonstration running line some ½ mile in length.
Gauge : 4ft 8½in.

D2325		0-6-0DM		RSHD	8184	1961		
				DC	2706	1961	(a)	
	EMPRESS (DEMELZA)	0-6-0ST	OC	WB	3061	1954	(b)	
	ELLAND No.1	0-4-0DM		HC	D1153	1959	(c)	
	AUSTIN WALKER							
	-	4wDM	R/R	S&H	7502	1966	(d)	
	MINNIE	0-6-0ST	OC	FW	358	1878	(e)	
	BROOKFIELD	0-6-0PT	OC	WB	2613	1940	(f)	
11104		0-6-0DM		VF	D78	1949		
				DC	2252	1949	(g)	
	ARMY 226	0-4-0DM		VF	5261	1945		
				DC	2180	1945	(h)	
	TOTO	0-4-0ST	OC	AB	1619	1919	(j)	
	ROF 8 No.8	0-4-0ST	OC	AB	2157	1943	(k)	
A34W		2w-2PMR		Wkm	8501	1960	(l)	(3)
DB 965080		2w-2PMR		Wkm	7595	1957	(m)	(1)
	No.15 HASTINGS	0-6-0ST	IC	HE	469	1888	(n)	
	BROOKES No.1	0-6-0T	IC	HE	2387	1941	(o)	(2)
(D2018 03018) 600 No.2		0-6-0DM		Sdn		1958	(p)	
03081 (D2081) LUCIE		0-6-0DM		Don		1960	(q)	
98401		4wBE		Perm	001	1987	(r)	
DS 1169 IDRIS		4wDM		RH	207103	1941	(s)	

(a) ex NCB, Norwich Coal Concentration Depot, Norwich, via Tannick Commercial Repairs, Norwich, Norfolk, 19/3/1989.
(b) ex store at CEGB, Rye House, Hoddesden, Hertfordshire, c8/1989.
(c) ex CEGB, Elland Power Station, Yorkshire (WR), via Booth Roe Metals Ltd, Rotherham, South Yorkshire, 13/12/1989.
(d) ex RFS Ltd, Kilnhurst, South Yorkshire, 23/11/1990.
(e) ex East Kent Light Railway, Shepherdswell, Kent, after 16/7/1993 by 20/8/1993.
(f) ex Pontypool & Blaenavon Railway, South Wales, after 25/7/1993 by 30/8/1993.
(g) ex Pontypool & Blaenavon Railway, South Wales, c/9/1993.
(h) ex East Kent Light Railway, Shepherdswell, Kent, after 12/8/1994 by 3/1995 , to Wickham Rail, Thorrington, by 6/1999, returned 3/9/2006.
(j) ex?, Bexley, Kent by 1/3/1998.
(k) ex Staffordshire Locomotive Co Ltd, Shropshire, 7/1999.
(l) ex East Anglian Railway Museum, Chappel & Wakes Colne Station, near Colchester, after 14/8/1999 by 15/9/1999.
(m) ex East Anglian Railway Museum, Chappel & Wakes Colne Station, near Colchester, 5/3/2000.
(n) ex Tenterden Railway Co Ltd, Kent, by 3/2003.

(o) ex Middleton Railway Trust, West Yorkshire, /2003 by 18/10/2003, rebuild of 0-6-0ST.
(p) ex Harry Needle Railroad Co Ltd, Lavender Line, Sussex, 3/2004.
(q) ex Gennape Sugar Factory, Belgium, 8/3/2004.
(r) ex Transplant Ltd, /2004 by 6/6/2004.
(s) ex Telford Horsehay Steam Trust, Horsehay, Telford, Shropshire, 14/2/2006.
(1) to private location 31/8/2003.
(2) to Middleton Railway Trust, Hunslet, Leeds, West Yorkshire, /2005.
(3) to Royal Deeside Railway, Banchory, Milton of Crathes, Aberdeen, Scotland, 25/11/2006.

Gauge : 3ft 6in

| 8 | | 4wRER | A.C. Cars | 1949 | (a) |

(a) ex Talycafn Tramway Museum, Talycafn, North Wales, by 7/2004, originally Southend Pier Railway.

NEWHAM BOROUGH COUNCIL
Winsor Terrace, Beckton AP17
earlier **Beckton Council** until 31/3/1998 TQ 440816
Gauge : 4ft 8½in

| ROBERT | | 0-6-0ST | OC | AE | 2068 | 1933 | (a) | (1) |

(a) ex Kew Bridge Steam Museum, Greater London, c/1994 by 6/7/1994.

(1) to Kew Bridge Steam, Museum, 7/3/2000.

Meridian Square, Stratford CP18

Gauge : 4ft 8½in

| ROBERT | | 0-6-0ST | OC | AE | 2068 | 1933 | (a) | (1) |

(a) ex Kew Bridge Steam Museum, Greater London, 4/7/2000.

(1) to East Anglian Railway Museum, Chappel & Wakes Colne Station, 9/3/2008.

NORTH WOOLWICH OLD STATION MUSEUM
Pier Road, North Woolwich AP19
TQ 433798
Museum using the restored station buildings at North Woolwich. Opened in 1984, but closed again when the site was needed for re-development in 2008.
Gauge : 4ft 8½in

	No 229	0-4-0ST	OC	N	2119	1876	(a)	(6)
		0-6-0ST	OC	P	2000	1942	(b)	(4)
No.14	DOLOBRAN	0-6-0ST	IC	MW	1762	1910	(c)	(3)
No.15	RHYL	0-6-0ST	IC	MW	2009	1921	(c)	(3)
No.29		0-6-0ST	IC	RSHN	7667	1950	(c)	(1)
BARKING POWER	DUDLEY	4wDM		FH	[?3294	1948]	(d)	(5)

No.35	RHIWNANT	0-6-0ST	IC	MW	1317	1895	(e)	(3)
8		0-6-0ST	IC	AE	2068	1933	(f)	(2)
5	ABERNANT	0-6-0ST	IC	MW	2015	1921	(g)	(3)

(a) ex Dean Forest Railway, Norchard, Gloucestershire, 12/9/1984.
(b) ex Nene Valley Railway, Cambridgeshire, c11/1984.
(c) ex Kent & East Sussex Railway, Tenterden, Kent; here by 5/1985.
(d) ex Stour Valley Railway, Chappel & Wakes Colne Station, here by 20/4/1987.
(e) ex Nene Valley Railway, Cambridgeshire, c7/1991.
(f) ex Peak Rail, Derbyshire, c9/1991 (by 12/10/1991).
(g) ex Birmingham Railway Museum, Tyseley, West Midlands, 8/1992 (by 18/9/1992).

(1) to Great Central (Nottingham) Ltd, Ruddington, Nottinghamshire, 28/1/1994.
(2) to Kew Bridge Steam Museum, Greater London, c/1994.
(3) to Peak Rail Ltd, Rowsley, Derbyshire, 13/6/2002.
(4) to Barrow Hill Engine Shed Society, Barrow Hill, Derbyshire, c31/8/2004.
(5) to Waltham Abbey Royal Gunpowder Mills, 30/4/2008
(6) to Bill Parker, The Flour Mill, Bream, Forest of Dean, Gloucestershire, 6/6/2008

SHARPES AUTOS (LONDON) LTD
GABLES SERVICE STATION, Rawreth, near Rayleigh GP20
TQ 784920

This locomotive stood in the open behind an Esso garage on the south side of the A129 road. There was also a collection of vintage cars at this site but the owner has been unable to obtain planning permission to open this display to the public.
Gauge : 4ft 8½in

| 139 | BEATTY | 0-4-0ST | OC | HL | 3240 | 1917 | (a) | (1) |

(a) ex British Steel Corporation, General Steels Division, Dock Street Foundry, Middlesbrough, Cleveland, 7/1972.

(1) to Telford & Horsehay Steam Trust, Horsehay, Telford, Shropshire, 3/12/2002.

L. J. SMITH
THE BUNGALOW, Rectory Lane, Battlesbridge GP21
TQ 783965

Private narrow gauge railway about 300yards long at this location served nursery gardens and as a pleasure line. Closed by /1994.
Gauge : 2ft 0in

	SMUDGE	4wDM		MR	8729	1941	(a)	(3)
	-	4wDM		RH	218016	1943	(b)	(2)
02	HAYLEY	0-4-0DM	S/O	Bg	3232	1947	(c)	(3)
	-	4wBE		CE	5667	1969	(d)	(1)

(a) ex Alan Keef Ltd, Cote Farm, Bampton, Oxfordshire, /1978.
(b) ex M. E. Engineering Ltd, Neasden, London, c3/1978.
(c) ex Alan Keef Ltd, Cote Farm, Bampton, Oxfordshire, c10/1978; earlier Butlins Ltd, Clacton Holiday Camp.
(d) ex Alan Keef Ltd, Cote Farm, Bampton, Oxfordshire, c2/1979.

(1) to Ayle Colliery Co, Cumbria, via Alan Keef Ltd, Bampton, Oxfordshire, 4/1985.
(2) to Leighton Buzzard Narrow Gauge Railway, Bedfordshire, 26/12/1988.
(3) to Linton & Barnstaple Railway Association, Lynbarn Railway, Downland Farm, near Clovelly, Devon, c3/1994, by 15/5/1994.

SOUTHEND MINIATURE RAILWAY
THE KURSAAL, Southend on Sea   JP22
N. G. Parkinson 1932-1938
W. H. Bond 1920-1932

Railway originally constructed as a 200yd end to end line but re laid in 1932 as a 400yd balloon loop which had two stations at Lakeside and Central, the latter with an overall roof, line closed 1938. It is thought that the test line built by Never Stop Railways prior to the 1924 British Empire Exhibition at Wembley was located at this site.

Gauge : 1ft 3in

		4-4-0	OC	Herschell Spellman		New?	(2)
GEORGE THE FIFTH		4-4-2	OC	BL	18 1911	(a)	(1)
			Reb	A. Barnes 99 1919			
41		4-4-2	OC	BL	30 1912	(b)	(2)
7		2-4-0PM	S/O				
				N.G. Parkinson	c1932	New	(2)
		4w-4w+4-4PER					
				N.G. Parkinson	1932	(c)	(2)

(a) ex Skegness Miniature Railway, Lincolnshire, /1928.
(b) ex unknown owner, Cricklewood, London, /1930.
(c) ex Yarmouth Miniature Railway, Great Yarmouth, Norfolk.

(1) to Belle Vue Miniature Railway, Manchester, /1936.
(2) to H Dunn Plant & Machinery Ltd, Bishop Auckland, Co. Durham, /1938.

SOUTHEND-ON-SEA BOROUGH COUNCIL

SOUTHEND BOROUGH PARKS DEPARTMENT, CENTRAL NURSERY,
Barling Road, Southend-on-Sea   JP23
                                                                        TQ 915873
Gauge : 3ft 6in

7	4wRER	A.C. Cars	1949	(a)

(a) ex Talycafn Tramway Museum, Talycafn, North Wales, c/2001, originally Southend Pier Railway.

SOUTHEND PIER RAILWAY, Southend-on-Sea JP24
Brent Walker Leisure Ltd from 2/5/1986 until c/1988 TQ 884850

Construction of the first pier began on 25/7/1829 by the Southend Pier Co, this was a wooden structure which, by 1846, had reached deep water and was 1¼ miles long making it the longest pier in Europe. The period after 1846 saw the laying of the first tramway with wooden rails to a gauge of approximately 3ft 6in, the single track had three hand propelled trucks for conveying the luggage of steamer passengers, there was also a special sail fitted truck for use when the wind was favourable. Ownership of the pier and tramway changed hands and in 1873 the new owner the Local Board decided to adapt the line for passenger transport and the wooden rails were replaced by flat bottomed iron rails spiked directly to the decking. Rolling stock consisted of three small enclosed carriages pulled by two horses.

In 1885 the decision was taken to rebuild the pier with iron piles and to install an electric railway. Construction began in 1888, the contractors being Messrs Arrol Bros of Glasgow, and took two years to complete. Although not finished the pier opened in 1889 and work began on the railway the following year with the line being laid to 3ft 6in gauge. Although the railway had not been completed, with only ¼ mile laid, the pier and railway were officially opened on 2/8/1890, the track reached the pier head in 1891 a total distance of 1¼ miles. The original car No.1 was supplied by Falcon Works of Loughborough with two trailer cars purchased in 1891 and a further set of three cars purchased in 1893, again from Falcon Works. A passing loop opened in 1898 when a further six cars were purchased from Brush Electrical Engineering Co Ltd, successors to Falcon. The stock was added to again in 1902 which allowed the operation of 4 four car trains.

Work commenced on doubling the track in 1928 and was completed in 1929 by Titan Trackwork Co of Sheffield. When war broke out in 1939 the pier was taken over by the Royal Navy and re-opened to the public in May 1945. By 1947 annual passenger figures had reached 3 million and with passenger numbers ever increasing and lack of maintenance on the rolling stock during the war years, the decision was taken to replace the entire passenger car fleet. An order was placed with A.C. Cars Ltd of Thames Ditton for four complete trains each of seven cars, the first being delivered in March 1949. In 1974 it was decided to single the line to reduce costs however the pier head was destroyed by fire on 29/7/1976 after which the line closed on 1/10/1978, within four years the entire stock had been disposed of.

In 1985 May Gurney (Colchester) Ltd were contracted to lay a new 3ft 0in gauge single line with passing loop which re opened on 2/5/1986.

Reference : Pier Railways & Tramways of the British Isles, Keith Turner, Oakwood Press, 1999.

Gauge : 3ft 0in

A	SIR JOHN BETJEMAN	4w-4wDH	SL	SE4	1986	New
B	SIR WILLIAM HEYGATE	4w-4wDH	SL	SE4	1986	New
No.1835		4wBER			1996	

Gauge : 3ft 6in

-		2w-2PMR	Wkm	10943	1976	New (1)

(1)   to Brecon Mountain Railway, Pant, Glamorgan.

SOUTHEND PIER RAILWAY MUSEUM
Southend on Sea   JP25
                                                                                                    TQ 884850

Gauge : 3ft 6in
11	4wRER	A.C. Cars	1949	(a)
22	4wRER	A.C. Cars	1949	(a)
6	2-2wRER	BE	1890	(a)

(a)   ex Southend Pier Railway.

Gauge : 2ft 8½in
8	4wRER	BE	1898	(a)	(1)
		reb BE	c1911		
		reb Btn Corp	1950		

(a)   ex Volks Electric Railway, Brighton, Sussex, 27/2/1999.

(1)   to unknown location in Essex.

WALTHAM ABBEY ROYAL GUNPOWDER MILLS CO LTD
Waltham Abbey   DP26
                                                                                                    TL 376013

Gauge : 4ft 8½in
BARKING POWER DUDLEY   4wDM   FH   [3294 1948]   (a)

(a)   ex North Woolwich Old Station Museum, London, 30/4/2008.

Gauge : 3ft 0in
BB 307   2w-2DE   GB   6099 1964   (a)

(a)   ex Health & Safety Laboratory, Harpur Hill, Buxton, Derbyshire, 9/3/2000.

Gauge : 2ft 6in
	4wDH	HE	8828	1979	(a)	
1	4wBEF	CE	B3482A	1988	(b)	Scr 7/2009
	4wDH	Ruhr	3920	1969	(c)	
	4wDH	BD	3755	1981	(d)	

(a)   ex BAE Systems, Bishopton, Ayrshire, 23/7/2003.
(b)   ex Mines Rescue Service, Selby, North Yorkshire, 28/2/2004.
(c)   ex Whipsnade Wild Animal Park Ltd, Whipsnade, Bedfordshire, 13/9/2007.
(d)   ex MoD Dean Hill, Hampshire, 23/7/2009.

Gauge : 1ft 6in
No.1 WOOLWICH	0-4-0T	OC	AE	1748	1916	(a)
BUDLEIGH	4wDM		RH	235624	1945	(b)
CARNEGIE	0-4-4-0DM		HE	4524	1954	(c)

(a)   ex Bicton Woodland Railway, Devon.
(b)   ex Bicton Woodland Railway, Devon.
(c)   ex Alan Keef Ltd, Lea Line, Herefordshire, 13/12/2004.

WALTON-ON-THE NAZE PIER RAILWAY
Walton-on-the Naze   HP27
                                                                                                                 TM 255214

Construction of the pier was promoted by the Walton-on-the-Naze Pier & Hotel Co Ltd and authorised by the Walton-on-the-Naze Pier Order of 1897. During construction this company became incorporated into the Coast Development Co Ltd under whose ownership the pier opened in 8/1898. The pier was 2,600ft long and included a single track electric railway of 3ft 6in gauge. Rolling stock consisted of a motor car and two trailers built by the Ashbury Railway Carriage & Iron Co Ltd of Manchester, a second identical set of cars also ran on the railway, whether these dated from 1898 or 1904 is uncertain. In 4/1930 Walton Pier Ltd placed an order with Baguley (Engineers) Ltd of Burton on Trent for the supply of a bogie tramcar chassis without electrical equipment or superstructure, this to be fitted by the Pier Co. The railway ran until the end of the 1935 summer season when it was closed and the track lifted. The railway was replaced by a wooden guide-way for a single battery powered carriage built by Electricars Ltd, in 1937 the pier and battery car were purchased by the **New Walton Pier Co** and remained in service until 30/5/1942 when both were destroyed by fire. Walton-on-the-Naze pier was rebuilt after the war by Walton-on-the-Naze Urban District Council. During reconstruction a 2ft gauge contractor's railway was laid along it's half mile length, this line was adapted to carry passengers and opened in 1948. Line closed and dismantled following severe storm damage in 12/1978.

Gauge : 3ft 6in

	4w-4wRE	BgE	1795	1930	New	(a)	s/s /1935

(a)   body and secondhand electrical equipment fitted by Pier Co.

Gauge : 2ft 0in

No.1 DREADNOUGHT	0-4-0PM	Bg	3024	1939	(a)	
	reb to 0-4-0DM			1952		(1)

(a)   ex Wilson's Pleasure Railway, Allhallows-on-Sea, Kent, c/1947-48.

(1)   to ARM (UK) Ltd, Enstone, Oxfordshire (dealers), /1982, thence to Camelot Theme Park, Charnock Richard, Lancashire, via overhaul at Steamtown, Carnforth, Lancashire, c/1982, by 4/1983.

WICKFORD NARROW GAUGE RAILWAY GROUP
Wickford   GP28

Following the purchase of MR 7374 and its subsequent restoration a short private running line in the form of a balloon loop was established in the grounds of 222 Southend Road, Wickford. The chassis of MR 8640 was also kept here and was used as a brake van. Following the sale of the grounds for development, the line was dismantled 12/1998 and the locomotives were placed into storage. L 3916, L 42494, RH 441951 and FH 3983 were kept and operated nearby at the home of Graham & Lesley Feldwick. The remaining locomotives were stored at a farm in Shotgate. MR 7374, L 3916 and L 42494 were moved to Bradfield Post Office & Stores, Manningtree after 6/1999, where a short private line was again established around the garden. All stock was moved to storage in Wareham, Dorset by 2002, and later moved to Twyford Waterworks, Winchester where it is currently operating in 2010.

Gauge : 2ft 0in

No.29	AYALA		4wDM		MR	7374	1939	(a)	(5)
	-		4wDM		MR	8640	1941	(b)	(4)
	DOE 3983		4wDM		FH	3983	1962	(c)	(3)
	3916		4wPM		L	3916	1931	(d)	(5)
			4wDM		RH	441951	1960	(e)	(2)
			4wDH		RH	283513	1949		
				reb	AK	20R	1986	(f)	(1)
			4wDH		AK	10	1983	(g)	(1)
			4wDM		L	42494	1956	(h)	(5)

(a) ex George Garside (Sand) Ltd, Leighton Buzzard, Bedfordshire, 10/11/1981.
(b) ex MOD(AD), Shoeburyness, 4/11/1985.
(c) ex Medway Port Authority, Hoo Ness, Kent, 28/4/1990.
(d) ex M.A.G. Jacob, Long Eaton, Derbyshire, 23/3/1991.
(e) ex Butterley Building Materials Ltd, 19/11/1992.
(f) ex Butterley Building Materials Ltd, 4/12/1992.
(g) ex Butterley Building Materials Ltd, 5/12/1992.
(h) ex FMB Engineering Co Ltd, Hampshire, 6/4/1996.

(1) to Alan Gartell, Somerset, 27/3/1993, per FMB Engineering.
(2) to FMB Engineering Co Ltd, Hampshire, 6/4/1996.
(3) to Wareham, Dorset, 6/7/2002.
(4) to Wareham, Dorset, 14/8/2002.
(5) to Wareham, Dorset, 23/8/2002.

Mr WOODS
private location   XP29

The locomotive listed below was reported as sold to Mr Woods and was still extant here in 2009, stored on private property, until moved for further restoration.

Gauge : 4ft 8½in

		4wVBT	VCG S	7492	1928	(a)	(1)

(a) ex R. Finbow, Bacton, Suffolk, c7/1990.

(1) to Avon Valley Railway, Bitton, Somerset, 4-5/9/2010.

SECTION 5
NON LOCO WORKED LINES

ADMIRALTY
HMS OSEA, Osea Island   GH1
TL 914063

HMS Osea was established in 1918 as headquarters of the Coastal Motor-Boat Service and was located on Osea Island in the River Blackwater, the OS map for 1922 show this Naval Establishment with a tramway which ran to piers on both the north and south sides of the island, closed 14/6/1921.

BENFLEET BRICKWORKS LTD
BENFLEET BRICKWORKS   JH2
Frank Butler, c1890 until 16/8/1899   TQ 782864
Benfleet Brick & Tile Co Ltd, incorporated 30/12/1879, wound up 5/3/1884

A brickworks established in 1879 and located on the south side of Vicarage Hill and south west of the vicarage, the works had a short tramway to the clay pit as shown on the OS map for 1895, company went into voluntary liquidation 21/8/1907.

BILLERICAY BRICK CO LTD
STATION BRICKWORKS, Billericay   GH3
Company registered 9/3/1922   TQ 671952

Brickworks on land at Charity Farm, Buttsbury later operated by Harris Bros from 1929 until 1946 at least, company went into liquidation 15/1/1929. The works had a 20in gauge tramway with a cable operated incline which ran from the clay pit to the works.

BOARD OF TRADE
SEA TRANSPORT STORES DEPOT No.79, Pitsea   KH4
TQ 735863

The Sea Transport Stores Depot was part of the Mercantile Marine Department of the Board of Trade. The depot had earlier been an Army Transport Stores Depot in operation by 5/1920. The land and property had been taken on lease (dated 21/7/1920) from the British Explosives Syndicate Ltd ; the Sea Transport Stores Depot opened in 8/1939 to supply provisions for troop ships including bedding and other stores. A light railway was installed to move supplies around the site; use of a locomotive unknown. The depot closed in 7/1964 when the remaining stores were transferred to Wapley Common depot near Chipping Sodbury, Bristol.

BRITISH EXPLOSIVES SYNDICATE LTD
PITSEA WORKS KH5
TQ 735863

Production commenced in 1894 and in 1895 there was a tramway in the form of a closed loop which linked the numerous small buildings (around TQ 736865) and terminated on a wharf on Vange Creek. The line was 2ft gauge, laid with ferrous rails was worked by horses. At least two explosions took place on the site; on 28/3/1913 three men were killed and several injured when about five tons of gun-cotton in a drying shed exploded destroying buildings and causing considerable damage in the surrounding area; the shock of the explosion was felt nearly twenty miles away. On 10/5/1916 an explosion took place in the laboratory when Nitro-glycerine exploded killing the chemist and his assistant, the resultant fire destroyed a whole block of buildings including the offices. In 1919 the British Explosives Syndicate and many other explosives companies amalgamated to form the **Explosives Trades Ltd** (incorporated 11/1918), this company changed it's name in 12/1920 to **Nobel Industries Ltd** which went into voluntary liquidation on 28/8/1928. The site (or part of it) is recorded as the Sea Transport Stores Depot. In WW2 at least part of the site was used by the War Department and is thought to have been occupied by No.114 REME workshop of Anti Aircraft Command; a railway of some sort was used, beside Pitseahall Fleet on the east side of the site.

"The Times" 16/8/1919, Fuller Horsey & Co will sell by auction on 16/9/1919 and the following days, on the premises of the British Explosives Syndicate Ltd at Pitsea Factory, modern chemical plant which included about 4 miles of 24in gauge tramway with steel sleepers and turntables, also 32 flat top, tank, side and end tip trolleys.

Reference : The Quarry, 2/1911.

JAMES BROWN
Company registered 21/5/1904 as **James Brown (London) Ltd**

James Brown established his business c1860 at Braintree and Chelmsford. He also operated a works at Hatfield Peverel and a small works at Boreham which made plain bricks A receiver was appointed by the High Court on 30/10/1917 and the company dissolved 6/7/1923.

CHELMSFORD BRICKWORKS GH6
James Wilkin early 1840's until 1887   Works TL 701064 Clay Pit TL 700064

Brickworks between New Writtle Road and the GER railway embankment. The works in operation from the early 1840s, had a tramway which ran under the viaduct which crossed the River Can to a clay pit on the west side of the embankment and is shown on the OS map for 1897. Works closed in 1904.

BRAINTREE BRICKWORKS FH7
TL 761228

Brickworks situated between Braintree Station and Lower Railway Street in operation from the mid 1870s. The works was served by a standard gauge siding from the Witham to Braintree branch of the GER and had a standard gauge tramway into the Brick Works. Works closed in 1903.

BRENTWOOD BRICKWORKS   EH8
James Brown (1919 Brentwood) Ltd until 1930s   TQ 587931
James Brown London Ltd, until 10/1919
James Brown Ltd until 1904

Brickworks established in 1899 on 18½ acres of land bounded by London Road and Kavanagh Lane. The works had a Brown's patent continuous kiln with a 100ft chimney, drying sheds and tunnel dryers and produced facing, rubber moulded and ornamental bricks. A tramway system and rolling stock connected the works to a siding on the GE Railway, siding agreement dated 7/3/1899. Works later taken over by the British Cavity Brick & Tile Works Ltd (registered 14/6/1928) until it's closure in late 1940s.

UPMINSTER BRICKWORKS, Bird Lane, Upminster   LH9
TQ 566888 & 566885

Matthew Howland Patrick built a circular brick kiln 70ft high about one and a half miles north of Upminster in 1774. Subsequent operators were Champion Branfill (1777-1792), Joseph Knight (1820's), Thomas Sandford (1840s-late 1850s) and Samuel Gardner (c1858-c1875). From c1875 to 1885 Thomas Lewis Wilson and Edward Hook trading as Wilson & Hook were the lessees and from 31/12/1886 it was in the hands of Ernest James Brown. The works was situated on 35 acres of land, the site of an earlier pottery, and had a Brown's patent continuous kiln with a 100ft high chimney and drying sheds; it produced hand made red facing bricks and tiles. There were two brickfields, the old brickfield (north of Bird Lane) had a 2ft 4in gauge tramway. The new brickfield (south of Bird Lane) had a 1ft 8in gauge tramway. Clay was ground in vertical pug mills driven by windlass & chain, steam was generated by a Field vertical boiler used to power a Shanks Caledonian engine of 16hp. The old brickfield had an engine house which contained a 10hp Ransome Sims & Jefferies portable engine. A narrow gauge tramway of 2ft 4in gauge, 1¼ miles long linked the works with Upminster Station (TQ 562869) on the London Tilbury & Southend Railway, and was worked by horses hauling wooden trucks with drop sides. Site later taken over by the Upminster Brick Co Ltd (registered 26/8/1912, and by 1920 the New Upminster Brickworks Ltd which went into voluntary liquidation 18/10/1922; later Upminster United Brick Works Ltd (registered 15/10/1927) & South Essex Brick & Tile Co Ltd (1927-1933). A section of track was still in situ in the 1970s.

THE BULMER BRICK & TILE COMPANY LTD
HOLE FARM BRICKWORKS, near Bulmer Tye, Sudbury   FH10
TL 832381

Brickworks operated by John English in the 1840s, and George English by 1873, later George William English from 1880 to 1920, thence Parker and Fry who in 1923 sold the works to George E. Gray Ltd. In 1936 the works was acquired by Lawrence Albert Minter and the Bulmer Brick & Tile Company Ltd was incorporated on 12/12/1936. There were three up draught Scotch kilns in use up to 1939, each holding 36,000 bricks (two still extant in 2009). A short 2ft gauge tramway extended from clay pits up an incline to a pug mill, these tracks were exposed in 2004 when a 'V' skip was recovered from the flooded pit. The brickworks continues to make high quality hand-made bricks for use in the restoration of old buildings using a down draught Beehive kiln which dates back to 1937. The works is one of only two brickworks in Essex still making hand-made bricks in 2009.

CARTER AND WOODGATE
WEST THURROCK BRICKWORKS   NH11
Gibbs & Co Ltd, by 1878 until 1893   TQ 595775

A brickworks located north east of St. Clement's Church, West Thurrock. It was established by Gibbs & Co Ltd and was connected to the tramway system of the nearby Thames Portland Cement Works. The brickworks closed c/1907.

ALFRED COOK & JOHN EPHRAIM POTTER
JUBILEE BRICKFIELD, Thorpe Bay   JH12
TQ 921854

A brickworks at Thorpe Bay operated by Cook and Potter and listed in the quarry directory for 1933 as operated by Alfred Cook. The works is shown on the OS map for 1898 with a tramway to a wharf of the River Thames where barges were used to transport bricks.

DAGENHAM (THAMES) DOCK CO
DAGENHAM BREACH   PH13

The River Thames broke through its banks in 1707, flooding a large area of the low-lying Dagenham Levels. The breach was eventually closed, leaving a large lake which was known as "The Gulf". Messrs Rigby, the Holyhead Harbour contractors were involved and Sir John Rennie directed the work. A standard gauge railway was built from a junction with the London Tilbury & Southend Railway to a timber pier on the west end of the former Breach. At this point the money ran out; further Acts of 1866 and 1870 were ineffective and the railway became derelict. The property was purchased by Samuel Williams, owner of a successful lighterage business, in 1877.

DIX, GREEN & CO
NORTH ESSEX PORTLAND CEMENT WORKS, near Saffron Walden   FH14
Works TL 548374 Clay Pit 558375

A brick works was started in 1877 about half a mile east of the Saffron Walden – Thaxted road, along a lane to Bears Hall. In a year or so the concern failed and was taken over by D. Dix of Saffron Walden, who re-organised it for cement production. A brick-built cement kiln was built at the west end of the lane and clay was brought from a pit near the brickworks by means of a tramway which ran in the lane. Dix, Green & Co, a partnership between Dick Dix and Richard Crafton Green, was dissolved 1/1/1896, a later partnership of Joseph Emmanuel Dix, William Richard Dix, and Andrew Robert Dix was dissolved 31/12/1904 when Andrew Robert Dix left the partnership, the partners were still operating the works in 1905, but closed down in 1907; in c1908 W. Bell & Sons took over. It is believed that the last owners were Ray's Cement Co Ltd. The works and tramway were closed between 1927 and 1930. In c1920 the tramway consisted of a track of about 2ft gauge on wooden sleepers, with four wheeled, drop sided wagons pulled by ex – army mules which could haul two wagons, a load of about two tons. Later steel side tipping wagons were used. There was a passing loop midway along the track, some of which were said to be buried still under the road in 1973.

EASTWOODS
SOUTH SHOEBURY BRICKFIELDS, Shoeburyness   JH15
John Francis Eastwood trading as **Eastwoods**   TQ 944857
Josiah Jackson trading as **J. Jackson & Co** until mid 1890s
These brickfields were on land which was part of Cherry Tree Farm at South Shoebury, the land acquired by agreement of 28/2/1895. The brickfield is shown on the OS map for 1898 with a tramway between the clay pits and the kiln.

EASTWOODS LTD
SHORE FIELD BRICKWORKS, Shoeburyness   JH16
Eastwood & Co Ltd, until 1920   TQ 944850
John Francis Eastwood trading as **Eastwoods**, until 6/8/1902
Josiah Jackson trading as **J. Jackson & Co** until mid 1890s
Dale Knapping c 1850s until 1878
Brickworks established c1850s by Dale Knapping and known as Shore Field. The brickfields were located along the shoreline at Shoeburyness and had wooden jetties down the beach. Barges would lie either side of these jetties to be loaded. The brickworks had a narrow gauge tramway which extended in a north north westerly direction to earth pits at TQ 939856 & 937862 as shown on the OS map for 1874. Tramways also ran down to the jetties and used wooden wagons hauled by horses. By 1897 the tramway had been diverted to the later Model Brickfield at TQ 937854.

ESSEX BRICK COMPANY LTD
Bridgemarsh Island   GH17
Company registered 10/7/1866   TQ 888970
Brickworks established in 1866 on Bridgemarsh Island; the works had a short tramway from the brickfield to a pier on the River Crouch as shown on the 25in OS map for 1875, company wound up 28/9/1869.

EVERETTS BRICKWORKS LTD
LAND LANE BRICKWORKS, Colchester   HH18
  TM 002257
Brickworks in existence from the 1870's later operated by Henry Everett and joined by his son John Everett in 1908 when they traded as Henry Everett & Son. Everetts Brickworks Ltd was formed in 1954 and went into voluntary liquidation 16/8/1971. The Kiln held 200,000 bricks and produced about 2,000,000 bricks a year, the site extended to about 60 acres and had a narrow gauge tramway to the clay pit; works closed in 1966.

JOSEPH FELS
MAYLAND SMALLHOLDINGS, Mayland   GH19
  TL 918020
A Tomato enterprise started about 1905 by American soap millionaire Joseph Fels, (said to be connected with the Fels Naptha Soap Co). A colony of smallholdings was constructed which employed the French Gardening technique and a dock known as Pigeon Dock was constructed in a creek off the River Blackwater. A light railway was laid down to facilitate transport of stable manure from London brought in by barge to the dock. The barges would

then take tomatoes to the nearest point, Althorne station on the GER Southminster branch, the enterprise failed and the railway and dock were abandoned.

FINGRINGHOE BRICK & TILE WORKS LTD
FINGRINGHOE BRICKWORKS   HH20
Company registered 1/3/1930
R.G. Ward & Co Ltd, registered 6/2/1919 until 1930   TM 043207
Colne Brick & Terra-Cotta Co Ltd, 31/12/1918 until 24/8/1927
Benjamin S. Barton, by 1859 until 1869
Joseph Barton & Sons, until late 1850s

Company with an address at 5 Laurence Pountney Hill, London, EC4 whose directors are listed as Robert George Ward and Clifford George Washington Ward. Works taken over on lease from R.G. Ward & Co Ltd by agreement dated 20/3/1930, the leasehold premises held for residue of a term of twenty one years commencing 1919; lease dated 5/5/1919 made between Thames Sand Dredging Co Ltd & Colne Brick & Terra Cotta Co Ltd. Company went into receivership 28/2/1938. The Colne Brick & Terra Cotta Co Ltd also had an address at 5 Laurence Pountney Hill and whose directors were John Henry West (director of the Thames Sand Dredging Co) and Robert George Ward (an Ilford Builders Merchant), a letter dated 9/7/1925 stated that the company had not traded for two years and the works closed down, liquidator appointed 24/8/1927. The 25in OS map for 1923 shows the brickworks adjacent to Ballast Quay Farm with a tramway within the works and sand pit which ran to Fingringhoe Quay on the River Colne, sand pits later taken over by Thames Sand Marketing Co Ltd .

W. & C. FRENCH LTD

HALLSFORD BRIDGE BRICKWORKS, High Ongar   EH21
Earlier F.M. Noble & Son Ltd   TL 562024
Frederic Miller Noble trading as Frederic Noble & Son, c1905 until 19/6/1924

Brickworks situated east of the road between Hallsford Bridge and High Ongar and in operation from c1905 until c1975 or later. The works had a tramway; track and wagons from the works were sold early c1940s to the Bulmer Brick & Tile Company.

RAY LODGE BRICKFIELDS, Woodford   DH22
TQ 415920

Brick Works north of Snakes Lane operated in 1850s by J.B. James. Subsequent operators were Edward Williams (early 1880s), Williams and Harris (late 1880s), Williams and Wells (partnership dissolved 26/3/1892), William French (late 1890s), William and Charles French (c1900 to c1930). The works produced hand made yellow stock bricks and red place bricks which were fired in a clamp. A tramway of about 2ft was used to transport clay from the pits, when the works temporarily closed during the First World War all the trucks were chained together and padlocked.

GRAYS CHALK QUARRIES CO LTD
SEABROOKE'S BRICKWORKS, Grays **NH23**
TQ 620780

Brickworks operated by Thomas Seabrooke from 1806 to c1835 when George Henry Errington took over until the mid 1840s. Then it became Meeson & Co and later the steam brickworks of Grays Chalk Quarries Co Ltd. The works was connected by a tramway to Meesons tramway which ran from a chalk quarry at TQ 612783 to Grays Wharf TQ 614774 and is shown on the OS map for 1873, tramway lifted by 1898.

HAROLD WOOD BRICK CO LTD
CHURCH ROAD BRICKWORKS, Harold Wood **EH24**
Company registered 20/4/1928
George King trading as **Harold Wood Brick Co** until c1904 TQ 552910
Alfred Rutley trading as **Harold Wood Brick Co** until c1894
Harold Wood Brickfield Co c1878 until 1887

Brickworks established c1878 by John Compton trading as Harold Wood Brickfield Co. An article in the British Clayworker for June 1931 shows this works to have a narrow gauge tramway and a cable operated incline from the works to the clay pit. The works covered an area of 16½ acres with standard gauge sidings connecting to the GER at Harold Wood Station, the company went into voluntary liquidation 5/5/1933. An advert by Henry Butcher & Co offered the works for sale by tender in 12/1933.

W.H. HOWLETT & CO LTD
FOBBING HORSE ISLAND RUBBISH SHOOT **KH25**
Company registered 30/9/1925 TQ 741844

Rubbish shoot in use from c/1922, about 600 tons of refuse was received each week in barges from Bermondsey which were unloaded at a wharf operated by a steam grab crane, the refuse being loaded into jubilee wagons which were man handled along a light track to the tipping face. In 9/1924 the firm employed twelve men at the site.

LANGWITH METAL FINISHERS
RAWRETH WORKS, Rawreth Industrial Estate, Rawreth **GH26**
TQ 795927

A 2ft 7½in gauge hand-worked line around 30 yards in length running from the yard into the shot-blasting area. The single wagon is converted from a WW2 bomb carrier, with rims partially removed to form flanges. Currently in regular use at time of writing (2010).

LONDON BRICK CO LTD
GREAT BENTLEY BRICKWORKS HH27
John Cathles Hill, trading as London Brick Co, until 1900 TM 112181
Arthur Solomon Went, c1880 until c1898

Brickworks located at Dines Farm east of Flag Creek and established c1880 by Arthur Solomon Went. The works on 70 acres of land had two Hoffmann kilns and two downdraft kilns; the chimney for the Hoffmann kilns was about 150ft high. The engine house contained two steam engines of high and low pressure built by Cole, Marchand & Morley and were of 120 hp. Tramways were used within the works; and a wharf on Flag Creek was used by barges for the transportation of bricks. The brick works which had been for sale since 1907 was sold in 1916 and closed down.

LOWER HOPE DEVELOPMENT CO LTD
MUCKING GRAVEL PITS KH28
Company registered 26/2/1930
Subsidiary of London & Thames Haven Oil Wharves. TQ 690813

Gravel pits ¼ mile east of Mucking village at Cabborns Farm and shown on the OS map for 1915. A narrow gauge tramway (presumably horse-worked) ran south-east for 400 yards to Stanford-le-Hope Wharf and was listed in HMSO list of quarries for 1931 and 1934 but not thereafter, the site was later taken over by Stuart (Thamesmouth) Sand & Shingle Co Ltd. The company went into voluntary liquidation 29/9/1950.

MILTON HALL (SOUTHEND) BRICK CO LTD
PRITTLEWELL BRICKWORKS JH29
Milton Hall Brick Co TQ 884863

Brickworks situated off Milton Street and Sutton Road and shown on the OS map for 1897 with a tramway and a standard gauge rail connection to Southend Station goods yard. Milton Hall (Southend) Brick Co Ltd was registered 24/5/1878 to take over the brickfields of John Ephraim Potter & James Hodges (established 1872). The original workings were on 18 acres of Brick Clay land, part of the Milton Hall Estate at Sutton Road, Prittlewell. The works produced yellow london stock bricks also red facing bricks on a small scale, production of these ceased in 1911 and the works were closed early in 1931 when the brick earth deposits were exhausted; the area eventually redeveloped for housing.

MINERS SAFETY EXPLOSIVES CO LTD
CURRY MARSH WORKS, Thames Haven KH30
Incorporated 1/5/1888 TQ 713813

The O.S. map surveyed in 1895 shows a tramway to a pier on the river from this factory. By 1919 when the map was revised, another tramway had been laid to the pier, these lines appear to be of narrow gauge, possibly hand or horse worked; no use of locomotives known. Company went into voluntary liquidation 1/1/1945.

OIKOS STORAGE LTD
HAVEN ROAD TANK FARM, Canvey Island
JH31
TQ 774821

Oil tank farm with two hand-worked jetty railways, both of 2ft 0in gauge and around 200 yards long. The main jetty has a long-wheelbase flat wagon which is used in connection with pipeline maintenance. The second jetty formerly had a plated Hudson flat wagon but the line is now disused.

PARAGON BRICK WORKS LTD
WEELEY BRICKWORKS, Weeley
HH32
TM 142217

Company registered 14/11/1936 to purchase the brick works owned by Clifford White. The brick works on 20 acres of land had five Scotch kilns each with a capacity of 50,000 bricks; the works had three pug mills a drying shed and engine & machine sheds together with tramways & tip wagons and a private siding to the GER goods yard at Weeley station. The company went into voluntary liquidation 21/10/1937.

PATTRICKS BRICKWORKS
LOWER DOVERCOURT BRICKWORKS
HH33
TM 252316

John Pattrick & Son until 1890s
John Pattrick until 1872
Pattrick & Daniels c1850 until 1858

Brick and Cement works at Lower Dovercourt with a tramway which ran to a quay on the river known as Phoenix Dock. The firm closed down by 1906 when the works and quay were sold to Groom & Sons, timber merchants.

PRIZEMAN & CO LTD
LITTLE MUSSELS RUBBISH SHOOT, Pitsea
KH34
Prizeman & Son until 13/4/1923
TQ 741851

Company formed to take over the business of a lighterman, wharfinger, barge & tug owner and general contractor carried on by Charles James Nelson Prizeman under style of Prizeman & Son. The company had a rubbish shoot, in operation in 9/1924, about 200 yards north of Fobbing Horse Island which received refuse in barges from Poplar and West Ham. The wharf had two grab cranes and used a horse to haul wagons to the tipping face. The company also operated a rubbish shoot at Barking; liquidator appointed 22/4/1925.

GEORGE HENRY RAYNER
MAIDEN LEY BRICKWORKS, Castle Hedingham
FH35
William John Rayner, until 1934
TL 785343
William Rayner & Son 1893 until 1905

Brickworks established in 1893 by William Rayner & Son and shown on the OS map for 1922 with a standard gauge siding to the CV&HR and a narrow gauge tramway to the clay pit; the works had six up draught Scotch kilns; works closed 1952. Derelict remains of tram lines and a pug mill still survived in 1997.

RETTENDON ROOFING TILE WORKS LTD
EAST HANNINGFIELD BRICK, TILE & PIPE WORKS
GH36
TQ 754983

Brick, tile & pipe works located near Rettendon Common established by Henry Rayner c1880. Subsequent operators were Mrs Susannah Davison Rayner (1894-5/1901), Hedley T. Rayner & Charles H. Moger trading as Hedley T. Rayner & Co, (5/1901-3/1903), Hedley T. Rayner trading as Hedley T. Rayner & Co, (3/1903-1905), D.B. McBride, (1905-c1909), Frederick Sparrow Wilkinson, (c1909-1919). In 5/1905 there was a serious fire at the works which destroyed the drying sheds and engine house. By 1906 the Rettendon Tile Works of which D.B. McBride was proprietor comprised 40 acres with two downdraft kilns, one was rectangular holding 75,000 tiles, the other circular holding 35,000 tiles. By 1933 there were at least three down draught rectangular kilns. The works had a tramway which ran to the clay pits and used bogie wagons to transport clay, the company became a subsidiary of the Southend Estates Group from 1945 until the works were sold in 1955.

D. & C. RUTTER LTD
LANDWICK BRICKWORKS, Shoeburyness
JH37
D.& C. Rutter c1890 until 27/4/1905
TQ 964882

Brickworks established c1890 by Daniel and Charles Rutter who operated a brick works at Hillingdon in Middlesex. The brickworks had a tramway to a wharf on Havengore Creek where barges were loaded with bricks, works closed c1914. Company struck off 30/1/1934.

SAW MILLS COMPANY LTD
COOKS ROAD SAW MILLS, Stratford
CH38
TQ 378832

Saw Mills and timber yard in Cooks Road, Stratford on the banks of Bow Creek. The company were wooden box and bottle box makers and wine case and packing case manufacturers. The OS map for 1916 shows the timber yard with an internal narrow gauge tramway. Company listed in directories until 1963, dissolved 30/9/1992.

SIBLE HEDINGHAM RED BRICK CO LTD
SIBLE HEDINGHAM BRICKWORKS
FH39
Company registered 22/12/1919
TL 790349

Brickworks effectively owned by Rippers Ltd who had purchased the former Highfield, Purls Hill and Sidings Brick Works which were conveyed to the Sible Hedingham Red Brick Co Ltd in 1920 and thereafter continued as one brick works. It is shown on the 1923 OS map with a standard gauge siding to the works and a narrow gauge tramway from a sand pit in the adjacent Purlshill Plantation to a tipping dock on the siding connected to the CV&HR. The initial directors were Reuben Hunt, Eli Cornish, Harry Tucker Ripper & William Charlton Ripper, production ceased in 1954.

THAMES BRICK & TILE WORKS
CROWSTONE BRICKWORKS, Southend

JH40
TQ 862853

Brickworks on the banks of the River Thames in operation in the 1890s and shown on the OS map for 1898 with two tramways to jetties on the river. An auction of brickmaking plant and machinery took place in 3/1897.

THUNDERSLEY BRICKFIELD LTD
MANOR BRICKWORKS, Thundersley

GH41
TQ 776893

Company registered 24/11/1933
Manor Brick & Tile Co c1921 until 1926
Manor Brickworks Ltd until 8/1/1915 when dissolved
Manor Brickworks 10/1898 until 22/12/1900

Brickworks established 10/1898 by Geo Stuck, taken over by John Nash in 4/1899 assisted in the promotion of the company by John Hughes Ellis & Frederick Charles Spanswick. The works covered 7½ acres, by 1900 his output was 100,000 bricks; in 1905 a disastrous fire destroyed buildings and machinery and 300,000 unfinished tiles. The works had by 1926 a 700 ft narrow gauge tramway between the clay pits and the works which closed in 1936. The company went into voluntary liquidation 6/10/1936.

TOLLESBURY BRICK, TILE AND DEVELOPMENT COMPANY LTD
TOLLESBURY BRICKWORKS

GH42
TL 959106 & TL 965105

Incorporated 26/7/1905

Brickworks established by George Henry Wombwell at Tollesbury, the brickworks was situated on 14 acres of land and included a 16hp compound Engine & Boiler by Davey Paxman, a patent "Eclipse All" brickmaking & pressing machine, wagons and a tramway, wash mills and a kiln. The tramway connected with a siding on the GER Tollesbury branch the works also had a dock where barges could be loaded. The company had ceased trading by 1919 and was dissolved 11/10/1927.

TUDOR BRICK & TILE CO LTD
HULLBRIDGE BRICKWORKS,

GH43
TQ 814955

Ames & Hunter, until 1912
Alfred Hobman, 1899 until 1904

Brickworks established c1899 by Alfred Hobman. The works had five kilns and a site which extended to 65 acres including a tramway system with a timber built jetty on the River Crouch for loading barges. An auction of the brick works and plant took place in 1904. Subsequent owners were R.Y. Ames and R.W. Hunter trading as Ames and Hunter until 1912, when the Tudor Brick & Tile Co Ltd was incorporated to take over the brick and tile business of Ames and Hunter who became directors of the new company which was dissolved 15/12/1916.

J.J. WAGSTAFF
ROYAL ALBERT BRICKFIELDS, Eastwood, Rayleigh   JH44
TQ 836890

Brickworks established by James John Wagstaff and in operation from 1898–1910. The works had an Osman continuous kiln with an initial output of about 10,000 bricks per day and by 1902 over one million a year. The brick works employed about 60 men and extended to 9¾ acres and latterly operated an Osman continuous a Crystal Palace and a Suffolk kiln. An advert for the Wolff Dryer Co in The British Clayworker of April 1899 shows the brickworks with a horse drawn tramway.

WAR DEPARTMENT

DENGIE FLAT   GH45
TM 051037

OS 1in Seventh Series map revised 1952-55, shows a tramway leaving the shoreline at TM 036052 and extending out onto the flats for about 300 yards where it splits into two branches heading north east and south east, each about 1 mile long. It is not known what the tramway was used for but possibly by the War Department for targets. The line was still present c/1963 with one flat wagon at the landward end, presumably hand-worked.

DUNTON INTERMEDIATE AMMUNITION DEPOT, Dunton, near Laindon   LH46
TQ 652871

An ammunition depot located between West Horndon and Laindon on the south side of the LT&SR line to Southend. The depot was connected by a short branch line with a run round loop to a fan of five sidings which ran between three rows of storage magazines, use of a locomotive unknown. Depot closed and demolished and land restored to agricultural use.

EAST BEACH, Shoeburyness   JH47

Barrage Balloon factory dating from c1914 and served by a 2ft 0in gauge tramway. All buildings removed and site has been used for boat and car parking for many years. A length of track 150 yards long with 9 pairs of points remains set in concrete.

FINGRINGHOE RANGES, Fingringhoe   HH48
TM 027188

Rifle Ranges at Fingringhoe Camp on Fingringhoe Marsh with a 2ft gauge tramway which ran from a point near South House Farm. The line ran in a southerly direction in an inverted 'Y' formation to TM 024183 with a branch to TM 029180. The line was cable hauled and used Hudson skip frames on which targets were mounted. The line is shown on the 25in OS map for 1960 and was still in use in 3/1967.

PURFLEET RIFLE RANGES, Aveley Marshes   LH49
TQ 548796

Tramways ran from a point adjacent to the LT&SR at TQ 548796 extending across the ranges to TQ 539792; also tramways on the banks of the Mar Dyke at Purfleet Garrison and Barracks used in connection with gunpowder magazines. Construction of five magazines started about 1760 and had a capacity of 10,400 barrels, the line on the west bank extended to a jetty on the Thames built about 1890. No.5 magazine is now preserved.

RAINHAM RIFLE RANGES, Rainham Marshes   LH50
TQ 521818

A tramway ran from a point near Rainham Station in a southerly direction across the ranges on Rainham Marshes.

TILBURY FORT, Tilbury   MH51
TQ 650753

The artillery fort was begun in 1672 under Charles II. The fort's artillery was designed to stop warships getting up river to London and in 1716 two magazines were built to store large quantities of gunpowder. The fort accommodated and supplied troops destined for the trenches in the First World War and remained in military hands until 1950. The fort had an extensive narrow gauge system of approximately 18in gauge which ran to a jetty on the Thames. The system was hand worked and was used to transport stores; some track remains in situ at time of writing.

ROBERT WARNER & CO (ENGINEERS) LTD
WALTON ON THE NAZE FOUNDRY   HH52
Robert Warner & Co   TM 257228

An iron foundry listed in directories by 1874 and shown on the OS map for that year as having a tramway from the works to a wharf on the nearby dock known as Port Said (later renamed Foundry Dock). The works was later closed and the dock filled in.

W.H. WHITBREAD
PURFLEET QUARRIES   NH53
TQ 557782

Quarries with a plateway of L section iron rails, probably 3ft 6in gauge, had branches to two quarries, early maps suggest it was double track. The line, which was horse worked, ran from a wharf on the Thames to chalk pits and lime kilns. It was opened by 1807 and possibly as early as 1805. In 1826 the pits were operated by Meeson Hinton & Co, later by Gibbs & Co Ltd. Said to be the first tramway in Essex.

GEORGE WHITE
WICKFORD BRICKWORKS   GH54
TQ 740942

A brick works in existence from c/1906 to c/1908, with a narrow gauge tramway to Wickford Station Yard and shown on the 25in OS map for 1922, further details unknown.

W.W. WILSON
STANFORD-LE-HOPE SAND PITS   KH55
TQ 681816

Sand pits located in Butts Lane, Stanford-le-Hope ceased production at the beginning of WW2. An auction of plant took place at St Clere's Hall on 13/5/1941 which included eighty 2ft gauge steel side tipping wagons and 1½ miles of light railway track, use of locomotives unknown; site possibly that formerly worked by Dobson Ellis & Co .

YORK INTERNATIONAL LTD
BASILDON FACTORY, Gardiner's Lane South, Basildon   KH56
TQ 723905

A 3ft 0in gauge line running around the inside of the factory building for 150 yards and used for conveying air conditioning units. Haulage was by a battery-electric hand controlled and steered traction unit. Flat wagons had flangeless wheels on one side and double-flanged wheels on the other. The line was constructed c1987. Factory closed by end of 2006.

LOCOMOTIVE INDEX

NOTES :   Information normally relates to the locomotive as built.
Column 1   Works Number (or original company running number for locomotives built in main line workshops without a works number).
Column 2   Date ex-works where known - this may be a later year than the year of building or the year recorded on the worksplate.
Column 3   Gauge.
Column 4   Wheel arrangement.

	Steam Locomotives :	Diesel Locomotives :
Column 5	Cylinder position	Horse power
Column 6	Cylinder size	Engine type #
Column 7	Driving wheel diameter	Weight in working order
Column 8	Either weight in working order and/or Manufacturers type designation	Page references
Column 9	Page references	

 # Manufacturers of petrol and diesel engines :

Ailsa	- Ailsa Craig Ltd, Salfords, Redhill, Surrey
Allen	- W.H. Allen Sons & Co Ltd, Queens Engineering Works, Bedford
Baguley	- Baguley Engineers Ltd, Burton-on-Trent, Staffordshire
Blackstone	- Blackstone & Co Ltd, Rutland Engine Works, Stamford, Lincolnshire
Caterpillar	- Caterpillar Tractor Co Ltd, Desford, Leicestershire
Cummins	- Cummins Engine Co Ltd, Shotts, Lanarkshire
Deutz	- Motorenfabrik Deutz AG, Kóln, Germany
Dorman	- W.H.Dorman & Co Ltd, Tixall Rd, Stafford
EE	- English Electric Co Ltd
Ford	- Ford Motor Co Ltd, Dearborn, Michigan, USA
Fowler	- John Fowler & Co (Leeds) Ltd, Hunslet, Leeds
Gardner	- L. Gardner & Sons Ltd, Barton Hall Engine Works, Patricroft
JAP	- J.A.Prestwich Industries Ltd, Northumberland Park, Tottenham, London
Leyland	- Leyland Motors Ltd, Leyland, Lancashire
Lister	- R.A. Lister & Co Ltd, Dursley, Gloucestershire
MAN	- Maschinenfabrik Augsurg-Nürnberg AG, Germany
McLaren	- J. & H. McLaren Ltd, Midland Engineering Works, Leeds
Meadows	- Henry Meadows Ltd, Fallings Park Engine Works, Wolverhampton
Morris	- Morris Engines Ltd, Gosford Street, Coventry
National	- National Gas & Oil Engine Co Ltd, Ashton-under-Lyne, Lancashire
O&K	- Orenstein & Koppel AG, Berlin & Nordhausen, Germany
Paxman	- Davey, Paxman & Co Ltd, Colchester, Essex
Perkins	- Perkins Engine Co Ltd, Peterborough, Northants
R-R	- Rolls-Royce Ltd, Oil Engine Division, Shrewsbury
Ruston	- Ruston & Hornsby Ltd, Lincoln
Saurer	- Armstron Saurer Ltd, Newcastle upon Tyne
Thornycroft	- John I. Thornycroft & Co Ltd, Basingstoke, Hampshire.
White & Poppe	- White & Poppe Ltd, Lockhurst Lane, Coventry, Warwickshire

ANDREW BARCLAY, SONS & CO LTD, Kilmarnock — AB

Works	Date	Gauge	Type		Cylinders	Wheels	Refs
185	30.5.1877	4ft 8½in	0-4-0ST	OC	10 x 17	3ft 0in	212
193	3.4.1878	4ft 8½in	0-4-0ST	OC	12 x 20	3ft 6in	132
262	4.2.1884	4ft 8½in	0-4-0ST	OC	11 x 18	3ft 3in	134
266	21.3.1884	4ft 8½in	0-4-0ST	OC	11 x 18	3ft 3in	65,138
279	19.2.1885	4ft 8½in	0-4-0ST	OC	11 x 18	3ft 3in	212
282	25.3.1886	4ft 8½in	0-4-0ST	OC	11 x 18	3ft 3in	168
636	29.8.1889	4ft 8½in	0-4-0ST	OC	11 x 18	3ft 0in	134
680	13.12.1890	4ft 8½in	0-4-0ST	OC	9 x 18	2ft 9in	53,96
681	13.12.1890	4ft 8½in	0-4-0ST	OC	9 x 18	2ft 9in	53
699	22.7.1891	4ft 8½in	0-4-0ST	OC	11 x 18	3ft 0in	168
714	24.3.1892	4ft 8½in	0-4-0ST	OC	12 x 20	3ft 2in	60
730	18.4.1893	4ft 8½in	0-4-0ST	OC	10 x 18	3ft 0in	137
747	21.11.1894	4ft 8½in	0-4-0ST	OC	12 x 20	3ft 2in	168,182
757	16.4.1895	4ft 8½in	0-4-0ST	OC	10 x 18	3ft 0in	137
761	3.5.1895	3ft 0in	0-4-0T	OC	6 x 12	1ft 10in	213
859	28.2.1900	4ft 8½in	0-4-0ST	OC	14 x 22	3ft 5in	89
887	23.2.1901	4ft 8½in	0-4-0ST	OC	12 x 20	3ft 2in	212,272
891	28.2.1901	4ft 8½in	0-4-0ST	OC	10 x 18	3ft 0in	212,272
957	4.2.1903	4ft 8½in	0-4-0ST	OC	10 x 18	3ft 0in	134
967	29.4.1903	4ft 8½in	0-4-0ST	OC	10 x 18	3ft 0in	80
992	20.8.1904	4ft 8½in	0-4-0ST	OC	10 x 18	3ft 0in	79,218,254,255
1018	23.1.1905	4ft 8½in	0-4-0ST	OC	12 x 20	3ft 2in	247
1047	8.9.1905	4ft 8½in	0-4-0ST	OC	14 x 22	3ft 5in	288
1129	17.10.1907	4ft 8½in	0-4-0ST	OC	15 x 22	3ft 5in	185,268
1158	7.7.1909	4ft 8½in	0-6-0ST	OC	14 x 22	3ft 5in	72
1236	12.3.1911	4ft 8½in	0-6-0T	OC	16 x 24	3ft 9in	143,147
1237	22.3.1911	4ft 8½in	0-6-0T	OC	16 x 24	3ft 9in	143,147
1238	3.4.1911	4ft 8½in	0-6-0T	OC	16 x 24	3ft 9in	147
1281	18.5.1912	4ft 8½in	0-4-0ST	OC	14 x 22	3ft 6in	60,168,182,275
1290	29.11.1912	4ft 8½in	0-4-0ST	OC	14 x 22	3ft 5in	71
1294	30.7.1912	4ft 8½in	0-6-0T	OC	16 x 24	3ft 9in	143
1297	31.5.1915	4ft 8½in	0-6-0ST	OC	12 x 20	3ft 2in	70,119
1300	13.9.1912	4ft 8½in	0-6-0T	OC	16 x 24	3ft 9in	143
1301	21.9.1912	4ft 8½in	0-6-0T	OC	16 x 24	3ft 9in	143
1302	21.9.1912	4ft 8½in	0-6-0T	OC	16 x 24	3ft 9in	143,147
1309	19.3.1913	4ft 8½in	0-4-0ST	OC	12 x 20	3ft 2in	76
1391	31.5.1915	4ft 8½in	0-4-0ST	OC	12 x 20	3ft 2in	59
1411	1.10.1915	4ft 8½in	0-4-0ST	OC	12 x 20	3ft 2in	59
1437	1.4.1916	4ft 8½in	0-4-0F	OC	15 x 20	3ft 0in	160
1471	14.6.1916	4ft 8½in	0-4-0F	OC	15 x 18	3ft 0in	160
1472	22.6.1916	4ft 8½in	0-4-0F	OC	15 x 18	3ft 0in	113,149
1492	27.12.1916	4ft 8½in	0-4-0F	OC	15 x 20	3ft 0in	175
1493	27.12.1916	4ft 8½in	0-4-0F	OC	15 x 20	3ft 0in	175
1551	11.10.1917	4ft 8½in	0-6-0F	OC	14½ x 18	3ft 0in	113
1553	13.11.1917	4ft 8½in	0-6-0F	OC	14½ x 18	3ft 0in	113
1574	14.3.1918	4ft 8½in	0-6-0F	OC	14½ x 18	3ft 0in	49
1577	12.4.1918	4ft 8½in	0-6-0ST	OC	12 x 20	3ft 2in	150,275
1603	10.7.1918	4ft 8½in	0-4-0F	OC	15 x 18	3ft 0in	85
1619	15.3.1919	4ft 8½in	0-4-0ST	OC	16 x 24	3ft 7in	293
1652	23.12.1919	4ft 8½in	0-4-0ST	OC	10 x 18	3ft 0in	80

1666	13.4.1920	4ft 8½in	0-4-0ST	OC	12 x 20	3ft2in		65,138,267
1674	13.4.1920	4ft 8½in	0-4-0ST	OC	12 x 20	3ft2in		138
1711	20.4.1921	4ft 8½in	0-4-0ST	OC	12 x 20	3ft2in		172
1720	3.2.1921	4ft 8½in	0-4-0T	OC	12 x 18	3ft0in		134
1721	3.2.1921	4ft 8½in	0-4-0T	OC	12 x 18	3ft0in		134
1722	3.2.1921	4ft 8½in	0-4-0T	OC	12 x 18	3ft0in		134
1741	3.6.1921	3ft 9in	0-4-0ST	OC	12 x 20	3ft2in		60
1816	13.11.1923	4ft 8½in	0-4-0ST	OC	12 x 20	3ft2in		60
1844	26.8.1924	4ft 8½in	0-6-0ST	OC	12 x 20	3ft2in		70,187,231,259
1865	15.3.1926	4ft 8½in	0-4-0ST	OC	12 x 20	3ft2in		288
1870	13.9.1925	4ft 8½in	0-4-0F	OC	12 x 18	3ft0in		84,85
1895	29.7.1926	4ft 8½in	0-4-0ST	OC	8¼ x 14	3ft0in		80
1927	4.2.1927	4ft 8½in	0-4-0ST	OC	14 x 22	3ft5in		60
1940	10.10.1927	4ft 8½in	0-4-0ST	OC	12 x 20	3ft2in		150,168,182
1979	8.1.1930	4ft 8½in	0-4-0ST	OC	14 x 22	3ft5in		89
1993	15.4.1932	4ft 8½in	0-4-0ST	OC	12 x 20	3ft2in		67,231
2028	29.1.1937	4ft 8½in	0-4-0ST	OC	14 x 22	3ft5in		103
2157	20.4.1943	4ft 8½in	0-4-0ST	OC	14 x 22	3ft5in		293
2167	23.8.1946	4ft 8½in	0-4-0ST	OC	10 x 18	3ft0in		80
2182	9.1.1945	4ft 8½in	0-6-0ST	IC	18 x 26	4ft3in		121,182
2184	14.3.1945	4ft 8½in	0-6-0ST	IC	18 x 26	4ft3in		122
2199	3.5.1945	4ft 8½in	0-4-0ST	OC	12 x 20	3ft2in		286
2350	11.1.1954	4ft 8½in	0-6-0ST	OC	15 x 22	3ft5in		288
2353	6.9.1954	4ft 8½in	0-4-0ST	OC	14 x 22	3ft5in		68,233
333	24.9.1938	4ft 8½in	0-4-0DM	150hp	Paxman-Ricardo 8cyl			288
349	25.1.1941	4ft 8½in	0-4-0DM	153hp	Gardner 6L3	14½T		286
354	7.7.1941	4ft 8½in	0-4-0DM	153hp	Gardner 6L3	21T		123
368	11.5.1945	4ft 8½in	0-4-0DM	153hp	Gardner 6L3	21T		122
370	22.6.1945	4ft 8½in	0-4-0DM	153hp	Gardner 6L3	21T		124
371	20.9.1945	4ft 8½in	0-4-0DM	153hp	Gardner 6L3	21T		122
372	23.11.1945	4ft 8½in	0-4-0DM	153hp	Gardner 6L3	21T		122
395	17.1.1956	4ft 8½in	0-4-0DM	153hp	Gardner 6L3	25T		57,176
419	2.8.1957	4ft 8½in	0-4-0DM	102hp	Gardner 4L3			121,182,262
506	15.11.1965	4ft 8½in	0-8-0DH	504hp	2xR-R C8NFL			58
506/1	1969	4ft 8½in	0-4-0DH	{rebuild of AB 506}				58
506/2	1969	4ft 8½in	0-4-0DH	{rebuild of AB 506}				58

A.C. CARS LTD, Thames Ditton, Surrey   A.C. Cars

	1949	3ft 6in	4wRER		292,294,296,298

AVONSIDE ENGINE CO LTD, Bristol   AE

804	1871	7ft 0in	2-4-0T	IC	9 x 16	3ft0in	85
1016	1875	4ft 8½in	0-6-0ST	IC	17 x 24	4ft6in	233
1460	7.1903	4ft 8½in	0-6-0ST	OC	14 x 20	3ft3in	89,187
1505	1906	4ft 8½in	0-6-0ST	OC	14 X 20	3ft3in	117
1518	1907	4ft 8½in	0-6-0ST	OC	14 X 20	3ft3in	70
1578	1910	4ft 8½in	0-4-0ST	OC	12 x 18	3ft0in	75,76
1610	1912	4ft 8½in	0-4-0ST	OC	14 x 20	3ft1in	206,275
1631	1912	4ft 8½in	0-4-0ST	OC	12 x 18	3ft0in	275

1655	1913	4ft 8½in	0-6-0ST	OC	14 x 20	3ft3in		275
1668	1913	3ft 6in	0-4-0T	OC	8½x12	1ft8in		110
1672	1914	4ft 8½in	0-6-0ST	OC	14 x 20	3ft3in	B3	58,117,247
1702	1915	4ft 8½in	0-4-0ST	OC	10 x 16	2ft9in		80
1748	1916	1ft 6in	0-4-0T	OC	8½ x 12	2ft1in	'2814'	298
1763	1917	4ft 8½in	0-6-0ST	OC	14½x20	3ft3in		178
1771	1917	4ft 8½in	0-6-0ST	OC	14 x 20	3ft3in		58,106
1819	1918	4ft 8½in	0-6-0ST	OC	14½x20	3ft3in		118,277
1874	1921	4ft 8½in	0-4-0ST	OC	12 x 18	2ft11in		206,274
1875	1921	4ft 8½in	0-4-0ST	OC	12 x 18	2ft11in		206,274,286,288
1876	1921	4ft 8½in	0-4-0ST	OC	12 x 18	2ft11in		274
1877	1921	4ft 8½in	0-4-0ST	OC	12 x 18	2ft11in		172,274
1878	1921	4ft 8½in	0-4-0ST	OC	12 x 18	2ft11in		274
1879	1921	4ft 8½in	0-4-0ST	OC	12 x 18	2ft11in		274
1880	1921	4ft 8½in	0-4-0ST	OC	12 x 18	2ft11in		274
1881	1921	4ft 8½in	0-4-0ST	OC	12 x 18	2ft11in		275
2002	1931	4ft 8½in	0-4-0ST	OC	14½x20	3ft6in		268
2003	1931	4ft 8½in	0-4-0ST	OC	14½x20	3ft6in		268
2037	1931	4ft 8½in	0-4-0ST	OC	12 x 20	3ft1in		115,231
2068	22.5.1933	4ft 8½in	0-6-0ST	OC	15 x 20	3ft6in		288,294,295

ALAN KEEF LTD, Ross-on-Wye, Herefords — AK

10	26.8.1983	2ft 0in	4wDH	K30	Lister ST2	3½T	63,300
20R	6.6.1986	2ft 0in	4wDH		Deutz F3L912	3½T	63,300
26	22.9.1988	2ft 0in	4wDM	40S	Deutz F3L912	3½T	63,64
28	19.6.1989	2ft 0in	4wDM	40S	Deutz F3L912	3½T	63
40SD530	6.8.1987	2ft 0in	4wDM	40S	Deutz F3L912	3½T	63

AVELING & PORTER LTD, INVICTA WORKS, Rochester, Kent — AP

151	7.1865	4ft 8½in	4wWT	G	10hp	96
167	11.1865	4ft 8½in	4wWT	G	10hp	96
212	6.1866	4ft 8½in	4wWT	G	10hp	96
508	8.1869	4ft 8½in	4wWT	G	10hp	96
524	13.1.1870	4ft 8½in	4wWT	G	6hp	53
1121	8.1875	4ft 8½in	4wWT	G	10hp	172
1602	4.1880	4ft 8½in	2-2-0WT	G	8hp	53
1780	8.1882	4ft 8½in	2-2-0WT	G	8hp	53
3888	4.1887	4ft 8½in	4wWT	G		75

ATKINSON WALKER WAGGONS LTD, FRENCHWOOD WORKS, Preston, Lancashire — Atw

102	1927	4ft 8½in	4wVBT	VCG	7x10	3ft0in	Class A	275
110	1928	4ft 8½in	6wVBT	VCG			Class D	50
117	1930	4ft 8½in	4wVBT	VCG			Class B	193

AUSTRO-DAIMLER, Wiener Naustadt, Austria — Austro-Daimler

		2ft 0in	4wPM					72,174,207,279

SIR W.G.ARMSTRONG, WHITWORTH & CO (ENGINEERS) LTD, Newcastle-upon-Tyne AW

D23	1933	4ft 8½in	0-4-0DE	85hp	Saurer 6BLD		118,137

BARCLAYS & CO, Kilmarnock B

272	1883	3ft 6in	0-4-0ST	OC	218

ARCHIBALD BAIRD & SON, Hamilton, Glasgow Baird

334		3ft 0in	0-4-0ST	213
335		3ft 0in	0-4-0ST	213

BALMFORTH BROS, PEEL INGS FOUNDRY, Rodley, Yorks Balmforth

	4ft 8½in	0-4-0VBT	OC 7 x	198

BURY CURTIS & KENNEDY, CLARENCE FOUNDRY, Liverpool BCK

	26.10.1845	4ft 8½in	0-4-0	IC	15 x 20 5ft0in	214

BAGULEY-DREWRY LTD, Burton-on-Trent, Staffordshire BD

3706	16.6.1975	4ft 8½in	4wDHR	75hp	Perkins 4236	6T6C	127,288
3745	18.11.1976	4ft 8½in	4wDHR	75hp	Perkins 4236	6T6C	127
3755	18.3.1982	2ft 6in	4wDH	60hp	Perkins 4236	6.5T	298

BRUSH ELECTRICAL ENGINEERING CO LTD, Loughborough, Leicestershire BE

	1898	2ft 8½in	4wRER		298
276	1898	4ft 8½in	0-4-0ST	OC	96,109
306	1904	4ft 8½in	0-4-0ST	OC	96

GAS LIGHT & COKE CO, Beckton Gas Works Beckton

1	1902	4ft 8½in	0-4-0T	OC	10 x 18 2ft10in	134
2	1902	4ft 8½in	0-4-0T	OC	10 x 18 2ft10in	134

BRITISH ELECTRICAL VEHICLES LTD, Southport BEV

194	10.3.1920	2ft 6in	4wBE	Type No1	2T	102
393	30.6.1922	2ft 6in	4wBE	Type No2	2¾T	102
622	4.11.1925	2ft 6in	4wBE	Type No2	2¾T	102

E.E. BAGULEY LTD, Burton-on-Trent, Staffs Bg

1386	30.9.1924	4ft 8½in	0-4-0PM	25/30hp	Baguley	4T16C	67
1387	30.9.1924	4ft 8½in	0-4-0PM	25/30hp	Baguley	4T16C	67
1805	31.3.1932	1ft 8in	4wPM	25hp	Ford 4cyl		250
1837	6.1937	4ft 8½in	2w-2DMR	27hp	Perkins 'Wolf'	3T4C	125
1895	18.1.1950	4ft 8½in	2w-2PMR	12/24hp	Morris	1T8C	126,289
2081	25.1.1934	3ft 6in	4wPM	25hp	Ford A	3T10C	110

2201	30.12.1943	4ft 8½in	4wPM	30hp	Baguley	6T		126
2395	2.1953	2ft 0in	0-6-0DM	150hp	Gardner 8LW	17½T		264
3024	1939	2ft 0in	0-4-0PM	24hp	Ford	BB		299
3227	14.3.1951	4ft 8½in	0-4-0DM	150hp	Paxman 6RW	22T		103
3232	31.3.1947	2ft 0in	0-4-0DM	34hp	Perkins P4			284,295
3589	9.11.1962	4ft 8½in	0-4-0DM	112hp	Gardner 6LW	20T		71

BAGULEY CARS LTD, Burton-on-Trent, Staffs   BgC

566	5.7.1916	4ft 8½in	0-4-0PM	60hp	White & Poppe	7T		113
702	19.4.1918	4ft 8½in	0-4-0PM	60hp	White & Poppe			52
790	22.7.1920	4ft 8½in	2w-2PMR	25hp	Baguley	3T16C		134
1023	29.7.1918	4ft 8½in	2w-2PMR	10hp	Baguley	1T4C		125
1335	5.10.1923	4ft 8½in	2w-2PMR	25hp	Baguley	3T16C		134

BAGULEY (ENGINEERS) LTD, Burton-on-Trent, Staffs   BgE

1795	1930	3ft 6in	4w-4wRE					299
2046	1930	2ft 0in	4wPM	25hp	Ford 4cyl A	3T		249
2051	1931	2ft 0in	4wPM	25hp	Ford 4cyl A	3T		91

BLACK, HAWTHORN & CO LTD, Gateshead   BH

order dates

496	5.1879	4ft 8½in	0-6-0ST	12 x 19	3ft2in	254,255
758	27.8.1883	4ft 8½in	0-6-0ST	12 x 19	3ft2in	247
774	10.10.1883	4ft 8½in	0-4-0ST	13 x 19	3ft2in	138
864	29.3.1886	4ft 8½in	0-4-0T	10 x 18	2ft9in	134
865	29.3.1886	4ft 8½in	0-4-0ST	10 x 18	2ft9in	134
889	14.9.1887	4ft 8½in	0-4-0ST	8 x 14	2ft6in	196
1038	9.6.1881	4ft 8½in	0-4-0ST	12 x 19	3ft2in	75,76
1105	8.10.1894	4ft 8½in	0-6-0ST	14 x 20	3ft7in	243

W.J. BASSETT LOWKE LTD, Northampton   BL

| 18 | 1911 | 1ft 3in | 4-4-2 | OC | 296 |
| 30 | 1912 | 1ft 3in | 4-4-2 | OC | 296 |

BALDWIN LOCOMOTIVE WORKS, Philadelphia, USA   BLW

45285	3.1917	4ft 8½in	0-4-0ST	OC	197,272
46407	9.1917	4ft 8½in	0-6-0PT	OC	118
46489	9.1917	4ft 8½in	0-6-0PT	OC	118
46956	11.1917	4ft 8½in	0-6-0PT	OC	118

BEYER, PEACOCK & CO LTD, GORTON FOUNDRY, Manchester   BP

| 2467 | 1885 | 4ft 8½in | | 17 x 24 | 5ft6in | 117 |

BRAITHWAITE MILNER & CO, London   Braithwaite

| | 1838 | 5ft 0in | 0-4-0 | IC | 201 |
| | 1839 | 5ft 0in | 0-4-0 | IC | 202 |

BRIGHTON WORKS, Sussex Bton
    c1865 4ft 8½in 0-6-0ST IC   111,220

BROWN & TAWSE (TUBES) LTD, WEST HORNDON WORKS, Essex
                                                                            Brown & Tawse
                4ft 8½in   4wPM   61
                4ft 8½in   4wDM   155

BRITISH THOMSON-HOUSTON CO LTD, Rugby, Warwickshire BTH
    1932 4ft 8½in Bo-BoDE 150hp Allen 6cyl 44T   89
        frames and bodywork supplied by Metropolitan-Vickers of Sheffield

CALEDON MOTORS LTD, GLASGOW Caledon
        4ft 8½in 4wPM   125

NEI MINING EQUIPMENT LTD, CLAYTON EQUIPMENT, Hatton, Derbys CE
Originally **CLARKE CHAPMAN LTD.**
Note that, to save space, batches of locomotives are indexed as single entries. For example, 5792A-D refers to locomotives 5792A, 5792B, 5792C and 5792D.

5339	5.1967	2ft 0in	4wBE			4T	204
5431	1.1968	1ft 6in	4wBE	7hp		1¾T	228,259
5667	9.1969	2ft 0in	4wBE			3½T	295
5827	10.1970	1ft 6in	4wBE	7hp		1¾T	217
Two locos built under 5827							
5858	1.1971	1ft 6in	4wBE	7hp		1¾T	259
5911A	1.1972	1ft 6in	4wBE	7hp		1¾T	199
5920	3.1972	1ft 6in	4wBE	7hp		1¾T	217
Two locos built under 5920							
5940A	6.1972	2ft 0in	4wBE	7hp		1¾T	251
5940B	6.1972	2ft 0in	4wBE	7hp		1¾T	199,251
5940C	7.1972	2ft 0in	4wBE	7hp		1¾T	251
5940D	7.1972	2ft 0in	4wBE	7hp		1¾T	251
5961A	11.1972	2ft 0in	4wBE	7hp		1¾T	251
5965B	1.1973	1ft 6in	4wBE	7hp		1¾T	202,229
B0109A	3.1973	1ft 6in	4wBE	7hp		1¾T	228
B0109B	3.1973	1ft 6in	4wBE	7hp		1¾T	228
B0483	12.1975	762mm	4wBEF	25hp		5½T	130
Three locos built under B0483							
B1534B	8.1977	2ft 0in	4wBE	25hp		5½T	238
B2200B	10.1979	1ft 6in	4wBE	7hp		1¾T	256
B3482A	1988	2ft 6in	4wBEF	17½hp	Pony		298
B4253	1998		{reb of B0483 TERMINATOR}				130
B4408	c11.2004		{reb of B0483 TERMINATOR}				130
B4408	4.7.2005		{reb of B0483 RAMBO}				130
B4427A	20.2.2006	4ft 8½in	4wBE	150hp		15T	194
B4427B	20.2.2006	4ft 8½in	4wBE	150hp		15T	194
B4427C	1.4.2006	4ft 8½in	4wBE	150hp		15T	194
B4427D	1.4.2006	4ft 8½in	4wBE	150hp		15T	194

CHAPMAN & FURNEAUX LTD, Gateshead — CF

1151	1897	3ft3$^{3}/_{8}$in	0-4-2WT	OC	7 x 12	2ft3in	218
1164	1898	4ft 8½in	0-4-0ST	OC	12 x 19	2ft10in	53,254

CHANCE MANUFACTURING CO INC, Wichita, Kansas, U.S.A. — Chance

76-50145-24	1976	2ft 0in	4w-2-4wPH S/O	291
76-50148-24	1976	2ft 0in	4-2-4DH S/O	284

ALEXANDER CHAPLIN & CO LTD, CRANTONHILL WORKS, Glasgow — Chaplin

140	9.3.1860	4ft 8½in	0-4-0VBT	VCG	12hp	6 x 13	96
188	2.3.1861	4ft 8½in	0-4-0VBT	VCG	21hp	8 x 14	96
1675	27.4.1874	4ft 8½in	0-4-0VBT	VCG	9hp	5¼x11	133
1756	29.9.1874	4ft 8½in	0-4-0VBT	VCG	9hp	5¼x11	133
1757	15.10.1874	4ft 8½in	0-4-0VBT	VCG	9hp	5¼x11	133

HENRY COLEY, WEST LONDON IRONWORKS — Coley

	1854	4ft 8½in	2-2-0WT	IC 6 x 12	96

CREWE WORKS, Cheshire (LNWR / LMSR / BR) — Crewe

		4ft 8½in	0-6-0T					118
[LNWR 1911]	1847	4ft 8½in	2-4-0T	OC	15 x 20	5ft0in		72,111
[LNWR 1819]	1849	4ft 8½in	2-4-0T	OC	15 x 20	5ft0in		72,111
[LNWR 1927]	1849	4ft 8½in	2-4-0T	OC	15 x 20	5ft0in		72,111
2726	1884	4ft 8½in	2-4-2T	IC	17 x 20	4ft6in		117
[08588]	1959	4ft 8½in	0-6-0DE	350hp	EE 6KT		48T	190

W. R. CUNIS, WOOLWICH, London — Cunis

	1963	4ft 8½in	0-4-0DH	{Rebuild of AB 2167}	80

DÜBS & CO, GLASGOW LOCOMOTIVE WORKS, Glasgow — D

1438	1881	4ft 8½in	0-6-0ST	OC	15 x 20	3ft6in	111,142
1439	1881	4ft 8½in	0-6-0ST	OC	15 x 20	3ft6in	111,142
1440	1881	4ft 8½in	0-6-0ST	OC	15 x 20	3ft6in	111,142
1441	1881	4ft 8½in	0-6-0ST	OC	15 x 20	3ft6in	111,142
3404	1896	4ft 8½in	4-4-0	OC	19 x 24	6ft3in	130

DICK & STEVENSON, AIRDRIE ENGINE WORKS, Bell Street, Airdrie — D&S

58	c1875	4ft 8½in	0-4-0ST	OC	246

DARLINGTON WORKS, Co Durham (NER) — Dar

548	1908	4ft 8½in	4-4-0	IC	19 x 26	6ft10in	130
[D3476]	1957	4ft 8½in	0-6-0DE	350hp	Blackstone ER6T		286

DAVENPORT LOCOMOTIVE WORKS, Davenport, USA — Dav

2534	1943	4ft 8½in	0-6-0T	OC	16½x24 4ft6in	146
2547	1943	4ft 8½in	0-6-0T	OC	16½x24 4ft6in	146

DREWRY CAR CO LTD, London (Suppliers only) — DC

1837	1937	see Bg 1837	125
1895	1950	see Bg 1895	126,289
2081	1934	see Bg 2081	110
2161	1941	see EE 1192	85
2175	1945	see VF 5256	123
2180	1945	see VF 5261	282,293
2201	1943	see Bg 2201	126
2251	1948	see VF D77	54,282
2252	1949	see VF D78	293
2269	1949	see VF D98	75
2270	1949	see VF D99	75
2395	1953	see Bg 2395	264
2566	1955	see VF D293	75,291
2567	1955	see VF D294	75,76
2583	1956	see VF D297	151
2589	1957	see RSHN 7921	151
2606	1957	see RSHD 7892	89
2611	1958	see RSHD 7897	89,288
2656	1960	see RSHD 8097	288
2657	1960	see RSHD 8098	89,288
2706	1961	see RSHD 8184	293
2714	1961	see RSHD 8192	89

DERBY LOCOMOTIVE WORKS — Derby

[2791]	1872	4ft 8½in	0-6-0	IC			117
[WD 878]	1945	4ft 8½in	0-6-0DE	350hp	EE 6KT	48T	124
[D3255]	6.1956	4ft 8½in	0-6-0DE	350hp	Blackstone ER6T		286
[08423]	8.1958	4ft 8½in	0-6-0DE	350hp	EE 6KT	48T	58

DICK, KERR & CO LTD, BRITTANIA ENGINEERING WORKS, Kilmarnock — DK

	c1887	3ft 6in	0-4-0ST		110
	1889	4ft 8½in	0-4-0ST	OC	275
6899	1901	4ft 8½in	0-4-0ST	OC	88
9976	1901	4ft 8½in	0-4-0ST	OC	197

DONCASTER WORKS — Don

[2051]	1.1959	4ft 8½in	0-6-0DM	204hp	Gardner 8L3	31T	89
[03081]	1.1960	4ft 8½in	0-6-0DM	204hp	Gardner 8L3	31T	293

MOTORENFABRIK DEUTZ AG, Koln, Germany — Dtz

57871	1965	4ft 8½in	4wDH	Type KG 230B	191

E. BORROWS & SONS, St Helens, Lancashire — EB

10	1880	4ft 8½in	0-4-0WT	OC	102

ENGLISH ELECTRIC CO LTD, DICK, KERR WORKS, Preston — EE

687	1925	4ft 8½in	4wBE	84hp			118
785	1930	4ft 8½in	4w-4wBE				118
807	1932	2ft 0in	2w-2-2-2wRER				284
809	1931	2ft 0in	2w-2-2-2wRER				284
1192	4.9.1941	4ft 8½in	0-4-0DM	153hp	Gardner 6L3	22½T	85

ENGLISH ELECTRIC CO LTD, VULCAN WORKS, Newton-le-Willows — EEV

3582	1966	4ft 8½in	Bo-BoDE/RE			190
3670	1967	4ft 8½in	Bo-BoDE			190
D911	1964	4ft 8½in	0-6-0DH	455hp	Dorman 8QAT	123
D1122	1966	4ft 8½in	0-4-0DH	305hp	Cummins NHRS 6B 1	182
D1124	1966	4ft 8½in	0-4-0DH	305hp	Cummins NHRS 6B 1	89
D1229	1967	4ft 8½in	0-6-0DH	380hp	Dorman 6QT	89
E352	see 3582					

ELECTROMOBILE LTD, PROSPECT WORKS, Otley, Yorkshire — Electromobile

W247	1928	4ft 8½in	0-4-0BE	118,276
	1927	4ft 8½in	4wBER	125
	1921		4wBE	163

WILLIAM JONES LTD, CHARLTON WORKS, London (Suppliers) — Excelsior

	2ft 0in	4wPM	83

FALCON ENGINE & CAR WORKS LTD, Loughborough — FE

150	1887	4ft 8½in	0-4-0T	OC	52
211	1892	4ft 8½in	0-4-0T	OC	52
		4ft 8½in	0-4-0ST	OC	111,143,219

F.C. HIBBERD & CO LTD, Park Royal, London — FH

1636	1930	2ft 0in	4wPM	6hp			74
1645	12.1929	2ft 0in	4wPM	20hp			250
1656	1.1930	2ft 0in	4wPM	20hp			250
1677	5.1930	2ft 0in	4wPM	40hp			198
1678	6.1930	2ft 0in	4wPM	40hp			198
1682	3.1931	2ft 0in	4wPM	10hp			165
1697	8.1930	2ft 0in	4wPM	16/25hp	Meadows EH	2½T	211
1708	21.10.1932	2ft 0in	4wDM	20hp	Lister 18/2	4T	164
1709	10.1930	2ft 0in	4wPM	20hp	Dorman 2JO		55
1731	3.1931	2ft 0in	4wPM	20hp	Dorman 2JO		217
1737	14.4.1931	2ft 0in	4wPM	16/25hp	Meadows EH	4T	183
1738	1.1931	1ft 11⅝in	4wPM	16/25hp	Meadows EH	4T	149
1763	24.6.1931	2ft 0in	4wPM	20hp	Dorman 2JOR		55

1771	31.8.1931	2ft 0in	4wPM	16/25hp	Meadows 4EH	4T		183
1779	27.10.1931	2ft 0in	4wPM	40hp	Dorman 4JO	6T		198
1830	6.12.1933	60cm	4wPM	8hp	Ford Y	1½T		154
1844	1.1934	4ft 8½in	4wDM	40hp	Blackstone BPVG4	8T		115
1853	13.3.1934	4ft 8½in	4wDM	30hp	National 3D	6T		57
1862	6.1934	4ft 8½in	4wDM	30hp	Fordson	5½T		89
1889	10.1934	2ft 0in	4wPM	20hp	Dorman 2JO	2½T		55
1894	11.1934	2ft 0in	4wPM	20hp	Dorman 2JO	2½T		55
1977	7.1936	4ft 8½in	4wDM	55hp	Paxman 5RQ	13T		57
2102	27.5.1938	4ft 8½in	4wDM	70hp	Paxman 6RQT	18T		57
2198	7.1939	2ft 0in	4wDM	20hp	National 2D	2½T		208
2199	7.1939	2ft 0in	4wDM	20hp	National 2D	2½T		208
2280	3.1940	2ft 0in	4wDM	20hp	National 2D	2½T		208
2281	3.1940	2ft 0in	4wDM	20hp	National 2D	2½T		208
2282	4.1940	2ft 0in	4wDM	20hp	National 2D	2½T		208
2283	5.1940	2ft 0in	4wDM	20hp	National 2D	2½T		208
2286	7.1940	2ft 0in	4wDM	20hp	National 2D	2½T		208
2287	7.1940	2ft 0in	4wDM	20hp	National 2D	2½T		208
2416	8.1941	1ft 11in	4wDM	20hp	National 2D	2½T		114
2496	12.8.1941	4ft 8½in	4wDM	50hp	Dorman 3DL	9T		49
2834	3.1943	2ft 0in	4wDM	20hp				158
2914	9.1944	4ft 8½in	4wDM			6T		141
3075	10.1945	1ft 11in	4wDM			1½T		154
3147	31.3.1947	4ft 8½in	4wDM		National DA	11T		65,286
3264	25.11.1947	4ft 8½in	4wDM	20hp	National 2D			141
3294	19.7.1948	4ft 8½in	4wDM		National DA4	11T		67,288,294
								298
3307	15.7.1948	2ft 0in	4wDM		Perkins PG	6T		139
3317	28.5.1948	2ft 0in	4wPM		Ford C	1½T		115
3491	1951	4ft 8½in	4wDM		Dorman 3DL	11T		154
3596	9.2.1953	4ft 8½in	4wDM	48hp	Dorman 3DL			174,291
3641	2.7.1953	4ft 8½in	4wDM		Dorman 4DL	18T		180,182,269
3700	20.4.1955	4ft 8½in	4wDM		Dorman 3DL	11T		208
3709	12.4.1954	2ft 0in	4wDM		Lister 18/2	3¼T		222
3722	20.1.1955	4ft 8½in	4wDM		Dorman 6DL	23T		185
3736	17.2.1955	4ft 8½in	4wDM		Dorman 6DL	23T		71,80
3768	13.10.1955	4ft 8½in	4wDM		Dorman 6DL	23T		185
3777	29.8.1956	4ft 8½in	4wDM		Foden FD6			286
3787	28.9.1955	2ft 0in	4wDM		Lister 18/2	3¼T		51
3791	28.3.1956	4ft 8½in	4wDM		Dorman 4DL	18T		80
3799	16.4.1956	4ft 8½in	4wDM		Dorman 6DL	23T		185
3813	27.9.1956	4ft 8½in	4wDM		Dorman 6DL	23T		185
3821	31.8.1956	4ft 8½in	4wDM		Dorman 4DL	18T		80
3885	29.9.1958	4ft 8½in	4wDM		Dorman 6DL	23T		80
3889	30.9.1958	4ft 8½in	4wDM		Dorman 6DL	23T		135
3900	1959	4ft 8½in	4wDM					180
3907	20.6.1959	4ft 8½in	4wDM		Dorman 6DL	23T		135
3908	20.6.1959	4ft 8½in	4wDM		Dorman 6DL	23T		135
3909	20.6.1959	4ft 8½in	4wDM		Dorman 6DL	23T		135
3910	21.9.1959	4ft 8½in	4wDM		Dorman 6DL	23T		80,135
3911	30.9.1959	4ft 8½in	4wDM		Dorman 6DL	23T		135

3912	20.10.1959	4ft 8½in	4wDM	Dorman 6DL	23T		135
3945	24.10.1960	4ft 8½in	4wDM	Dorman 6LC	24T		185
3949	1960	4ft 8½in	4wDH				185
3959	25.3.1961	4ft 8½in	4wDM	Dorman 6DL			135
3960	27.3.1961	4ft 8½in	4wDM	Dorman 6DL			135
3961	30.3.1961	4ft 8½in	4wDM	Dorman 6DL			135
3983	20.3.1962	2ft 0in	4wDM	Perkins 3152			300
3994	31.8.1962	4ft 8½in	4wDM	Dorman 6LC	23T		80,138
3997	22.11.1962	4ft 8½in	4wDM	Ford 590E			185
4008	19.4.1963	2ft 0in	4wDM	Lister FR1	1½T		158

FLETCHER JENNINGS & CO, LOWCA ENGINE WORKS, Whitehaven — FJ

168	1879	4ft 8½in	0-4-0T	OC	8 x 16	59

FOX, WALKER & CO LTD, ATLAS ENGINE WORKS, Bristol — FW

149		1872	4ft 8½in	0-6-0ST	OC	13 x 20 3ft6in	111
160		1872	4ft 8½in	0-4-0ST	OC	10 x 18 2ft8in	103
263		1875	4ft 8½in	0-6-0ST	OC	13 x 20 3ft6in	111,143
288	30.11.1875		4ft 8½in	0-6-0ST	OC	13 x 20 3ft6in	111
343	22.8.1877		4ft 8½in	0-6-0ST	OC	13 x 20 3ft6in	210
358		1878	4ft 8½in	0-6-0ST	OC	13 x 20 3ft6in	293

GREENWOOD & BATLEY LTD, Armley, Leeds — GB

1602	22.12.1938	2ft 0in	4wBE	5hp		239
1668	19.4.1940	1ft 6in	4wBE	5hp		131
1669	19.4.1940	1ft 6in	4wBE	5hp		131
1670	19.4.1940	1ft 6in	4wBE	5hp		131
1671	19.4.1940	1ft 6in	4wBE	5hp		131
1672	19.4.1940	1ft 6in	4wBE	5hp		131
1673	19.4.1940	1ft 6in	4wBE	5hp		131
1851	16.11.1942	1ft 6in	4wBE	5hp		131
1852	16.11.1942	1ft 6in	4wBE	5hp		131
1861	30.11.1942	1ft 6in	4wBE	5hp		131
1862	30.11.1942	1ft 6in	4wBE	5hp		131
3586	1948	2ft 6in	4wBE			262
6099	7.4.1964	3ft 0in	2w-2DE	84hp	Ruston 6YDA	298
420461/21	1981	2ft 0in	2w-2-2-2wRER		{see HE 9120}	284

GEORGE ENGLAND & CO LTD, HATCHAM IRONWORKS, London — GE

	4ft 8½in	2-4-0T	OC	11 x 16	236,237

GREENSBURG MACHINE CO, Greensburg, Pennsylvania, USA — Greensburg

1ft 6in	4wBE	199

GILKES WILSON & CO, TEESSIDE ENGINE WORKS, Middlesborough — GW

160	6.1863	4ft 8½in	0-4-0T	195

JAMES & FREDK HOWARD LTD, BRITANNIA IRONWORKS, Bedford — H

939	26.10.1928	3ft 0in	4wPM	25hp	Dorman 4JUL	3T	65
940	19.7.1928	2ft 0in	4wPM	31hp	Dorman 4JUL	4T	98
964	31.8.1929	3ft 0in	4wPM	25hp	Dorman 4JU	3T	65
985	20.3.1931	50cm	4wPM	20hp	Morris 1M	3T	61

HUDSWELL, CLARKE & CO LTD, RAILWAY FOUNDRY, Leeds — HC

195	4.11.1879	4ft 8½in	0-4-0ST	OC	8 x 16	2ft 4in	246
220	16.6.1882	4ft 8½in	0-4-0ST	OC	10 x 16	2ft 9in	246
237	28.3.1883	4ft 8½in	0-6-0ST	OC	13 x 20	3ft 3in	247
287	12.11.1887	4ft 8½in	0-4-0ST	OC	10 x 16	2ft 9in	134
309	4.10.1888	4ft 8½in	0-4-0ST	OC	10 x 16	2ft 9in	250
313	27.3.1888	4ft 8½in	0-6-0ST	IC	13 x 20	3ft 3in	206
402	3.7.1893	4ft 8½in	0-4-0ST	OC	10 x 16	2ft 9in	70,150,254,258
435	11.4.1895	4ft 8½in	0-4-0ST	OC	10 x 16	2ft 9in	79,258
440	27.3.1896	4ft 8½in	0-6-0ST	OC	13 x 20	3ft 3in	70,258
442	2.9.1895	4ft 8½in	0-6-0ST	IC	12 x 18	3ft 0in	254,255
477	8.12.1897	4ft 8½in	0-4-0ST	OC	10 x 16	2ft 9in	243
522	29.3.1899	4ft 8½in	0-4-0ST	OC	9 x 15	2ft 9in	135
532	18.12.1899	4ft 8½in	0-4-0ST	OC	10 x 16	2ft 9in	224
534	31.1.1900	4ft 8½in	0-4-0ST	OC	10 x 16	2ft 8in	224
560	20.3.1900	4ft 8½in	0-6-0ST	IC	13 x 20	3ft 3in	243
604	10.2.1902	4ft 8½in	0-4-0ST	OC	11 x 16	2ft 9½in	72,168
653	4.5.1903	4ft 8½in	0-6-0ST	IC	12 x 18	3ft 1in	224
656	17.7.1903	4ft 8½in	0-4-0ST	OC	10 x 16	2ft 9½in	70,258
657	11.9.1903	4ft 8½in	0-4-0ST	OC	10 x 16	2ft 9½in	134
664	8.3.1905	4ft 8½in	0-6-0ST	IC	12 x 18	3ft 1in	224
696	31.3.1904	3ft 0in	0-4-0ST	OC	8 x 12	2ft 0½in	77,78,159
805	10.9.1907	4ft 8½in	0-4-0	OC	12 x 16	3ft 6in	72
823	4.6.1908	4ft 8½in	0-6-0ST	OC	14 x 20	3ft 7in	150,277
833	18.12.1910	4ft 8½in	0-6-0ST	IC	12 x 18	3ft 1in	243
845	26.5.1909	4ft 8½in	0-6-0ST	IC	14 x 20	3ft 3½in	119
857	31.8.1908	4ft 8½in	0-4-0ST	OC	8 x 12	2ft 0½in	77
859	30.9.1908	4ft 8½in	0-4-0ST	OC	8 x 12	2ft 0½in	77
888	30.8.1909	4ft 8½in	0-6-0ST	IC	15 x 20	3ft 7in	224
911	14.3.1910	4ft 8½in	0-4-0ST	OC	12 x 18	3ft 1in	76
1011	26.11.1912	4ft 8½in	0-6-0ST	IC	15 x 20	3ft 7in	224
1026	14.4.1913	4ft 8½in	0-6-0ST	IC	15 x 20	3ft 7in	224,226
1028	14.5.1913	4ft 8½in	0-6-0ST	IC	12 x 18	3ft 1in	224
1029	30.5.1913	4ft 8½in	0-6-0ST	IC	12 x 18	3ft 1in	224
1034	7.6.1913	2ft 0in	0-4-0WT	OC	5 x 8	1ft 8in	93
1061	10.3.1914	4ft 8½in	0-6-0ST	IC	12 x 18	3ft 1in	187
1101	8.2.1915	4ft 8½in	0-6-0T	OC	16 x 24	3ft 9in	143,147
1102	22.2.1915	4ft 8½in	0-6-0T	OC	16 x 24	3ft 9in	147
1103	15.3.1915	4ft 8½in	0-6-0T	OC	16 x 24	3ft 9in	143,147

1132	13.6.1916	2ft 0in	0-4-0WT	OC	6 x 9	1ft 8in	72,236
1153	20.9.1915	4ft 8½in	0-6-0T	OC	16 x 24	3ft 9in	143
1154	27.9.1915	4ft 8½in	0-6-0T	OC	16 x 24	3ft 9in	143,147
1155	30.9.1915	4ft 8½in	0-6-0T	OC	16 x 24	3ft 9in	143
1164	24.2.1915	4ft 8½in	0-6-0T	OC	16 x 24	3ft 9in	117
1170	4.3.1918	2ft 0in	0-4-0WT	OC	5 x 8	1ft 8in	176
1174	2.11.1915	4ft 8½in	0-4-0ST	OC	8 x 12	2ft 0½in	78
1183	8.7.1918	2ft 0in	0-4-0WT	OC	5 x 8	1ft 8in	266
1206	15.3.1916	4ft 8½in	0-4-0ST	OC	10 x 16	2ft 9½in	77,78
1244	23.4.1917	4ft 8½in	0-6-0T	OC	16 x 24	3ft 9in	143,147
1245	3.5.1917	4ft 8½in	0-6-0T	OC	16 x 24	3ft 9in	143,147
1254	25.6.1917	4ft 8½in	0-6-0T	OC	16 x 24	3ft 9in	143,147
1255	27.7.1917	4ft 8½in	0-6-0T	OC	16 x 24	3ft 9in	143
1318	30.11.1918	60cm	0-6-0WT	OC	6½ x 12	1ft 11in	179
1319	30.11.1918	60cm	0-6-0WT	OC	6½ x 12	1ft 11in	179
1323	24.4.1918	4ft 8½in	0-6-0T	OC	16 x 24	3ft 9in	143
1324	14.10.1918	4ft 8½in	0-6-0T	OC	16 x 24	3ft 9in	143
1326	28.2.1919	4ft 8½in	0-4-0ST	OC	12 x 18	3ft 1in	76
1336	31.5.1919	2ft 6in	0-4-0T	OC	7 x 10	2ft 0in	281
1337	29.4.1918	4ft 8½in	0-4-0ST	OC	14 x 20	3ft 3½in	180,272
1373	28.2.1919	60cm	0-6-0WT	OC	6½ x 12	1ft 11in	179
1374	28.2.1919	60cm	0-6-0WT	OC	6½ x 12	1ft 11in	179
1375	9.4.1919	60cm	0-6-0WT	OC	6½ x 12	1ft 11in	179
1376	15.4.1919	60cm	0-6-0WT	OC	6½ x 12	1ft 11in	179
1414	5.11.1920	4ft 8½in	0-6-0T	OC	16 x 24	3ft 9in	143,147
1415	25.11.1920	4ft 8½in	0-6-0T	OC	16 x 24	3ft 9in	143
1429	17.12.1920	4ft 8½in	0-6-0ST	IC	13 x 20	3ft 3½in	61,273
1442	23.3.1921	4ft 8½in	0-4-0ST	OC	8 x 12	2ft 0½in	77,78,262
1453	20.12.1921	4ft 8½in	0-6-0T	OC	16 x 24	3ft 9in	143
1454	20.12.1921	4ft 8½in	0-6-0T	OC	16 x 24	3ft 9in	143,147
1455	29.12.1921	4ft 8½in	0-6-0T	OC	16 x 24	3ft 9in	143
1466	29.9.1921	4ft 8½in	0-6-0ST	IC	13 x 20	3ft 3½in	273
1494	18.9.1923	4ft 8½in	0-6-0ST	IC	12 x 18	3ft 1in	224
1508	3.4.1924	4ft 8½in	0-6-0ST	OC	14 x 20	3ft 3½in	89
1510	26.10.1923	4ft 8½in	0-4-0ST	OC	12 x 18	3ft 1in	182
1511	14.12.1923	4ft 8½in	0-6-0ST	IC	12 x 18	3ft 1in	224
1513	20.3.1924	4ft 8½in	0-6-0ST	IC	13 x 20	3ft 3½in	224
1524	30.6.1924	4ft 8½in	0-6-0T	IC	18 x 24	4ft 0in	224
1525	11.4.1924	4ft 8½in	0-6-0ST	IC	12 x 18	3ft 1½in	224
1526	26.4.1924	4ft 8½in	0-6-0ST	IC	12 x 18	3ft 1½in	72,185,224
1528	23.5.1924	4ft 8½in	0-6-0ST	IC	12 x 18	3ft 3½in	224
1529	2.6.1924	4ft 8½in	0-6-0ST	IC	13 x 20	3ft 3½in	119
1534	29.5.1924	3ft 0in	0-4-0ST	OC	9 x 15	2ft 6½in	225
1535	2.7.1924	3ft 0in	0-4-0ST	OC	9 x 15	2ft 6½in	227
1536	8.7.1924	3ft 0in	0-4-0ST	OC	9 x 15	2ft 6½in	225,227
1537	16.7.1924	3ft 0in	0-4-0ST	OC	9 x 15	2ft 6½in	225
1538	16.7.1924	4ft 8½in	0-6-0ST	IC	13 x 20	3ft 3½in	224
1539	23.7.1924	4ft 8½in	0-6-0ST	IC	13 x 20	3ft 3½in	224
1561	30.6.1925	4ft 8½in	0-4-0ST	OC	10 x 16	2ft 9½in	76,77,78
1563	21.7.1925	4ft 8½in	0-4-0ST	OC	10 x 16	2ft 9½in	259
1564	28.7.1925	4ft 8½in	0-4-0ST	OC	10 x 16	2ft 9½in	97,254,259

1583	16.12.1926	4ft 8½in	0-6-0ST	IC	12 x 18	3ft1½in	187,231
1585	11.7.1927	4ft 8½in	0-6-0ST	IC	13 x 20	3ft3½in	224
1586	30.8.1927	4ft 8½in	0-6-0ST	IC	13 x 20	3ft3½in	224
1593	23.8.1927	4ft 8½in	0-6-0ST	OC	14 x 20	3ft7in	187,231
1596	23.7.1927	4ft 8½in	0-6-0T	OC	16 x 24	3ft9in	143
1597	29.8.1927	4ft 8½in	0-6-0T	OC	16 x 24	3ft9in	143
1598	19.9.1927	4ft 8½in	0-6-0T	OC	16 x 24	3ft9in	144
1601	23.7.1927	4ft 8½in	0-6-0ST	IC	12 x 18	3ft1½in	224
1602	24.8.1927	4ft 8½in	0-6-0ST	IC	12 x 18	3ft1½in	224,226
1676	5.4.1937	4ft 8½in	0-6-0ST	IC	13 x 20	3ft3½in	72,185
1719	3.2.1943	4ft 8½in	0-6-0T	OC	16 x 24	3ft9in	144
1720	23.3.1943	4ft 8½in	0-6-0T	OC	16 x 24	3ft9in	144
1725	28.7.1941	4ft 8½in	0-6-0ST	OC	14 x 22	3ft3½in	172
1731	2.11.1942	4ft 8½in	0-6-0T	OC	17 x 24	3ft9in	286
1748	27.9.1943	4ft 8½in	0-6-0ST	IC	18 x 26	4ft3in	144
1762	27.3.1944	4ft 8½in	0-6-0ST	IC	18 x 26	4ft3in	119,182
1873	7.7.1954	4ft 8½in	0-6-0T	OC	16 x 24	3ft9in	144,147
1874	30.7.1954	4ft 8½in	0-6-0T	OC	16 x 24	3ft9in	144,147
P251	9.1925	2ft 0in	4wPM	40hp	Dorman 4cyl		156,281
D610	18.1.1939	4ft 8½in	0-4-0DM	150hp	Davey Paxman 8RWT		61
D611	14.3.1939	1ft 9in	4-6-2DM	32½hp	Dorman 2DL		284
D612	19.4.1939	1ft 9in	4-6-2DM	32½hp	Dorman 2DL		284
D680	2.1949	4ft 8½in	0-6-0DM	200hp	Gardner 8L3		185
D701	13.5.1949	4ft 8½in	0-6-0DM	204hp	Gardner 8L3		185
D702	13.7.1949	4ft 8½in	0-6-0DM	204hp	Gardner 8L3		185
D760	11.5.1951	4ft 8½in	0-6-0DM	200hp	Gardner 8L3		271
D894	30.9.1954	4ft 8½in	0-4-0DM	100hp	Gardner 6LW		271
D915	3.8.1956	4ft 8½in	0-6-0DM	200hp	Gardner 8L3		271
D916	30.8.1956	4ft 8½in	0-6-0DM	200hp	Gardner 8L3		271
D917	10.1956	4ft 8½in	0-6-0DM	200hp	Gardner 8L3		80,271
D918	31.10.1956	4ft 8½in	0-6-0DM	200hp	Gardner 8L3		80,271
D963	16.3.1956	4ft 8½in	0-4-0DM	200hp	Gardner 8L3		271
D964	10.5.1956	4ft 8½in	0-4-0DM	200hp	Gardner 8L3		271
D1009	30.11.1956	4ft 8½in	0-4-0DM	153hp	Gardner 6L3		180
D1153	23.3.1959	4ft 8½in	0-4-0DM	204hp	Gardner 8L3		293
D1291	9.9.1964	4ft 8½in	0-4-0DH	191hp	Cummins NH220		180
D1373	1.1.1966	4ft 8½in	0-6-0DH	260hp	Gardner 8L3B		84,85
D1376	15.4.1966	4ft 8½in	0-6-0DH	307hp	Cummins NHRS6IP		89
D1377	9.2.1966	4ft 8½in	0-6-0DH	307hp	Cummins NHRS6IP		89
D1378	31.5.1966	4ft 8½in	0-6-0DH	307hp	Cummins NHRS6IP		89
D1396	20.4.1967	4ft 8½in	0-6-0DH	307hp	Cummins NHRS6IP		89

HUNSLET ENGINE CO LTD, Hunslet, Leeds   HE

1	18.7.1865	4ft 8½in	0-6-0ST	IC	14 x 18	3ft4in	185,212
4	13.12.1865	4ft 8½in	0-6-0ST	IC	12 x 18	3ft1in	254
45	1.11.1870	4ft 8½in	0-6-0ST	IC	12 x 18	3ft1in	254,255
98	4.7.1873	4ft 8½in	0-4-0ST	OC	10 x 15	2ft9in	59
103	30.6.1874	4ft 8½in	0-4-0ST	OC	10 x 15	2ft9in	133
215	1.12.1879	4ft 8½in	0-4-0ST	OC	10 x 15	2ft9in	70,257,258

223	1.5.1879	4ft 8½in	0-4-0ST	OC	12 x 18	3ft1in	168
229	18.9.1879	4ft 8½in	0-4-0ST	OC	9 x 14	2ft8½in	210
234	26.2.1880	4ft 8½in	0-6-0ST	IC	14 x 18	3ft1in	112,147,214,221
235	26.2.1880	4ft 8½in	0-6-0ST	IC	14 x 18	3ft1in	214,221
288	30.4.1884	4ft 8½in	0-6-0ST	IC	12 x 18	3ft1in	230
343	12.9.1884	4ft 8½in	0-4-0ST	OC	13 x 18	3ft1in	111,219
360	31.12.1884	4ft 8½in	0-4-0ST	OC	9 x 14	2ft8½in	79,212
365	18.9.1885	4ft 8½in	0-4-0ST	OC	10 x 15	2ft9in	210
366	24.4.1886	4ft 8½in	0-4-0ST	OC	10 x 15	2ft9in	210
387	25.8.1886	4ft 8½in	0-4-0ST	OC	9 x 14	2ft8½in	170
388	1.3.1887	4ft 8½in	0-4-0ST	OC	9 x 14	2ft8½in	230
401	8.7.1886	4ft 8½in	0-6-0T	IC	15 x 20	3ft4in	185,212
420	25.5.1887	4ft 8½in	0-4-0ST	OC	10 x 15	2ft9in	243
437	21.12.1887	4ft 8½in	0-6-0ST	IC	14 x 18	3ft1in	212
457	8.6.1888	4ft 8½in	0-6-0ST	IC	13 x 18	3ft1in	255
469	21.12.1888	4ft 8½in	0-6-0ST	IC	13 x 18	3ft1in	293
515	1.7.1890	4ft 8½in	0-4-0ST	OC	12 x 18	3ft1in	89,250
550	2.5.1892	4ft 8½in	0-6-0ST	IC	13 x 18	3ft1in	172,184,243,244
574	20.2.1893	4ft 8½in	0-6-0ST	OC	11 x 15	2ft6in	193
593	19.9.1893	4ft 8½in	0-6-0ST	OC	11 x 15	2ft6in	254,255
620	20.4.1895	4ft 8½in	0-4-0ST	OC	10 x 15	2ft9in	98,243
629	17.12.1895	4ft 8½in	0-4-0ST	OC	10 x 15	2ft10in	59,79,80,159
630	28.6.1895	4ft 8½in	0-6-0T	IC	15 x 20	3ft4in	243
640	5.11.1895	3ft 4in	0-4-0ST	OC	6 x 8	1ft6½in	161
716	12.2.1900	4ft 8½in	0-6-0ST	IC	12 x 18	3ft2½in	242
717	5.3.1900	4ft 8½in	0-6-0ST	IC	12 x 18	3ft2½in	243
718	21.3.1900	4ft 8½in	0-6-0ST	IC	12 x 18	3ft2½in	243
882	19.1.1906	4ft 8½in	0-6-0T	OC	13 x 18	3ft7in	272
1252	9.1.1917	2ft 0in	4-6-0T	OC	9½x12	2ft0in	278
1322	23.11.1918	1ft11⁵/₈	4-6-0T	OC	9½x12	2ft0in	179
1323	2.12.1918	1ft11⁵/₈	4-6-0T	OC	9½x12	2ft0in	179
1335	12.1919	4ft 8½in	0-4-0ST	OC	12 x 18	3ft1in	134
1499	20.9.1920	4ft 8½in	0-6-0ST	IC	14 x 20	3ft4in	70,259
1644	4.11.1929	4ft 8½in	0-6-0ST	OC	14 x 20	3ft4in	172,188,276
1647	15.6.1931	4ft 8½in	0-6-0ST	OC	14 x 20	3ft4in	89,187,231
1648	29.5.1931	4ft 8½in	0-6-0ST	OC	14 x 20	3ft4in	231,232
1686	13.7.1931	4ft 8½in	0-6-0ST	OC	14 x 20	3ft4in	172,188,240
1688	20.7.1931	4ft 8½in	0-6-0ST	OC	14 x 20	3ft4in	172,240
1689	27.7.1931	4ft 8½in	0-6-0ST	OC	14 x 20	3ft4in	188,231
1690	27.7.1931	4ft 8½in	0-6-0ST	OC	14 x 20	3ft4in	184,187
1697	15.5.1933	4ft 8½in	0-4-0DM	150hp McLaren MR6			113
1710	29.9.1932	2ft 0in	4wDM	20hp	Lister CS		166
1711	6.10.1932	2ft 0in	4wDM	20hp	Lister CS		166
1712	6.10.1932	2ft 0in	4wDM	20hp	Lister CS		166
1713	6.10.1932	2ft 0in	4wDM	20hp	Lister CS		166
1714	9.11.1932	2ft 0in	4wDM	20hp	Lister CS		166
1715	9.11.1932	2ft 0in	4wDM	20hp	Lister CS		166
1735	26.10.1933	1ft 6in	4wDM	20hp	Lister 18/2	4T	164
1764	28.12.1934	60cm	4wDM	20hp	Lister CS		281
1846	4.3.1937	4ft 8½in	0-6-0DM	150/170hp Gardner 8L3			119
1872	10.11.1937	4ft 8½in	0-4-0ST	OC	14 x 22		67

1974	17.11.1939	2ft 0in	4wDM	20hp	Ailsa RF2	127
1975	17.11.1939	2ft 0in	4wDM	20hp	Ailsa RF2	127,177
2067	11.6.1940	4ft 8½in	0-4-0DM	153hp	Gardner 6L3	84,85
2387	29.7.1941	4ft 8½in	0-6-0ST	IC 14 x 20 3ft4in		293
2409	20.5.1942	4ft 8½in	0-6-0ST	IC 15 x 20 3ft7in		288
2413	30.12.1941	4ft 8½in	0-6-0ST	IC 18 x 26 4ft0½in		288
2414	27.1.1942	4ft 8½in	0-6-0ST	IC 18 x 26 4ft0½in		144
2591	11.5.1942	60cm	4wDM	20hp	Ailsa	114
2700	13.8.1942	2ft 0in	4wDM	25hp	McLaren	166
2701	13.8.1942	2ft 0in	4wDM	25hp	McLaren	166
2702	13.8.1942	2ft 0in	4wDM	25hp	McLaren	166
2868	27.8.1943	4ft 8½in	0-6-0ST	IC 18 x 26 4ft3in		119
2876	14.10.1943	4ft 8½in	0-6-0ST	IC 18 x 26 4ft3in		144
2878	28.10.1943	4ft 8½in	0-6-0ST	IC 18 x 26 4ft3in		144,147
2881	5.11.1943	4ft 8½in	0-6-0ST	IC 18 x 26 4ft3in		144
2884	17.11.1943	4ft 8½in	0-6-0ST	IC 18 x 26 4ft3in		121
2893	17.1.1944	4ft 8½in	0-6-0ST	IC 18 x 26 4ft3in		121
2896	25.1.1944	4ft 8½in	0-6-0ST	IC 18 x 26 4ft3in		182
2898	31.1.1944	4ft 8½in	0-6-0ST	IC 18 x 26 4ft3in		121
2924	17.4.1944	60cm	4wDM	20hp	Ailsa RF2	280
2946	9.6.1944	2ft 0in	4wDM	20hp	Ailsa RF2	280
3157	22.3.1944	4ft 8½in	0-6-0ST	IC 18 x 26 4ft3in		119
3163	24.4.1944	4ft 8½in	0-6-0ST	IC 18 x 26 4ft3in		119
3166	12.5.1944	4ft 8½in	0-6-0ST	IC 18 x 26 4ft3in		144,147
3172	15.6.1944	4ft 8½in	0-6-0ST	IC 18 x 26 4ft3in		146
3212	30.4.1945	4ft 8½in	0-6-0ST	IC 18 x 26 4ft3in		120
3213	10.5.1945	4ft 8½in	0-6-0ST	IC 18 x 26 4ft3in		121,182
3214	18.5.1945	4ft 8½in	0-6-0ST	IC 18 x 26 4ft3in		120
3790	22.1.1953	4ft 8½in	0-6-0ST	IC 18 x 26 4ft3in		123,286,288
3791	22.1.1953	4ft 8½in	0-6-0ST	IC 18 x 26 4ft3in		123
3792	27.1.1953	4ft 8½in	0-6-0ST	IC 18 x 26 4ft3in		123
3794	10.2.1953	4ft 8½in	0-6-0ST	IC 18 x 26 4ft3in		123
3796	4.3.1953	4ft 8½in	0-6-0ST	IC 18 x 26 4ft3in		124
3800	22.4.1953	4ft 8½in	0-6-0ST	IC 18 x 26 4ft3in		123
3801	30.4.1953	4ft 8½in	0-6-0ST	IC 18 x 26 4ft3in		122
3802	12.5.1953	4ft 8½in	0-6-0ST	IC 18 x 26 4ft3in		123
3809	29.1.1954	4ft 8½in	0-6-0ST	IC 18 x 26 4ft3in		268
4208	29.11.1948	4ft 8½in	0-6-0DM	186/204hp Gardner 8L3		172
4250	2.7.1951	4ft 8½in	0-4-0DM	139/153hp Gardner 6L3		113
4300	12.12.1951	60cm	4wDM	21hp	Ailsa RFS2	201
4345	3.3.1952	2ft 0in	4wDM	21hp	Ailsa RFS2	201
4394	27.10.1952	2ft 0in	4wDM	35hp	Gardner 2LW	171
4395	24.11.1952	2ft 0in	4wDM	35hp	Gardner 2LW	171
4396	18.12.1952	2ft 0in	4wDM	35hp	Gardner 2LW	171
4524	16.3.1954	1ft 6in	0-4-4-0DM 80hp	McLaren M4 13¾T		263,298
4525	30.5.1953	4ft 8½in	0-4-0DM	153hp	Gardner 6L3	113
6285	30.1.1968	2ft 6in	4wDH	40hp	Perkins 4203	158
6660	17.12.1965	2ft 6in	4wDH	60hp	Gardner 4LW	287
6950	15.8.1967	4ft 8½in	0-6-0DH	325hp	RR C8SFL	58
8828	14.3.1979	2ft 6in	4wDH	42hp	Ford 2401E	298
8829	14.3.1979	2ft 6in	4wDH	42hp	Ford 2401E	287

8900	19.7.1977	4ft 8½in	0-6-0DH	308hp	Cummins NHRS 6			89
9079	14.3.1985	762mm	4wDH	42hp	Perkins D3152			287
9080	14.3.1985	762mm	4wDH	42hp	Perkins D3152			287
9120	10.1981	2ft 0in	2w-2-2-2wRER		{see GB 420461/21}			284
9282	18.3.1988	900mm	4wDH	150hp	Caterpillar 3306			292
9346	1994	610mm	4wDH	80hp	Deutz F6L912W 12T			238
9348	1994	610mm	4wDH	80hp	Deutz F6L912W 12T			238
9351	1994	610mm	4wDH	80hp	Deutz F6L912W 12T			238

HOPKINS GILKES & CO, TEES IRON WORKS, Middleborough   HG

246	9.1866	4ft 8½in	0-4-0ST	OC	12 x	228

HENRY HUGHES & COMPANY, FALCON WORKS, Loughborough   HH

		4ft 8½in	0-4-0ST	OC	59
		4ft 8½in	0-4-0ST	OC	168
		4ft 8½in	0-4-0ST	OC	221
	c1868	4ft 8½in	0-4-0ST	OC	168
	1883	4ft 8½in	0-4-0ST	OC	95,254,255

R & W HAWTHORN, LESLIE & CO LTD, FORTH BANK WORKS, Newcastle-upon-Tyne   HL

2345	6.1896	4ft 8½in	0-4-0ST	OC	14 x 20	3ft0in	103
2402	10.1898	4ft 8½in	0-4-0CT	OC	12 x 15	2ft10in	247
2502	11.6.1901	4ft 8½in	0-4-0ST	OC	14 x 20	3ft6in	89
2513	19.6.1902	4ft 8½in	0-4-0ST	OC	14 x 20	3ft6in	103
2800	1.12.1909	4ft 8½in	0-4-0ST	OC	10 x 15	2ft10in	68
2839	31.12.1910	4ft 8½in	0-4-0ST	OC	14 x 22	3ft6in	180
3083	20.1.1915	4ft 8½in	0-4-0ST	OC	14 x 22	3ft6in	72
3091	29.1.1915	4ft 8½in	0-4-0ST	OC	14 x 22	3ft6in	72
3177	6.5.1916	4ft 8½in	0-4-0ST	OC	14 x 22	3ft6in	147
3240	16.7.1917	4ft 8½in	0-4-0ST	OC	12 x 18	3ft0½in	295
3247	14.5.1917	4ft 8½in	0-4-0ST	OC	14 x 22	3ft4in	177
3308	7.3.1918	4ft 8½in	0-4-0ST	OC	12 x 18	3ft0½in	137
3424	3.9.1920	4ft 8½in	0-4-0ST	OC	14 x 22	3ft6in	273
3425	3.9.1920	4ft 8½in	0-4-0ST	OC	14 x 22	3ft6in	273
3467	30.10.1920	4ft 8½in	0-4-0ST	OC	14 x 22	3ft6in	273
3468	30.10.1920	4ft 8½in	0-4-0ST	OC	14 x 22	3ft6in	273
3469	30.10.1920	4ft 8½in	0-4-0ST	OC	14 x 22	3ft6in	274
3477	6.5.1921	4ft 8½in	0-4-0ST	OC	14 x 22	3ft6in	274
3478	6.5.1921	4ft 8½in	0-4-0ST	OC	14 x 22	3ft6in	274
3479	6.5.1921	4ft 8½in	0-4-0ST	OC	14 x 22	3ft6in	274
3529	16.10.1922	4ft 8½in	0-6-0T	OC	16 x 24	3ft9in	143,147
3530	16.10.1922	4ft 8½in	0-6-0T	OC	16 x 24	3ft9in	143,147
3539	9.3.1923	4ft 8½in	0-4-0ST	OC	14 x 22	3ft6in	103
3540	9.6.1923	4ft 8½in	0-4-0ST	OC	14 x 22	3ft6in	103
3595	19.8.1924	4ft 8½in	0-4-0F	OC	17 x 16	2ft11in	67,137
3596	1.10.1924	4ft 8½in	0-4-0F	OC	17 x 16	2ft11in	67
3653	20.1.1927	4ft 8½in	0-4-0ST	OC	14 x 22	3ft6in	67,68
3683	5.12.1927	4ft 8½in	0-4-0ST	OC	12 x 20	3ft1in	262
3715	22.6.1928	4ft 8½in	0-4-0ST	OC	15 x 22	3ft5in	286

3721	14.4.1928	4ft 8½in	0-4-0ST	OC	14 x 22	3ft6in		89,102
3742	27.11.1929	4ft 8½in	0-4-0ST	OC	14 x 22	3ft6in		135
3760	10.2.1932	4ft 8½in	0-4-0ST	OC	12 x 20	3ft1in		168,182
3791	19.1.1932	4ft 8½in	0-4-0ST	OC	14 x 22	3ft6in		67,68
3794	19.11.1931	4ft 8½in	0-4-0ST	OC	14 x 22	3ft4in		135
3901	10.3.1937	4ft 8½in	0-4-0ST	OC	16 x 24	3ft8in		54

HORWICH WORKS, Lancs (BR)   Hor

1097	5.1910	4ft 8½in	0-4-0ST	OC	13x18	3ft0in		71
1111	1910	4ft 8½in	0-4-0ST	OC	13x18	3ft0in		67
[08764]	1961	4ft 8½in	0-6-0DE	350hp	EE 6KT		48T	171

HIGHLAND RAILWAY, Lochgorm   HR

	1899	4ft 8½in	4-4-0	IC	18½x26 6ft0in	130

ROBERT HUDSON LTD, Leeds   HU

HT169	c1928	4ft 8½in	4wDM	20hp	Fordson	56
36459	1929	2ft 0in	4wPM			249

JEFFREY MANUFACTURING CO, Columbus, Ohio, U.S.A.   Jeffrey

5136	1919	4ft 8½in	4wBE	151

JENBACHWERKE A G, Jenbach, Austria   Jenbach

4ft 8½in	4wDH	190

JOHN FOWLER & CO (LEEDS) LTD, Hunslet, Leeds   JF

1347	1.11.1870	4ft 8½in	2-2-0WT	8hp		82
2850	3.1876	4ft 8½in	0-4-0ST	8½ x		246
19024	9.1930	4ft 8½in	0-4-0DM	65/70hp	MAN	69,71,75,77
19231	4.1931	2ft 0in	4wPM	10hp		70
19424	5.1931	4ft 8½in	0-4-0DM	70hp	MAN	170
19351	7.1933	4ft 8½in	0-4-0DM	70hp	MAN	75,77
20550	11.1934	4ft 8½in	0-4-0DM	85hp	Fowler 6cyl	170
21293	8.4.1936	2ft 0in	4wDM	40hp	Fowler 4B	249
21294	15.4.1936	2ft 0in	4wDM	40hp	Fowler 4B	249
21295	23.4.1936	2ft 0in	4wDM	40hp	Fowler 4B	249
21408	30.10.1936	2ft 0in	4wDM	40hp	Fowler 4B	249
21455	24.9.1936	4ft 8½in	0-4-0DM	80hp	Fowler 6A	50
22077	31.1.1938	4ft 8½in	0-4-0DM	80hp	Fowler 6A	168
22885	31.10.1940	4ft 8½in	0-6-0DM	200hp	Fowler 6C	276
22889	12.9.1939	4ft 8½in	0-4-0DM	150hp	Fowler 4C	121
22890	20.9.1939	4ft 8½in	0-4-0DM	150hp	Fowler 4C	119
22934	23.1.1941	4ft 8½in	0-4-0DM	150hp	Fowler 4C	180
22947	31.10.1941	4ft 8½in	0-4-0DM	150hp	Fowler 4C	277

22973	23.3.1942	4ft 8½in	0-4-0DM	150hp	Fowler 4C		57
22976	11.5.1942	4ft 8½in	0-4-0DM	150hp	Fowler 4C		121
22978	30.6.1942	4ft 8½in	0-4-0DM	150hp	Fowler 4C		277
22987	20.10.1942	4ft 8½in	0-4-0DM	150hp	Fowler 4C		277
23003	29.7.1943	4ft 8½in	0-4-0DM	150hp	Fowler 4C		85,182
4160002	31.10.1952	4ft 8½in	0-4-0DM	100hp	McLaren M4		170
4200018	8.12.1947	4ft 8½in	0-4-0DM	150hp	Fowler 4C		89
4200035	19.1.1949	4ft 8½in	0-4-0DM	150hp	Fowler 4C		162
4210003	21.9.1949	4ft 8½in	0-4-0DM	150hp	McLaren		57,176,180
4210005	21.11.1949	4ft 8½in	0-4-0DM	150hp	McLaren		160
4210007	6.12.1949	4ft 8½in	0-4-0DM	150hp	McLaren		160
4210072	16.5.1952	4ft 8½in	0-4-0DM	150hp	McLaren		182
4210076	17.6.1952	4ft 8½in	0-4-0DM	150hp	McLaren		180
4210130	30.9.1957	4ft 8½in	0-4-0DM	150hp	McLaren		160
4210140	1.4.1958	4ft 8½in	0-4-0DM	150hp	McLaren		150
4210143	28.8.1958	4ft 8½in	0-4-0DM	150hp	McLaren		150
4210144	29.8.1958	4ft 8½in	0-4-0DM	150hp	McLaren		85
4220001	26.2.1959	4ft 8½in	0-4-0DM	176hp	Leyland EN900		87
4220008	26.11.1959	4ft 8½in	0-4-0DH	185hp	Leyland EN900		282
4220009	29.8.1960	4ft 8½in	0-4-0DH	185hp	Leyland EN900		172
4220031	10.11.1964	4ft 8½in	0-4-0DH	203hp	Leyland EN900		160
4220039	29.9.1965	4ft 8½in	0-4-0DH	203hp	Leyland EN900		160,288
4240016	8.9.1964	4ft 8½in	0-6-0DH	275hp	Leyland		113,160

KITSON & CO, AIRDALE FOUNDRY, Leeds K

T109	1884	4ft 8½in	0-4-0T	8x12	3ft0in		58,106

KENT CONSTRUCTION & ENGINEERING CO LTD, Ashford, Kent KC

		2ft 0in	4wPM				97

KERR, STUART & CO LTD, CALIFORNIA WORKS, Stoke-on-Trent KS

124	15.5.1897	4ft 8½in	0-4-2T	OC	8½x15			53
692	23.3.1901	4ft 8½in	0-4-2T	OC	9½x15	2ft9in	Waterloo	58,106
850	24.12.1903	2ft 1¾in	0-4-0ST	OC	6 x 9	1ft8in	Wren	93
876	27.10.1904	3ft 0in	0-4-2ST	OC	7¼ x 12	2ft0in	Tattoo	282
1099	23.3.1910	2ft 1¾in	0-4-0ST	OC	6½ x 9	1ft8in	Wren	93
1142	7.2.1911	2ft 0in	0-4-2ST	OC	7 x 12	2ft0in	Tattoo	248
1144	30.9.1911	2ft 0in	0-4-2ST	OC	7 x 12	2ft0in	Tattoo	248
1269	29.7.1912	4ft 8½in	4wWE	70hp				288,292
1283	28.4.1915	4ft 8½in	0-4-0ST	OC	11 x 16	2ft9in	Huxley	106
1291	8.12.1915	2ft 0in	0-4-2ST	OC	7 x 12	2ft0in	Tattoo	248
2459	20.8.1915	2ft 0in	0-4-0ST	OC	6 x 9	1ft8in	Wren	234,278
2460	23.8.1915	2ft 0in	0-4-0ST	OC	6 x 9	1ft8in	Wren	234,278
2469	21.6.1916	2ft 0in	0-4-0ST	OC	6 x 9	1ft8in	Wren	234,164
3048	27.3.1917	4ft 8½in	0-4-0WT	OC	14½x20	3ft4in	Priestley	102,181
3078	11.12.1917	4ft 8½in	0-6-0T	OC	14 x 20	3ft9in	Argentina	275
3129	10.10.1918	4ft 8½in	0-4-2ST	OC	8 x 12	2ft3in	Darwin	100,275
4001	20.12.1918	2ft 0in	0-4-0ST	OC	6 x 9	1ft8in	Wren	248
4002	20.12.1918	2ft 0in	0-4-0ST	OC	6 x 9	1ft8in	Wren	234,278

4005	31.12.1918	2ft 0in	0-4-0ST	OC	6 x 9	1ft8in	Wren	234,278
4020	14.6.1919	2ft 0in	0-4-0ST	OC	6 x 9	1ft8in	Wren	93,202
4022	14.6.1919	2ft 0in	0-4-0ST	OC	6 x 9	1ft8in	Wren	202,203
4065	13.3.1920	4ft 8½in	0-4-2ST	OC	9 x 15	2ft6in	Brazil	182
4154	7.6.1920	2ft 1¾in	0-4-0ST	OC	6 x 9	1ft8in	Wren	93
4162	19.10.1921	2ft 0in	0-4-0ST	OC	6 x 9	1ft8in	Wren	203,234,236
4163	21.10.1921	2ft 0in	0-4-0ST	OC	6 x 9	1ft8in	Wren	234
4168	17.9.1920	4ft 8½in	0-4-0ST	OC	15 x 20	3ft6in	Moss Bay	68
4199	23.7.1920	4ft 8½in	0-4-0WT	OC	14½x20	3ft4in	Priestley	102
4246	31.1.1922	2ft 0in	0-4-0ST	OC	6 x 9	1ft8in	Wren	234
4247	31.1.1922	2ft 0in	0-4-0ST	OC	6 x 9	1ft8in	Wren	234,241
4248	2.2.1922	2ft 0in	0-4-0ST	OC	6 x 9	1ft8in	Wren	234
4249	2.2.1922	2ft 0in	0-4-0ST	OC	6 x 9	1ft8in	Wren	234
4250	24.2.1922	2ft 0in	0-4-0ST	OC	6 x 9	1ft8in	Wren	234
4251	24.6.1922	2ft 0in	0-4-0ST	OC	6 x 9	1ft8in	Wren	234
4252	30.6.1922	2ft 0in	0-4-0ST	OC	6 x 9	1ft8in	Wren	234
4253	30.6.1922	2ft 0in	0-4-0ST	OC	6 x 9	1ft8in	Wren	234
4254	4.3.1922	2ft 0in	0-4-0ST	OC	6 x 9	1ft8in	Wren	234
4255	4.3.1922	2ft 0in	0-4-0ST	OC	6 x 9	1ft8in	Wren	234
4256	4.3.1922	2ft 0in	0-4-0ST	OC	6 x 9	1ft8in	Wren	234,278
4257	8.3.1922	2ft 0in	0-4-0ST	OC	6 x 9	1ft8in	Wren	234
4258	9.3.1922	2ft 0in	0-4-0ST	OC	6 x 9	1ft8in	Wren	234
4259	9.3.1922	2ft 0in	0-4-0ST	OC	6 x 9	1ft8in	Wren	234
4260	27.3.1922	2ft 0in	0-4-0ST	OC	6 x 9	1ft8in	Wren	234,278
4261	28.3.1922	2ft 0in	0-4-0ST	OC	6 x 9	1ft8in	Wren	235
4262	30.3.1922	2ft 0in	0-4-0ST	OC	6 x 9	1ft8in	Wren	235
4263	31.3.1922	2ft 0in	0-4-0ST	OC	6 x 9	1ft8in	Wren	235
4264	14.4.1922	2ft 0in	0-4-0ST	OC	6 x 9	1ft8in	Wren	235
4265	6.4.1922	2ft 0in	0-4-0ST	OC	6 x 9	1ft8in	Wren	235
4267	21.7.1922	2ft 0in	0-4-0ST	OC	6 x 9	1ft8in	Wren	235
4268	27.7.1922	2ft 0in	0-4-0ST	OC	6 x 9	1ft8in	Wren	235
4269	28.7.1922	2ft 0in	0-4-0ST	OC	6 x 9	1ft8in	Wren	235
4270	31.7.1922	2ft 0in	0-4-0ST	OC	6 x 9	1ft8in	Wren	235
4273	6.12.1922	2ft 0in	0-4-0ST	OC	6 x 9	1ft8in	Wren	202,203
4274	18.12.1922	2ft 0in	0-4-0ST	OC	6 x 9	1ft8in	Wren	202,203
4290	7.2.1923	2ft 0in	0-4-0ST	OC	6 x 9	1ft8in	Wren	249
4291	6.4.1923	2ft 0in	0-4-0ST	OC	6 x 9	1ft8in	Wren	248
4421	2.12.1929	4ft 8½in	6wDM	90hp	McLaren			154

R. & A. LISTER & CO LTD, Dursley, Gloucestershire   L

3145	1930	2ft 0in	4wPM		JAP		249
3270	25.7.1930	2ft 0in	4wPM	6hp	JAP	Type R 1½T	101
3289	1930	2ft 0in	4wPM		JAP		249
3834	24.4.1931	2ft 0in	4wPM	6hp	JAP	Type R 1½T	249
3854	22.5.1931	2ft 0in	4wPM	6hp	JAP	Type R 1½T	189
3916	19.6.1931	2ft 0in	4wPM	6hp	JAP	Type R 1½T	300
4449	1932	2ft 0in	4wPM		JAP		51
7622	9.4.1936	2ft 0in	4wPM	6hp	JAP	Type R 1½T	81
9256	1937	2ft 0in	4wPM	9.8hp	JAP	Type RT 2T	74
13258	20.2.1940	2ft 0in	4wPM	6hp	JAP	Type R 1T 17cwt	110

25919	20.9.1944	2ft 0in	4wPM	6hp	JAP	Type R 1T 8cwt		158
33937	13.5.1949	1ft 8in	4wPM	6hp	JAP	Type R		73
34521	19.10.1949	2ft 0in	4wPM	6hp	JAP	Type R		69,203
37658	7.2.1952	2ft 0in	4wPM	6hp	JAP	Type R		74
42494	16.3.1956	2ft 0in	4wPM	6hp	JAP	Type R		300
50265	29.3.1958	2ft 0in	4wPM	6hp	JAP	Type R		86
52031	10.11.1960	2ft 0in	4wDM	12¾hp	Lister SL3 Type RM3 2T			74

LAKE & ELLIOT LTD, ALBION WORKS, Braintree, Essex — Lake & Elliot

	c1923	4ft 8½in	4wPM		107,152
	c1924	4ft 8½in	4wPM		107
		4ft 8½in	4wDM		152,286

LISTER BLACKSTONE TRACTION LTD, Dursley, Gloucestershire — LB

54183	29.9.1964	1ft 8in	4wDM	12hp	Lister Type RM2 1.52T		73

STEPHEN LEWIN, DORSET FOUNDRY. Poole, Dorset — Lewin

	c1876	4ft 8½in	0-4-0WT	OC	52
	c1877	4ft 8½in	0-4-0T	OC	52

LILLESHALL CO, St George's, Okengates — Lilleshall

187	1873	4ft 8½in	0-4-0ST	OC	53

LOGAN MACHINERY & MINING CO LTD — LMM

1049	1950	2ft 0in	4wBE		3T	229
1053	1950	2ft 0in	4wBE		3T	229
1066	1950	2ft 0in	4wBE		3T	229
1072	1950	2ft 0in	4wBE		3T	229

R. B. LONGRIDGE & CO, Bedlington, Northumberland — Longridge

	1847	4ft 8½in	0-6-0ST	IC	111,220

MATISA MATERIAL INDUSTRIEL SA, Lausanne, Switzerland — Matisa

D8.005	1971	4ft 8½in	2w-2wDMR		286
2654	1975	4ft 8½in	4wDE		282
2655	1975	4ft 8½in	4wDE		282

McEWAN PRATT & CO LTD, Wickford, Essex — McP

		2ft 0in	0-4-0PM		52,113,234

MUIR-HILL (ENGINEERS) LTD, Trafford Park, Manchester — MH

L103	29.7.1930	4ft 8½in	4wPM		89
		4ft 8½in	4wPM		61

S A MOTEURS MOËS, Waremme, Belgium — Moës

600mm	4wDM		287
750mm	4wDM		287

MOTOR RAIL LTD, SIMPLEX WORKS, Bedford — MR

No.	Date	Gauge	Type	Power	Engine	T	Page
234	1916	60cm	4wPM	20hp	Dorman 2JO		127
320	4.4.1917	60cm	4wPM	20hp	Dorman 2JO	2½T	127
476	5.4.1918	60cm	4wPM	40hp	Dorman 4JO	6¾T	155
856	7.5.1918	60cm	4wPM	20hp	Dorman 2JO		253
867	7.5.1918	60cm	4wPM	20hp	Dorman 2JO		253
915	10.6.1918	60cm	4wPM	20hp	Dorman 2JO		196
1234	6.1.1919	60cm	4wPM	20hp	Dorman 2JO		187
1313	3.9.1918	60cm	4wPM	40hp	Dorman 4JO		125,127
1643	30.4.1918	60cm	4wPM	20hp	Dorman 2JO	2½T	127
1740	14.8.1918	60cm	4wPM	20hp	Dorman 2JO	2½T	141
1744	14.8.1918	60cm	4wPM	20hp	Dorman 2JO	2½T	127
1985	19.5.1920	2ft 0in	4wPM	20hp	Dorman 2JO	2½T	253
2029	13.9.1920	4ft 8½in	4wPM	40hp	Dorman 4JO	8T	288
2034	16.12.1920	4ft 8½in	4wPM	40hp	Dorman 4JO	8T	65
2210	26.6.1923	60cm	4wPM	20hp	Dorman 2JO	2½T	242
2214	26.6.1923	60cm	4wPM	20hp	Dorman 2JO	2½T	242
3790	26.3.1926	2ft 0in	4wPM	20hp	Dorman 2JO	2½T	139
3896	1928	4ft 8½in	4wPM	40hp	Dorman 4JO	8T	84
3983	c1936	2ft 0in	4wPM	20hp	Dorman 2JO	2½T	139
4028	31.3.1926	2ft 0in	4wPM	20hp	Dorman 2JO	2½T	207
4032	7.7.1926	2ft 0in	4wPM	20hp	Dorman 2JO	2½T	98
4048	11.1.1927	2ft 0in	4wPM	20hp	Dorman 2JO	2½T	98
4502	9.6.1927	2ft 0in	4wPM	20hp	Dorman 2JO	2½T	98
4510	11.2.1928	60cm	4wPM	20hp	Dorman 2JO	2½T	98
4516	13.4.1928	60cm	4wPM	20hp	Dorman 2JO	2½T	98
4521	18.8.1928	60cm	4wPM	20hp	Dorman 2JO	2½T	98
4587	21.11.1928	1ft11⁵⁄₈in	4wPM	20hp	Dorman 2JO	2½T	279
4592	19.6.1930	2ft 0in	4wPM	20hp	Dorman 2JO	2½T	226
4593	19.6.1930	2ft 0in	4wPM	20hp	Dorman 2JO	2½T	226
4594	19.6.1930	2ft 0in	4wPM	20hp	Dorman 2JO	2½T	226
4596	3.2.1932	2ft 0in	4wPM	20hp	Dorman 2JO	2½T	226
4597	11.2.1932	2ft 0in	4wPM	20hp	Dorman 2JO	2½T	226
4598	11.2.1932	2ft 0in	4wPM	20hp	Dorman 2JO	2½T	226
4599	23.2.1932	2ft 0in	4wPM	20hp	Dorman 2JO	2½T	226
4600	26.2.1932	2ft 0in	4wPM	20hp	Dorman 2JO	2½T	226
4601	27.2.1932	2ft 0in	4wPM	20hp	Dorman 2JO	2½T	226
4602	27.2.1932	2ft 0in	4wPM	20hp	Dorman 2JO	2½T	226
4603	2.3.1932	2ft 0in	4wPM	20hp	Dorman 2JO	2½T	226
4604	5.3.1932	2ft 0in	4wPM	20hp	Dorman 2JO	2½T	226
4605	7.3.1932	2ft 0in	4wPM	20hp	Dorman 2JO	2½T	226
4606	9.3.1932	2ft 0in	4wPM	20hp	Dorman 2JO	2½T	226
4607	11.3.1932	2ft 0in	4wPM	20hp	Dorman 2JO	2½T	226
4608	18.3.1932	2ft 0in	4wPM	20hp	Dorman 2JO	2½T	226
4609	19.3.1932	2ft 0in	4wPM	20hp	Dorman 2JO	2½T	226
4610	23.3.1932	2ft 0in	4wPM	20hp	Dorman 2JO	2½T	226

4611	24.3.1932	2ft 0in	4wPM	20hp	Dorman 2JO	2½T	226
4612	1.4.1932	2ft 0in	4wPM	20hp	Dorman 2JO	2½T	226
4613	5.4.1932	2ft 0in	4wPM	20hp	Dorman 2JO	2½T	226
4614	5.4.1932	2ft 0in	4wPM	20hp	Dorman 2JO	2½T	226
4615	7.4.1932	2ft 0in	4wPM	20hp	Dorman 2JO	2½T	226
4616	8.4.1932	2ft 0in	4wPM	20hp	Dorman 2JO	2½T	226
4617	13.4.1932	2ft 0in	4wPM	20hp	Dorman 2JO	2½T	226
4618	14.4.1932	2ft 0in	4wPM	20hp	Dorman 2JO	2½T	226
4619	21.4.1932	2ft 0in	4wPM	20hp	Dorman 2JO	2½T	226
4620	21.4.1932	2ft 0in	4wPM	20hp	Dorman 2JO	2½T	226
4621	30.4.1932	2ft 0in	4wPM	20hp	Dorman 2JO	2½T	227
4622	3.5.1932	2ft 0in	4wPM	20hp	Dorman 2JO	2½T	227
4711	14.4.1936	2ft 0in	4wPM	20/26hp	Dorman 2JO	2½T	187
4721	12.7.1937	2ft 0in	4wPM	20/26hp	Dorman 2JO	2½T	187
4722	29.7.1937	2ft 0in	4wPM	20/26hp	Dorman 2JO	2½T	187
4723	8.1.1938	2ft 0in	4wPM	20/26hp	Dorman 2JO	2½T	62
5005	13.5.1929	60cm	4wPM	20hp	Dorman 2JO	2½T	187
5027	10.9.1929	2ft 0in	4wPM	20hp	Dorman 2JO	2½T	97
5061	29.3.1930	60cm	4wPM	20hp	Dorman 2JO	2½T	98
5063	8.4.1930	2ft 0in	4wPM	20hp	Dorman 2JO	2½T	211
5075	22.5.1930	2ft 0in	4wPM	20hp	Dorman 2JO	2½T	193
5076	22.5.1930	2ft 0in	4wPM	20hp	Dorman 2JO	2½T	193
5077	26.5.1930	2ft 0in	4wPM	20hp	Dorman 2JO	2½T	193
5078	26.5.1930	2ft 0in	4wPM	20hp	Dorman 2JO	2½T	193
5086	17.9.1930	2ft 0in	4wPM	20hp	Dorman 2JO	4T	55
5204	5.8.1930	2ft 0in	4wPM	20/35hp	Dorman4MRX	2½T	98
5223	21.10.1930	2ft 0in	4wPM	20/35hp	Dorman4MRX	2½T	193
5224	21.10.1930	2ft 0in	4wPM	20/35hp	Dorman4MRX	2½T	193
5258	14.1.1931	2ft 0in	4wPM	20/35hp	Dorman4MRX	2½T	105
5277	23.2.1931	2ft 0in	4wPM	20/35hp	Dorman4MRX	2½T	104
5289	20.3.1931	2ft 0in	4wPM	20/35hp	Dorman4MRX	2½T	157
5294	20.3.1931	2ft 0in	4wPM	20/35hp	Dorman4MRX	2½T	279
5301	14.4.1931	2ft 0in	4wPM	20/35hp	Dorman4MRX	2½T	98
5319	18.5.1931	2ft 0in	4wPM	20/35hp	Dorman4MRX	2½T	104
5322	1.6.1931	2ft 0in	4wPM	20/35hp	Dorman4MRX	4T	203
5340	11.8.1931	2ft 0in	4wPM	20/35hp	Dorman4MRX	4T	203
5344	6.8.1931	2ft 0in	4wPM	20/35hp	Dorman4MRX	4T	203,242
5345	20.8.1931	2ft 0in	4wPM	20/35hp	Dorman4MRX	2½T	157
5414	1934	2ft 0in	4wPM	20/35hp	Dorman4MRX	4T	203
5415	1934	2ft 0in	4wPM	20/35hp	Dorman4MRX	4T	203
5416	1934	2ft 0in	4wPM	20/35hp	Dorman4MRX	4T	203
5454	1935	2ft 0in	4wPM	20/35hp	Dorman4MRX	2½T	279
5612	20.8.1931	2ft 0in	4wDM	20hp	Dorman 2RB	4T	157,211
5613	1.9.1931	2ft 0in	4wDM	20hp	Dorman 2RB	4T	157,211
5614	25.5.1932	2ft 0in	4wDM	20hp	Dorman 2RB	2½T	226
5615	29.2.1932	2ft 0in	4wDM	20hp	Dorman 2RB	4T	264,281
5620	25.5.1932	2ft 0in	4wDM	20hp	Dorman 2RB	2½T	226
5621	25.5.1932	2ft 0in	4wDM	20hp	Dorman 2RB	2½T	226
5638	22.2.1933	2ft 0in	4wDM	20/36hp	Dorman 2RB	2½T	199
5640	8.4.1933	2ft 0in	4wDM	20/36hp	Dorman 2RB	2½T	199
5645	18.5.1933	2ft 0in	4wDM	20/36hp	Dorman 2RB	2½T	98

5646	18.5.1933	2ft 0in	4wDM	20/36hp Dorman 2RB	2½T	98
5752	25.3.1935	4ft 8½in	4wDM	65/85hp Dorman 4DL	20T	276
5755	31.1.1949	4ft 8½in	4wDM	65/85hp Dorman 4DL	15T	84,182
5853	18.1.1934	2ft 0in	4wDM	20/28hp Dorman 2HW	3½T	162
5855	24.1.1934	2ft 0in	4wDM	20/28hp Dorman 2HW	2½T	157
5875	26.3.1935	2ft 0in	4wDM	20/28hp Dorman 2HW	2½T	157
5876	26.1.1935	2ft 0in	4wDM	20/28hp Dorman 2HW	3½T	199
5877	1935	2ft 0in	4wDM	20/28hp Dorman 2HW	3½T	199
5878	1935	2ft 0in	4wDM	20/28hp Dorman 2HW	3½T	199
5909	4.4.1934	2ft 0in	4wDM	32/42hp Dorman 2RBL	4T	155
5933	30.3.1936	2ft 0in	4wDM	32/42hp Dorman 2RBL	6T	153
6011	11.11.1930	2ft 0in	4wPM	12/20hp Dorman 4MVR	2T	183
6014	15.4.1931	2ft 0in	4wPM	12/20hp Dorman 4MVR	2T	198
6017	21.5.1931	2ft 0in	4wPM	12/20hp Dorman 4MVR	2T	198
7001	5.5.1932	60cm	4wPM	25hp Ford	2½T	91
7035	1936	2ft 0in	4wPM	20/26hp Dorman 2JO	2½T	157
7040	1937	2ft 0in	4wPM	20/26hp Dorman 2JO	2½T	104,153
7046	23.3.1937	2ft 0in	4wPM	20/26hp Dorman 2JO	2½T	104
7061	7.9.1938	2ft 0in	4wPM	20/26hp Dorman 2JO	2½T	257
7073	1940	2ft 0in	4wPM	20/26hp Dorman 2JO	2½T	153
7107	1936	2ft 0in	4wDM	20/28hp Dorman 2HW	3½T	104
7128	1936	2ft 0in	4wDM	20/28hp Dorman 2HW	3½T	109
7143	1936	2ft 0in	4wDM	20/28hp Dorman 2HW	2½T	153
7146	1936	2ft 0in	4wDM	20/28hp Dorman 2HW	2½T	153
7214	1938	2ft 0in	4wDM	20/28hp Dorman 2HW	2½T	109
7225	1938	2ft 0in	4wDM	20/28hp Dorman 2HW	2½T	280
7226	1938	2ft 0in	4wDM	20/28hp Dorman 2HW	2½T	104
7228	1938	2ft 0in	4wDM	20/28hp Dorman 2HW	2½T	157
7305	5.9.1938	2ft 0in	4wDM	20/28hp Dorman 2DW	2½T	200
7306	1938	2ft 0in	4wDM	20/28hp Dorman 2DW	2½T	153
7308	19.10.1938	2ft 0in	4wDM	20/28hp Dorman 2DW	2½T	55
7309	1938	2ft 0in	4wDM	20/28hp Dorman 2DW	2½T	218
7331	1938	2f 0in	4wDM	20/28hp Dorman 2DW	2½T	140
7358	31.12.1938	2ft 0in	4wDM	20/28hp Dorman 2DWD	2½T	153
7374	22.2.1939	2ft 0in	4wDM	20/28hp Dorman 2DWD	2½T	300
7377	20.3.1939	2ft 0in	4wDM	20/28hp Dorman 2DWD	2½T	55
7380	1939	2ft 0in	4wDM	20/28hp Dorman 2DWD	3½T	153
7398	1939	2ft 0in	4wDM	20/28hp Dorman 2DWD	2½T	153
7403	17.8.1939	2ft 0in	4wDM	20/28hp Dorman 2DWD	2½T	109
7441	24.11.1939	2ft 0in	4wDM	20/28hp Dorman 2DWD	3½T	153
7454	7.12.1939	2ft 0in	4wDM	20/28hp Dorman 2DWD	2½T	200
7456	27.12.1939	2ft 0in	4wDM	20/28hp Dorman 2DWD	2½T	153
7457	16.12.1939	2ft 0in	4wDM	20/28hp Dorman 2DWD	2½T	104
7465	1.1.1940	2ft 0in	4wDM	20/28hp Dorman 2DWD	3½T	203
7467	1940	2ft 0in	4wDM	20/28hp Dorman 2DWD	2½T	153
7472	12.2.1940	2ft 0in	4wDM	20/28hp Dorman 2DWD	3½T	203,204
7473	12.2.1940	2ft 0in	4wDM	20/28hp Dorman 2DWD	3½T	91,204
7486	22.4.1940	2ft 0in	4wDM	20/28hp Dorman 2DWD	2½T	92,203
7487	22.4.1940	2ft 0in	4wDM	20/28hp Dorman 2DWD	3½T	203
7902	3.11.1939	2ft 0in	4wDM	32/42hp Dorman 2DL	6T	132
7927	12.3.1941	2ft 0in	4wDM	32/42hp Dorman 2DL	6T	132

7928	12.6.1941	2ft 0in	4wDM	32/42hp	Dorman 2DL		6T	109
7981	19.8.1946	2ft 0⅛in	4wDM	32/42hp	Dorman 2DL		5T	132
7994	2.7.1947	2ft 0in	4wDM	32/42hp	Dorman 2DL		7T	261
8565	27.9.1940	60cm	4wDM	20/28hp	Dorman 2DWD	2½T		157
8588	14.1.1941	60cm	4wDM	20/28hp	Dorman 2DWD	2½T		140
8598	1940	2ft 0in	4wDM	20/28hp	Dorman 2DWD	2½T		104
8602	26.12.1940	2ft 0in	4wDM	28/28hp	Dorman 2DWD	2½T		101
8614	8.4.1941	60cm	4wDM	20/28hp	Dorman 2DWD	2½T		63
8615	8.4.1941	60cm	4wDM	20/28hp	Dorman 2DWD	2½T		128
8626	16.5.1941	60cm	4wDM	20/28hp	Dorman 2DWD	2½T		84
8640	25.6.1941	60cm	4wDM	20/28hp	Dorman 2DWD	2½T		128,300
8641	25.6.1941	60cm	4wDM	20/28hp	Dorman 2DWD	2½T		128
8675	6.10.1941	2ft 0in	4wDM	20/28hp	Dorman 2DWD	2½T		104,153
8696	23.12.1941	2ft 0in	4wDM	20/28hp	Dorman 2DWD	2½T		216
8729	1941	2ft 0in	4wDM	20/28hp	Dorman 2DWD	2½T		295
8745	6.2.1942	2ft 0in	4wDM	20/28hp	Dorman 2DWD	2½T		128
8788	26.3.1943	2ft 0in	4wDM	20/28hp	Dorman 2DWD	2½T		157
8813	17.3.1943	60cm	4wDM	20/28hp	Dorman 2DWD	2½T		129
8820	25.5.1943	60cm	4wDM	20/28hp	Dorman 2DWD	2½T		128
8857	20.10.1943	60cm	4wDM	20/28hp	Dorman 2DWD	2½T		128
8882	21.4.1944	60cm	4wDM	20/28hp	Dorman 2DWD	2½T		216
8886	21.4.1944	60cm	4wDM	20/28hp	Dorman 2DWD	2½T		129
8887	21.4.1944	60cm	4wDM	20/28hp	Dorman 2DWD	2½T		128
8903	30.6.1944	60cm	4wDM	20/28hp	Dorman 2DWD	2½T		128
9102	10.2.1941	2ft 0in	4wDM	20/26hp	Dorman 2JO		2½T	183
9263	16.5.1947	2ft 0in	4wDM	20/28hp	Dorman 2DWD	3½T		216
9409	7.12.1948	2ft 0in	4wDM	20/28hp	Dorman 2DWD	3½T		208,209
9410	8.12.1948	2ft 0in	4wDM	20/28hp	Dorman 2DWD	3½T		208,209
9411	13.12.1948	2ft 0in	4wDM	20/28hp	Dorman 2DWD	3½T		208,209
9412	6.1.1949	2ft 0in	4wDM	20/28hp	Dorman 2DWD	3½T		208,209
9413	18.1.1949	2ft 0in	4wDM	20/28hp	Dorman 2DWD	3½T		208,209
9414	12.4.1949	2ft 0in	4wDM	20/28hp	Dorman 2DWD	3½T		208,209
9415	27.4.1949	2ft 0in	4wDM	20/28hp	Dorman 2DWD	3½T		208,209
9416	18.5.1949	2ft 0in	4wDM	20/28hp	Dorman 2DWD	3½T		208,209
9417	24.5.1949	2ft 0in	4wDM	20/28hp	Dorman 2DWD	3½T		208,209
9418	14.6.1949	2ft 0in	4wDM	20/28hp	Dorman 2DWD	3½T		208,209
10030	8.6.1948	2ft 0in	4wDM	32/42hp	Dorman 2DL		6T	109
10031	9.7.1948	2ft 0in	4wDM	32/42hp	Dorman 2DL		6T	261
10114	18.5.1949	2ft 0in	4wDM	32/42hp	Dorman 2DL		6T	109
10130	16.8.1949	2ft 0in	4wDM	32/42hp	Dorman 2DL		6T	132
10154	31.1.1950	2ft 0in	4wDM	32/42hp	Dorman 2DL		6T	109
10155	6.2.1950	2ft 0in	4wDM	32/42hp	Dorman 2DL		6T	109
10160	7.2.1950	2ft 11in	4wDM	32/42hp	Dorman 2DL		5T	223
10342	6.11.1952	600mm	4wDM	32/42hp	Dorman 2DL		6T	109
10362	23.4.1953	600mm	4wDM	32/42hp	Dorman 2DL		6T	261
11090	19.9.1957	600mm	4wDM	50hp	Dorman 3LA		6T	153
11091	20.9.1957	600mm	4wDM	50hp	Dorman 3LA		6T	153
11110	1.6.1959	600mm	4wDM	50hp	Dorman 3LA		6T	153
11111	28.4.1959	600mm	4wDM	50hp	Dorman 3LA		6T	63
21286	11.6.1959	600mm	4wDM	20/28hp	Dorman 2DWD	2½T		209
21287	11.6.1959	600mm	4wDM	20/28hp	Dorman 2DWD	2½T		209

21520	1.11.1955	2ft 0in	4wDM	20/28hp	Dorman 2DWD	2½ T		63,64,97
21620	20.11.1957	600mm	4wDM	20/28hp	Dorman 2DWD	2½ T		204
22031	26.6.1959	600mm	4wDM	40hp	Dorman 2LB	4½ T		216
22032	26,6,1959	600mm	4wDM	40hp	Dorman 2LB	4½ T		216
22070	6.5.1960	600mm	4wDM	30hp	Dorman 2LB	3½ T		284
26009	18.1.1965	2ft 6in	4wDM	12hp				261
40S343	31.10.1969	2ft 0in	4wDM	40hp	Dorman 2LD	2½ T		63,64

MANNING, WARDLE & CO LTD, BOYNE ENGINE WORKS, Leeds   MW

44	17.3.1862	4ft 8½in	0-6-0ST	IC	11 x 17	3ft 1½in	Old Class I	205
78	9.5.1863	4ft 8½in	0-4-0ST	OC	9 X 14	2ft 9in	E	214,221
143	14.1.1865	4ft 8½in	0-4-0ST	OC	9 x 14	2ft 9in	E	207
212	10.7.1866	4ft 8½in	0-6-0ST	IC	12 x 17	3ft 1$^{3}/_{8}$in	K	121,182,211
221	30.11.1866	4ft 8½in	0-4-0ST	OC	9 x 14	2ft 9in	E	88,192,221
225	20.3.1867	4ft 8½in	0-4-0ST	OC	9 x 14	2ft 9in	E	133,214,221 255
318	11.10.1870	4ft 8½in	0-6-0ST	IC	13 x 18	3ft 0in	M	210
336	28.12.1871	4ft 8½in	0-4-0ST	OC	12 x 18	3ft 0in	H	172
457	5.4.1875	4ft 8½in	0-4-0ST	OC	10 x 16	2ft 9in	F	133
554	31.1.1876	4ft 8½in	0-4-0ST	OC	10 x 16	2ft 9in	F	214,221,257
559	27.5.1875	4ft 8½in	0-4-0ST	OC	8 x 14	2ft 8in	D	133
585	3.7.1876	4ft 8½in	0-4-0ST	OC	10 x 16	2ft 9in	F	185,214
588	13.3.1876	4ft 8½in	0-6-0ST	IC	12 x 17	3ft 1½in	K	220
589	3.4.1876	4ft 8½in	0-6-0ST	IC	12 x 17	3ft 1½in	K	220
595	5.4.1876	4ft 8½in	0-6-0ST	IC	12 x 17	3ft 1½in	K	70,254,258
606	19.4.1876	4ft 8½in	0-4-0ST	OC	9 x 14	2ft 9in	E	220
607	21.6.1876	4ft 8½in	0-6-0ST	IC	12 x 17	3ft 1½in	K	220
618	30.8.1876	4ft 8½in	0-4-0ST	OC	10 x 16	2ft 9in	F	215,221,257
619	6.10.1876	4ft 8½in	0-4-0ST	OC	10 x 16	2ft 9in	F	220
621	11.8.1876	4ft 8½in	0-6-0ST	IC	12 x 17	3ft 1½in	K	220
624	3.8.1876	4ft 8½in	0-4-0ST	OC	8 x 14	2ft 8in	D	220
625	9.8.1876	4ft 8½in	0-4-0ST	OC	8 x 14	2ft 8in	D	220
631	31.10.1876	4ft 8½in	0-4-0ST	OC	10 x 16	2ft 9in	F	220
635	27.4.1877	4ft 8½in	0-4-0ST	OC	10 x 16	2ft 9in	F	181,185
636	11.12.1876	4ft 8½in	0-6-0ST	IC	12 x 17	3ft 1½	K	210
641	20.3.1877	4ft 8½in	0-6-0ST	IC	12 x 17	3ft 1$^{3}/_{8}$in	K	185
663	28.6.1877	4ft 8½in	0-6-0ST	IC	12 x 17	3ft 1$^{3}/_{8}$in	K	112,214,221
672	5.11.1877	4ft 8½in	0-4-0ST	OC	8 x 14	2ft 8in	D	214
690	26.2.1878	4ft 8½in	0-4-0ST	OC	10 x 16	2ft 9in	F	220
694	30.9.1878	4ft 8½in	0-6-0ST	IC	12 x 17	3ft 1½in	K	112,147,214 221
704	30.7.1878	4ft 8½in	0-4-0ST	OC	8 x 14	2ft 8in	D	246
706	30.8.1878	4ft 8½in	0-4-0ST	OC	9 x 14	2ft 9in	E	77
718	29.3.1880	4ft 8½in	0-4-0ST	OC	8 x 14	2ft 8in	D	257
738	11.4.1881	4ft 8½in	0-6-0ST	IC	12 x 17	3ft 1$^{3}/_{8}$in	K	233
779	3.3.1881	4ft 8½in	0-4-0ST	OC	12 x 18	3ft 0in	H	172
780	8.9.1881	4ft 8½in	0-4-0ST	OC	12 x 18	3ft 0in	H	177
786	13.8.1881	4ft 8½in	0-4-0ST	OC	8 x 14	2ft 9in	E	115
809	9.9.1881	4ft 8½in	0-4-0ST	OC	8 x 14	2ft 8in	D	88
810	14.12.1881	4ft 8½in	0-4-0ST	OC	8 x 14	2ft 8in	D	161

814	7.2.1881	4ft 8½in	0-4-0ST	OC	8 x 14	2ft8in	D	152
825	23.2.1882	4ft 8½in	0-4-0ST	OC	8 x 14	2ft8in	D	80
841	7.8.1882	4ft 8½in	0-6-0ST	IC	13 x 18	3ft0in	M	184,221,243 244
847	17.11.1882	4ft 8½in	0-6-0ST	IC	12 x 17	3ft1^{3}/$_{8}$in	K	116
856	13.2.1883	4ft 8½in	0-4-0ST	OC	9 x 14	2ft9in	E	185,214
858	18.12.1882	4ft 8½in	0-6-0ST	IC	13 x 18	3ft0in	M	59,221,243 244
861	22.1.1883	4ft 8½in	0-4-0ST	OC	10 x 16	2ft9in	F	214,221
862	23.1.1883	4ft 8½in	0-4-0ST	OC	10 x 16	2ft9in	F	214,221
867	26.12.1883	4ft 8½in	0-4-0ST	OC	8 x 14	2ft8in	D	215,221
872	6.6.1883	4ft 8½in	0-6-0ST	IC	13 x 18	3ft0in	M	112,147,215 221
873	13.6.1883	4ft 8½in	0-6-0ST	IC	13 x 18	3ft0in	M	112,147,215 221
882	30.3.1883	4ft 8½in	0-4-0ST	OC	10 x 16	2ft9in	F	215,221
883	2.4.1883	4ft 8½in	0-4-0ST	OC	10 x 16	2ft9in	F	185,215,221 277
886	9.3.1883	4ft 8½in	0-4-0ST	OC	8 x 14	2ft8in	D	214
887	24.6.1884	4ft 8½in	0-4-0ST	OC	8 x 14	2ft8in	D	230
888	12.3.1883	4ft 8½in	0-6-0ST	IC	12 x 17	3ft1½in	K	214
889	6.4.1883	4ft 8½in	0-6-0ST	IC	12 x 17	3ft1½in	K	116
893	2.12.1884	4ft 8½in	0-4-0ST	OC	8 x 14	2ft8in	D	111,219
898	21.12.1883	4ft 8½in	0-6-0ST	IC	13 x 18	3ft0in	M	210
900	6.3.1884	4ft 8½in	0-4-0ST	OC	9 x 14	2ft9in	E	214
901	19.1.1885	4ft 8½in	0-4-0ST	OC	9 x 14	2ft9in	E	134,138,180
902	8.9.1884	4ft 8½in	0-6-0ST	IC	12 x 17	3ft1½in	K	230
905	27.8.1884	4ft 8½in	0-4-0ST	OC	12 x 18	3ft0in	H	111,143,219
927	14.1.1885	4ft 8½in	0-6-0ST	IC	13 x 18	3ft0in	M	214
941	28.3.1885	4ft 8½in	0-4-0ST	OC	8 x 14	2ft8in	D	79
944	27.2.1885	4ft 8½in	0-6-0ST	IC	12 x 17	3ft1½in	K	116
951	7.1.1885	4ft 8½in	0-6-0ST	IC	14 x 20	3ft6in	Q	185,212,277
958	17.3.1885	4ft 8½in	0-6-0ST	IC	12 x 17	3ft1½	K	116
969	1.5.1885	4ft 8½in	0-6-0ST	IC	12 x 17	3ft1½in	K	116
983	6.8.1886	4ft 8½in	0-4-0ST	OC	10 x 16	2ft9in	F	110,177
1008	25.2.1887	4ft 8½in	0-4-0ST	OC	14 x 18	3ft0in	P	143,168
1014	21.4.1887	4ft 8½in	0-4-0ST	OC	10 x 16	2ft9in	F	230
1017	2.11.1887	4ft 8½in	0-4-0ST	OC	9 x 14	2ft9in	E	258
1039	19.12.1887	4ft 8½in	0-4-0ST	OC	10 x 16	2ft9in	F	181,185
1069	26.10.1888	4ft 8½in	0-6-0ST	IC	12 x 17	3ft0in	K	254,255
1088	19.12.1888	4ft 8½in	0-4-0ST	OC	10 x 16	2ft9in	F	252
1106	20.9.1888	4ft 8½in	0-4-0ST	OC	14 x 18	3ft0in	P	143,247
1119	23.2.1889	4ft 8½in	0-6-0ST	IC	12 x 17	3ft0in	K	252
1135	23.4.1891	4ft 8½in	0-4-0ST	OC	9 x 14	2ft9in	E	59,102,243 244
1141	29.10.1889	4ft 8½in	0-4-0ST	OC	12 x 18	3ft0in	H	95,254
1190	2.9.1890	4ft 8½in	0-6-0ST	IC	12 x 18	3ft0in	L	185
1196	21.8.1890	4ft 8½in	0-6-0T	IC	14 x 22	3ft9in	Special	59,243
1208	29.12.1890	4ft 8½in	0-6-0ST	IC	12 x 17	3ft0in	K	243
1241	22.1.1892	4ft 8½in	0-4-0ST	OC	10 x 16	2ft9in	F	53,230
1242	2.8.1894	4ft 8½in	0-4-0ST	OC	10 x 16	2ft9in	F	80

1279	5.4.1895	4ft 8½in	0-4-0ST	OC	9 x 14	2ft9in	E	218
1303	3.8.1895	4ft 8½in	0-4-0ST	OC	10 x 16	2ft9in	F	88
1306	29.4.1895	4ft 8½in	0-4-0ST	OC	9 x 14	2ft9in	E	50,59,109 244
1317	30.10.1895	4ft 8½in	0-6-0ST	IC	15 x 22	3ft6in	Special	295
1318	15.10.1895	4ft 8½in	0-4-0ST	OC	8 x 14	2ft8in	D	80,218
1383	29.3.1898	4ft 8½in	0-4-0ST	OC	10 x 16	2ft9in	F	170
1406	21.12.1898	4ft 8½in	0-4-0ST	OC	10 x 16	2ft9in	F	79
1427	8.2.1899	4ft 8½in	0-4-0ST	OC	14 x 18	3ft0in	P	135
1442	12.4.1899	4ft 8½in	0-4-0ST	OC	12 x 18	3ft0in	H	150,178
1446	29.3.1899	4ft 8½in	0-6-0ST	IC	12 x 18	3ft0in	L	258,259
1449	29.10.1900	4ft 8½in	0-6-0ST	IC	14 x 20	3ft6in	Q	70
1450	19.6.1899	4ft 8½in	0-4-0ST	OC	8 x 14	2ft8in	D	88
1488	29.5.1900	4ft 8½in	0-6-0ST	IC	12 x 18	3ft0in	L	72,185
1539	17.2.1902	4ft 8½in	0-6-0ST	IC	12 x 17	3ft0in	K	231
1552	21.1.1902	3ft 6in	0-4-0ST	OC	9½ x 14	2ft6in	Special	218
1560	16.5.1902	4ft 8½in	0-6-0ST	IC	12 x 18	3ft0in	L	224
1561	2.6.1902	4ft 8½in	0-6-0ST	IC	12 x 17	3ft0in	K	247
1576	26.1.1903	4ft 8½in	0-6-0ST	IC	12 x 17	3ft0in	K	195
1581	14.12.1903	4ft 8½in	0-4-0ST	OC	12 x 18	3ft0in	H	172
1590	27.2.1903	4ft 8½in	0-6-0ST	IC	12 x 17	3ft0in	K	185,268
1617	10.12.1903	4ft 8½in	0-6-0ST	IC	12 x17	3ft0in	K	72,185
1619	21.4.1904	4ft 8½in	0-4-0ST	OC	12 x 18	3ft0in	H	150,178
1642	28.11.1904	4ft 8½in	0-6-0ST	IC	12 x 18	3ft0in	L	195,196
1665	29.8.1905	4ft 8½in	0-6-0ST	IC	13 x 18	3ft0in	M	231
1671	16.12.1905	4ft 8½in	0-4-0ST	OC	10 x 16	2ft9in	F	167,181
1674	3.1.1906	4ft 8½in	0-6-0ST	IC	12 x 18	3ft0in	L	72,185,259
1756	4.1.1910	4ft 8½in	0-4-0ST	OC	10 x 16	2ft9in	F	76,78
1762	20.10.1910	4ft 8½in	0-6-0ST	IC	16 x 22	3ft6in	T	294
1832	17.10.1913	4ft 8½in	0-4-0ST	OC	10 x 16	2ft9in	F	134
1846	29.4.1914	4ft 8½in	0-4-0ST	OC	12 x 18	3ft0in	H	59,70,168 254
1913	9.3.1917	4ft 8½in	0-4-0ST	OC	16 x 24	3ft6in	Special	178
1951	29.4.1918	4ft 8½in	0-4-0PM		180hp	3ft0in	Thornycroft	98
1973	8.4.1919	4ft 8½in	0-4-0ST	OC	12 x 18	3ft0in	H	76
2004	4.6.1921	4ft 8½in	0-6-0ST	IC	13 x 18	3ft0in	Special	273
2005	18.6.1921	4ft 8½in	0-6-0ST	IC	13 x 18	3ft0in	Special	100,172 187,273
2006	18.6.1921	4ft 8½in	0-6-0ST	IC	13 x 18	3ft0in	Special	273
2009	1.11.1921	4ft 8½in	0-6-0ST	IC	16 x 22	3ft6in	Special	294
2015	7.12.1921	4ft 8½in	0-6-0ST	IC	13 x 20	3ft0in	Special	295
2044	8.2.1926	4ft 8½in	0-4-0ST	OC	10½x16	2ft9in	F	79,80
2045	17.6.1926	4ft 8½in	0-6-0ST	IC	13 x 18	3ft0in	Special	72,185

NEILSON & CO LTD, SPRINGBURN WORKS, Glasgow   N

1561	1870	4ft 8½in	0-4-0WT	OC	10 x 18		133
1562	1870	4ft 8½in	0-4-0WT	OC	10 x 18		133,134
1659	1872	4ft 8½in	0-4-0WT	OC	10 x 18		133
2119	1876	4ft 8½in	0-4-0ST	OC	12 x 20	3ft7in	294
2151	1876	4ft 8½in	0-4-0WT	OC	10 x 18		133
2227	1877	4ft 8½in	0-4-0WT	OC	10 x 18		133

2228	1877	4ft 8½in	0-4-0WT	OC	10 x 18		133
2380	1878	4ft 8½in	0-4-0WT	OC	10 x 18		133
2382	1878	4ft 8½in	0-4-0T	OC	10 x 18		133
2465	1879	4ft 8½in	0-4-0T	OC	10 x 18		133
2466	1879	4ft 8½in	0-4-0T	OC	10 x 18		133
2597	1880	4ft 8½in	0-4-0T	OC	12 x 20		134
2598	1880	4ft 8½in	0-4-0T	OC	10 x 18		134
3097	1883	4ft 8½in	0-4-0ST	OC	10 x 18		134
3345	1884	4ft 8½in	0-4-0ST	OC	10 x 18		134
3451	1885	4ft 8½in	0-4-0ST	OC	10 x 18		134
3789	1888	4ft 8½in	0-4-0T	OC	12 x 20		134
4249	1890	4ft 8½in	0-4-0T	OC	10 x 18		134
4250	1890	4ft 8½in	0-4-0ST	OC	10 x 18		134
4397	1891	4ft 8½in	0-4-0T	OC	11 x 19	3ft0in	138,267
4408	1892	4ft 8½in	0-4-0T	OC	10 x 18		134
4414	1892	4ft 8½in	0-4-0T	OC	12 x 20		134
4444	1892	4ft 8½in	0-4-0T	OC	12 x 20		137
4445	1892	4ft 8½in	0-4-0T	OC	12 x 20		137
4571	1892	4ft 8½in	0-4-0T	OC	10 x 18		135
4572	1892	4ft 8½in	0-4-0T	OC	10 x 18		137
5086	1896	4ft 8½in	0-4-0T	OC	10 x 18		134
5087	1896	4ft 8½in	0-4-0ST	OC	10 x 18		134
5228	1897	4ft 8½in	0-4-0T	OC	10 x 18		134
5229	1897	4ft 8½in	0-4-0T	OC	10 x 18		134
5230	1897	4ft 8½in	0-4-0T	OC	10 x 18		134
5231	1897	4ft 8½in	0-4-0T	OC	10 x 18		134,137

NORTH BRITISH LOCOMOTIVE CO LTD, Glasgow — NB
HYDE PARK WORKS — NBH

16032	1903	4ft 8½in	2-4-2ST	OC	14 x 20	3ft8in	116
16601	1905	4ft 8½in	2-4-2ST	OC	14 x 20	3ft8in	116
17223	1906	4ft 8½in	2-4-2ST	OC	14 x 20	3ft8in	116,181
19019	1909	4ft 8½in	2-4-2ST	OC	14 x 20	3ft8in	116

QUEENS PARK WORKS — NBQ

27646	1959	4ft 8½in	0-4-0DH		275hp National M4AAU5	123

NASMYTH GASKELL & CO, BRIDGEWATER FOUNDRY, Patricroft — NG

64	4.1847	4ft 8½in	0-6-0	IC	15 x 24	4ft6in	214

NOORD NEDERLANDSCHE MACHINEFABRIEK BV, Winschoten, Holland — NNM

73511	1979	4ft 8½in	4wDH	R/R	Type 45TMC	66
80504	1980	4ft 8½in	4wDH	R/R	Type 45TMC	163
80505	1980	4ft 8½in	4wDH	R/R	Type 45TMC	163
80508	1980	4ft 8½in	4wDH	R/R	Type 45TMC	163
81514	1983	4ft 8½in	4wDH	R/R	Type 45TMC	163

NEILSON REID & CO, Glasgow NR

5348	1898	4ft 8½in	0-4-0T	OC	10x18	2ft9in	137
5936	1902	4ft 8½in	0-4-0ST	OC	14x20	3ft8in	118
6302	1902	4ft 8½in	0-4-0T	OC	10x18	2ft9in	137

ORENSTEIN & KOPPEL AG, Berlin & Nordhausen, Germany OK

[Note that the ex-works dates for the i/c locos below are those of despatch in kit form from the Nordhausen factory. Final assembly was undertaken by William Jones at Greenwich, under the 'Montania' brand name, and so the dates to traffic were considerably later.]

4031	22.3.1930	605mm	4wDM			Type RL1a	249
4241	25.9.1930	605mm	4wDM			Type RL1a	203
4501	17.6.1931	605mm	4wDM			Type RL1a	253
							263
5129	18.11.1933	605mm	4wDM			Type RL1a	91
							203,285
6194	28.6.1935	605mm	4wDM	11hp	O&K 1cyl 2¼T	Type MD	158,216
6504	11.1.1936	605mm	4wDM	11hp	O&K 1cyl 3T	Type RL1b	228
6707	11.5.1936	605mm	4wDM			Type RL1a	292
6711	27.5.1936	605mm	4wDM			Type RL1a	55
7268	16.11.1936	605mm	4wDM			Type RL1c	83
8141	8.1916	4ft 8½in	0-4-0F	OC			175
10253	4.2.1939	605mm	4wDM	14hp	O&K 1cyl 3T	Type RL1c	222
20463	30.8.1934	600mm	4wDM			Type RL2b	166

PERMANENT WAY EQUIPMENT CO LTD, Bulwell, Nottingham Perm

011	1986	4ft 8½in	4wDHR	286
001	1987	4ft 8½in	4wBE	293
001	1987	4ft 8½in	4wDHR	190

PECKETT & SON LTD, ATLAS ENGINE WORKS, Bristol P

442	1.9.1885	4ft 8½in	0-4-0ST	OC	10 x 14	2ft6in	M3	210
596	20.12.1894	4ft 8½in	0-4-0ST	OC	12 x 18	3ft0in	R1	158
633	2.12.1896	4ft 8½in	0-4-0ST	OC	12 x 18	3ft0in	R1	60
665	19.11.1897	4ft 8½in	0-4-0ST	OC	10 x 14	2ft6in	M4	53,59
697	12.10.1897	4ft 8½in	0-6-0ST	OC	14 x 20	3ft7in	B1	117
720	8.7.1898	4ft 8½in	0-6-0ST	OC	14 x 20	3ft7in	B1	70,243
771	18.4.1899	4ft 8½in	0-4-0ST	OC	10 x 14	2ft6in	M4	59
798	21.8.1899	4ft 8½in	0-4-0ST	OC	10 x 14	2ft6in	M4	59,242,277
800	30.8.1899	4ft 8½in	0-4-0ST	OC	10 x 14	2ft6in	M4	59,242
806	28.12.1899	4ft 8½in	0-6-0ST	OC	14 x 20	3ft7in	B1	243
936	6.9.1902	4ft 8½in	0-4-0ST	OC	10 x 14	2ft6in	M4	68,268
939	12.11.1902	4ft 8½in	0-6-0ST	OC	14 x 20	3ft7in	B1	243
958	6.10.1902	4ft 8½in	0-6-0ST	IC	16 x 22	3ft10in	X	243
1067	29.7.1905	4ft 8½in	0-6-0ST	IC	14 x 20	3ft2in	C	71
1114	12.6.1907	4ft 8½in	0-4-0ST	OC	14 x 20	3ft2½in	W5	74
1142	27.7.1908	4ft 8½in	0-4-0ST	OC	14 x 20	3ft2½in	W5	103
1241	8.4.1911	4ft 8½in	0-6-0ST	IC	16 x 22	3ft10in	X2	243
1287	8.8.1912	4ft 8½in	0-4-0T	OC	8 x 12	2ft3in	8X12	172

No.	Date	Gauge	Type	Cyl.	Size	Wheels	Class	Page
1314	17.3.1913	4ft 8½in	0-4-0ST	OC	10 x 15	2ft9in	M5	59
1318	17.3.1913	4ft 8½in	0-4-0ST	OC	12 x 18	3ft0in	R2	172,268
1319	25.8.1913	4ft 8½in	0-4-0ST	OC	12 x 18	3ft0in	R2	172
1369	28.5.1914	4ft 8½in	0-4-0ST	OC	12 x 18	3ft0in	R2	172
1438	10.7.1916	4ft 8½in	0-4-0ST	OC	14 x 20	3ft2½in	W5	288
1439	31.7.1916	4ft 8½in	0-4-0ST	OC	14 x 20	3ft2½in	W5	61
1571	3.1.1921	4ft 8½in	0-4-0ST	OC	12 x 18	3ft0in	R2	85,150
1574	24.11.1920	4ft 8½in	0-4-0ST	OC	10 x 15	2ft9in	M5	135,137
1575	6.12.1920	4ft 8½in	0-4-0ST	OC	10 x 15	3ft0in	M5	137
1576	15.11.1921	4ft 8½in	0-4-0ST	OC	12 x 18	3ft0in	R2	137
1606	16.4.1923	4ft 8½in	0-4-0ST	OC	12 x 18	3ft0in	R2	72,185
1616	27.4.1923	4ft 8½in	0-6-0ST	OC	14 x 20	3ft7in	B2	277
1617	17.5.1923	4ft 8½in	0-6-0ST	OC	14 x 20	3ft7in	B2	277
1689	11.5.1925	4ft 8½in	0-4-0ST	OC	12 x 18	3ft0in	R2	172
1701	8.3.1926	4ft 8½in	0-4-0ST	OC	14 x 22	3ft2½in	W6	60
1707	5.8.1926	4ft 8½in	0-4-0ST	OC	14 x 22	3ft2½in	W6	60,184
1725	20.4.1927	4ft 8½in	0-4-0ST	OC	14 x 22	3ft2½in	W6	61
1734	11.7.1927	4ft 8½in	0-4-0ST	OC	12 x 18	3ft0in	R2	168,182
1735	31.10.1927	4ft 8½in	0-4-0ST	OC	12 x 18	3ft0in	R2	85,150
1739	13.2.1928	4ft 8½in	0-4-0ST	OC	14 x 22	3ft2½in	W6	74,276
1741	30.11.1927	4ft 8½in	0-4-0ST	OC	14 x 22	3ft2½in	W6 Special	150
1742	29.12.1927	4ft 8½in	0-4-0ST	OC	14 x 22	3ft2½in	W6 Special	187
1746	29.12.1927	4ft 8½in	0-4-0ST	OC	14 x 22	3ft2½in	W6	168
1806	18.2.1930	4ft 8½in	0-4-0ST	OC	12 x 20	3ft0½in	R4	172,268
1811	16.6.1930	4ft 8½in	0-4-0ST	OC	14 x 22	3ft2½in	W6 Special	135
1821	19.9.1931	4ft 8½in	0-4-0ST	OC	12 x 20	3ft0½in	R4	172
1837	20.10.1931	4ft 8½in	0-4-0ST	OC	14 x 22	3ft2½in	W6	135
1861	15.10.1934	4ft 8½in	0-6-0ST	OC	14 x 22	3ft7in	B3	89
1890	20.1.1936	4ft 8½in	0-6-0ST	OC	14 x 22	3ft7in	B3	89
1908	7.6.1937	4ft 8½in	0-4-0ST	OC	12 x 20	3ft0½in	R4	89
1919	27.8.1936	4ft 8½in	0-6-0ST	OC	14 x 22	3ft7in	B3	172
1920	14.12.1936	4ft 8½in	0-6-0ST	OC	14 x 22	3ft7in	B3	172
1923	14.4.1937	4ft 8½in	0-6-0ST	OC	14 x 22	3ft7in	B3	172
1932	7.7.1937	4ft 8½in	0-4-0ST	OC	14 x 22	3ft2½in	W6 Special	135
1933	9.8.1937	4ft 8½in	0-4-0ST	OC	14 x 22	3ft2½in	W6 Special	135
1938	13.9.1937	4ft 8½in	0-6-0ST	OC	14 x 22	3ft7in	B3	89
1966	15.3.1939	4ft 8½in	0-4-0ST	OC	12 x 18	3ft0in	R2 Special	135
1994	19.11.1940	4ft 8½in	0-4-0ST	OC	14 x 22	3ft2½in	W7	172,268
2000	7.12.1942	4ft 8½in	0-6-0ST	OC	14 x 22	3ft7in	B3	294
2015	28.7.1941	4ft 8½in	0-4-0ST	OC	7 x 12	2ft0in	Yorktown	268
2025	17.8.1942	4ft 8½in	0-4-0ST	OC	14 x 22	3ft2½in	W7	144
2039	15.11.1943	4ft 8½in	0-4-0ST	OC	10 x 15	2ft9in	M5	288
2047	21.9.1943	4ft 8½in	0-4-0ST	OC	12 x 20	3ft0½in	R4	276
2049	28.8.1944	4ft 8½in	0-4-0ST	OC	12 x 20	3ft0½in	R4	277
2083	14.7.1947	4ft 8½in	0-4-0ST	OC	12 x 20	3ft0½in	R4	137
2099	31.8.1948	4ft 8½in	0-4-0ST	OC	12 x 20	3ft0½in	R4	137
2100	17.1.1949	4ft 8½in	0-4-0ST	OC	12 x 20	3ft0½in	R4	288
2123	30.4.1952	4ft 8½in	0-4-0ST	OC	12 x 18	3ft0in	R2 Special	135
2135	23.12.1953	4ft 8½in	0-4-0ST	OC	12 x 20	3ft0½in	R4	138
2154	10.1954	4ft 8½in	0-6-0ST	OC	14 x 22	3ft7in	B3	89
5001	1956	4ft 8½in	0-4-0DM	200hp				89

PLASSER RAILWAY MACHINERY (GB) LTD, West Ealing, London   **Plasser**

52465A	1982	4ft 8½in	4wDMR				282

PENDERSHAAB MASHINFABRIC A/S   **PMA**

13403		2ft 6in	4wDM				196

R.F.S. ENGINEERING LTD, KILNHURST WORKS,
Kilnhurst, South Yorkshire   **RFSK**

L106	20.12.1989	900mm	4wDH	112kw	Cummins		15T	292

T.D. RIDLEY, Middlesborough   **Ridley**

13	1899	4ft 8½in	0-4-0ST	OC			53,59

ROBEL & CO, MASCHINENFABRIK, München, Germany   **Robel**

56.27-10-AG39	1983	4ft 8½in	2w-2DMR				127
21 11 RK1	1966	4ft 8½in	4wDM				89

RUSTON & HORNSBY LTD, Lincoln   **RH**

164333	12.1.1932	2ft 0in	4wDM	10hp	Lister 10/2	2½T	92
164336	1.2.1932	2ft 0in	4wDM	10hp	Lister 10/2	2½T	82
164337	19.12.1931	2ft 0in	4wDM	10hp	Lister 10/2	2½T	62,84
164338	15.2.1932	2ft 0in	4wDM	10hp	Lister 10/2	2½t	227
164342	3.3.1932	2ft 0in	4wDM	10hp	Lister 10/2	2½T	158
164347	13.3.1933	2ft 0in	4wDM	10hp	Lister 10/2	2½T	92,183,262
165365	14.3.1932	2ft 0in	4wDM	16hp	Lister 18/2	2¾T	227
166009	8.6.1932	2ft 0in	4wDM	16hp	Lister 18/2	2¾T	227,280
166010	3.9.1934	2ft 0in	4wDM	16hp	Lister 18/2	2¾T	92,114
166011	1.11.1932	2ft 0in	4wDM	16hp	Lister 18/2	2¾T	165
166048	21.2.1934	2ft 0in	4wDM	10hp	Lister 10/2	2¾T	56,82,83
166050	4.4.1934	2ft 0in	4wDM	10hp	Lister 10/2	2¾T	165
168437	29.4.1933	2ft 0in	4wDM	22/28hp Lister 3JP		4T	174
172889	23.8.1934	2ft 0in	4wDM	22/28hp Lister 3JP		4T	156,282
173394	7.11.1934	2ft 0in	4wDM	18/21hp Lister 18/2		2¾T	114
174526	16.9.1935	2ft 0in	4wDM	18/21hp Lister 18/2		2¾T	255
174528	19.12.1935	2ft 0in	4wDM	10hp	Lister 10/2	2½T	65,107,108
174945	19.3.1935	2ft 0in	4wDM	36/42hp Lister 4JP			108
174946	19.3.1935	2ft 0in	4wDM	36/42hp Lister 4JP			108
175399	8.10.1935	2ft 0in	4wDM	18/21hp Lister 18/2		2¾T	155,156,282
175408	17.12.1935	2ft 0in	4wDM	18/21hp Lister 18/2		2¾T	92,183,262
177531	20.5.1936	2ft 0in	4wDM	36/42hp Lister 4JP			50
177606	26.5.1936	2ft 0in	4wDM	27/32hp Lister 3JP		4T	174
177641	22.7.1936	2ft 0in	4wDM	10hp	Lister 10/2	2½T	280
178003	18.6.1936	2ft 0in	4wDM	36/42hp Lister 4JP			50
178004	18.6.1936	2ft 0in	4wDM	36/42hp Lister 4JP			50
178256	13.3.1936	2ft 0in	4wDM	18/21hp Lister 18/2		2¾T	156,282

178994	8.4.1936	2ft 0in	4wDM	18/21hp Lister 18/2	2¾T	200
179004	5.8.1936	2ft 0in	4wDM	18/21hp Lister 18/2	2¾T	139
179012	2.9.1936	2ft 0in	4wDM	18/21hp Lister 18/2	2¾T	204
179866	12.11.1936	2ft 0in	4wDM	36/42hp Lister 4JP		50
179872	20.8.1936	2ft 0in	4wDM	18/21hp Lister 18/2	2¾T	279
179880	16.9.1936	2ft 0in	4wDM	27/32hp Lister 3JP	4T	64,162
179881	16.9.1936	2ft 0in	4wDM	27/32hp Lister 3JP	4T	63,64,162
179884	17.10.1936	2ft 0in	4wDM	40hp Lister 4JP		108
181815	27.10.1936	2ft 0in	4wDM	20hp Lister 18/2	2¾T	204
181824	27.1.1937	2ft 6in	4wDM	27/32hp Lister 3JP	4T	50
183073	20.3.1937	2ft 0in	4wDM	36/42hp Lister 4JP		108
183741	14.4.1937	2ft 0in	4wDM	16/20hp Ruston 2VSO	3¼T	139
183744	16.4.1937	2ft 0in	4wDM	44/48hp Ruston 4VRO	5½T	50
183757	26.5.1937	2ft 0in	4wDM	33/40hp Ruston 3VRO	5T	108
186327	10.8.1937	2ft 0in	4wDM	25/30hp Ruston 3VSO	3¼T	100,280
187046	4.9.1937	2ft 0in	4wDM	16/20hp Ruston 2VSO	3¼T	200
187064	7.10.1937	2ft 0in	4wDM	25/30hp Ruston 3VSO	3¼T	99,100
187083	30.10.1937	2ft 0in	4wDM	16/20hp Ruston 2VSO	3¼T	114,280
187099	21.2.1938	2ft 0in	4wDM	11/13hp Ruston 2VTO	2½T	51,188
187101	11.11.1937	2ft 0in	4wDM	16/20hp Ruston 2VSO	3¼T	156,282
187111	8.12.1937	2ft 0in	4wDM	25/30hp Ruston 3VSO	3¼T	239
187113	17.12.1937	2ft 0in	4wDM	25/30hp Ruston 3VSO	4T	239
187120	11.3.1938	2ft 0in	4wDM	16/20hp Ruston 2VSO	3¼T	203,204
187123	14.3.1938	2ft 0in	4wDM	16/20hp Ruston 2VSO	3¼T	204
189941	31.12.1937	2ft 0in	4wDM	25/30hp Ruston 3VSO	4T	240
189948	22.1.1938	2ft 0in	4wDM	25/30hp Ruston 3VSO	4T	239
189952	18.1.1938	2ft 0in	4wDM	25/30hp Ruston 3VSO	4T	240
189993	14.3.1938	2ft 0in	4wDM	16/20hp Ruston 2VSO		204
189997	16.3.1938	2ft 0in	4wDM	16/20hp Ruston 2VSO		204
191646	16.3.1938	2ft 0in	4wDM	16/20hp Ruston 2VSO	2½T	129
191682	19.7.1938	2ft 0in	4wDM	16/20hp Ruston 2VSO	3¼T	91
192325	27.6.1938	4ft 8½in	4wDM	80/88hp Ruston 4VPO	18T	172
192861	2.6.1939	2ft 0in	4wDM	44/48hp Ruston 4VRO	6½T	50
192876	19.11.1938	2ft 0in	4wDM	25/30hp Ruston 3VSO	4T	99
193966	23.3.1939	2ft 0in	4wDM	25/30hp Ruston 3VSO	4T	99
193972	30.12.1938	2ft 0in	4wDM	16/20hp Ruston 2VSO	2¾T	101
195856	7.7.1939	4ft 8½in	4wDM	44/48hp Ruston 4VRO	7½T	217
198265	28.9.1939	2ft 0in	4wDM	16/20hp Ruston 2VSO	3½T	280
200513	13.5.1940	2ft 0in	4wDM	44/48hp Ruston 4VRO	7½T	265
200771	11.3.1941	2ft 0in	4wDM	25/30hp Ruston 3VSO	3¼T	99
200780	21.7.1941	2ft 0in	4wDM	25/30hp Ruston 3VSO	4T	239
200781	23.8.1941	2ft 0in	4wDM	25/30hp Ruston 3VSO	4T	239
200782	23.8.1941	2ft 0in	4wDM	25/30hp Ruston 3VSO	4T	240
200794	27.8.1940	2ft 0in	4wDM	44/48hp Ruston 4VRO	7¼T	61
200800	1.4.1941	2ft 0in	4wDM	44/48hp Ruston 4VRO	7¼T	265
201999	9.7.1940	60cm	4wDM	20DL Ruston 2VSO	2¾T	129
202000	12.7.1940	60cm	4wDM	20DL Ruston 2VSO	2¾T	129
202005	24.6.1940	2ft 2¾in	4wDM	20DL Ruston 2VSO	2¾T	261
202967	29.10.1940	60cm	4wDM	20DL Ruston 2VSO	2¾T	127
202969	7.12.1940	60cm	4wDM	20DL Ruston 2VSO	2¾T	128,131
202970	7.12.1940	60cm	4wDM	20DL Ruston 2VSO	2¾T	128

202980	17.1.1941	2ft 0in	4wDM	20DL	Ruston 2VSO	2¾T	280
202998	11.2.1941	60cm	4wDM	20DL	Ruston 2VSO	2¾T	128
203012	14.2.1941	2ft 0in	4wDM	20DL	Ruston 2VSO	3¼T	279
203014	26.7.1941	2ft 0in	4wDM	20DL	Ruston 2VSO	3¼T	263
207103	29.4.1941	4ft 8½in	4wDM	44/48hp	Ruston 4VRO	7½T	49,293
209429	17.4.1943	2ft 0in	4wDM	13DL	Ruston 2VTH	2½T	158
210481	9.12.1941	4ft 8½in	4wDM	48DS	Ruston 4VRO	7½T	89
210483	25.4.1942	4ft 8½in	4wDM	44/48hp	Ruston 4VRO	7½T	49
211608	31.5.1941	60cm	4wDM	20DL	Ruston 2VSO	2¾T	92,183,262
211626	6.10.1941	2ft 0in	4wDM	20DL	Ruston 2VSO	2¾T	279
211632	6.11.1941	60cm	4wDM	20DL	Ruston 2VSO	2¾T	281
211635	6.11.1941	60cm	4wDM	20DL	Ruston 2VSO	2¾T	281
211646	18.2.1942	60cm	4wDM	20DL	Ruston 2VSO	2¾T	281
211649	27.2.1942	2ft 0in	4wDM	20DL	Ruston 2VSO	2¾T	279
211662	3.6.1942	60cm	4wDM	20DL	Ruston 2VSO	2¾T	179
211685	2.1.1943	2ft 0in	4wDM	30DL	Ruston 3VSO	3¼T	99
211686	5.1.1943	2ft 0in	4wDM	30DL	Ruston 3VSO	3¼T	99
213824	28.8.1942	2ft 0in	4wDM	20DL	Ruston 2VSO	2¾T	280
217963	6.11.1942	60cm	4wDM	20DL	Ruston 2VSO	2¾T	201,280
217967	30.10.1942	60cm	4wDM	20DL	Ruston 2VSO	2¾T	128,131
217968	18.11.1942	60cm	4wDM	20DL	Ruston 2VSO	2¾T	179
217999	7.12.1942	2ft 0in	4wDM	20DL	Ruston 2VSO	2¾T	87,285
218007	27.2.1943	2ft 0in	4wDM	20DL	Ruston 2VSOL	2¾T	179
218014	8.3.1943	2ft 0in	4wDM	20DL	Ruston 2VSOL	2¾T	281
218016	11.3.1943	60cm	4wDM	20DL	Ruston 2VSO	2¾T	295
218032	12.2.1943	2ft 0in	4wDM	33/40hp	Ruston 3VRO	4½T	139
218047	2.2.1943	4ft 8½in	4wDM	48DS	Ruston 4VROL	7½T	180
221604	25.5.1943	60cm	4wDM	20DL	Ruston 2VSO	2¾T	280
221639	30.8.1943	4ft 8½in	4wDM	48DS	Ruston 4VRO	7½T	49,61,269 286
221648	1.12.1943	4ft 8½in	4wDM	48DS	Ruston 4VRO	7½T	119
223696	21.1.1944	60cm	4wDM	20DL	Ruston 2VSO	2¾T	129
223716	21.1.1944	2ft 0in	4wDM	20DL	Ruston 2VSO	2¾T	255
224342	17.10.1944	4ft 8½in	4wDM	48DS	Ruston 4VRO	7½T	120
224343	20.11.1944	4ft 8½in	4wDM	48DS	Ruston 4VRO	7½T	120
224344	18.12.1944	4ft 8½in	4wDM	48DS	Ruston 4VRO	7½T	120
224347	6.3.1945	4ft 8½in	4wDM	48DS	Ruston 4VRO	7½T	121
226264	8.6.1944	60cm	4wDM	20DL	Ruston 2VSO	2¾T	261
226267	8.6.1944	60cm	4wDM	20DL	Ruston 2VSO	2¾T	281
226288	12.8.1944	60cm	4wDM	20DL	Ruston 2VSO	2¾T	242
226294	22.8.1944	60cm	4wDM	20DL	Ruston 2VSO	2¾T	158
226298	30.8.1944	60cm	4wDM	20DL	Ruston 2VSO	2¾T	261
235622	22.1.1945	60cm	4wDM	20DL	Ruston 2VSH	2¾T	281
235624	26.1.1945	60cm	4wDM	20DL	Ruston 2VSH	2¾T	298
235712	26.10.1945	60cm	4wDM	20DL	Ruston 2VSO	3¼T	261
235743	7.8.1945	3ft 6in	4wDM	48DL	Ruston 4VRO	5½T	54,110
237916	22.7.1946	2ft 0in	4wDM	30DL	Ruston 3VSH	4T	63,64
237918	27.12.1945	4ft 8½in	4wDM	48DS	Ruston 4VRO	7½T	71
237929	8.12.1946	4ft 8½in	4wDM	48DS	Ruston 4VRO	7½T	180,217
252687	27.7.1949	4ft 8½in	0-4-0DM	165DS	Ruston 6VPH	28T	172,265
256194	2.2.1948	2ft 0in	4wDM	30DL	Ruston 3VSH	4T	64,261

265611	19.7.1948	4ft 8½in	4wDM	48DS	Ruston 4VRH	7½T	69,71
277273	31.3.1949	2ft 0in	4wDM	30DL	Ruston 3VSH	3¼T	216
277278	31.3.1949	2ft 0in	4wDM	30DL	Ruston 3VSH	3¼T	100
281266	31.8.1950	4ft 8½in	0-4-0DM	165DS	Ruston 6VPH	28T	286
281269	27.11.1950	4ft 8½in	0-4-0DM	165DS	Ruston 6VPH	28T	61
283513	22.8.1949	2ft 0in	4wDM	30DL	Ruston 3VSH	3¼T	63,300
287665	24.1.1952	2ft 0in	4wDM	40DL	Ruston 3VRH	4½T	261
299103	28.10.1950	4ft 8½in	4wDM	88DS	Ruston 4VPH	20T	102
305314	9.4.1951	4ft 8½in	4wDM	88DS	Ruston 4VPH	20T	68,180
306092	8.6.1950	4ft 8½in	4wDM	88DS	Ruston 4VPH	17T	68
310081	24.8.1951	4ft 8½in	0-4-0DM	165DS	Ruston 6VPH	28T	172
338438	7.5.1953	2ft 6in	4wDM	LBU	Ruston 3VSH	3½T	287
338440	27.6.1955	2ft 0in	4wDM	LBT	Ruston 3VSH	3¼T	99,100
349041	15.6.1954	4ft 8½in	4wDM	48DS	Ruston 4VRH	7½T	277
354028	21.9.1953	2ft 0in	4wDM	20DL	Ruston 2VSH	2¾T	97,250
373359	1.11.1954	2ft 0in	4wDM	20DL	Ruston 2VSH	2¾T	51
375354	6.11.1954	2ft 0in	4wDM	LAT	Ruston 2VSH	3½T	215
375361	1.2.1955	2ft 0in	4wDM	LAT	Ruston 2VSH	3½T	215
375693	25.8.1954	2ft 6in	4wDM	LBU	Ruston 3VSH	3½T	287
381703	20.6.1955	2ft 0in	4wDM	20DL	Ruston 2VSH	3½T	261
386871	4.3.1955	4ft 8½in	4wDM	48DS	Ruston 4VRH	7½T	58
387819	10.6.1955	2ft 0in	4wDM	LBT	Ruston 3VSH	3¾T	215
393327	26.6.1956	2ft 0in	4wDM	LBT	Ruston 3VSH	3¾T	99,282
398099	26.6.1956	2ft 0in	4wDM	LBT	Ruston 3VSH	3¾T	99
398100	26.6.1956	2ft 0in	4wDM	LBT	Ruston 3VSH	3¾T	99,100
398611	16.6.1956	4ft 8½in	4wDM	88DS	Ruston 4VPH	20T	180
398616	1.10.1956	4ft 8½in	4wDM	88DS	Ruston 4VPH	17T	291
411319	21.6.1957	4ft 8½in	4wDM	48DS	Ruston 4VRH	7½T	123
412715	16.8.1957	4ft 8½in	0-4-0DE	200DE	Ruston 6RPH	30T	89
412716	2.11.1957	4ft 8½in	0-4-0DE	200DE	Ruston 6RPH	30T	151,172,265
414301	8.7.1957	4ft 8½in	0-4-0DM	165DS	Ruston 6VPH	28T	49
416211	16.10.1957	4ft 8½in	0-4-0DE	165DE	Ruston 6VPH	28T	269
418791	14.4.1958	4ft 8½in	0-4-0DM	165DS	Ruston 6VPH	28T	58
418803	31.12.1957	3ft 0in	4wDM	48DLG	Ruston 4VRH	7T	287
421416	20.3.1958	4ft 8½in	4wDM	88DS	Ruston 4VPH	20T	135,139,265
425478	9.7.1959	4ft 8½in	0-4-0DE	165DE	Ruston 6VPH	28T	207
432664	24.8.1959	2ft 0in	4wDM	LBU	Ruston 2YDA	4½T	252
433493	18.11.1958	2ft 0in	4wDM	48DLU	Ruston 4YC	7T	261
433676	27.3.1961	4ft 8½in	0-4-0DE	200DE	Ruston 6RPH	30T	172,265
437362	21.1.1960	4ft 8½in	0-4-0DH	165DH	Ruston 6YEX	26T	150,264,266
437364	5.4.1961	4ft 8½in	0-4-0DH	165DH	Ruston 6YEX	28T	185
441951	28.1.1960	2ft 0in	4wDM	LB	Ruston 2YDA	3½T	63,64,162 300
457303	23.4.1964	4ft 8½in	0-4-0DH	165DH	Ruston 6YEX	28T	150,266
459518	20.6.1961	4ft 8½in	0-6-0DH	LSSH	Ruston 6RPH	36T	124
459519	5.7.1961	4ft 8½in	0-6-0DH	LSSH	Ruston 6RPH	36T	124
459520	28.6.1961	4ft 8½in	0-6-0DH	LSSH	Ruston 6RPH	36T	124
466621	4.12.1961	4ft 8½in	0-6-0DH	LSSH	Ruston 6RPH	36T	124
466622	10.1.1962	4ft 8½in	0-6-0DH	LSSH	Ruston 6RPH	36T	124
466625	22.1.1962	4ft 8½in	4wDM	88DS	Ruston 4VPH	20T	291
468043	16.10.1963	4ft 8½in	0-6-0DH	LSSH	Ruston 6RPH	36T	124

468048	20.11.1963	4ft 8½in	0-4-0DH	LSSH	Ruston 6RPH	45T	89
476142	24.9.1963	4ft 8½in	4wDM	88DS	Ruston 4VPH	20T	269
476143	31.12.1963	4ft 8½in	4wDM	88DS	Ruston 4VPH	17T	89
512463	18.1.1965	4ft 8½in	0-4-0DH	LPSH	Ruston 6YE	28T	150,266
512464	5.2.1965	4ft 8½in	0-4-0DH	LPSH	Ruston 6YE	28T	150
512572	18.2.1965	4ft 8½in	4wDM	88DS	Ruston 4VPH		291
512842	8.6.1965	4ft 8½in	0-4-0DE	LPSE	Ruston 6YE	28T	264,266
513142	22.9.1967	4ft 8½in	4wDM	88DS	Ruston 4VPH	17T	69
7002/0566/3	3.11.1965	2ft 6in	4wDM	LBT	Ruston 2YDA	3½T	196
7002/0566/4	3.11.1965	2ft 6in	4wDM	LBT	Ruston 2YDA	3½T	196

RUSTON, PROCTOR & CO LTD, SHEAF IRON WORKS, Lincoln RP

	c1870	4ft 8½in	0-4-0ST	OC			111,143,219
13111	1888	4ft 8½in	0-4-0ST	OC			102
51697	30.1.1917	1ft 6in	4wPM	10hp	Ruston Proctor ZLH		131
51707	16.3.1917	1ft 6in	4wPM	10hp	Ruston Proctor ZLH		131
51901	28.9.1917	1ft 6in	4wPM	10hp	Ruston Proctor ZLH		131
51927	29.10.1917	1ft 6in	4wPM	10hp	Ruston Proctor ZLH		131

ROLLS-ROYCE LTD, SENTINEL WORKS, Shrewsbury RR

10189	21.7.1964	4ft 8½in	0-4-0DH	325hp	R-R C8SFL	40T	180
10214	27.11.1964	4ft 8½in	0-6-0DH	375hp	R-R C8TFL	48T	264
10230	18.2.1965	4ft 8½in	4wDH	230hp	R-R C6SFL	34T	172
10232	27.5.1965	4ft 8½in	4wDH	255hp	R-R C6SFL	34T	60
10235	21.6.1965	4ft 8½in	4wDH	256hp	R-R C6SFL	34T	172
10247	30.12.1965	4ft 8½in	4wDH	256hp	R-R C6SFL	34T	106
10248	27.1.1966	4ft 8½in	4wDH	256hp	R-R C6SFL	34T	106
10249	10.2.1966	4ft 8½in	4wDH	256hp	R-R C6SFL	34T	106
10276	2.1.1968	4ft 8½in	4wDH	256hp	R-R C6SFL	34T	172

RANSOMES & RAPIER LTD, RIVERSIDE WORKS, Ipswich R&R

	1938	2ft 0in	4wDM				105

ROBERT STEPHENSON & CO LTD, Forth Street, Newcastle-upon-Tyne RS

162	11.1836	4ft 8½in	0-4-2	IC	14 x 15	4ft6in	260
2844	2.10.1896	4ft 8½in	0-6-0ST	OC	16 x 20	3ft6in	71,111,142
2845	6.10.1896	4ft 8½in	0-6-0ST	OC	16 x 20	3ft6in	111,142
2981	23.5.1900	4ft 8½in	0-6-0ST	OC	16 x 20	3ft6in	111,142
2982	28.5.1900	4ft 8½in	0-6-0ST	OC	16 x 20	3ft6in	111,112,147
2983	31.5.1900	4ft 8½in	0-6-0ST	OC	16 x 20	3ft6in	111,143
3050	26.4.1901	4ft 8½in	0-6-0ST	OC	16 x 20	3ft6in	111,143
3053	1901	4ft 8½in	0-6-0ST	OC	16 x 20	3ft6in	112,147
3070	1901	4ft 8½in	0-6-0ST	OC	16 x 20	3ft6in	111,143
3094	1902	4ft 8½in	0-6-0ST	OC	16 x 20	3ft6in	70,111,143
3120	30.1.1904	4ft 8½in	0-6-0ST	OC	16 x 20	3ft6in	70,112,147
3170	27.6.1905	4ft 8½in	0-6-0ST	OC	16 x 20	3ft6in	112,147
3296	28.10.1907	4ft 8½in	0-6-0ST	OC	16 x 20	3ft6in	112,147

ROBERT STEPHENSON & HAWTHORNS LTD — RSH
D DARLINGTON WORKS — RSHD
N NEWCASTLE-UPON-TYNE WORKS — RSHN

Works	No.	Date	Gauge	Type	Cyl	Bore × Stroke	Wheel	Notes	Page
N	6991	26.12.1940	4ft 8½in	0-4-0DM	Fowler 150hp				276
N	7031	22.5.1941	4ft 8½in	0-6-0ST	IC	16 x 22	3ft6in		288
D	7040	3.12.1941	4ft 8½in	0-4-0ST	OC	14 x 22	3ft6in		276
N	7058	18.4.1942	4ft 8½in	0-4-0ST	OC	14 x 22	3ft6in		288
N	7096	8.8.1943	4ft 8½in	0-6-0ST	IC	18 x 26	4ft3in		119
N	7098	19.8.1942	4ft 8½in	0-6-0ST	IC	18 x 26	4ft3in		119
N	7099	26.8.1943	4ft 8½in	0-6-0ST	IC	18 x 26	4ft3in		119
N	7103	23.9.1943	4ft 8½in	0-6-0ST	IC	18 x 26	4ft3in		144
N	7104	30.9.1943	4ft 8½in	0-6-0ST	IC	18 x 26	4ft3in		144
N	7105	6.10.1943	4ft 8½in	0-6-0ST	IC	18 x 26	4ft3in		144
N	7107	20.10.1943	4ft 8½in	0-6-0ST	IC	18 x 26	4ft3in		144
N	7108	27.10.1943	4ft 8½in	0-6-0ST	IC	18 x 26	4ft3in		182
N	7111	18.11.1943	4ft 8½in	0-6-0ST	IC	18 x 26	4ft3in	{WB 7077}	181
N	7112	29.11.1943	4ft 8½in	0-6-0ST	IC	18 x 26	4ft3in		122
N	7113	9.12.1943	4ft 8½in	0-6-0ST	IC	18 x 26	4ft3in		144
N	7130	30.3.1944	4ft 8½in	0-6-0ST	IC	18 x 26	4ft3in		122
N	7141	8.6.1944	4ft 8½in	0-6-0ST	IC	18 x 26	4ft3in		120
N	7150	2.7.1945	4ft 8½in	0-4-0ST	OC	14 x 22	3ft6in		67
N	7210	5.4.1945	4ft 8½in	0-6-0ST	IC	18 x 26	4ft3in		120
N	7286	31.5.1945	4ft 8½in	0-6-0ST	IC	18 x 26	4ft3in		120
N	7309	19.8.1946	4ft 8½in	0-4-0ST	OC	12 x 20	3ft1in		138,267
N	7386	18.5.1948	4ft 8½in	0-4-0ST	OC	14 x 22	3ft6in		288
N	7474	24.1.1949	4ft 8½in	0-4-0ST	OC	12 x 20	3ft0in		135
N	7540	30.5.1949	4ft 8½in	0-4-0ST	OC	14 x 22	3ft6in		68
N	7597	12.9.1949	4ft 8½in	0-6-0T	OC	18 x 24	3ft8in		288
N	7665	1.5.1951	4ft 8½in	0-4-0F	OC	17 x 26	2ft11in		135
N	7667	29.8.1950	4ft 8½in	0-6-0ST	IC	18 x 26	4ft2½in		294
N	7671	10.11.1950	4ft 8½in	0-6-0ST	IC	18 x 26	4ft0½in		286,288
N	7765	11.9.1954	4ft 8½in	0-6-0T	OC	18 x 24	4ft6in		286
N	7803	26.7.1954	4ft 8½in	0-4-0F	OC	17 x 16	2ft11in		66,135
N	7817	17.8.1954	4ft 8½in	0-4-0ST	OC	16 x 24	3ft8in		286
N	7846	19.10.1955	4ft 8½in	0-6-0T	OC	18 x 24	3ft8in		288
D	7892	19.11.1957	4ft 8½in	0-6-0DM	204hp	Gardner 8L3		{DC 2606}	89
D	7897	16.1.1958	4ft 8½in	0-6-0DM	204hp	Gardner 8L3		{DC 2611}	89,288
D	7922	19.3.1957	4ft 8½in	0-4-0DM	153hp	Gardner 6L3		{DC 2589}	151
D	8097	15.2.1960	4ft 8½in	0-6-0DM	204hp	Gardner 8L3		{DC 2656}	288
D	8098	25.2.1960	4ft 8½in	0-6-0DM	204hp	Gardner 8L3		{DC 2657}	89,288
D	8184	29.4.1961	4ft 8½in	0-6-0DM	204hp	Gardner 8L3		{DC 2706}	293
D	8192	20.7.1961	4ft 8½in	0-6-0DM	204hp	Gardner 8L3		{DC 2714}	89
D	8343	9.1962	4ft 8½in	0-6-0DH	311hp	R-R C8SFL			176
D	8366	1962	4ft 8½in	0-4-0DH	179hp	R-R C6NFL		{WB 3211}	251
D	8367	1962	4ft 8½in	0-4-0DH	266hp	Dorman 6QA		{WB 3212}	182

RUHRTHALER MASCHINENFABRIK SCHWARTZ & DYCKERHOFF AG — Ruhr

	3920	28.11.1969	750mm	4wDH	100hp				298

R & W HAWTHORN & CO, FORTH BANK WORKS, Newcastle-upon-Tyne RWH

217	1836	4ft 8½in	0-4-2ST	IC	14 x 15		260

SENTINEL (SHREWSBURY) LTD, BATTLEFIELD WORKS, Shrewsbury   S

5735	1926	4ft 8½in	4wVBT	VCG	6¾x9	2ft6in	100hp	72,185
6310CH	1926	4ft 8½in	0-4-0VBT	VCG	6¾x9			168
6893	1927	4ft 8½in	4wVBT	VCG	6¾x9	2ft6in	80hp	102
6951	1927	4ft 8½in	0-4-0VBT	VCG	6¾x9			134
6994	1927	4ft 8½in	4wVBT	VCG	6¾x9	2ft6in	80hp	65,70
7060	1927	4ft 8½in	4wVBT	VCG	6¾x9	2ft6in	100hp	75
7297	1928	4ft 8½in	4wVBT	VCG	6¾x9	2ft6in	100hp	102
7492	1928	4ft 8½in	4wVBT	VCG	6¾x9	2ft6in	100hp	300
7696	1929	4ft 8½in	4wVBT	VCG	6¾x9	2ft6in	100hp	75
8796	1933	4ft 8½in	4wVBT	VCG	6¾x9	2ft6in	100hp	75
9365	1945	4ft 8½in	4wVBT	VCG	6¾x9	2ft6in	100hp	75
9398	1.1.1950	4ft 8½in	4wVBT	VCG	6¾x9	2ft6in	100hp	67,135
10127	19.4.1963	4ft 8½in	0-4-0DH	272hp	R-R C8NFL	37T		89
10143	11.12.1963	4ft 8½in	0-8-0DH	621hp	2xR-R C8SFL	54T		124

CRISTOPH SCHÖTTLER MASCHINENFABRIK GmbH, Germany   Schöma

4415	1980	2ft 0in	4wDH	185hp	Deutz	24T	CFL-180 DCL	239
5424	2.1995	900mm	4wDH		Deutz		CFL-180 DCL	239
5554	1998	900mm	4wDH	230hp	Deutz	30T	CFL-180 DCL	200
5555	1998	900mm	4wDH	230hp	Deutz	30T	CFL-180 DCL	200
5556	1998	900mm	4wDH	230hp	Deutz	30T	CFL-180 DCL	200
5557	1998	900mm	4wDH	230hp	Deutz	30T	CFL-180 DCL	200
5558	1998	900mm	4wDH	230hp	Deutz	30T	CFL-180 DCL	200
5559	1998	900mm	4wDH	230hp	Deutz	30T	CFL-180 DCL	200
5560	1998	900mm	4wDH	230hp	Deutz	30T	CFL-180 DCL	241
5561	1998	900mm	4wDH	230hp	Deutz	30T	CFL-180 DCL	241
5562	1998	900mm	4wDH	230hp	Deutz	30T	CFL-180 DCL	241
5610	6.1999	900mm	4wDH	230hp	Deutz	27T	CFL-180 DCL	241
5611	6.1999	900mm	4wDH	230hp	Deutz	27T	CFL-180 DCL	241
5615	8.1999	900mm	4wDH		Deutz		CFL-180 DCL	241
5713	2001	750mm	4wDH		Deutz		CHL-60 G	238
5714	2001	750mm	4wDH		Deutz		CHL-60 G	238
5715	2001	750mm	4wDH		Deutz		CHL-60 G	238
5724	2002	900mm	4wDH		Deutz		CFL-180 DCL	239
5725	2002	900mm	4wDH		Deutz		CFL-180 DCL	239
5726	2002	900mm	4wDH		Deutz		CFL-180 DCL	239
5727	2002	900mm	4wDH		Deutz		CFL-180 DCL	239
5728	2002	900mm	4wDH		Deutz		CFL-180 DCL	239
5729	2002	900mm	4wDH		Deutz		CFL-180 DCL	239

SWINDON WORKS, Wilts (GWR / BR)   Sdn

[D2018]	4.1958	4ft 8½in	0-6-0DM	204hp	Gardner 8L3	30¾T	293
[D2041]	4.1959	4ft 8½in	0-6-0DM	204hp	Gardner 8L3	30¾T	67,286
[D2170]	11.1960	4ft 8½in	0-6-0DM	204hp	Gardner 8L3	30¾T	291

[D2184]	4.1962	4ft 8½in	0-6-0DM	204hp	Gardner 8L3	30¾T		162,286
[D9504]	7.1964	4ft 8½in	0-6-0DH	650hp	Paxman 6TJX	50T		190
[D9529]	1.1965	4ft 8½in	0-6-0DH	650hp	Paxman 6TJX	50T		190
[D9538]	3.1965	4ft 8½in	0-6-0DH	650hp	Paxman 6TJX	50T		160

SLAUGHTER, GRUNING & CO, Bristol — SG

	1863	4ft 8½in	0-4-0ST OC	9 x 14	96

ALEXANDER SHANKS & SON LTD, DENS IRON WORKS, Arbroath — Shanks

1870	4ft 8½in	0-4-0ST	OC	111
1872	4ft 8½in	0-4-0ST	OC	111,219

STRACHAN & HENSHAW LTD, Ashton, Bristol — S&H

7502	1966	4ft 8½in	4wDM	R/R	293
7509	1968	4ft 8½in	4wDM	R/R	160

SEVERN-LAMB UK LTD, Stratford-Upon-Avon, Warwickshire — SL

SE4	1986	3ft 0in	4w-4wDH	297

SWEDISH RAIL SYSTEMS EUROC, Solna, Sweden — SRS

	4ft 8½in	2w-4DMR	R/R	190
	4ft 8½in	4wDM	R/R	190

SHARP STEWART & CO LTD, ATLAS WORKS, Manchester & Glasgow — SS

		4ft 8½in	0-6-0ST	IC			185,221
3472	1888	4ft 8½in	0-6-0T	IC	13x20	3ft0in	236
3474	1888	4ft 8½in	0-6-0T	IC	13x20	3ft0in	212,272
4114	1895	2ft11½in	0-4-0WT	OC			285
4115	1895	2ft11½in	0-4-0WT	OC			285

STRATFORD WORKS, Essex (GER / BR) — Str

[68574]	1896	4ft 8½in	0-6-0T	IC	16½x22 4ft0in	146
[68578]	1896	4ft 8½in	0-6-0T	IC	16½x22 4ft0in	146
[68644]	1912	4ft 8½in	0-6-0T	IC	16½x22 4ft0in	146

T.GREEN & SON LTD, Leeds — TG

227	1900	4ft 8½in	0-4-0ST	OC	10 x 16		98
366	1904	2ft 0in	0-6-0ST	OC	9½ x 14	2ft4in	192
440	1907	4ft 8½in	0-4-0ST	OC	5½ x 10		77

THOMAS HILL (ROTHERHAM) LTD, VANGUARD WORKS, Kilnhurst — TH

144v	2.10.1964	4ft 8½in	4wDH	179hp	R-R C6NFL	30T	104,149
176v	28.12.1966	4ft 8½in	4wDH	256hp	R-R C6SFL	34T	180
187v	26.9.1967	4ft 8½in	4wDH	272hp	R-R C8NFL	37T	113,160
239v	9.2.1972	4ft 8½in	0-4-0DH	272hp	R-R C8NFL	45T	58,160
272v	11.7.1977	4ft 8½in	4wDH	300hp	R-R C6TFL	35T	125
279v	20.11.1978	4ft 8½in	4wDH	272hp	R-R C8NFL	40T	160
280v	14.12.1978	4ft 8½in	4wDH	272hp	R-R C8NFL	40T	160
281v	20.12.1978	4ft 8½in	4wDH	272hp	R-R C8NFL	40T	160
282v	10.4.1979	4ft 8½in	4wDH	272hp	R-R C8NFL	40T	160
285v	25.3.1980	4ft 8½in	0-6-0DH	370hp	R-R C8TFL	50T	89
291v	26.11.1980	4ft 8½in	0-6-0DH	400hp	R-R DV8N	70T	58
294v	10.8.1981	4ft 8½in	0-6-0DH	427hp	R-R DV8N	60T	58
295v	6.10.1981	4ft 8½in	0-6-0DH	427hp	R-R DV8N	60T	58
299v	10.11.1981	4ft 8½in	4wDH	300hp	R-R C6TFL	35T	124
301v	16.6.1982	4ft 8½in	4wDH	300hp	R-R C6TFL	35T	125
306v	22.8.1983	4ft 8½in	4wDH	300hp	R-R C6TFL	35T	125
309v	13.12.1983	4ft 8½in	4wDH	300hp	R-R C6TFL	35T	125

TRACK SUPPLIES & SERVICES LTD, Wolverton, Bucks — TS&S

NO/1023	1985	4ft 8½in	2w-2BER	289

THORNEWILL WARHAM LTD, Burton-on-Trent — TW

373	1873	4ft 8½in	0-4-0ST	OC	178
420	1876	4ft 8½in	0-4-0ST	OC	178

UCA, Antwerp, Belgium — UCA

	2004	4ft 8½in	4wDM	R/R	190

UNILOKOMOTIVE LTD, Galway, Ireland — Unilok

2109	1980	4ft 8½in	4wDM	R/R	286

MERCEDES BENZ AG, Germany — Unimog

		4ft 8½in	4wDM	R/R	251
092692	1982	4ft 8½in	4wDM	R/R	66
166200	1991	4ft 8½in	4wDM	R/R	251
166228	1991	4ft 8½in	4wDM	R/R	251

VULCAN FOUNDRY LTD, Newton-le-Willows, Lancs — VF

1236	1888	4ft 8½in	0-6-0ST	IC	12 x 17	3ft1in	258
1436	1895	4ft 8½in	0-6-0CT	OC	16 x 20	3ft3in	116
5256	2.6.1945	4ft 8½in	0-4-0DM	153hp	Gardner 6L3 {DC 2175}		123
5261	3.9.1945	4ft 8½in	0-4-0DM	153hp	Gardner 6L3 {DC 2180}		282,293
5272	1945	4ft 8½in	0-6-0ST	IC	18 x 26	4ft3in	120
5274	1945	4ft 8½in	0-6-0ST	IC	18 x 26	4ft3in	120
5275	1945	4ft 8½in	0-6-0ST	IC	18 x 26	4ft3in	121
5276	1945	4ft 8½in	0-6-0ST	IC	18 x 26	4ft3in	122

5284	1945	4ft 8½in	0-6-0ST	IC	18 x 26		4ft3in	122
5309	1945	4ft 8½in	0-6-0ST	IC	18 x 26		4ft3in	286
D77	4.3.1948	4ft 8½in	0-4-0DM	153hp	Gardner 6L3		{DC 2251}	54,282
D78	29.7.1949	4ft 8½in	0-6-0DM	204hp	Gardner 8L3		{DC 2252}	293
D98	1949	4ft 8½in	0-4-0DM	153hp	Gardner 6L3		{DC 2269}	75
D99	1949	4ft 8½in	0-4-0DM	153hp	Gardner 6L3		{DC 2270}	75
D293	8.10.1955	4ft 8½in	0-4-0DM	153hp	Gardner 6L3		{DC 2566}	75,291
D294	22.10.1955	4ft 8½in	0-4-0DM	153hp	Gardner 6L3		{DC 2567}	75,76
D297	1956	4ft 8½in	0-4-0DM	153hp	Gardner 6L3		{DC 2583}	151
D1065	1967	4ft 8½in	Bo BoDE	1000hp	EE 8SVT		{20189}	190

CHARLES WALL LTD, GLOBE WORKS, Grays, Essex — Ch Wall

	2ft 4in	0-4-2T	OC		256
	2ft 4in	0-4-0VBT	HCG		256
	2ft 4in	2-4-0WT	OC		256
	2ft 4in	2-4-0			256
	2ft 4in	4wTG			256
	2ft 4in	4wTG			256

W.G. BAGNALL LTD, CASTLE ENGINE WORKS, Stafford — WB

840	3.1887	3ft 0in	0-4-0IST	OC	8 x 12		159
1116	5.1889	3ft 0in	0-4-0IST	OC	9 x 13½	2ft3in	213
1423	15.7.1893	3ft 0in	0-4-0IST	OC	8 x 12	2ft0in	213
1424	9.12.1893	3ft 6in	0-4-0T	OC	7 x 12	2ft0in	110
1434	17.11.1894	3ft 0in	0-4-0IST	OC	8 x 12	2ft0in	213
1444	21.7.1897	2ft 9in	0-4-0ST	OC	4½ x 7½	1ft1¾in	138
1480	7.9.1897	3ft 0in	0-4-0IST	OC	8 x 12	2ft0½in	213
1504	22.5.1897	3ft 0in	0-4-0IST	OC	10 x 15	2ft7½in	213
1553	29.11.1898	2ft 9in	0-4-0ST	OC	6 x 9	1ft7in	138
1567	18.12.1899	3ft 0in	0-4-0ST	OC	8 x 12	2ft0½in	213
1583	28.12.1899	4ft 8½in	0-4-0ST	OC	10 x 15	2ft6in	96
1613	24.9.1900	4ft 8½in	0-4-0ST	OC	10 x 15	2ft10in	72
1637	26.6.1901	4ft 8½in	0-4-0ST	OC	10 x 15	2ft6in	96
1657	17.5.1902	3ft 0in	0-4-0ST	OC	8 x 12	2ft0½in	213
1663	11.1902	1ft 11in	0-4-0ST	OC	6 x 9	1ft8in	235
1734	18.10.1903	2ft 9in	0-4-0ST	OC	7 x 12	1ft9½in	138
1855	13.1.1909	2ft 9in	0-4-0ST	OC	7 x 12	1ft9½in	138
1964	9.4.1914	2ft 9in	0-4-0ST	OC	7 x 12	1ft9½in	138
2047	21.3.1918	2ft 0in	0-4-0ST	OC	6 x 9	1ft7in	235
2048	27.3.1918	2ft 0in	0-4-0ST	OC	6 x 9	1ft7in	235
2073	29.6.1918	2ft 0in	0-4-0ST	OC	6 x 9	1ft7in	235
2103	28.7.1919	2ft 6in	0-4-0ST	OC	7 x 12	1ft9½in	281
2133	10.7.1924	3ft 6in	0-4-0ST	OC	7 x 12	1ft9½in	290
2135	15.4.1926	3ft 6in	0-4-0ST	OC	7 x 12	1ft9½in	290
2168	13.2.1921	4ft 8½in	0-6-0ST	OC	13 x 18	2ft9½in	187
2170	22.12.1921	4ft 8½in	0-4-0ST	OC	9 x 14	2ft3½in	79,80
2172	24.12.1921	4ft 8½in	0-4-0ST	OC	12 x 18	3ft0½in	277
2178	22.9.1922	4ft 8½in	0-4-0ST	OC	12 x 18	3ft0½in	275
2370	6.5.1929	4ft 8½in	0-6-0F	OC	20 x 18	2ft9¼in	149
2542	14.2.1936	4ft 8½in	0-4-0ST	OC	12 x 18	3ft0½in	288

2613	17.9.1940	4ft 8½in	0-6-0PT	OC	13 x 18	2ft 11½in		293
2657	24.3.1942	4ft 8½in	0-4-0ST	OC	12 x 18	2ft 9in		135
2658	24.3.1942	4ft 8½in	0-4-0ST	OC	12 x 18	2ft 9in		135
2739	29.5.1944	4ft 8½in	0-6-0ST	IC	18 x 26	4ft 3in		182
2740	29.5.1944	4ft 8½in	0-6-0ST	IC	18 x 26	4ft 3in		120
2753	9.1944	4ft 8½in	0-6-0ST	IC	18 x 26	4ft 3in		122
2764	16.1.1945	4ft 8½in	0-6-0ST	IC	18 x 26	4ft 3in		121
2767	13.2.1945	4ft 8½in	0-6-0ST	IC	18 x 26	4ft 3in		122
2773	19.3.1945	4ft 8½in	0-6-0ST	IC	18 x 26	4ft 3in		122
2775	3.5.1945	4ft 8½in	0-6-0ST	IC	18 x 26	4ft 3in		120,182
2851	22.9.1947	4ft 8½in	0-4-0F	OC	18½x18	3ft 0½in		149
2879	26.10.1948	4ft 8½in	0-4-0ST	OC	14½x22	3ft 6½in		168
3061	13.10.1954	4ft 8½in	0-6-0ST	OC	16 x 24	3ft 6½in		293
3160	29.10.1959	4ft 8½in	0-6-0DM	304hp	Dorman 6QAT			58
3211	1962	4ft 8½in	0-4-0DH	179hp	R-R C6NFL	{RSHD 8366}		251
3212	1962	4ft 8½in	0-4-0DH	262hp	Dorman 6QA	{RSHD 8367}		182
7077	25.4.1955	{returned after repairs}			see RSH 7111			181

WHITING CORPORATION, Harvey, Illinois, U.S.A   WhC

4001	1966	4ft 8½in	4wDM	R/R			66

D. WICKHAM & CO LTD, Ware, Hertfordshire   Wkm

626	16.8.1932	4ft 8½in	2w-2PMR	1100cc	JAP	Type 17	289
1583	19.7.1934	4ft 8½in	2w-2PMR	1100cc	JAP	Type 17	289
1946	28.6.1935	4ft 8½in	2w-2PMR	1323cc	JAP	Type 17A	286,289
3030	10.7.1941	60cm	2w-2PM	1323cc	JAP	R1389	129
3031	10.7.1941	60cm	2w-2PM	1323cc	JAP	R1389	129
3032	10.7.1941	60cm	2w-2PM	1323cc	JAP	R1389	129
3033	10.7.1941	60cm	2w-2PM	1323cc	JAP	R1389	129
3034	10.7.1941	60cm	2w-2PM	1323cc	JAP	R1389	129
3096	8.4.1942	4ft 8½in	2w-2PMR	1323cc	JAP	Type 17A	125
3227	21.4.1943	4ft 8½in	2w-2PMR	1323cc	JAP	Type 17A Spl	125
3235	7.5.1943	2ft 0in	2w-2PM	1323cc	JAP	R1389	290
3287	7.9.1943	60cm	2w-2PM	1323cc	JAP	R1389	129
3403	5.11.1943	2ft 0in	2w-2PMR	1323cc	JAP	Type 17A	265
3414	7.9.1943	60cm	2w-2PM	1323cc	JAP	R1389	129
3860	5.10.1945	4ft 8½in	2w-2PMR	1323cc	JAP	Type 17a Spl	126
5008	27.9.1949	4ft 8½in	2w-2PMR	10hp	Ford	Type 27A	289
6896	28.10.1954	4ft 8½in	2w-2PMR	600cc		JAP Type 4B	289
6963	10.1.1955	4ft 8½in	2w-2PMR	30hp	Ford	Type 40	126
7390	11.6.1956	4ft 8½in	2w-2PMR	30hp	Ford	Type 40	126
7391	11.6.1956	4ft 8½in	2w-2PMR	30hp	Ford	Type 40	126
7397	16.1.1957	4ft 8½in	2w-2PMR	Perkins P4		Type 40	126
7595	21.1.1957	4ft 8½in	2w-2PMR	1323cc	JAP	Type 17A	289,293
8086	22.10.1958	4ft 8½in	2w-2PMR	10hp	Ford	Type 27	126
8089	22.10.1958	4ft 8½in	2w-2PMR	10hp	Ford	Type 27	126
8197	20.11.1958	4ft 8½in	2w-2PMR	10hp	Ford	Type 27	126
8501	28.7.1960	4ft 8½in	2w-2PMR	600cc		JAP Type 4B	289,293
8502	2.9.1960	4ft 8½in	2w-2PMR	600cc	JAP	Type 4B	289
10943	9.2.1976	3ft 6in	2w-2PMR		Ford	Type 27A	297

WINGROVE & ROGERS LTD, Kirkby, Liverpool — WR

778	27.11.1930	2ft 6in	4wBE	Type No2		2¾T	102
1060	30.6.1937	2ft 0in	4wBE	4.8hp	W417	2½T	231
1061	30.6.1937	2ft 0in	4wBE	4.8hp	W417	2½T	231
1062	30.6.1937	2ft 0in	4wBE	4.8hp	W417	2½T	231
1063	30.6.1937	2ft 0in	4wBE	4.8hp	W417	2½T	231
1064	30.6.1937	2ft 0in	4wBE	4.8hp	W417	2½T	231
1065	30.6.1937	2ft 0in	4wBE	4.8hp	W417	2½T	231
1066	30.6.1937	2ft 0in	4wBE	4.8hp	W417	2½T	231
1067	30.6.1937	2ft 0in	4wBE	4.8hp	W417	2½T	231
1071	15.7.1937	2ft 0in	4wBE	12hp	W227	3T	216
1199	17.3.1938	2ft 0in	4wBE	4.8hp	W417	2½T	203,204
1200	17.3.1938	2ft 0in	4wBE	4.8hp	W417	2½T	203
1211	17.3.1938	2ft 0in	4wBE	4.8hp	W417	2½T	203
1212	17.3.1938	2ft 0in	4wBE	4.8hp	W417	2½T	203,204
2063	22.7.1941	1ft 6in	0-4-0BE	4hp	W217	1½T	142
2064	30.7.1941	1ft 6in	0-4-0BE	4hp	W217	1½T	142
2065	15.9.1941	1ft 6in	0-4-0BE	4hp	W217	1½T	142
2066	15.9.1941	1ft 6in	0-4-0BE	4hp	W217	1½T	142
2067	15.9.1941	1ft 6in	0-4-0BE	4hp	W217	1½T	142
2068	15.9.1941	1ft 6in	0-4-0BE	4hp	W217	1½T	142
2352	12.8.1942	1ft 6in	0-4-0BE	4hp	W217	1½T	142
2353	12.8.1942	1ft 6in	0-4-0BE	4hp	W217	1½T	142
2354	21.8.1942	1ft 6in	0-4-0BE	4hp	W217	1½T	142
2355	28.8.1942	1ft 6in	0-4-0BE	4hp	W217	1½T	142
2356	17.9.1942	1ft 6in	0-4-0BE	4hp	W217	1½T	142
2357	17.9.1942	1ft 6in	0-4-0BE	4hp	W217	1½T	142
3219	20.9.1945	2ft 0in	0-4-0BE	4hp	W217	1½T	216
3788	1.7.1948	1ft 6in	0-4-0BE	4hp	W217	1½T	217
4352	9.2.1950	2ft 0in	4wBE	12hp	W227	3T	216
4475	13.10.1950	2ft 0in	0-4-0BE	4hp	W217	1½T	216
4476	13.10.1950	2ft 0in	0-4-0BE	4hp	W217	1½T	216
4579	29.12.1950	2ft 0in	0-4-0BE	4hp	W217	1½T	217
4580	29.12.1950	2ft 0in	0-4-0BE	4hp	W217	1½T	217
4818	19.11.1951	2ft 0in	4wBE	4.8hp	W417	2½T	206
4897	3.3.1952	2ft 0in	4wBE	12hp	W227	3T	216
5659	31.5.1957	1ft 6in	4wBE	4hp	W217	1½T	204
6097	28.9.1959	2ft 0in	4wBE	12hp	W227	3T	216
6600	25.7.1962	1ft 6in	0-4-0BE	4hp	W217	1½T	205,259
6702	25.7.1962	1ft 6in	0-4-0BE	4hp	W217	1½T	205,259
6711	27.3.1963	1ft 6in	0-4-0BE	4hp	W217	1½T	205
D6878	26.5.1964	2ft 0in	0-4-0BE	4hp	W217	1½T	216
D6879	26.5.1964	2ft 0in	0-4-0BE	4hp	W217	1½T	216
J7206	30.5.1969	2ft 0in	4wBE	12hp	W227	3T	227
J7208	7.5.1969	2ft 0in	4wBE	12hp	W227	3T	227
J7272	19.5.1969	2ft 0in	4wBE	12hp	W227	3T	227
J7273	30.5.1969	2ft 0in	4wBE	12hp	W227	3T	227
J7274	30.5.1969	2ft 0in	4wBE	12hp	W227	3T	227
J7282	27.2.1970	2ft 0in	4wBE	12hp	W227	3T	227
M7550	15.12.1972	2ft 0in	0-4-0BE	5hp	WR5L	1½T	238
M7556	16.3.1972	2ft 0in	4wBE	12hp	W227	3T	197

L801	1983	1ft 6in	2w-2wBE	3hp	WR3	229
544901	5.9.1986	1ft 6in	2w-2wBE	3hp	WR3	229

WELLMAN SMITH OWEN ENGINEERING CORPORATION LTD
Darlaston, Staffordshire   **WSO**

1408	1929	4ft 8½in	0-4-0WE	80hp	15T	135
1529	1930	4ft 8½in	0-4-0WE	80hp	18T	89
4191	1945	4ft 8½in	0-4-0WE	80hp		135
5785	16.7.1954	4ft 8½in	0-4-0WE	80hp		135

YORKSHIRE ENGINE CO LTD, MEADOW HALL WORKS, Sheffield   **YE**

284	1876	4ft 8½in	0-4-0ST	OC	12 x 20 3ft 3in	111,143
1923	1933	4ft 8½in	0-4-0ST	OC	{reb of BH 1038}	75,76
2619	12.11.1956	4ft 8½in	0-6-0DE	400hp	2 x R-R C6SFL	144,147
2620	26.11.1956	4ft 8½in	0-6-0DE	400hp	2 x R-R C6SFL	144,147
2630	13.5.1957	4ft 8½in	0-6-0DE	400hp	2 x R-R C6SFL	144,147
2633	24.5.1957	4ft 8½in	0-6-0DE	400hp	2 x R-R C6SFL	144,147
2640	3.6.1957	4ft 8½in	0-6-0DE	400hp	2 x R-R C6SFL	144,147
2641	1.7.1957	4ft 8½in	0-6-0DE	400hp	2 x R-R C6SFL	57,147
2686	10.5.1958	4ft 8½in	0-4-0DE	200hp	R-R C6SFL	84,85,282
2690	7.5.1959	4ft 8½in	0-6-0DE	400hp	2 x R-R C6SFL	144
2691	14.5.1959	4ft 8½in	0-6-0DE	400hp	2 x R-R C6SFL	144
2739	27.5.1959	4ft 8½in	0-6-0DE	400hp	2 x R-R C6SFL	144
2740	4.6.1959	4ft 8½in	0-6-0DE	400hp	2 x R-R C6SFL	144
2755	30.9.1959	4ft 8½in	0-6-0DE	300hp	R-R C8SFL	147
2756	14.10.1959	4ft 8½in	0-6-0DE	300hp	R-R C8SFL	147
2757	7.10.1959	4ft 8½in	0-6-0DE	300hp	R-R C8SFL	147
2758	25.11.1959	4ft 8½in	0-6-0DE	300hp	R-R C8SFL	144
2759	26.10.1959	4ft 8½in	0-6-0DE	300hp	R-R C8SFL	144
2760	25.11.1959	4ft 8½in	0-6-0DE	300hp	R-R C8SFL	144
2761	14.12.1959	4ft 8½in	0-6-0DE	300hp	R-R C8SFL	144
2762	21.12.1959	4ft 8½in	0-6-0DE	300hp	R-R C8SFL	144
2763	21.12.1959	4ft 8½in	0-6-0DE	300hp	R-R C8SFL	144
2769	6.1.1960	4ft 8½in	0-6-0DE	300hp	R-R C8SFL	144
2770	11.1.1960	4ft 8½in	0-6-0DE	300hp	R-R C8SFL	144
2853	5.5.1961	4ft 8½in	0-4-0DE	220hp	R-R C6SFL	147
2854	5.5.1961	4ft 8½in	0-4-0DE	220hp	R-R C6SFL	147,172
2856	1.6.1961	4ft 8½in	0-4-0DE	220hp	R-R C6SFL	172

NINE ELMS WORKS, London   **9E**

781	1906	4ft 8½in	2-2-0T	OC	10 x 14 3ft 0in	117
795	1906	4ft 8½in	2-2-0T	OC	10 x 14 3ft 0in	117
798	1906	4ft 8½in	2-2-0T	OC	10 x 14 3ft 0in	117

INDEX OF LOCOMOTIVE NAMES

A.E.CLOW	89	BETA	52
ABBERTON	249	BEXLEY	76
ABBEY	243	BIM	93
ABERNANT	295	BIRKENHEAD	288
ACCRINGTON	185, 212	BLACK DIAMOND	221
ADJUTANT	178	BLACK	102, 181
AGENOR	112, 147	BLUEBIRD	61
AJAX	110, 112, 147, 177, 211	BOBBY	187, 231
ALBERT	111, 168, 230	BOMBAY	72, 185, 259
ALBION	275	BORDON	117
ALICE	53, 89, 187, 254	BOUNTY	269
ALICIA	210	BRAMLEY No.6	172, 188, 276
ALICIE	210	BRANCKER	213
ALLAN BRUCE	53, 96	BRISTOL	96
ALPHA	52	BROOKFIELD	293
ALSTON	213	BRUCE	214, 221
ALWILDA	59, 242, 277	BUDLEIGH	298
AMBERGATE	254, 255	BUNTY	213
AMOS	96	BURNHOPE	248
ANGLOCO	85	BURTON	89
ANGLO-DANE	172, 268		
ANIK	273	C.H. STANTON	247
ANNE	194	C.P. HUNTINGTON	284, 291
ANNIE	59, 243, 244, 254, 255	CAESAR	60
ANT	112, 214, 221	CALDEW	133, 214, 221, 255
ARAB	59, 242	CAM	236, 237
ASHBURY	275	CAMEL	247
AUNTY	53	CANADA	243
AUSTIN WALKER	293	CARNEGIE	263, 298
AVON	233	CASTLE HEDINGHAM	286
AVONSIDE	172	CECIL LEVITA	70, 259
AYALA	300	CEMENT	172
		CENTURY	292
BAID	215	CHANNEL DRIVER	256
BALDER	60	CHARLETON	96
BANBURY	243	CHARLTON	75
BARKING POWER	294, 298	CHELSEA	111, 220
BARKING	59, 77, 218, 230, 243, 244	CHINGFORD	243
BARRINGTON	286	CHRISTOPHER	185
BARRY	70, 258	CHURCHILL	150
BASSETT	71	CIRCE	71, 75
BAT	112, 147	CLARENCE	53, 59
BEATRICE	53, 138, 210	CLITHEROE	214
BEATTY	295	COFFEE STALL	256
BECKTON	243	COLNE	236, 237
BECONTREE	259	COLUMBIA	274
BELVEDERE	75, 80, 218	COMET	168
BELVOIR	288	CORDITE	58, 106
BEN RINNES	130	CORONATION	172

CORSTON	59, 243
COVINGTON	80
CROOKES	102
CROSTON No.2	277
CRYMLYN	70
CUNARDER	184, 187
CWM TÂF	273
DAGENHAM	97, 254, 259
DARENTH	150
DELTA	52
DEMELZA	293
DENIS	256
DENMARK	150, 178
DEPTFORD	75, 76
DEVON	254, 255
DEVONPORT	185, 212
DIANA	72, 236
DOLOBRAN	294
DOVER	212, 272
DRAPER	125
DREADNOUGHT	299
DUDLEY	294, 298
DUNHAM No.1	277
DUNRAVEN	185
DUNSTON	258, 259
DUVAL	220
DUVALS	96
EARL OF CARYSFORT	117
EAST RIGGS No.1	277
ECLIPSE	103
EDDIE	89
EDEN	260
EDGWARE	72, 185
EDWARD	80
EILEEN	166
ELIZABETH II	71, 80
ELIZABETH	60
ELLESMERE	72
ELSIE	85, 167, 181
EMPRESS	293
ERICA	256
ESSEX	76, 78, 201
ETTRICK	89, 102
FELSPAR	59, 70, 168, 254
FERRO	77
FIREFLY	79, 80
FLEETWOOD	206
FLORENCE	171

FLY	112, 147
FOLA	172, 268
FORTH	70, 187, 231, 259
FORWARD	275
FOX	168
FRANCES	79, 258
FRANK	195, 196
GALLIONS	76
GAMMA	52
GENERAL LORD ROBERTSON	124
GEORGE THE FIFTH	296
GEORGE	60, 168, 182
GIPSY	170
GLADYS	166
GLASGOW	212
GLENCOE	213
GLOBE No.1	53
GLOBE No.2	53
GNAT	112, 248
GOLIATH	50, 60
GORDON	223
GOWY	252
GREENHITHE	187
GREENWICH	75, 230
GROSVENOR	215, 221, 230
GUERNSEY	53, 230
GUINNION	80
GUNBY	288
GYP	213
HAMBLE-LE-RICE	58
HAMPTON	214, 221
HARBORO	180
HARTHOPE	248
HASSALL	102
HASTINGS	293
HAYLEY	295
HECTOR	112, 147
HENGIST	182
HENRY APPLEBY	254, 255
HILTON	60
HON. HOWE BROWN	133
HORNBLOWER	269
HORNCHURCH	77
IDA	98
IDRIS	293
IOTA	50, 52
IRESHOPE	248
IRIS	138

Name	Pages
ITCHEN	115, 231
J.B.GANDY	102
JACOB WILSON	254, 255
JAMES	61
JASON	112, 147
JEAN	166
JEFF	153
JEFFREY	288
JENNIFER	286
JETTY	75, 76
JOAN	164
JOE	98, 243
JOHN PEEL	288
JOYCE	166
JUBILEE	159, 172, 288
JUMBO	95, 221, 254, 255
JUPITER	286, 288
KAPPA	53, 254
KENT	75
KENTISH MAID	58
KENYA	79, 80
KILINDINI	187
KILMARNOCK	168
KING ARTHUR	210
KING GEORGE	180, 288
KINGSTON	192
KIRKBY	88
KITTY	194
KOLVADA	274
KYNITE	58, 106
LADYSMITH	243
LAUREL	72
LAURIE	172, 273
LAYER	249
LEA	218
LIMERICK	255
LION	59, 198
LIZ	174
LLWYN-ON	273
LOCH LEVEN	213
LOCH SHIN	130
LOCKWOOD	242
LONDONDERRY	243
LONG WIND	111
LOOE	111, 143
LORD MAYOR	70, 133, 150, 254, 258
LORD NELSON	77
LOU	194

Name	Pages
LOUGHBOROUGH	168
LUCIE	293
LUMPY TOM	256
LYDIA	138
MABEL	138
MALCOM	89
MANCHESTER	212
MARFLEET	172, 184, 243, 244
MARGAM	50, 59, 109, 244
MARY	292
MARYHILL	72, 185
MASHAM	192
MATUNGA	274
MAX	180, 272
MAYBURY	197
MEESON	96, 109
MET	68
METEOR	237
MIDDLESEX	201
MIDGE	249
MILDRED	178
MILLBROOK	231, 232
MINNIE	293
MONARCH	172
MONMOUTH	79
MORNINGTON	150, 178
MOTH	112, 147
MYFANWY	251
NANCY	71
NAPIER	272
NAPOLEON	177
NESTOR	112, 147
NEWCASTLE	247
NEWPORT	231
NICHOLSON	116
NOAH	89
NORFOLK	201
NORSEMAN	172
NORTHERN OUTFALL	205
NORTHFLEET	168
NOTTINGHAM	212
NUTTALL	188
OBERON	75
ODIN	172
OILER	75, 77
OLWEN	288
OPHIR	96
ORMSBY	168

OSCAR	103
P.H.B.	168, 182
PARANA	95, 254
PARK	275
PARKER	213
PARKESTON	246
PARTINGTON	258
PAULINE	93
PAXMAN	288
PEGASUS	75
PEGGY	65, 138
PELDON	249
PENN GREEN	288
PENN	231
PERFECTION	85
PERSEUS	75
PETROS	59, 159
PHILIP	275
PHOENIX	70
PHYLLIS	210
PLANET	168, 182
PLUMSTEAD	78
PLUTO	286
POLAND	172, 268
PORTLAND	172
PORTSMOUTH	254, 255
PREDATOR	130
PRIAM	75
PRINCESS ELIZABETH	284
PRINCESS MARGARET ROSE	284
PRINCESS	88
PURFLEET	76, 150, 178
QUEEN	243
QUEEN'S COUNSEL	133
R.A. GRAY	133
RAINHAM	77, 78, 88
RALPH	80
RAMBO	130
RANNOCH	213
RHIWNANT	295
RHYL	294
RIBBLESDALE	214, 221
ROBERT	288, 294
ROCKET	291
RODNEY	268
ROMFORD	76, 77, 78, 249
ROSE	138
ROYAL ENGINEER	118
ROYAL	150, 275

RUSHTON	220
RUTH	138
SANKEY	252
SCHOOL OF GUNNERY	118
SCOT	76
SHARPNESS	236
SHEPPERTON	277
SHIRLEY	67, 231
SHOEBURYNESS	116
SIMON ADAM BECK	133
SIR JOHN BETJEMAN	297
SIR TOM	290
SIR WILLIAM CRUNDALL	243
SIR WILLIAM HEYGATE	297
SMUDGE	295
SOLVAY	102
SOMERFORD	70, 243
SOMERTON	70, 258
SOUTHAMPTON	89, 187, 231
SOUTHERN	172, 240
SOUTHFLEET	168
SOUTHSEA	70, 257, 258
ST MONANS	212
STANLEY	59
STANLOW	258
STAR	150, 168, 182
STOREFIELD	288
STRATFORD	79, 218, 254, 255
STUART	53
SUFFOLK	201
SUNBURY	277
SUSAN	180
SWANSCOMBE	168
SWANSEA	70, 185, 254, 258
SYLVIA	236
TALBOT	218
TAMAR	79, 212
TERMINATOR	130
TERRIER	72
TEUCER	75, 76
THAMES	180, 218
THE AUDITOR	243
THE BEAR	117
THE CAPTAIN	79
THE NIDD	213
THE NIPPER	170
THOR	59, 79, 159, 172
THURROCK	77, 78
THURWHIT	168, 182

TILBURY	214, 221	WALTER	102, 194
TIM	93	WALTHAMSTOW	243
TOTO	293	WARWICK	59
TRAFFORD PARK	188, 231	WASP	112, 147, 248
TRIBRUIT	80	WEAR	181, 212, 254, 272
TRUDY	193	WEST THURROCK	60, 184
TUNNEL	172	WESTERHAM	257
TYNE	260, 276	WESTMINSTER	212, 272
		WHEELOCK	102
UNDAUNTED	72	WILLIAM MURDOCK	288
UPNEY	193	WILLIAM	88
UTILITY	77, 78	WILLIE	243
		WINSTON	172
VERA	167	WITHAM	59
VICTORIA	111, 212, 220, 272	WOLF	198
VICTORY	286	WOODCROFT-	187, 231
VIKING	172	WOOLWICH	75, 184, 243, 244, 298
		WOTO	290
WADALA	273	WOULDHAM	59, 60
WALLACE	214, 221	WYE	243
WALLASEY	172		
WALSHAW DEAN	213	YORK	275

INDEX OF OWNERS and LOCATIONS

Abberton Reservoir Construction	248
Abbey Mills Chemical Works	56
Adcock, William John	191
Admiralty	49,301
John Aird & Sons	192
Albion Works	106
Allen John S. & Son	261
Alpha Cement Ltd	49
Alresford Sand & Ballast Co	51
AMEC Construction	192
Anglian Water Authority	51
Anglo American Oil Co	84,85
Argall Works	263
Harold Arnold & Son	192
Sir William Arrol & Co	193
Associated Equipment Co	52
Associated Portland Cement Manufacturers	49,52
Aveley Clay Pit	174
Aveley Plant Depot	217
Baker, Ronald L.	54
Baker Street Gravel Pits	154
Balfour Beatty	193,194
Band, John	194
Barham & Tait	262
Barking & Ilford Joint Sewage Board	97
Barking Contract	192,196,198,206,218,251,258
Barking Jetty Co	54
Barking Power Station	66
Barking Sewage Works	110
Barkingside Plant Depot	233
Barton, Joseph & Sons	306
Barwyke Sand & Ballast Co	55
Basildon Contract	242
Beach Pit	183
Beach, W. & Son	55
Beach's Brickfields	55
Beckton By-Products Works	137
Beckton Contract	192,199,200,201,208,217,230,239
Beckton Gas Works	133
Becontree Contract	258
Bellhouse Brickworks	74
Benfleet Brickworks	301
Benfleet Contract	202,211
Bent-Marshall	55
Bentons Pit	83
Berk Ltd	56
Berk, F.W. & Co	56
Berk Spencer Acids	56
Bidder Street Scrap Yard	70
Billericay Brick Co	301
Birch Brook Pits	155
Bishops Stortford Contract	195
Blyth S.J.	56
Boot, Henry & Sons	194
Booth, Alfred & Co	154
Boreham Sand Pits	101
Bow Power Station	67
BP Oil	57
Braintree Brickworks	302
Braintree Contract	207
Bramble Island Works	86
Brand, Charles & Son	194
Brassey & Mackenzie	195
Brassey, Ogilvie & Harrison	195
Brentwood Brickworks	303
Brentwood Contract	197,229
Bretts Farm Clay Pits	50
Brick & Tile Manufacturers Corporation	107
Bridge Metals	262
Briggs Motor Bodies	88
Brightlingsea Contract	237
Brightlingsea Pits	99
British Coal	162
British Electricity Authority	66
British Explosives Syndicate	301,302
British Ferries	163
British Portland Cement Manufacturers	59
British Postal Museum	284
British Sugar Corporation	61
Bromley-By-Bow Gas Works	138
Brooks Shoobridge & Co	53
Brooks Works	53
Brown, James	302
Brown & Tawse	61
Broxbourne Pits	62
Broxbourne Sand & Ballast	62
B.R.T. Securities	262
Bruce, Alexander	62
Bruces Wharf	62
Brunner Mond & Co	102
Buckhurst Hill Brickworks	91

Buckhurst Hill Plant Depot	203	Collier, W.H. & Co	73
Bulmer Brick & Tile Co	303	Colliers Ltd	73,74
Burt Boulton & Haywood	149	Colne Brick & Terra Cotta Co	306
Butlins Ltd	284	Colne Valley Railway	286
Butterley Building Materials	63	Columbia Wharf	182
		Concrete Piling	198
Caffin & Co	195	Cook Alfred	304
Canewdon Brick Sand & Ballast Co	65	Co-operative Wholesale Society	74,162
Canning Town Contract	194	Cornish, Daniel	74,108
Canning Town Glass Works	65	Cornish, Edward	74
Canvey Casino	285	Cornish & Haylock	108
Carless Solvents	66	Cornish, Orbell	93
Carless Capel & Leonard	66	Cory Bros & Co	57
Cater Bros (Scotts Hall)	65	Coryton Bulk Terminal	57
Carter, J.	285	Cory, William & Son	75
Carter & Woodgate	304	Costain Civil Engineering	199
Central Electricity Authority	66	Costain, Richard	199
Central Electricity Generating Board	66	Costain, Skanska, Bachy,	
Central Line Contract	194	Soletanche J. V.	200
Chalkwell Contract	227	County and General Consumers	
Chamberlain Industries	263	Company	139
Channel Tunnel Rail Link	190,200, 239,241	County of London Electric Supply Co	66
Charing Cross West End & City Electricity Supply Co	67	Covington, H. & Sons	78
		Craven, Roger	287
Charring Cross & Strand Electricity Supply Corpn	67	Crowley Russell & Co	200
		Crowstone Brickworks	311
Chelmsford Brickworks	302	Cunis, W.R.	80
Chelmsford Contract	223,257	Custom House Contract	216
Chelmsford Works	152	Curtis, Robert L.	175
Cherry Orchard Brickworks	63		
Chigwell Urban District Council	285	Dagenham Contract	222,231,250,253
Chingford Contract	228,231,255	Dagenham Dock	132,184
Christiani & Nielsen	196	Dagenham (Thames) Dock Co	304
Clacton Contract	229,252,192	Dagenham Works	88
Clancy, M.J. & Sons	197	Dannatt, B.	81
Clark and Standfield	69	Dartford Contract	195
Clark, S.A.	197	Dartnell, F.J.	263
Cleveland Bridge and Engineering Co	197	Davey Paxman & Co	267
		Demolition & Construction Co	201
Clover, William	107	Dengie Flat	312
Clyde Wharf Sugar Refinery	82	Dix Green & Co	304
Coal Conversion	54	Dobson Ellis & Co	81
Coalhouse Fort	176	Docklands Light Railway	82
Cochrane & Sons	198	Duncan Bell & Elott	82
Cohen, George Sons & Co	69,70	Duncan, James	82
Colchester By Pass Contract	198	Dunton Ammunition Depot	312
Colchester Contract	192,195,260,		
Colchester Council	198	East & West India Dock Co	112
Colchester Plant Depot	223	East Anglian Railway Museum	288
Colconite Works	54	Eastbrook Farm Pits	99

Eastern Counties Railway	201	Gascoine Road Sewage	
East Essex (Corringham) Sand Co	82	Pumping Station	97
East Essex Pits	82	Gas Light & Coke Co	93,133,137,138
East Essex Reclamation Scheme	165	Gas Lighting Improvement Company	85
East London Transport	83	Gentry, Mark	93
East London Water Works Co	115	Gevertz, Jack	265
East Tilbury Contract	191	Gibbs & Co	52,304
East Tilbury Rubbish Shoot	166	Gibbs Wharf	184
Eastwood & Co	84,305	Gill, John	206
Eastwoods	305	Glendale Forge	291
Eastwoods Flettons	84	Glico Petroleum	85
Ebano Oil Co	84	Globe Cement, Brick, Whiting	
Eldonwall Industrial Estates	106	& Chalk Co	94
Elms, P.	290	Globe Wharf Rubbish Shoot	176
Engineering Services	264	Globe Works	253
Epping & Ongar Railway	291	Glossop W. & J.	206
Essex Brick Co	305	Goldsmith, E.J. & W.	95
Essex County Council	202	Grantrail	207
Essex Iron & Steel Co	264	Gray, J.G.	207
Essex Welding Co	262	Grays Chalk Quarries Co	96,307
Esso Petroleum Co	84	Grays Coal Depot	97
Everetts Brickworks	305	Grays Contract	194,242
Explosives & Chemical Products	86	Grays Co-operative Society	97
		Grays Engineering Works	69
Fairlop Gravel Pits	153	Grays Plant Depot	196,200,212
Featherby, George	86	Grays Rubbish Shoot	95
Featherby's Brickworks	86	Grays Works	96,109
Fels, Joseph	305	Great Coldharbour Rubbish Shoot	80
Felsted Works	61	Greater London Council	97
Ferro Works	241	Great Sampford (RAF) Contract	240
Fineturret	202	Great Wakering Brick Co	97
Fingringhoe Brick & Tile Works	306	Great Wakering Sand Pits	162
Fingringhoe Ranges	312	Griffiths Wharf Co	98
Fingringhoe Sand Pits	168	Groves Farm Sand & Gravel Pits	140
Fisons	87		
Flint Works	169	Hall & Co	98
Flower & Everett	87,108	Hall & Ham River	99
Fogdens	264	Hallsford Bridge Brickworks	306
Fondu Works	106	Halstead Contract	236
Ford Motor Co	88	Ham River Grit Co	99
French W.& C.	91,202,306	Hancock, Walter	265
Freshwater Sand & Ballast Co	92	Hanson, William	207
Freshwater Gravel Sand & Ballast Co	92	Harlow Development Corporation	291
Fresh Wharf	105	Harlow New Town Plant Depot	204
Frinton-on-Sea Contract	207	Harrisons (London)	100
Frog Island Timber Yard	141	Harrison's Wharf	84
Furness, George	205	Harold Wood Brick Co	307
		Harold Wood Brickfield Co	307
Gabbutt, Edmund	205	Harvey Road Plant Depot	216
Gallions Jetty	75	Harvey, Roger	292

Harwich Contract	260	Keyman Pearson Parker	265
Harwich Refinery	66	Key Works	174
Hatfield Peverel Works	81	Kirk & Parry	213
Haverhill Contract	236	Kirk & Randall	214
Hawkes, G.J. & Sons	101	Knapping, Dale	84,305
Hawkwell Brickfields	101	Kynoch	105
Hawkwell Brickworks	101	Kynoch, G. & Co	105
Heath T.W.	207	Kynochtown Works	105
Hedingham Brick & Tile Works	93		
Hedingham Brick, Tile		Lafarge Aluminous Cement Co	106
& Terra Cotta Works	93	Lake & Co	106
Hedley Thomas & Co	149	Lake & Elliot	106
Herts Plant Hire	101	Laing, John & Son	215
Higgs & Hill	207	Lamb, W.T.	65,107,108
Highways Construction	101	Land Reclamation Co	108
Hill, John Cathles	308	Landwick Brickworks	310
Hilton Anderson Brooks & Co	53	Langford Reservoir Contract	247
Hoffmann Manufacturing Co	152	Langham Valley Contract	202
Holloway Brothers (London)	208	Langthorne Brickworks	93
Holme & King	210	Langwith Metal Finishers	307
Hornchurch Contract	199	Lea Bridge District Gas Co	139
Hornchurch Rubbish Shoot	77	Lea Bridge Gas Works	139
Howlett, W.H. & Co	307	Lea Bridge Works	115
Hussey, Egan & Pickmere	211	Lee Conservancy Board	211,216
Hutton Contract	228	Lee Tunnel Contract	230
		Lee Valley Contract	227
ICI (Alkali)	102	Leeman, P.W.	265
Ilford Contract	216,236,239,250	Leigh-on-Sea Contract	196
Ilford Factory	141	Leslie & Co	216
Ilford Gas Co	93	Leyton Contract	231
Ilford Gas Light and Coke Co	93	Leytonstone Contract	259
Ilford Gas Works	93	Light Expanded Clay Aggregates	109
Ind Coope & Allsopp	103	Lilley–Waddington Ltd	216
Ind Coope & Co	103	Lind, Peter & Co	217
Industrial Chemicals	103	Lion Cement and Chalk Co	59
Inns & Co	104,108,152	Lion Cement Co	59
Intrade	105	Little Mardyke Pits	104
Imperial Chemical Industries	102	Little Thurrock Marshes	
Imperial Gas Light and Coke Co	138	Rubbish Shoot	78
Industrial Constructions	211	Lockington & Mander	94
		London Brick Co	63,64,308
Jackaman, A. & Son	211	London Electricity Joint Committee	67
Jackson, J. & Co	84,305	London & Continental Railways	190
Sir John Jackson	212	London & Grays Seaborne Coal Co	110
Jaywick Miniature Railway	292	London & India Docks Co	111
Jetty Works	69	London & St Katharine Dock Co	111
Jubilee Brickfield	304	London & Thames Haven	
Jubilee Line Extension	251	Oil Wharves	113,308
Jurgens	175	London & Tilbury Lighterage Co	169
Keyman & Pearson (1963) Ltd	265		

London & Tilbury Lighterage Contracting & Dredging Co	169
London & West India Docks Joint Committee	111
London County Council	110,218
London FerroConcrete Co	110
London Power Co	67
London Tilbury Southend Contract	246
Loughton Contract	207
Lower Hope Development Co	308
Low Street Rubbish Shoot	170
Lucas & Aird	219
Lunnon	266
Luxborough Lane Brickworks	91
Luxborough Lane Sewage Works	69
Lyle Abram & Sons	114
Lynn Tait Gallery	292
Maiden Ley Brickworks	309
Mallandain, Albert E.	86
Manor Brick & Tile Co	311
Manor Brickworks	311
Mardyke Ballast Pits	83
Mardyke Works	261
Marks Tey Brickworks	73
Marley Tile (Aveley) Co	114
Mangapps Farm Railway Museum	293
Marples Ridgeway & Partners	222
Marsh & Sons	114
Marsh Farm Sewage Works	51
Martin, Henry	223
May & Butcher	266
May Gurney & Co	223
Mayer-Newman & Co	267
Maynard Reservoir	115
McAlpine & Sons	224
McCormick & Sons	228
McEwan Pratt & Co	267
Meads Lane Brickfields	163
Meadgate Farm Pits	156
Meeson & Co	96
Metals & Ropes	69
Metropolitan Cement Company	50
Metropolitan Water Board	115,228
Metropolitan Works	50
Mid Essex Pits	165
Millar's Karri & Jarrah Forests	150
Miller Construction Ltd	228
Millhead Brickworks	97
Milton Hall (Southend) Brick Co	63,64,308
Mineral Oil Corporation	161
Miners Safety Explosives Co	308
Ministry of Defence	115,130
Ministry of Munitions	132
Ministry of Public Building & Works	131
Minoco Wharf	161
Mitchell Bros Sons & Co	229
Mitchell, Fred & Son	229
Mobil Oil Co	57
Morgan Sindall/Vinci Construction/ Bachy Soletanche JV	230
Mowlem John & Co	230,231
Mucking Hall Gravel Pit	164
Mucking Rubbish Shoot	166
Muirhead Macdonald Wilson & Co	233
Muirhead, William & Co	233
Munro, Wm	236
Murex	132
Murphy, J. & Sons	238
Murray Works	267
Murrell, W.C.	132
National Coal Board	162
Nazeing Pits	152
Nevendon Treatment Works	51
Newham Borough Council	294
New Walton Pier Co	299
Nishimatsu-Cementation, Skanska JV	239
Noble, F.M. & Son	306
Noble, Frederick & Son	306
North Essex Portland Cement Works	304
Northfleet Coal & Ballast Co	168
North Thames Gas Board	133
North Woolwich Old Station Museum	294
North Woolwich Wharf	98
Norton, F.A.	139
Nuttall, Edmund Sons & Co (London)	239
Nuttall, Wayss & Freytag-Kier JV	241
Odell R.G.	139
Oikos Storage	309
Olley, Chas T. & Sons	140
Organ & Co	241
Oxford & Shipton Cement	49
Pacific Wharf	85
Parkeston Quay Contract	198,211,246
Parkinson, J. & Sons (Blackpool)	242

Sir Lindsay Parkinson & Co	242
Paragon Brick Works	309
Pattricks Brickworks	309
Pearson, S. & Son	59,242
Perry & Co	245
Peters & Barham	140
Peto Betts & Brassey	246
Petroplus Holdings	141
Phoenix Timber Co	141
Pierce, A.D.	246
Pitsea Contract	193
Pitsea Rubbish Tip	108
Pitsea Wharf	54
Pitt E.L.& Co (Coventry)	267
Place Farm Pits	156
Plaistow & Barking Contract	205
Plaistow Wharf Sugar Refinery	114
Plessey Co	141
Plowman, Eli	107
Ponds Farm Sand Pits	174
Port of London Authority	142
Potter, John Ephraim	161,304
Premier Mills	151
Pressed Brick & Tile Co Ltd	107
Prince Regent Tar Co	149
Printar Industries	149
Prior, J.J.	168
Prittlewell Brickworks	308
Prittlewell Contract	211
Prizeman & Co	309
Proctor & Gamble	149
Purfleet Chalk Quarries	149
Purfleet Chalk Quarry	149
Purfleet Contract	240
Purfleet Deep Wharf & Storage Co	150
Purfleet Quarries	313
Purfleet Rifle Ranges	313
Purfleet Tank Farm	85
Purfleet Thames Terminal	150
Purfleet Wharf	76,100,150
Purfleet Wharf & Sawmills	150
Purfleet Works	57,175
Rainham Contract	252
Rainham Ferry Works	132
Rainham Pits	55
Rainham Rifle Ranges	313
Rainham Rubbish Shoot	77,87,159,166
Rainsford End Brickworks	55
Rank, Joseph	151
Ransome Hoffmann Pollard	152
Rayleigh Brick & Tile Co	107
Ray Lodge Brickfields	306
Rayner, George Henry	309
Rayner, William	309
Read & Company	110
Read, P.T. (Fairlop)	153
Redland-Inns Gravel	152
Redline Glico	85
Redline Motor Spirit Co	85
Reedham Wharves	113
Rettendon Roofing Tile Works	110
Ridley, Samuel Coote	246
Ridley, Thomas D.	246
River Lee Flood Relief Scheme	211
Rochford, Edmund	154
Roding Pits	104
Romford Brewery	103
Romford Contract	226
Romford Field Pits	141
Romford Gravel Pits	98
Romford Scrap & Salvage Co	271
Rom River Co	154
Rom River Sand & Gravel Co	154
Rookery Sand & Gravel Pit	155
Rotary Hoes	155
Rotavators	155
Rowhedge Pits	156
Rowhedge Sand & Ballast Co	155
Royal Albert Brickfield	312
Royal Albert Dock	111
Royal Albert Dock Contract	193,219,243,248
Royal Docks	142
Royal Engineers	177,178,179
Royal Gunpowder Factory	130
Royal Victoria Dock	111
Royal Victoria Dock Contract	227,240
Royal Victoria Dock Drainage Scheme	199,202,228,251
Rutter, D. & C.	97,310
Saffron Walden Contract	207
Sail & Steam Engineering	268
Salvation Army	157
Sanders & Forster	158
Sandon Hall Gravel Pit	139
Sandon Pits	165
Saunders, H. & J.R. & Co	268
Saw Mills Co	310
Seabrooke & Sons	159
Seabrooke's Brickworks	307

Sealink	163
Sea Transport Stores Depot	301
Second Anglo-Scottish Beet Sugar Corpn	61
Sharpes Autos (London)	295
Shelbourne, John & Co	159
Shellabear, G. & Son	247
Shell Haven Contract	232
Shell Haven Refineries	159
Shell-Mex & BP	57
Shell Refining Co	159
Shell Refining & Marketing Co	159
Shell U.K. Ltd	159
Shenfield Brickworks	74
Shenfield Contract	210, 226
Shenfield & Cranham Brick & Tile Co	74
Shoeburyness Brickworks	84
Shoeburyness Contract	195
Shoeburyness Depot	115, 131
Sible Hedingham Red Brick Co	310
Silver End Development Co	248
Silvertown Flour Mills	74
Silvertown Lubricants Ltd	161
Silvertown Machinery Works	180
Silvertown Works	102
Simmons, W. & Son	269
Smith, L.J.	295
Southchurch Brickfields Co	161
Southchurch Brickworks	161
South Eastern Gas Corporation	139
Southend Borough Council	250
Southend Coal Concentration Depot	162
Southend Gas Co	93
Southend Miniature Railway	296
Southend-on-Sea Borough Council	296
Southend-on-Sea & District Gas Co	93
Southend-on-Sea Estates Co	162
Southend-on-Sea Gas Works	93
Southend Pier Contract	223
Southend Pier Railway	297
Southend Pier Railway Museum	298
Southend Road Contract	234
Southend Sand & Gravel Co	162
Southern Depot Co	162
South Essex Gas Light and Coke Co	139
South Essex Waterworks Company	163, 248
South Hornchurch Gravel Pits	55
South Hornchurch Plant Depot	253
South Ockendon Pits	99
Speight, L.J.	250
Spencer Chapman and Messel	56
St. Albans Sand & Gravel Co	156
Stanford-Le-Hope Sand Pits	314
Stanstead Contract	238
Stanstead Nurseries	154
Star Lane Brickworks	64
Steamship Owners Coal Association Ltd	76
Stena Sealink	163
Stott S.S. & Sons	250
Stour Scheme	249
Stour Valley Railway Preservation Society	288
Stratford Contract	203
Streeter, A. & Co	251
Stroud, Robert	163
Stuart (Thamesmouth) Sand & Shingle Co	164
Stubbings, Clifford	165
Surridge, F.W.	165
Tank Storage & Carriage Co	85
Tarmac/Montcocol JV	251
Tate & Lyle	167
Tate, Henry & Sons	167
Tawse, Wm	252
Taylor Woodrow Construction	252
Temple Mills Contract	207
Thames Brick & Tile Works	311
Thames Haven Contract	215, 242, 246
Thames Iron Works & Shipbuilding Co	269
Thames Land Co Ltd	184
Thames Matex Ltd	175
Thames Portland Cement Works	52
Thames Sand Marketing Co	168
Thames Sugar Refinery	167
Thames Works	52
Thornback Brick Co	63
Thorpe Bay Brickworks	162
Thundersley Brickfield	311
Thurrock Chalk & Whiting Co	168
Thurrock Contract	192
Thurrock Brewery	159
Thurrock Flint Co	169
Tilbury Contract	208, 229, 253
Tilbury Contracting & Dredging Co	169
Tilbury Docks	112, 147
Tilbury Docks Contract	214, 220, 224
Tilbury Fort	313
Tilbury Machinery & Ironware Co	270

Tilbury Power Station	68
Tilbury Riverside Terminal	171
Tilbury Rubbish Shoot	170
Titan Works	103, 271
Tollesbury Brick, Tile & Development Co	311
Topham Jones & Railton	252
Tottenham-Forest Gate Contract	222
Towler, W.T.	270
Transfesa UK	171
Tredegar Works	245
Trentham, G. Percy	253
Trevethick, F.H.	171
Trollope & Colls	253
Tudor Brick & Tile Co	311
Tunnel Cement	171
Tunnel Holdings	171
Tunnel Industrial Services	171
Tunnel Portland Cement Co	171
Tunnel Portland Cement Works Co	171
United Glass (England)	174
United States Army	179
Universal Steam TramCar Construction Co	270
Upminster Brickworks	303
Upminster Contract	230
Upton, E.W.J.	174
Vacuum Oil Co	57
Van den Berghs & Jurgens	175
Vange Brick Fields	175
Van Ommeren Tank Terminal Purfleet	175
Victoria Dock Co	111
Victoria Dock Extension Contract	219
Vopak Terminal Purfleet	175
Wagstaff, J.J.	312
Wakefield, T. (Scrap Metals)	271
Wall, Charles	176,253
Walsh Bros (Tunnelling)	256
Walter Scott & Middleton	247
Waltham Abbey Royal Gunpowder Mills Co	298
Walthamstow Contract	206
Walthamstow Reservoirs Contract	242
Walton Contract	237
Walton-on-Naze Contract	195
Walton-on-the-Naze Pier Railway	299
War Department	115,177,312
War Office	176
Ward, Henry	257
Ward, R.G.	306
Ward, Thos. W.	180,271
Ward Ferrous Metals	180
Warner, Robert & Co	313
Watts, W.J. & Co	63
Wennington Marshes Rubbish Shoot	79
Wennington Sand & Ballast Co	183
West Ham Contract	238
West Ham Power Station	68
West Horndon Works	61,155
West, Samuel	183
West Thurrock Ballast Pits	252
West Thurrock Brickworks	304
West Thurrock Contract	209,215
West Thurrock Rubbish Shoot	78
West Thurrock Works	149,168,171
Westwick Rubbish Shoot	139
Wheeler, S.P. & Sons	282
Whitaker Ellis	257
Whitbread, W.H.	313
White Colne Pits	105
White, George	314
Whitehall Securities Corporation	184
Wickford Brickworks	108,314
Wickford Narrow Gauge Group	299
Wickham Rail Engineering Ltd	282
Williams, Samuel & Sons	184
Williams, Samuel (Dagenham Dock)	184
Willment Bros	259
Wills, C.J.	257
Wills, C.J. & Sons	258
Wilmot, Sidney	188
Wilson, W.W.	314
Wimpey, George & Co	260
Witham Works	154
Wivenhoe Sand, Stone & Gravel Co	189
Woodford Brickworks	92
Woodford Contract	205,238,257
Woods	300
Wouldham Cement Co	59
Wouldham Works	59
Wythes, George	260
York International	314

Fig 1, HL3760 1932 0-4-0ST OC P.H.B.
Thurrock Chalk & Whiting Co Ltd, 31-8-1964

Roger Hateley

Fig 2, HU c1928 4wDM
F.W. Berk & Co Ltd, West Ham, 25-5-1959
S.A. Leleux

Fig 3, YE2641 1957 0-6-0DE
BP Oil Ltd, Purfleet Works, 13-5-1981
Robin Waywell

Fig 4, AE1771 1917 0-6-0ST OC
Cory Bros, Corringham, 12-6-1948

IRS: K.J. Cooper Collection

Fig 5, WB3160 1959 0-6-0DM
Mobil Oil, Coryton Refinery, 26-9-1979

Robin Waywell

Fig 6, AB506/1 1969 0-4-0DH　　　　　　　　　　　　　　　　　　　　　Robin Waywell
Mobil Oil, Coryton Refinery, 14-5-1981

Fig 7, TH291v 1980 0-6-0DH　　　　　　　　　　　　　　　　　　　　　Robin Waywell
Mobil Oil, Coryton Refinery, 14-5-1981

Fig 8, P771 1899 0-4-0ST OC WOULDHAM IRS: K.J. Cooper Collection
BPCM Wouldham Works, 6-8-1950

Fig 9, MW1306 1895 0-4-0ST OC MARGAM
R.A. Wheeler photograph, IRS: Jim Peden Collection
BPCM Wouldham Works, 13-9-1937

Fig 10, P1314 1913 0-4-0ST OC STANLEY
BPCM Wouldham Works, 6-8-1950

IRS: K.J. Cooper Collection

Fig 11, AB1391 1915 0-4-0ST OC THOR
BPCM Wouldham Works, 17-3-1956

Jack Faithfull, RCTS

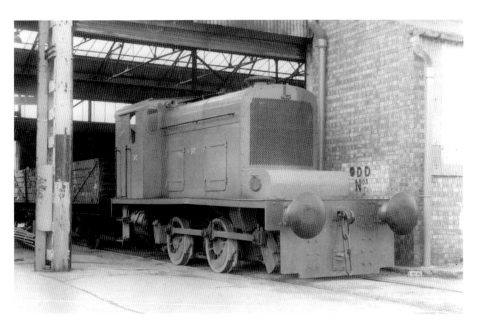

Fig 12, HC D610 1939 0-4-0DM
Brown & Tawse (Tubes) Ltd, 5-4-1975
Kevin Lane

Fig 13, RH221639 1943 4wDM
Brown & Tawse (Tubes) Ltd, 5-4-1975
Kevin Lane

Fig 14, FH3885 1958 4wDM
Alexander Bruce (Grays) Ltd, 26-9-1979

IRS: Howard Earl Collection

Fig 15, RH441951 1960 4wDM
Milton Hall (Southend) Brick Co Ltd, Star Lane Brickworks, 8-1978

Kevin Lane

Fig 16, RH441951 1960 4wDM   Kevin Lane
Milton Hall (Southend) Brick Co Ltd, Star Lane Brickworks, 10-1980

Fig 17, AK26 1988 4wDM   Robin Waywell
Butterley Building Materials Ltd, Star Lane Brickworks, 21-7-1989

Fig 18, MR2034 1920 4wPM S.A. Leleux
Canning Town Glass Works Ltd, 25-5-1959

Fig 19, FH3147 1947 4wDM S.A. Leleux
Canning Town Glass Works Ltd, 25-5-1959

Fig 20, HL3653 1927 0-4-0ST OC Jack Faithfull RCTS
CEA Barking Power Station 14-4-1956

Fig 21, FH3294 1948 4wDM Roger Hateley
CEGB Barking Power Station 2-9-1964

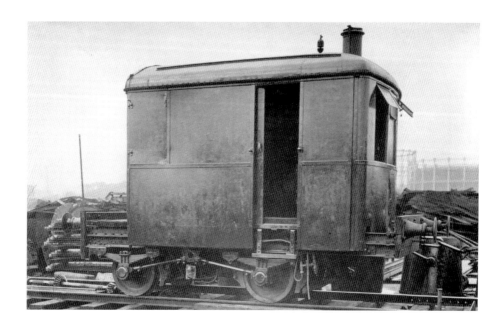

Fig 22, S6994 1927 4wVBT VCG ILS: Frank Jones Collection
George Cohen, Sons & Co Ltd, Canning Town

Fig 23, MW1756 1910 0-4-0ST OC ESSEX IRS: Jim Peden Collection
Steamship Owners Coal Association Ltd, Purfleet Wharf

Fig 24, MW1973 1919 0-4-0ST OC PURFLEET ILS: Frank Jones Collection
William Cory & Son Ltd, 1953

Fig 25, HC1326 1919 0-4-0ST OC GALLIONS Jack Faithfull RCTS
William Cory & Son Ltd, Purfleet, 18-4-1959

Fig 26, TG440 1907 0-4-0ST OC HORNCHURCH IRS: Jim Peden Collection
William Cory & Son Ltd, Rainham Rubbish Shoot

Fig 27, HC696 1904 0-4-0ST OC RAINHAM
Morton Middleditch, John Hutchings Collection
William Cory & Son Ltd, Rainham Rubbish Shoot, 31-3-1934

Fig 28, HC1206 1916 0-4-0ST OC UTILITY   Les Hallett, John Hutchings Collection
Willian Cory & Son Ltd, Rainham Rubbish Shoot 5-12-1948

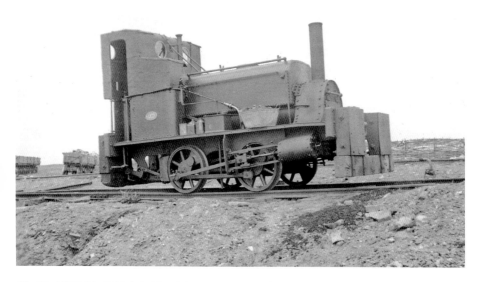

Fig 29, MW941 1885 0-4-0ST OC   Morton Middleditch, John Hutchings Collection
H. Covington & Sons Ltd, Wennington Marshes, 19-5-1934

Fig 30, HE629 1896 0-4-0ST OC   Morton Middleditch, John Hutchings Collection
H. Covington & Sons Ltd, Wennington Marshes, 19-5-1934

Fig 31, WB2170 1921 0-4-0ST OC KENYA   Jack Faithfull RCTS
H. Covington & Sons Ltd, Wennington Marshes, 29-5-1949

Fig 32, AB1652 1919 0-4-0ST OC   Morton Middleditch, John Hutchings Collection
W.R. Cunis Ltd, Great Coldharbour Rubbish Shoot, 19-5-1934

Fig 33, AB1895 1926 0-4-0ST OC RALPH
                    Morton Middleditch, John Hutchings Collection
W.R. Cunis Ltd, Great Coldharbour Rubbish Shoot, 19-5-1934

Fig 34, MW2044 1925 0-4-0ST OC FIREFLY   ILS: Frank Jones Collection
W.R. Cunis Ltd, Great Coldharbour Rubbish Shoot

Fig 35, AB992 1904 0-4-0ST OC   Morton Middleditch, John Hutchings Collection
W.R. Cunis Ltd, Great Coldharbour Rubbish Shoot, 19-5-1934

Fig 36, AB1870 1925 0-4-0F OC   Jack Faithfull RCTS
Esso Petroleum Co Ltd, Pacific Wharf, West Ham, 26-10-1957

Fig 37, AB1603 1918 0-4-0F OC   Ken Plant, IRS: Jim Peden Collection
Esso Petroleum Co Ltd, used at Purfleet Tank Farm

Fig 38, JF4220001 1959 0-4-0DM Robin Waywell
Fisons Ltd, Stanford-le-Hope, 26-9-1979

Fig 39, AE1460 1903 0-6-0ST OC ALICE John Hutchings Collection
Ford Motor Co Ltd, Dagenham

Fig 40, P1938 1937 0-6-0ST OC Roger Hateley
Ford Motor Co Ltd, Dagenham, 2-9-1964

Fig 41, HC1508 1924 0-6-0ST OC   Morton Middleditch, John Hutchings Collection
Ford Motor Co Ltd, Dagenham, 12-6-1948

Fig 42, P1908 1937 0-4-0ST OC Roger Hateley
Ford Motor Co Ltd, Dagenham, 2-9-1964

Fig 43, BTH 1932 Bo-BoDE IRS: K.J. Cooper Collection
Ford Motor Co Ltd, Dagenham, 12-6-1948

Fig 44, HC D1377 1966 0-6-0DH Jack Faithfull RCTS
Ford Motor Co Ltd, Dagenham, 25-5-1968

Fig 45, RSH/DC 2611 1958 0-6-0DM Robin Waywell
Ford Motor Co Ltd, Dagenham, 19-8-1978

Fig 46, EEV D1229 1967 0-6-0DH
Ford Motor Co Ltd, Dagenham, 19-8-1978
Robin Waywell

Fig 47, S10127 1963 0-4-0DH MALCOLM
Ford Motor Co Ltd, Dagenham, 2-2-2009
Roy Etherington

Fig 48, MW1141 1889 0-4-0ST OC PARANA ILS: Frank Jones Collection
E.J. & W. Goldsmith Ltd, Grays Rubbish Shoot 3-3-1905

Fig 49, BE276 1898 0-4-0ST OC MEESON Jack Faithfull RCTS
Grays Chalk Quarries Co Ltd, 28-6-1952

Fig 50, 0-4-0ST ICG   IRS: Jim Peden Collection
Grays Chalk Quarries Co Ltd, later I.W. Boulton, PUGSEY

Fig 51, KS4199 1920 0-4-0WT OC CROOKES   Jack Faithfull RCTS
Imperial Chemical Industries Ltd, Silvertown, 14-3-1959

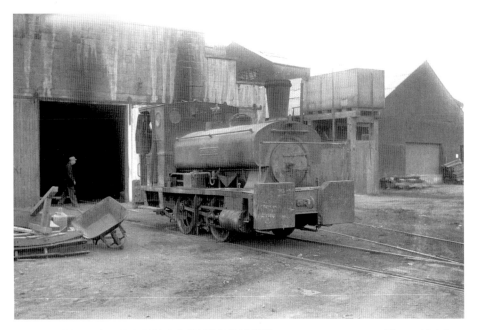

Fig 52, AB699 1891 0-4-0ST OC SWANSCOMBE Roger Hateley
Thurrock Chalk & Whiting Co Ltd, 31-8-1964

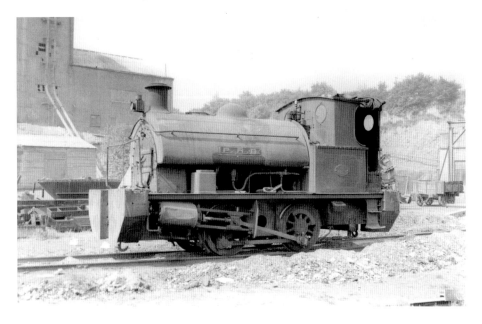

Fig 53, HL3760 1932 0-4-0ST OC P.H.B. Jack Faithfull RCTS
Thurrock Chalk & Whiting Co Ltd, 29-5-1965

Fig 54, WB2879 1948 0-4-0ST OC COMET Jack Faithfull RCTS
Thurrock Chalk & Whiting Co Ltd, 31-8-1957

Fig 55, RH192325 1939 4wDM Roger Hateley
Thurrock Chalk & Whiting Co Ltd, 31-8-1964

Fig 56, RR10249 1966 4wDH Kevin Lane
Lafarge Aluminous Cement Co Ltd, 10-1977

Fig 57, WB1424 1894 0-4-0IST OC IRS: Jim Peden Collection
London County Council, Northern Outfall, Beckton, 5-7-1947

Fig 58, RP c1870 0-4-0ST OC
London & India Docks Co, Royal Docks

IRS: Jim Peden Collection

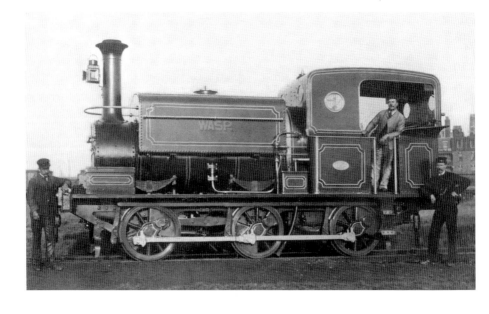

Fig 59, MW872 1883 0-6-0ST IC WASP
London & India Docks Co, Tilbury Docks

IRS Jim Peden Collection

Fig 60, HE234 1880 0-6-0ST IC MOTH
London & India Docks Co, Tilbury Docks

IRS Jim Peden Collection

Fig 61, AB1551 1917 0-6-0F OC
London & Thames Haven Oil Wharves Ltd, 16-8-1952

Jack Faithfull RCTS

Fig 62, VF1436 1895 0-6-0CT OC SHOEBURYNESS Shoeburyness Archive
War Department, Shoeburyness

Fig 63, VF1436 1895 0-6-0CT OC IRS: Jim Peden Collection
War Department, Shoeburyness

Fig 64, HC1164 1916 0-6-0T OC
War Department, Shoeburyness

Shoeburyness Archive

Fig 65, NBL16601 1905 2-4-2ST OC
War Department, Shoeburyness

Shoeburyness Archive

Fig 66, NR5936 1902 0-4-0ST OC
War Department, Shoeburyness

Shoeburyness Archive

Fig 67, TH309v 1983 4wDH
Ministry of Defence, Shoeburyness, 1-4-2005

Shoeburyness Archive

Fig 68, Chaplin built 1874 0-4-0VBT VCG
Gas Light & Coke Co, Beckton Gas Works

John Hutchings Collection

Fig 69, N3789 1888 0-4-0T OC
North Thames Gas Board, Beckton Gas Works, 22-4-1959

S.A. Leleux

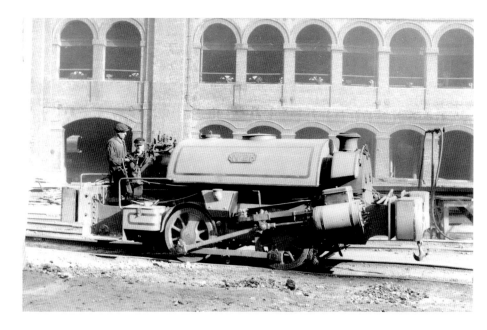

Fig 70, N4250 1890 0-4-0ST OC Jack Faithfull RCTS
North Thames Gas Board, Beckton Gas Works, 30-10-1954

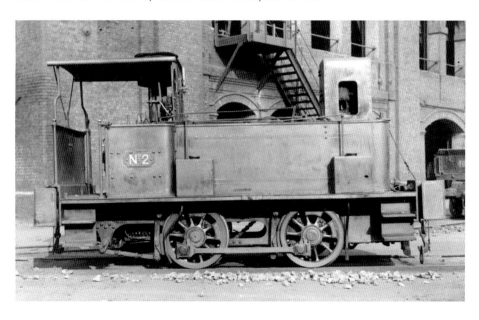

Fig 71, S6951 1927 0-4-0VBT VCG IRS: Jim Peden Collection
North Thames Gas Board, Beckton Gas Works

Fig 72, AB730 1893 0-4-0ST OC Jack Faithfull RCTS
North Thames Gas Board, Beckton By-Products Works, 27-4-1952

Fig 73, HL3308 1918 0-4-0ST OC Roger Hateley
North Thames Gas Board, Beckton By-Products Works 2-9-1964

Fig 74, HL3595 1924 0-4-0F OC Alec Swain, Kevin Lane Collection
North Thames Gas Board, Beckton By-Products Works

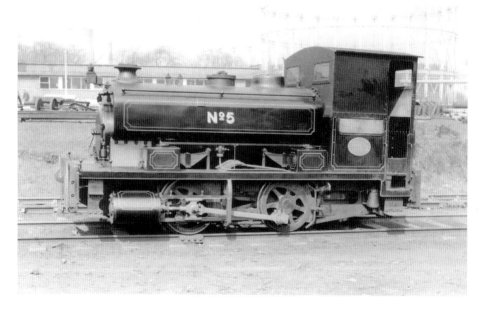

Fig 75, AB1674 1920 0-4-0ST OC Jack Faithfull RCTS
North Thames Gas Board, Bromley-By-Bow Gas Works, 4-4-1959

Fig 76, N4397 1891 0-4-0T OC ILS: Frank Jones Collection
Gas Light & Coke Co, Bromley-By-Bow Gas Works, 1948

Fig 77, P2135 1953 0-4-0ST OC Jack Faithfull RCTS
North Thames Gas Board, Bromley-By-Bow Gas Works, 4-4-1959

Fig 78, Port of London Authority, Custom House Loco Shed   Jack Faithfull RCTS
4-4-1959

Fig 79, HL3529 1922 0-6-0T OC   Jack Faithfull RCTS
Port of London Authority, Custom House 4-4-1959

Fig 80, Port of London Authority, Custom House Workshops Roger Hateley
31-8-1964

Fig 81, YE2640 1957 0-6-0DE IRS: Rodney Weaver Collection
Port of London Authority, Custom House

Fig 82, YE2760 1959, YE2619 1956, YE2759 1959   Roger Hateley
Port of London Authority, Custom House 31-8-1964

Fig 83, YE2758 1959 0-6-0DE   IRS: Rodney Weaver Collection
Port of London Authority, Custom House

Fig 84, RS3170 1905 0-6-0ST OC JASON
Port of London Authority, Tilbury Docks, 1931

J.M.Hutchings Collection

Fig 85, HC1101 1915 0-6-0T OC
Port of London Authority, Tilbury Docks 6-1950

IRS: Harry Townley Collection

Fig 86, HC1874 1954 0-6-0T OC Jack Faithfull RCTS
Port of London Authority, Tilbury Docks 17-3-1956

Fig 87, AB1238 1911 0-6-0T OC IRS: Jim Peden Collection
Port of London Authority, Tilbury Docks

Fig 88, WB2851 1947 0-4-0F OC Robin Waywell Collection
Prince Regent Tar Co Ltd, Silvertown, 16-3-1957

Fig 89, AB1472 1916 0-4-0F OC IRS: Howard Earl Collection
Proctor & Gamble Ltd, West Thurrock, 20-6-1978

Fig 90, WB2370 1928 0-4-0F OC
Proctor & Gamble Ltd, West Thurrock, 20-6-1978

IRS: Howard Earl Collection

Fig 91, TH144v 1964 4wDH
Proctor & Gamble Ltd, West Thurrock, 26-9-1979

Robin Waywell

Fig 92, MW1619 1904 0-4-0ST OC DENMARK Jack Faithfull RCTS
Purfleet Deep Wharf & Storage Co Ltd, 22-8-1953

Fig 93, AB1577 1918 0-6-0ST OC Jack Faithfull RCTS
Purfleet Deep Wharf & Storage Co Ltd, 31-8-1957

Fig 94, HC823 1908 0-6-0ST OC Jack Faithfull RCTS
Purfleet Deep Wharf & Storage Co Ltd, 22-8-1953

Fig 95, P1571 1921 0-4-0ST OC Jack Faithfull RCTS
Purfleet Deep Wharf & Storage Co Ltd, 22-8-1953

Fig 96, RH437362 1960 0-4-0DH Robin Waywell
Purfleet Deep Wharf & Storage Co Ltd, 26-9-1979

Fig 97, RH512463 1965 0-4-0DH Robin Waywell
Purfleet Deep Wharf & Storage Co Ltd, 26-9-1979

Fig 98, VF D297 1956 0-4-0DM
Purfleet Deep Wharf & Storage Co Ltd, 26-9-1979

Robin Waywell

Fig 99, FH3491 1951 4wDM
Rom River Co Ltd, Witham, 31-3-1977

Ian Peaty

Fig 100, FH4008 1963 4wDM   Kevin Lane
Sanders & Forster Ltd, Stratford, 21-6-1975

Fig 101, HE6285 1968 4wDM   Kevin Lane
Sanders & Forster Ltd, Stratford, 1-1980

Fig 102, JF4210130 1957 0-4-0DM　　　　　　　　　　　　John Wilkins
Shell U.K. Ltd, Thames Haven Refinery

Fig 103, JF4220031 1964 0-4-0DH　　　　　　　　　　　　John Wilkins
Shell U.K. Ltd, Thames Haven Refinery

Fig 104, JF4240016 1964 0-6-0DH
Shell U.K. Ltd, Thames Haven Refinery
John Wilkins

Fig 105, TH187v 1967 4wDH
Shell U.K. Ltd, Thames Haven Refinery
John Wilkins

Fig 106, TH239v 1971 0-4-0DH
Shell U.K. Ltd, Thames Haven Refinery, 14-5-1981

IRS: Howard Earl Collection

Fig 107, TH279v 1978 4wDH
Shell U.K. Ltd, Thames Haven Refinery, 13-6-1990

Alec Swain, Kevin Lane Collection

Fig 108, TH280v 1978 4wDH
Shell U.K. Ltd, Thames Haven Refinery, 13-6-1990

Alec Swain, Kevin Lane Collection

Fig 109, One of the batch HE1710 – 1715 built 1932 4wDM

Collection of Hunslet Engine Co

F.W. Surridge, Mucking Rubbish Shoot

Fig 110, P1806 1930 0-4-0ST OC FOLA Jack Faithfull RCTS
Tunnel Portland Cement Co Ltd, West Thurrock Works, 29-5-1965

Fig 111, P1994 1940 0-4-0ST OC POLAND Jack Faithfull RCTS
Tunnel Portland Cement Co Ltd, West Thurrock Works, 29-5-1965

Fig 112, P1920 1937 0-6-0ST OC MONARCH   Jack Faithfull RCTS
Tunnel Portland Cement Co Ltd, West Thurrock Works, 29-5-1965

Fig 113, P1919 1936 0-6-0ST OC & P1369 1914 0-4-0ST OC   Jack Faithfull RCTS
Tunnel Portland Cement Co Ltd, West Thurrock Works, 29-5-1965

Fig 114, RR10235 1965 4wDH Kevin Lane
Tunnel Industrial Services Ltd, West Thurrock Works, 10-1975

Fig 115, AB1493 1916 0-4-0F OC IRS: Jim Peden Collection
Van Den Berghs & Jurgens Ltd, Purfleet

Fig 116, OK8141 1916 0-4-0F OC Jack Faithfull RCTS
Van Den Berghs & Jurgens Ltd, Purfleet, 28-6-1952

Fig 117, RSHD/WB8343 1962 0-6-0DH Kevin Lane
Thames Matex Ltd, West Thurrock, 2-1981

Fig 118, MW 1913 1917 0-6-0ST OC ADJUTANT Miles McNair Collection
New to War Department, Royal Engineers Purfleet

Fig 119, MW1671 1905 0-4-0ST OC Robin Waywell Collection
Thos W. Ward Ltd, Silvertown Works

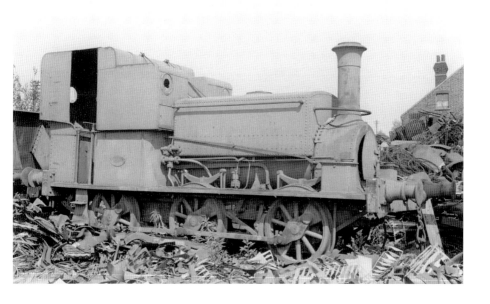

Fig 120, MW212 1866 0-6-0ST IC
Thos W. Ward Ltd, Columbia Wharf, Grays, 1949

ILS: Frank Jones Collection

Fig 121, HC1510 1923 0-6-0ST IC
Thos W. Ward Ltd, Columbia Wharf, Grays, 18-4-1959

Jack Faithfull RCTS

Fig 122, WB2739 1944 0-6-0ST IC ILS: Frank Jones Collection
Thos W. Ward Ltd, Columbia Wharf, Grays, 10-1959

Fig 123, RSHD/WB 8367 1962 0-4-0DH HENGIST Ian Peaty
Thos W. Ward Ltd, Columbia Wharf, Grays, 7-1983

Fig 124, AB419 1957 0-4-0DM Ian Peaty
Thos W. Ward Ltd, Columbia Wharf, Grays, 7-1983

Fig 125, MW841 1882 0-6-0ST IC WOOLWICH   M.J. Lee, Robin Waywell Collection
Whitehall Securities Corporation Ltd, 23-10-1954

Fig 126, HE550 1892 0-6-0ST IC MARFLEET
Whitehall Securities Corporation Ltd, 6-1951

ILS: Frank Jones Collection

Fig 127, MW635 1877 0-4-0ST OC
Samuel Williams & Sons Ltd, Dagenham Dock

IRS: Jim Peden Collection

Fig 128, HE1 1865 0-6-0ST OC
Samuel Williams & Sons Ltd, Dagenham Dock

ILS: Frank Jones Collection

Fig 129, MW641 1877 0-6-0ST IC
Samuel Williams & Sons Ltd, Dagenham Dock, 6-6-1949

IRS: K.J. Cooper Collection

Fig 130, S5735 1926 4wVBT VCG ILS: Frank Jones Collection
Samuel Williams & Sons Ltd, Dagenham Dock

Fig 131, MW1590 1903 0-6-0ST IC Jack Faithfull RCTS
Samuel Williams & Sons Ltd, Dagenham Dock, 18-10-1958

Fig 132, P1606 1923 0-4-0ST OC MARYHILL IRS: K.J. Cooper Collection
Samuel Williams & Sons Ltd, Dagenham Dock, 6-6-1949

Fig 133, HC1526 1924 0-6-0ST IC IRS: K.J. Cooper Collection
Samuel Williams & Sons Ltd, Dagenham Dock, 6-6-1949

Fig 134, AB1129 1907 0-4-0ST OC Jack Faithfull RCTS
Samuel Williams & Sons Ltd, Dagenham Dock, 18-10-1958

Fig 135, MW1674 1906 0-6-0ST IC BOMBAY IRS: K.J. Cooper Collection
Samuel Williams & Sons Ltd, Dagenham Dock, 6-6-1949

Fig 136, HC D680 1949 0-6-0DM IRS: Jim Peden Collection
Samuel Williams & Sons Ltd, Dagenham Dock, 16-8-1957

Fig 137, FH3997 1963 4wDM Roger Hateley
Samuel Williams & Sons Ltd, Dagenham Dock, 2-9-1964

Fig 138, FH3813 1956 & FH3722 1955 4wDM   Kevin Lane
Samuel Williams & Sons Ltd, Dagenham Dock, 12-2-1976

Fig 139, FH3949 1960 4wDM   Robin Waywell
Samuel Williams (Dagenham Dock) Ltd, Dagenham Dock, 26-9-1979

Fig 140, MW1576 1903 0-6-0ST IC R.G.Pratt, John Hutchings Collection
Caffin & Co Ltd, Walton-on-Naze contract

Fig 141, Simplex 20hp loco R.G. Pratt, John Hutchings Collection
Colchester Council, Colchester By Pass contract, 18-6-1930

Fig 142, FH 40hp 4wPM R.G. Pratt, John Hutchings Collection
Colchester Council, Colchester By Pass contract, 18-6-1930

Fig 143, MW143 1865 0-4-0ST OC LITTLE EASTERN IRS Jim Peden Collection
William Hanson, Saffron Walden contract

Fig 144, MW 1539 1902 0-6-0ST IC PENN Frank Jones, Robin Waywell Collection
John Mowlem & Co Ltd, Chingford contract

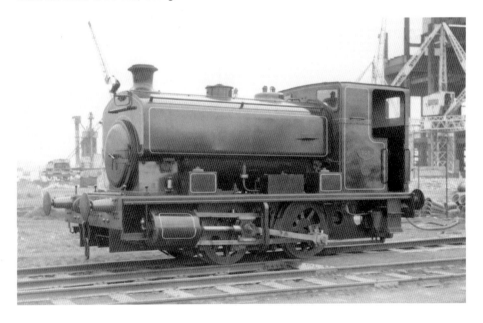

Fig 145, AB2353 1954 0-4-0ST OC ILS: Frank Jones Collection
John Mowlem & Co Ltd, West Thurrock Power Station contract

Fig 146, MW1135 1891 0-4-0ST OC ANNIE IRS: Jim Peden Collection
S. Pearson & Son Ltd, Walthamstow Reservoirs contract

Fig 147, P798 1899 0-4-0ST OC ALWILDA IRS: Jim Peden Collection
S. Pearson & Son Ltd, Walthamstow Reservoirs contract, 24-11-1900

Fig 148, OK 4wDM R.G. Pratt, John Hutchings Collection
South Essex Waterworks Company, Stour Scheme

Fig 149, MW1119 1889 0-6-0ST IC GOWY IRS: Jim Peden Collection
Topham Jones & Railton Ltd, Tilbury Dock contract

Fig 150, ERICA 2-4-0WT OC Ch. Wall　　　　　　IRS: Jim Peden collection
Charles Wall Ltd, Chingford contract

Fig 151, 4wTG Ch. Wall　　　　　　　　　　　Robin Waywell Collection
Charles Wall Ltd, Chingford contract

Fig 152, HC440 1896 0-6-0ST IC BARRY
C.J. Wills & Sons Ltd, Becontree contract

IRS: Jim Peden Collection

Fig 153, HE1499 1926 0-6-0ST IC CECIL LEVITA
C.J. Wills & Sons Ltd, Becontree contract

IRS: Jim Peden Collection

Fig 154, HE215 1879 0-4-0ST OC SOUTHSEA
C.J. Wills & Sons Ltd, Becontree contract

ILS: Frank Jones Collection

Fig 155, MW595 1876 0-6-0ST IC SWANSEA
C.J. Wills & Sons Ltd, Becontree contract

ILS: Frank Jones Collection

Fig 156, HC1336 1918 0-4-0T OC
Thos W. Ward Ltd, Titan Works, Grays, 19-7-1936
Morton Middleditch, John Hutchings Collection

Fig 157, HE1644 1929 0-6-0ST OC BRAMLEY No6   ILS: Frank Jones Collection
Thos W. Ward Ltd, Titan Works, Grays, 5-1960

Fig 158, KS3129 1918 0-4-2ST OC   ILS: Frank Jones Collection
Thos W. Ward Ltd, Titan Works, Grays

Fig 159, MW2005 1921 0-6-0ST IC   ILS: Frank Jones Collection
Thos W. Ward Ltd, Titan Works, Grays 2-1952